An Introduction to
Applied and
Environmental
Geophysics

An Introduction to Applied and Environmental Geophysics

JOHN M. REYNOLDS
Reynolds Geo-Sciences Ltd, UK

JOHN WILEY & SONS
Chichester · New York · Weinheim · Brisbane · Singapore · Toronto

Copyright © 1997 John Wiley & Sons Ltd,
          The Atrium, Southern Gate, Chichester,
          West Sussex PO19 8SQ, England

          Telephone   (+44) 1243 779777

Email (for orders and customer service enquiries): cs-books@wiley.co.uk
Visit our Home Page on www.wileyeurope.com or www.wiley.com

Reprinted 1998 (twice), January 2000, November 2001, October 2002, July 2003

Other Wiley Editorial Offices

John Wiley & Sons Inc., 111 River Street, Hoboken, NJ 07030, USA

Jossey-Bass, 989 Market Street, San Francisco, CA 94103-1741, USA

Wiley-VCH Verlag GmbH, Boschstr. 12, D-69469 Weinheim, Germany

John Wiley & Sons Australia Ltd, 33 Park Road, Milton, Queensland 4064, Australia

John Wiley & Sons (Asia) Pte Ltd, 2 Clementi Loop #02-01, Jin Xing Distripark, Singapore 129809

John Wiley & Sons Canada Ltd, 22 Worcester Road, Etobicoke, Ontario, Canada M9W 1L1

British Library Cataloguing in Publication Data

A catalogue record for this book is available from the British Library

ISBN: 978-0-471-95555-9

Contents

Preface

The idea for this book originated in 1987 whilst I was preparing for lectures on courses in applied geology and environmental geophysics at Plymouth Polytechnic (now the University of Plymouth), Devon, England. Students who had only very basic mathematical skills and little if any physics background found most of the so-called 'introductory' texts difficult to follow owing to the perceived opacity of text and daunting display of apparently complex mathematics. To junior undergraduates, this is immediately offputting and geophysics becomes known as a 'hard' subject and one to be avoided at all costs.

I hope that the information on the pages that follow will demonstrate the huge range of applications of modern geophysics – some now very well established, others very much in the early stages of implementation. It is also hoped that the book will provide a foundation on which to build if the reader wishes to take the subject further. The references cited, by no means exhaustive, have been included to provide pointers to more detailed discussions.

The aim of this book is to provide a basic introduction to geophysics, keeping the mathematics and theoretical physics to a minimum and emphasising the applications. Considerable effort has been expended in compiling a representative set of case histories that demonstrate clearly the issues being discussed.

This book is different from other introductory texts in that it pays attention to a great deal of new material, or topics not previously discussed in detail: for example, geophysical survey design and line optimisation techniques, image-processing of potential field data, recent developments in high-resolution seismic reflection profiling, electrical resistivity Sub-Surface Imaging (tomography), Spectral Induced Polarisation, and Ground Penetrating Radar, amongst many other subjects, which until now have never featured in detail in such a book. Many new and previously unpublished case histories from commercial projects have been included along with recently published examples of applications.

The subject material has been developed over a number of years, firstly while I was at Plymouth, and secondly and more recently while

I have been working as a geophysical consultant. Early drafts of the book have been tried out on several hundred second- and third-year students who have been unwitting 'guinea pigs' – their comments have been very helpful. While working in industry, I have found the need for an introductory book all the more evident. Many potential clients either appear unaware of how geophysics could possibly be of help to them, or have a very dated view as to the techniques available. There has been no suitable book to recommend to them that explained what they needed and wanted to know or that provided real examples.

While I have been writing this book, the development of new instruments, improved data-processing and interpretation software and increased understanding of physical processes have been unparalleled in the history of geophysical sciences. It has been difficult to keep abreast of all the new ideas, especially with an ever-increasing number of scientific publications. What is exciting is that the changes are still occurring and we can expect to see yet more novel developments over the next few years. We may see new branches of the science develop, particularly in environmental geophysics and applications to contaminated-land mapping, for example.

It is my hope that this book will be seen as providing a broad overview of applied and environmental geophysics methods, illustrating the power and sophistication of the various techniques, as well as their limitations. If this book helps in improving the acceptance of geophysical methods and in increasing the awareness of the methods available, then it will have met its objective. There is no doubt that applied and environmental geophysics have an important role to play, and that the potential for the future is enormous.

It is inevitable with a book of this kind that brand names, instrument types, and specific manufacturers are named. Reference to such information does not constitute an endorsement of any product and no preference is implied, nor should any inference be drawn over any omissions. In books of this type the material covered tends to be flavoured by the interests and experiences of the author, and I am sure that this one is no exception. I hope that what is included is a fair reflection of the current state of applied and evironmental geophysics. Should any readers have any case histories that they feel are of particular significance, I should be most interested to receive them for possible inclusion at a later date. Also, any comments or corrections that readers might have would also be gratefully received.

ACKNOWLEDGEMENTS

Drafts of the manuscript have been read by many colleagues, and their help, encouragement and advice have been most beneficial. I would particularly like to thank Professor Don Tarling for his continual encouragement to complete this book and for having commented on substantial parts of the manuscript. Thanks are also due to companies that have very kindly supplied material, and to colleagues around the world for permitting extracts of their work to be reproduced. I must also thank Richard Baggaley (formerly of the Open University Press) for commissioning me to write the book in the first place, and to Helen Bailey, Abi Hudlass and Louise Portsmouth at John Wiley for their patience in waiting so long for the final manuscript.

My final acknowledgement must be to my wife, Moira, and sons, Steven and David, for their support, encouragement and longsuffering patience while I have been closeted with 'The Book'. Without their help and forbearance, this project would have been abandoned long ago.

John M. Reynolds
Mold, Clwyd, North Wales, UK
June 1995

Chapter 1
Introduction

1.1   WHAT ARE 'APPLIED' AND 'ENVIRONMENTAL' GEOPHYSICS?

In the broadest sense, the science of *Geophysics* is the application of physics to investigations of the Earth, Moon and planets. The subject is thus related to astronomy. Normally, however, the definition of 'Geophysics' is used in a more restricted way, being applied solely to the Earth. Even then, the term includes such subjects as meteorology and ionospheric physics, and other aspects of atmospheric sciences.

To avoid confusion, the use of physics to study the interior of the Earth, from land surface to the inner core, is known as *Solid Earth Geophysics*. This can be subdivided further into *Global Geophysics*, or alternatively *Pure Geophysics*, which is the study of the whole or substantial parts of the planet, and *Applied Geophysics* which is concerned with investigating the Earth's crust and near-surface to achieve a practical and, more often than not, an economic aim.

'Applied geophysics' covers everything from experiments to determine the thickness of the crust (which is important in hydrocarbon exploration) to studies of shallow structures for engineering site investigations, exploring for groundwater and for minerals and other economic resources, to trying to locate narrow mine shafts or other forms of buried cavities, or the mapping of archaeological remains, or locating buried pipes and cables – but where in general the total depth of investigation is usually less than 100 m. The same scientific principles and technical challenges apply as much to shallow geophysical investigations as to pure geophysics. Sheriff (1991; p. 139) has defined '*applied geophysics*' thus:

> "Making and interpreting measurements of physical properties of the earth to determine sub-surface conditions, usually with an economic objective, e.g., discovery of fuel or mineral depositions."

'*Engineering geophysics*' can be described as being:

> "The application of geophysical methods to the investigation of sub-surface materials and structures which are likely to have (significant) engineering implications."

As the range of applications of geophysical methods has increased, particularly with respect to derelict and contaminated land investigations, the sub-discipline of '*environmental geophysics*' has developed (Greenhouse 1991; Steeples 1991). This can be defined as being:

> "The application of geophysical methods to the investigation of near-surface physico-chemical phenomena which are likely to have (significant) implications for the management of the local environment."

The principal distinction between engineering and environmental geophysics is more commonly that the former is concerned with structures and types of materials, whereas the latter can also include, for example, mapping variations in pore–fluid conductivities to indicate pollution plumes within groundwater. Chemical effects are equally as important as physical phenomena. Since the mid-1980s in the UK, geophysical methods have been used increasingly to investigate derelict and contaminated land, with a specific objective of locating polluted areas prior to direct observations using trial pits and boreholes (e.g. Reynolds and Taylor 1992). Geophysics is also being used much more extensively over landfills and other waste repositories (e.g. Reynolds and McCann 1992). One of the advantages of using geophysical methods is that they are largely environmentally benign – there is no disturbance of sub-surface materials. An obvious example is the location of a corroded steel drum containing toxic chemicals. To probe for it poses the real risk of puncturing it and creating a much more significant pollution incident. By using modern geomagnetic surveying methods, the drum's position can be isolated and a careful excavation instigated to remove the offending object without damage. Such an approach is cost-effective and environmentally safer.

A further major advantage of the use of environmental geophysics in investigating contaminated sites is that large areas can be surveyed quickly at relatively low cost. This provides information to aid the location of trial pits and boreholes. The alternative and more usual approach is to use a statistical sampling technique (e.g. Ferguson 1992). The disadvantage of this is that key areas of contamination can easily be missed, reducing the value substantially of such direct investigation. By targeting direct investigations by using a preliminary geophysical survey to locate anomalous areas, there is a much higher certainty that the trial pits and boreholes constructed will yield useful results. Instead of seeing the geophysical survey as a cost, it should be viewed as adding value by making the entire site investigation more cost-effective.

There are obviously situations where a specific site investigation contains aspects of engineering as well as environmental geophysics and there may well be considerable overlap. Indeed, if each sub-discipline of applied geophysics is considered, they may be represented as shown in Figure 1.1, as overlapping. Also included are three other sub-disciplines whose names are largely self-explanatory: namely, *Archaeo-geophysics* (geophysics in archaeology), *hydro-geophysics* (geophysics in groundwater investigations), and *Glacio-geophysics*

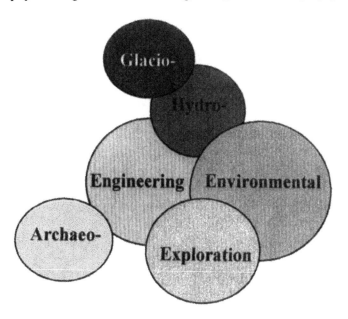

Figure 1.1 Inter-relationships between various sub-disciplines of applied geophysics

(geophysics in glaciology). The last one is the least well known, despite the fact that it has been in existence for far longer than either archaeo- or environmental geophysics, and is particularly well established within the polar scientific communities and has been since the 1950s.

The general orthodox education of geophysicists to give them a strong bias towards the hydrocarbon industry has largely ignored these other areas of our science. It may be said that this restricted view has delayed the application of geophysics more widely to other disciplines. Geophysics has been taught principally in Earth Science departments of universities. There is an obvious need for it to be introduced to engineers and archaeologists much more widely than at present. Similarly, the discipline of environmental geophysics needs to be brought to the attention of policy-makers and planners, to the insurance and finance industries (Doll 1994).

The term 'environmental geophysics' has been interpreted by some to mean geophysical surveys undertaken with environmental sensitivity – that is, ensuring that, for example, marine seismic sur- veys are undertaken sympathetically with respect to the marine environment (Bowles 1990). With growing public awareness of the environment and the pressures upon it, the geophysical community has had to be able to demonstrate clearly its intentions to minimise environmental impact (Marsh 1991). By virtue of scale, the greatest likely impact on the environment is from hydrocarbon and some mineral exploration, and the main institutions involved in these activities are well aware of their responsibilities. In small-scale surveys the risk of damage is much lower; but all the same, it is still important that those undertaking geophysical surveys should be mindful of their responsibilities to the environment and to others whose livelihoods depend upon it.

While the term 'applied geophysics' covers a wide range of applica- tions, the importance of 'environmental' geophysics is particularly highlighted within this book. The growth of the discipline, which appears to be expanding exponentially, is such that this subject may outstrip the use of geophysics in hydrocarbon exploration during the early part of the next century and provide the principal area of employment for geophysicists. Whether this proves to be the case is for history to decide. What is clear, however, is that even in the last decade of this century, environmental geophysics is becoming in- creasingly important in the management of our environment. Ignore it at your peril!

1.2   GEOPHYSICAL METHODS

Geophysical methods respond to the physical properties of the sub-surface media (rocks, sediments, water, voids, etc.) and can be classified into two distinct types.

- **Passive** methods are those that detect variations within the natural fields associated with the Earth, such as the gravitational and magnetic fields.
- In contrast are the **active** methods, such as those used in exploration seismology, in which artificially generated signals are transmitted into the ground, which then modifies those signals in ways that are characteristic of the materials through which they travel. The altered signals are measured by appropriate detectors whose output can be displayed and ultimately interpreted.

Applied geophysics provides a wide range of very useful and powerful tools which, when used correctly and in the right situations, will produce useful information. All tools, if misused or abused, will not work effectively. One of the aims of this book it to try to explain how applied geophysical methods can be employed appropriately, and to highlight the advantages and disadvantages of the various techniques.

Geophysical methods may form part of a larger survey, and thus geophysicists should always try to interpret their data and communicate their results clearly to the benefit of the whole survey team and particularly to the client. An engineering site investigation, for instance, may require the use of seismic refraction to determine how easy it would be to excavate the ground (e.g. the 'rippability' of the ground). If the geophysicist produces results that are solely in terms of seismic velocity variations, the engineer is still none the wiser. The geophysicist needs to translate the velocity data into a rippability index with which the engineer would be familiar.

Few, if any, geophysical methods provide a *unique* solution to a particular geological situation. It is possible to obtain a very large number of geophysical solutions to some problems, some of which may be geologically nonsensical. It is necessary, therefore, always to ask the question: "Is the geophysical model geologically plausible?" If it is not, then the geophysical model has to be rejected and a new one developed which does provide a reasonable geological solution. Conversely, if the geological model proves to be inconsistent with the geophysical interpretation, then it may require the geological information to be re-evaluated.

> It is of paramount importance that geophysical data are interpreted within a physically constrained or geological framework.

1.3 MATCHING GEOPHYSICAL METHODS TO APPLICATIONS

The various geophysical methods rely on different physical properties and it is important that the appropriate technique be used for a given type of application.

For example, gravity methods are sensitive to density contrasts within the sub-surface geology and so are ideal for exploring for major sedimentary basins where there is a large density contrast between the lighter sediments and the denser underlying rocks. It would be quite inappropriate to try to use gravity methods to search for localized near-surface sources of groundwater where there is a negligible density contrast between the saturated and unsaturated rocks. It is even better to use methods that are sensitive to different physical properties and are able to complement each other and thereby provide an integrated approach to a geological problem. Gravity and magnetic methods are frequently used in this way.

Case histories for each geophysical method are given in each chapter along with some examples of integrated applications where appropriate. The basic geophysical methods are listed in Table 1.1 with the physical properties to which they relate and their main uses. Table 1.1 should only be used.as a guide. More specific information about the applications of the various techniques is given in the appropriate chapters.

Some methods are obviously unsuitable for some applications but novel uses may yet be found for them. One example is that of ground

Table 1.1 Geophysical methods and their main applications

Geophysical method	Chapter number	Dependent physical property	Applications (see key below)									
			1	2	3	4	5	6	7	8	9	10
Gravity	2	Density	P	P	s	s	s	s	!	!	s	!
Magnetic	3	Susceptibility	P	P	P	s	!	m	!	P	P	!
Seismic refraction	4,5	Elastic moduli; density	P	P	m	P	s	s	!	!	!	!
Seismic reflection	4,6	Elastic moduli; density	P	P	m	s	s	m	!	!	!	!
Resistivity	7	Resistivity	m	m	P	P	P	P	P	s	P	m
Spontaneous potential	8	Potential differences	!	!	P	m	P	m	m	m	!	!
Induced polarization	9	Resistivity; capacitance	m	m	P	m	s	m	m	m	m	m
Electromagnetic (EM)	10	Conductance; inductance	s	P	P	P	P	P	P	P	P	m
EM–VLF	11	Conductance; inductance	m	m	P	m	s	s	s	m	m	!
EM – ground penetrating radar	12	Permitivity; conductivity	!	!	m	P	P	P	s	P	P	P
Magneto-telluric	11	Resistivity	s	P	P	m	m	!	!	!	!	!

P = primary method; s = secondary method; m = may be used but not necessarily the best approach, or has not been developed for this application; (!) = unsuitable

Applications
1 Hydrocarbon exploration (coal, gas, oil)
2 Regional geological studies (over areas of 100s of km^2)
3 Exploration/development of mineral deposits
4 Engineering site investigations
5 Hydrogeological investigations
6 Detection of sub-surface cavities
7 Mapping of leachate and contaminant plumes
8 Location and definition of buried metallic objects
9 Archaeogeophysics
10 Forensic geophysics

radar being employed by police in forensic work (see Chapter 12 for more details). If the physical principles upon which a method is based are understood, then it is less likely that that technique will be misapplied or the resultant data misinterpreted. This makes for much better science.

Furthermore, it must also be appreciated that the application of geophysical methods will not necessarily produce a unique geological solution. For a given geophysical anomaly there may be many possible solutions each of which is equally valid geophysically, but which may make geological nonsense. This has been demonstrated

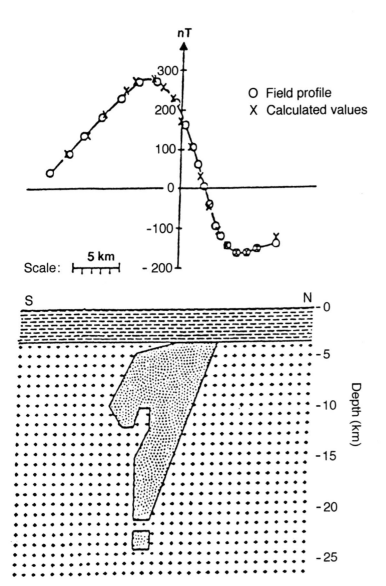

Figure 1.2   A magnetic anomaly over Lausanne, Switzerland, with a hypothetical and unreal model for which the computed anomaly still fits the observed data. After Meyer de Stadelhofen and Juillard (1987)

O Field profile

X Calculated values

very clearly in respect to a geomagnetic anomaly over Lausanne in Switzerland (Figure 1.2). While the model with the form of a question mark satisfies a statistical fit to the observed data, the model is clearly and quite deliberately geological nonsense in order to demonstrate the point. However, geophysical observations can also place stringent restrictions on the interpretation of geological models. While the importance of understanding the basic principles cannot be over-emphasised, it is also necessary to consider other factors that affect the quality and usefulness of any geophysical survey, or for that matter of *any* type of survey whether it is geophysical, geochemical or geotechnical. This is done in the following few sections.

1.4   PLANNING A GEOPHYSICAL SURVEY

1.4.1   General philosophy

Any geophysical survey tries to determine the nature of the subsurface, but it is of paramount importance that the prime objective of the survey be clear right at the beginning. The constraints on a commercial survey will have emphases different from those on an academic research investigation and, in many cases, there may be no *ideal* method. The techniques employed and the subsequent interpretation of the resultant data tend to be compromises, practically and scientifically.

There is no short-cut to developing a good survey style; only by careful survey planning backed by a sound knowledge of the geophysical methods and their operating principles, can cost-effective and efficient surveys be undertaken within the prevalent constraints. However, there have been only a few published guidelines – e.g. British Standards Institute BS 5930 (1981), Hawkins (1986), Geological Society Engineering Group Working Party Report on Engineering Geophysics (1988). Scant attention has been paid to survey design, yet a badly thought-out survey rarely produces worthwhile results. Indeed, Darracott and McCann (1986, p. 85) said that:

> "dissatisfied clients have frequently voiced their disappointment with geophysics as a site investigation method. However, close scrutiny of almost all such cases will show that the geophysical survey produced poor results for one or a combination of the following reasons: inadequate and/or bad planning of the survey; incorrect choice or specification of technique, and insufficiently experienced personnel conducting the investigation."

It is hoped that this chapter will provide at least a few pointers to help construct cost-effective and technically sound geophysical field programmes.

1.4.2 Planning strategy

Every survey must be planned according to some strategy, or else it will become an uncoordinated muddle. *The mere acquisition of data does not guarantee the success of the survey.* Knowledge (by way of masses of data) does not automatically increase our *understanding* of a site; it is the latter we are seeking, and knowledge is the means to this.

One less-than-ideal approach is the 'blunderbus' approach – take along a sufficient number of different methods and try them all out (usually inadequately owing to insufficient testing time per technique) to see which ones produce something interesting. Whichever method yields an anomaly, then use that technique. This is a crude statistical approach, such that if enough techniques are tried then at least one must work! This is hardly scientific or cost-effective.

The success of geophysical methods can be very site-specific and *scientifically-designed* trials of adequate duration may be very worthwhile to provide confidence that the techniques chosen will work or that the survey design needs modifying in order to optimise the main survey. It is in the interests of the client that suitably experienced geophysical consultants are employed for the vital survey design, site supervision and final reporting.

So what are the constraints that need to be considered by both clients and geophysical survey designers? An outline plan of the various stages in designing a survey is given in Figure 1.3. The remainder of this chapter discusses the relationships between the various components.

1.4.3 Survey constraints

The first and most important factor is that of *finance*. How much is the survey going to cost and how much money is available? The cost will depend on where the survey is to take place, how accessible the proposed field site is, and on what scale the survey is to operate. An airborne regional survey is a very different proposition to, say, a local, small-scale ground-based investigation. The more complex the survey in terms of equipment and logistics, the greater the cost is likely to be.

It is important to remember that the geophysics component of a survey is usually only a small part of an exploration programme and thus the costs of the geophysics should be viewed in relation to those of the whole project. Indeed, the judicious use of geophysics can save large amounts of money by enabling the effective use of resources (Reynolds 1987a). For example, a reconnaissance survey can identify smaller areas where much more detailed investigations ought to be undertaken – thus removing the need to do saturation surveying. The factors that influence the various components of a budget also vary

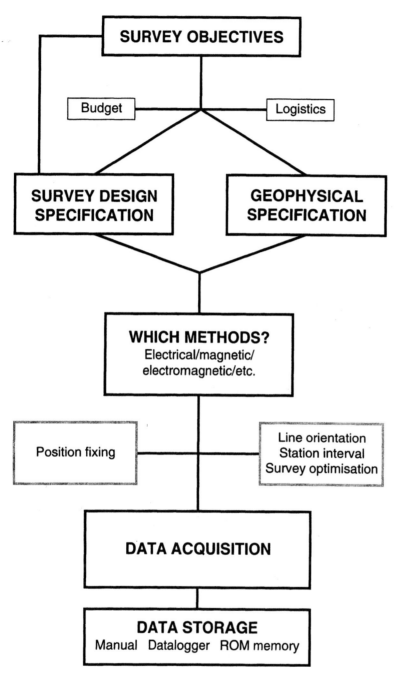

Figure 1.3 Schematic flow diagram to illustrate the decision-making process leading to the selection of geophysical and utility software. From Reynolds (1991a), by permission

from country to country, and from job to job, and there is no magic formula to guarantee success.

Some of the basic elements of a survey budget are given in Table 1.2. This list is not exhaustive but serves to highlight the most

Table 1.2 Basic elements of a survey budget

Staffing	Management, technical, support, administration, etc.
Operating costs	Including logistics
Cashflow	Assets versus usable cash
Equipment	For data acquisition and/or for data reduction/analysis – computers and software; whether or not to hire or buy
Insurance	To include liability insurance, as appropriate
Overheads	Administration; consumables; etc.
Development costs	Skills, software, etc.
Contingencies	Something is bound to go wrong at some time, usually when it is most inconvenient!

common elements of a typical budget. Liability insurance is especially important if survey work is being carried out as a service to others. If there is any cause for complaint, then this may manifest itself in legal action (Sherrell 1987).

It may seem obvious to identify *logistics* as a constraint but there have been far too many surveys ruined by a lack of even the most basic needs of a survey. It is easy to think of the main people to be involved in a survey – i.e. geologists, geophysicists, surveyors – but there are many more tasks to be done to allow the technical staff the opportunity to concentrate on the tasks in hand. Vehicles and equipment will need maintaining, so skilled technicians and mechanics may be required. Everybody has to eat and it is surprising how much better people work when they are provided with well-prepared food: a good cook at base camp can be a real asset. Due consideration should be paid to health and safety and any survey team should have staff trained in First Aid. Admittedly it is possible for one person to be responsible for more than one task, but on large surveys this can prove to be a false economy. Apart from the skilled and technical staff, local labour may be needed as porters, labourers, guides, translators, etc., or even as armed guards!

It is all too easy to forget what field conditions can be like in remote and inaccessible places. It is thus important to remember that in the case of many countries, access in the dry season may be possible whereas during the rains of the wet season, the so-called roads (which often are dry river beds) may be totally impassable. Similarly, access to land for survey work can be severely hampered during the growing season with some crops reaching 2–3 metres high and consequently making position fixing and physical access extremely difficult. There is then the added complication that some surveys, such as seismic refraction and reflection, may cause a limited amount of damage for which financial compensation may be sought. In some cases, claims may be made even when no damage has been caused! If year-round access is necessary the provision of all-terrain vehicles and/or helicopters may prove to be the only option, and these are never cheap to operate.

Where equipment has to be transported, consideration has to be given not only to its overall weight but to the size of each container. It can prove an expensive mistake to find that the main piece of equipment will not pass through the doorway of a helicopter so that alternative overland transport has to be provided at very short notice; or to find that many extra hours of flying time are necessary to airlift all the equipment. It may even be necessary to make provision for a bulldozer to excavate a rough road to provide access for vehicles. If this is accounted for inadequately in the initial budgeting, the whole success of the survey can be jeopardised. Indeed, the biggest constraint in some developing countries, for example, is whether the equipment can be carried by a porter or will fit on the back of a pack-horse.

Other constraints that are rarely considered are those associated with *politics*, *society* and *religion*. Let us take these in turn.

Political constraints   This can mean gaining permission from land-owners and tenants for access to land, and liaison with clients (which often requires great diplomacy). The compatibility of staff to work well together also needs to be considered, especially when working in areas where there may be conflicts between different factions of the local population – such as tribal disputes or party political disagreements. It is important to remember to seek permission from the *appropriate* authority to undertake geophysical fieldwork. For example, in Great Britain it is necessary to liaise with the police and local government departments if survey work along a major road is being considered, so as to avoid problems with traffic jams. In other cases it may be necessary to have permission from a local council, or in the case of marine surveys, from the local harbour master so that appropriate marine notices can be issued to safeguard other shipping. All these must be found out well before the start of any fieldwork. Delays cost money!

Social constraints   For a survey to be successful it is always best to keep on good terms with the local people. Treating other people with respect will always bring dividends (eventually). Each survey should be socially and environmentally acceptable and not cause a nuisance. An example is in not choosing to use explosives as a seismic source for reflection profiling through urban areas or at night. Instead, the seismic vibrator technique should be used (see Chapter 4). Similarly, an explosive source for marine reflection profiling would be inappropriate in an area associated with a lucrative fishing industry because of possibly unacceptably high fish-kill. In designing the geophysical survey, the question must be asked: "Is the survey technique socially and environmentally acceptable?"

Religious constraints The survey should take into account local social customs which are often linked with religion. In some Muslim countries, for example, it is common in rural areas for women to be the principal water-collectors. It is considered inappropriate for the women to have to walk too far away from the seclusion of their homes. Thus there is no point in surveying for groundwater for a tubewell several kilometres from the village (Reynolds 1987a). In addition, when budgeting for the provision of local workers, it is best to allow for their 'sabbath'. Muslims like to go to their mosques on Friday afternoons and are thus unavailable for work then. Similarly, Christian workers tend not to like being asked to work on Sundays, or Jews on Saturdays. Religious traditions must be respected to avoid difficulties.

However, problems may come if local workers claim to be Muslims on Fridays and Christians on Sundays – and then that it is hardly worth anyone's while to have to work only on the Saturday in between so they end up not working Friday, Saturday or Sunday! Such situations, while sounding amusing, can cause unacceptable delays and result in considerably increased survey costs.

1.5 GEOPHYSICAL SURVEY DESIGN

1.5.1 Target identification

Geophysical methods locate boundaries across which there is a marked contrast in physical properties. Such a contrast can be detected remotely because it gives rise to a *geophysical anomaly* (Figure 1.4) which indicates variations in physical properties relative to some background value (Figure 1.5). The physical source of each anomaly is termed the *geophysical target*. Some examples of targets are trap structures for oil and gas, mineshafts, pipelines, ore lodes, cavities, groundwater, buried rock valleys, and so on.

In designing a geophysical survey, the type of target is of great importance. Each type of target will dictate to a large extent the appropriate geophysical method(s) to be used, and this is where an understanding of the basic geophysical principles is important. The physical properties associated with the geophysical target are best detected by the method(s) most sensitive to those same properties.

Consider the situation where saline water intrudes into a near-surface aquifer; saline water has a high conductivity (low resistivity) in comparison with freshwater and so is best detected using electrical resistivity or electromagnetic conductivity methods; gravity methods would be inappropriate because there would be virtually no density contrast between the saline and freshwater. Similarly, seismic methods would not work as there is no significant difference in seismic wave velocities between the two saturated zones. Table 1.1

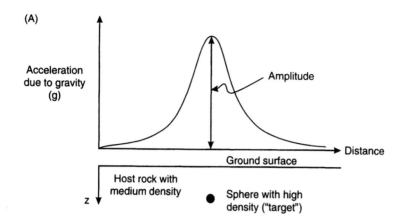

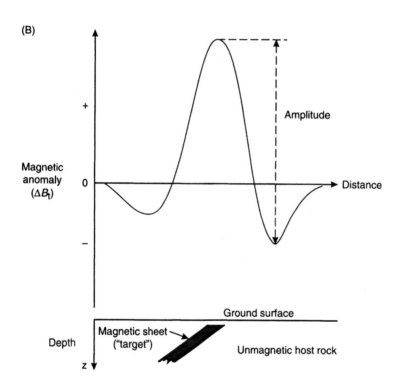

Figure 1.4 Examples of (A) a gravity anomaly over a buried sphere, and (B) a magnetic anomaly over an inclined magnetic sheet. For further details of gravity and magnetic methods, see Chapters 2 and 3 respectively

provides a ready means of selecting an appropriate technique for the major applications.

Although the physical characteristics of the target are important, so are its shape and size. In the case of a metallic ore lode, a mining company might need to know its lateral and vertical extent. An examination of the amplitude of the anomaly (i.e. its maximum peak-to-peak value) and its shape may provide further information about where the target is below ground and how big it is.

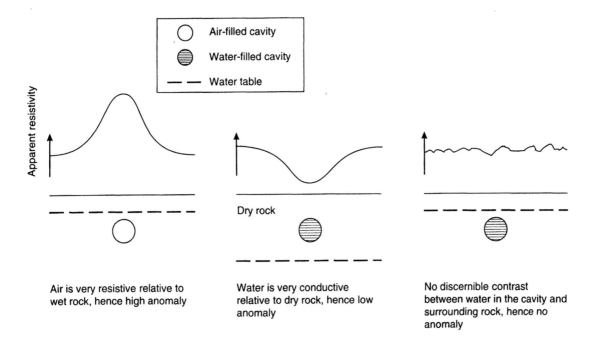

Air-filled cavity

Water-filled cavity

— — Water table

Apparent resistivity

Dry rock

Air is very resistive relative to
wet rock, hence high anomaly

Water is very conductive
relative to dry rock, hence low
anomaly

No discernible contrast
between water in the cavity and
surrounding rock, hence no
anomaly

1.5.2   Optimum line configuration

So far only the types of geological target and the selection of the most
appropriate geophysical methods have been discussed. In order to
complete a technically competent survey several other factors need to
be given very careful thought. How are the data to be collected in
order to define the geophysical anomaly? Two concepts need to be
introduced, namely *profiling* and *mapping*.

Profiling is a means of measuring the variation in a physical parameter along the surface of a two-dimensional cross-section (Figure
1.6A). Consideration needs to be given to the correct orientation and
length of profile (see below). Data values from a series of parallel lines
or from a grid can be contoured to produce a *map* (Figure 1.6B) on
which all points of equal value are joined by *isolines* (equivalent to
contours on a topographic map). However, great care has to be taken
over the methods of contouring or else the resultant map can be
misleading (see Section 1.5.3). There are many other ways of displaying geophysical data (Figure 1.5C), especially if computer graphics
are used (e.g. shaded relief maps as in Figure 1.6D), and examples are
given throughout the book.

The best orientation of a profile is normally at right-angles to the
strike of the target. A provisional indication of geological strike may
be obtained from existing geological maps, mining records, etc.
However, in many cases, strike direction may not be known at all and
test lines may be necessary to determine strike direction prior to

Figure 1.5 Contrasts in physical
properties from different geological
targets give rise to a geophysical
target. When there is no contrast, the
target is undetectable geophysically

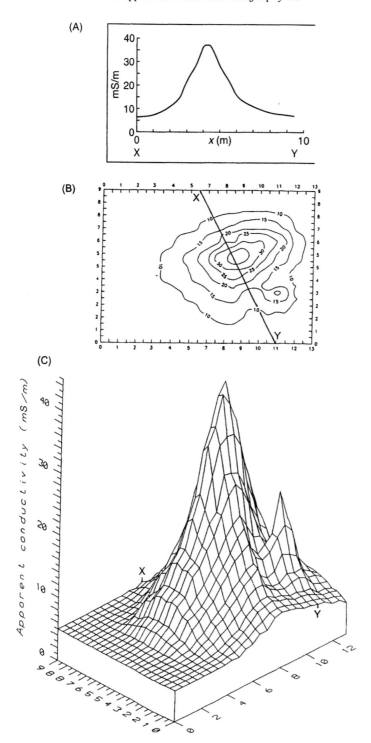

Figure 1.6 Geophysical anomaly plots: (A) profile, (B) map, and (C) isometric projection. All three plots are from the same set of electromagnetic ground-conductivity data (see Chapter 11). (D) A shaded relief/grey-scale shadow display can enhance features that otherwise would be hard to visualise – in this case the display is of magnetic data over an area in which complex faulting appears as a series of concentric features that possibly may be part of a meteorite impact crater. Photos courtesy of Geosoft Europe Ltd

(D)

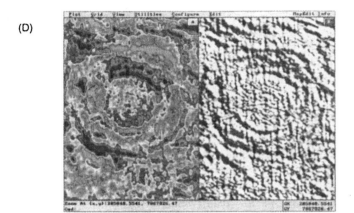

the main survey. The length of the profile should be greater than the width of the expected geophysical anomaly. If it is not, then it may be impossible to define a background value to determine the true anomaly amplitude and the value of the survey would be reduced greatly. The choice of line orientation also has to take into account sources of noise (see Section 1.5.4). If a map is required then it is advisable to carry out 'tie-lines' (cross-cutting profiles), the intersections (*nodes*) of which should have identical values. If the data are not the same at the nodes then the values need to be checked in case there has been a simple misprint in data entry, or there might have been an error in position-fixing or in instrumental calibration. When such data are compared, make sure all necessary data corrections have been made (see the individual chapters for details and examples) so that like is compared with like. Nodal values are vital for data quality control.

1.5.3   Selection of station intervals

The point at which a discrete geophysical measurement is made is called a *station* and the distances between successive measurements are *station intervals*.

   It is fundamental to the success of a survey that the correct choice of station intervals be made. It is a waste of time and money to record too many data and equally wasteful if too few are collected. So how is a reasonable choice to be made? This requires some idea of the nature and size of the geological target. Any geophysical anomaly found will always be larger than the feature causing it. Thus, to find a mineshaft, for example, with a diameter of, say, 2 m, an anomaly with a width of at least twice this might be expected. Therefore, it is necessary to choose a station interval that is sufficiently small to be able to resolve the anomaly, yet not too small as to take far too long to be practicable.

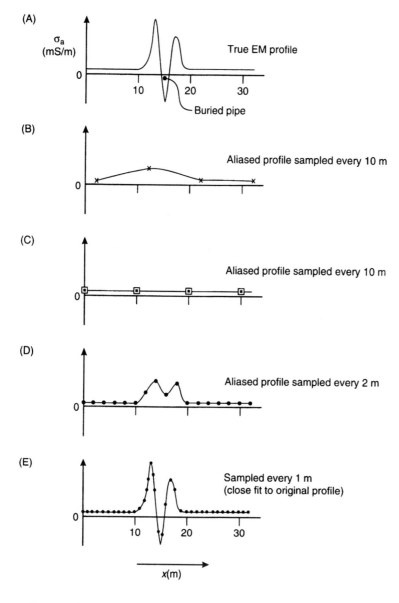

Figure 1.7 Examples of various degrees of spatial aliasing using different sampling intervals. (A) shows a continuously sampled profile. (B) and (C) show sampling every 10 m, but at different points along the profile. (D) shows sampling every 2 m: the profile is still aliased. (E) shows sampling every 1 m: this profile is the closest to that in (A)

Reconnaissance surveys tend to have coarser station intervals in order to cover a large area quickly and to indicate zones over which a more detailed survey should be conducted with a reduced station interval and a more closely spaced set of profiles.

Consider Figure 1.7A in which a typical electromagnetic anomaly for a buried gas pipe is shown. The whole anomaly is 8 m wide. If a 10 m sampling interval is chosen, then it is possible either to clip the anomaly, as in Figure 1.7B, or to miss it entirely (Figure 1.7C). The resultant profiles with 2 m and 1 m sampling intervals are shown in Figures 1.7D and 1.7E respectively. The smaller the sampling interval, the better the approximation is to the actual anomaly (compare with Figure 1.7B or C). The loss of high-frequency information, as in Figures 1.7B and C, is a phenomenon known as *spatial aliasing* and should be avoided.

Another form of spatial aliasing may occur when gridded data are contoured, particularly by computer software. If the grid network is too coarse, higher-frequency information may be smeared artificially and appear as lower-frequency anomalies. A common characteristic of spatially aliased gridded data is the 'bullseye' effect (see Figure 1.8) where the contouring program has had too little information to work on and so has contoured around individual data points or has linked data together unjustifiably (Cameron *et al.* 1976; Hood *et al.* 1979; Reid 1980; Wu 1990). This kind of problem can be created by an inadequately detailed or an inappropriately designed field programme.

Figure 1.8 shows a hypothetical aeromagnetic survey. The map in Figure 1.8A was compiled from contouring the original data at a line spacing of 150 m. Figures 1.8B and C were recontoured with line spacings of 300 m and 600 m respectively. The difference between the three maps is very marked, with a significant loss of information between Figures 1.8A and C. Noticeably the higher-frequency anomalies have been aliased out, leaving only the longer-wavelength (lower-frequency) features. In addition, the orientation of the major anomalies has been distorted by the crude contouring in Figure 1.8C.

Spatial stretching occurs on datasets acquired along survey lines separated too widely with respect to along-line sampling. This spatial aliasing can be removed or reduced using mathematical functions, such as the Radon Transform (Yuanxuan 1993). This method provides a means of developing a better gridding scheme for profile line-based surveys. The specific details of the method are beyond the scope of this chapter and readers are referred to Yuanxuan's paper for more information.

Similar aliasing problems associated with contouring can arise from radial survey lines and/or too few data points, as exemplified by Figure 1.9. Figure 1.9A and B both have 64 data points over the same area, and two effects can be seen very clearly: in Figure 1.9A the orientation of the contours (one marked 47 500 nT) artificially follows that of the line of data points to the top left-hand corner, whereas the orientation is more north–south in Figure 1.9B. The even grid in Figure 1.9B highlights the second effect (even more pronounced in Figure 1.9C), which is the formation of bullseyes around individual data points. The inadequacy of the number of data points is further demonstrated in Figure 1.9C, which is based on only 13 data values, by the formation of concentric contours that are artificially rounded in the top left and both bottom corners. For comparison, Figure 1.9D has been compiled on the basis of 255 data points, and exposes the observed anomalies much more realistically.

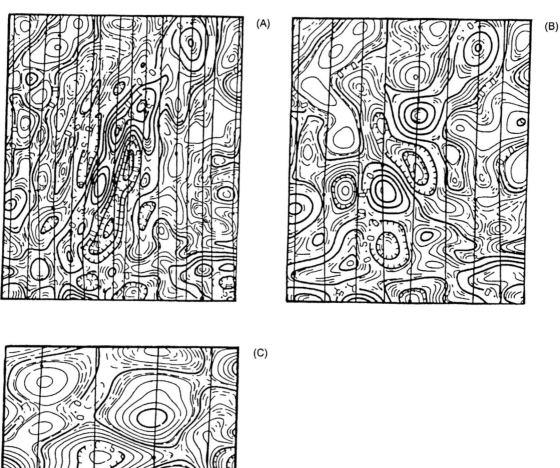

Figure 1.8 Example of spatial aliasing on aeromagnetic data, showing the loss of higher-frequency anomalies with increasing separation between flight lines and the increased 'bullseye' effect caused by stretching the data too far. From Hood *et al.* (1979), by permission

1.5.4 Noise

When a field survey is being designed it is important to consider what extraneous data (*noise*) may be recorded. There are various sources of noise, ranging from man-made sources ('cultural noise') as diverse as electric cables, vehicles, pipes and drains, to natural sources of noise

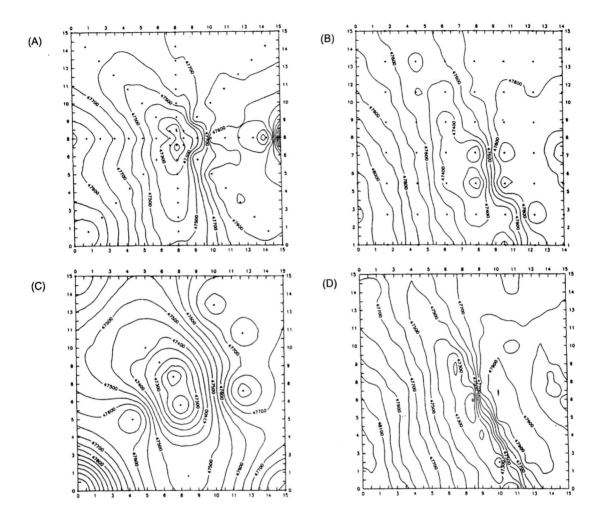

such as wind and rain, waves, and electric and magnetic storms (Figure 1.10).

Some aeromagnetic and electrical methods can suffer badly from cathodic currents that are used to reduce corrosion in metal pipes (Gay 1986). Electrical resistivity surveys should not be conducted close to or parallel to such pipes, nor parallel to cables as power lines will induce unwanted voltages in the survey wires. Before a survey starts, it is always advisable to consult with public utility companies which should, given enough time, provide maps of their underground and overhead facilities. It is important to check on the location of water mains, sewers, gas pipes, electricity cables, telephone cables and cable-television wires. In many cases such utilities may mask any anomalies caused by deeper-seated natural bodies. Furthermore, should direct excavation be required, the utilities underground may be damaged if their locations are not known.

Figure 1.9 Examples of contouring different patterns of data. (A) shows a set of radial lines, and (B) an even grid of data, both with 114 points per square kilometre. (C) has too few data points unevenly spread over the same area (23 data points per square kilometre). (D) shows an even grid of 453 data points per square kilometre. The contours are isolines of total magnetic field strength (units: nanoteslas); the data are from a ground magnetometer investigation of north-west Dartmoor, England

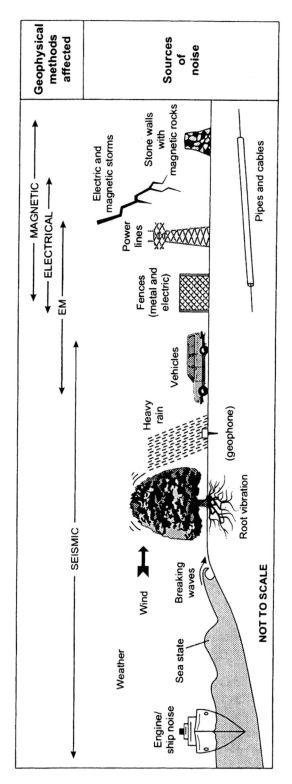

Figure 1.10 Schematic illustrating some common sources of geophysical noise

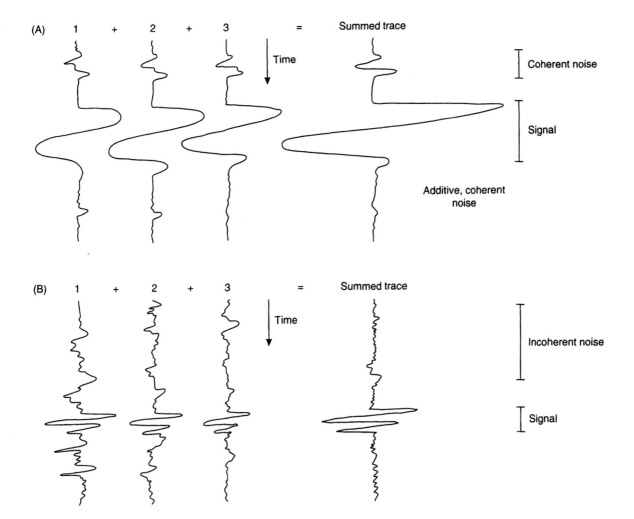

Figure 1.11 The effect of summing three traces with (A) coherent and (B) incoherent noise

It is also worth checking on the type of fencing around the survey area. Wire mesh and barbed wire fences, and metal sheds can play havoc with electromagnetic and magnetic surveys and will restrict the area over which sensible results can be obtained. It also pays to watch out for types of walling around fields as in many areas wire fences may be concealed by years of growth of the local vegetation. In addition, when undertaking a magnetic survey, be on the lookout for stone walls built of basic igneous rocks as these can give a noticeable magnetic anomaly.

There are two forms of noise (Figure 1.11).

- *Coherent noise*, such as that produced by power lines, occurs systematically (Figure 1.11A) and may degrade or even swamp the wanted signals. As coherent noise usually occurs with a definable frequency (e.g. mains electricity at 50–60 Hz), appropriate filters can be used to remove or reduce it.

- In contrast, *incoherent noise*, such as that due to waves breaking on a seashore or to traffic, is random. When summed together it tends to cancel to some extent, so reducing its overall effect (Figure 1.11B).

High, but incoherent, noise levels are often associated with surveys along road verges. Metal-bodied vehicles passing by during an electromagnetic survey can cause massive but brief disturbances. Vehicles, particularly heavy lorries, and trains can set up short-lived but excessive acoustic noise which can ruin a seismic survey. So, too, can the effects of waves washing onto beaches or the noise of turbulent riverwater close to geophone spreads on a seismic survey. In exposed areas, geophones that have not been planted properly may pick up wind vibration acting on the geophones themselves and on the connecting cable, but also from trees blowing in the breeze, as the motion transmits vibrations into the ground via their root systems. Similar effects can be observed close to man-made structures. Unprotected geophones are very sensitive to the impact of rain drops which can lead to the curtailment of a seismic survey during heavy rain.

Cultural and unnecessary natural noise can often be avoided or reduced significantly by careful survey design. Increasingly, modern technology can help to increase the *signal-to-noise ratio* so that, even when there is a degree of noise present, the important geophysical signals can be enhanced above the background noise levels (Figure 1.12). Details of this are given in the relevant sections of later chapters. However, it is usually better to use a properly designed field technique to optimise data quality in the first instance rather than relying on post-recording filtering. Further details of field methods are given, for example, by Milsom (1989).

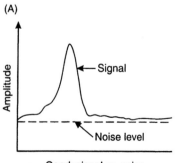

(A)

Good: signal ≫ noise

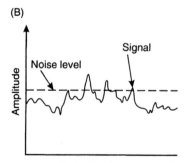

(B)

Bad: signal ≤ noise

Figure 1.12   Signal-to-noise ratio. In (A) the signal has a much larger amplitude than that of the background noise, so the signal can be resolved. In (B) the signal amplitude is less than, or about the same as, that of the noise and thus the signal is lost in the noise

1.5.5 Data analysis

All too often data are acquired without regard for how they are to be
processed and analysed. This oversight can lead to inadequate data
collection or the recording of data in such a way that vast amounts of
tedious transcribing or typing in of measurements has to be undertaken. Not only is this unproductive in terms of the person who has to
do all the 'number crunching', but it often allows the introduction of
errors into the datasets. The consequent back-checking to find the
bad data takes up valuable time and money. It therefore pays
dividends to think through how the data are to be collected in relation
to the subsequent methods of data reduction and analysis. A scheme
is presented in Figure 1.13.

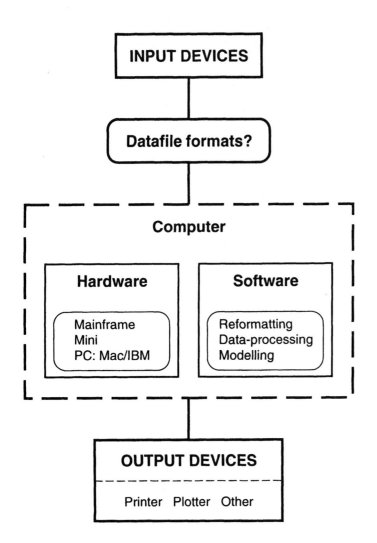

Figure 1.13 Schematic to show the relationship between various input devices, through data-file formats to the computer, and subsequently to some form of hardcopy device. From Reynolds (1991a), by permission

As automatic data-logging and computer analysis are becoming more commonplace (e.g. Sowerbutts and Mason 1984) it is increasingly important to standardise the format in which the data are recorded (Reeves and MacLeod 1986) to ease the portability of information transfer between computer systems. This also makes it easier to download the survey results into data-processing software packages. To make computer analysis much simpler it helps to plan the survey well before going into the field to ensure that the collection of data and the survey design are appropriate for the type of analyses anticipated. Even here, there are many pitfalls awaiting the unwary. How reliable is the software? Has it been calibrated against proven manual methods, if appropriate? What are the assumptions on which the software is based, and under what conditions are these no longer valid, and when will the software fail to cope and then start to produce erroneous results? (For an example of this, see Section 7.5.3.)

The danger with computers is that their output (especially if in colour) can have an apparent credibility that may not be justified by the quality of the data input or of the analysis. Unfortunately there are no guidelines or accepted standards for much geophysical software (Reynolds 1991a) apart from those for the major seismic data-processing systems. However, the judicious use of computers and of automatic data-logging methods can produce excellent and very worthwhile results (e.g. Sowerbutts and Mason 1984). Comments on some of the computer methods available with different geophysical techniques are made in the relevant chapters of this book, and some have been discussed more fully elsewhere (Reynolds 1991a).

For users of personal computers, there are two main software houses generating commercially available geophysical computer packages, namely Geosoft Ltd in Canada and Interpex Ltd in the USA. Geosoft also produces gridding and contouring packages, as does Golden Software (USA), producers of SURFER. Commercial products vary widely in their ranges of applications, flexibility and portability between different computers. Intending users of any software package should evaluate the software prior to purchase if possible.

BIBLIOGRAPHY

General geophysics texts

Beck, A.E. (1981) *Physical Principles of Exploration Methods*. London: Macmillan.
Dohr, G. (1981) *Applied Geophysics*. New York: Halstead.
Griffiths, D.H. and King, R.F. (1981) *Applied Geophysics for Geologists and Engineers*. Oxford: Pergamon.
Kearey, P. and Brooks, M. (1991) *An Introduction to Geophysical Exploration*, 2nd edn. Oxford: Blackwell Scientific.

Milsom, J. (1989) *Field Geophysics*. Milton Keynes: Open University Press.
Parasnis, D.S. (1986) *Principles of Applied Geophysics*, 4th edn. London: Chapman & Hall.
Robinson, E.S. and Coruh, C. (1988) *Basic Exploration Geophysics*. New York: John Wiley.
Sharma, P.V. (1986) *Geophysical Methods in Geology*. New York: Elsevier Science.
Telford, W.M., Geldart, L.P., Sheriff, R.E. and Keys, D.A. (1990) *Applied Geophysics*, 2nd edn. Cambridge: Cambridge University Press.

Further reading

See also monographs and special publications produced by the Society for Exploration Geophysicists (SEG), and by the Environmental and Engineering Geophysical Society (EEGS). The latter holds an annual Symposium on the Application of Geophysics to Engineering and Environmental Problems (SAGEEP) and publishes the proceedings. Other organisations of note are the Australian Society of Exploration Geophysics (ASEG), the Canadian Exploration Geophysics Society, the South African Geophysical Association, and the European Association of Geoscientists and Engineers (EAGE), among others.

ASEG publishes the quarterly journal *Exploration Geophysics*; SEG publishes the journals *Geophysics* and *Geophysics: The Leading Edge*, and books, monographs and audiovisual materials (slides, videos, etc.). In July 1995, the EEGS published an inaugural volume of the *Journal of Environmental and Engineering Geophysics*, and in January 1996 the European Section of the EEGS launched the first issue of the *European Journal of Environmental and Engineering Geophysics*. The EAGE produces *Geophysical Prospecting*.

The list above gives a general idea of what is available. For those interested particularly in archaeological geophysics, very useful guidelines have been produced by the English Heritage Society (David 1995). During 1995, John Wiley & Sons Ltd produced the first two issues of another new journal entitled *Archaeological Prospection*.

The rapid growth in the number of journals and other publications in environmental and engineering geophysics demonstrates the growing interest in the subject and the better awareness of the applicability of modern geophysical methods.

Section 1
POTENTIAL FIELD METHODS

Chapter 2
Gravity methods

2.1 INTRODUCTION

Gravity surveying measures variations in the Earth's gravitational field caused by differences in the density of sub-surface rocks. Although known colloquially as the 'gravity' method, it is in fact the variation of the *acceleration* due to gravity that is measured. Gravity methods have been used most extensively in the search for oil and gas, particularly in the early twentieth century. While such methods are still employed very widely in hydrocarbon exploration, many other applications have been found (Table 2.1), some examples of which are described in more detail in Section 2.7.

Micro-gravity surveys are those conducted on a very small scale – of the order of hundreds of square metres – and which are capable of detecting cavities, for example, as small as 1 m in diameter within 5 m of the surface.

Perhaps the most dramatic change in gravity exploration in the 1980s has been the development of instrumentation which now permits *airborne* gravity surveys to be undertaken routinely and with a high degree of accuracy (see Section 2.5.7). This has allowed aircraft-borne gravimeters to be used over otherwise inaccessible terrain and has led to the discovery of several small but significant areas with economic hydrocarbon potentials.

Table 2.1 Applications of gravity surveying

Hydrocarbon exploration
Regional geological studies
Isostatic compensation determination
Exploration for, and mass estimation of, mineral deposits
Detection of sub-surface cavities (micro-gravity)
Location of buried rock-valleys
Determination of glacier thickness
Tidal oscillations
Archaeogeophysics (micro-gravity); e.g. location of tombs
Shape of the earth (geodesy)
Military (especially for missile trajectories)
Monitoring volcanoes

2.2 PHYSICAL BASIS

2.2.1 Theory

The basis on which the gravity method depends is encapsulated in two laws derived by Sir Isaac Newton, which he described in *Principia Mathematica* (1687) – namely his Universal Law of Gravitation, and his Second Law of Motion.

The first of these two laws states that the force of attraction between two bodies of known mass is directly proportional to the product of the two masses and inversely proportional to the square of the distance between their centres of mass (Box 2.1). Consequently, the greater the distance separating the centres of mass, the smaller is the force of attraction between them.

Box 2.1   Newton's Universal Law of Gravitation

$$\text{Force} = \text{gravitational constant} \times \frac{\text{mass of Earth } (M) \times \text{mass } (m)}{(\text{distance between masses})^2}$$

$$F = \frac{G \times M \times m}{R^2} \qquad \text{(equation (1))}$$

where the gravitational constant $(G) = 6.67 \times 10^{-11}\,\text{N m}^2\,\text{kg}^{-2}$

Newton's law of motion states that a force (F) is equal to mass (m) times acceleration (Box 2.2). If the acceleration is in a vertical direction, it is then due to gravity (g).

Box 2.2   Newton's Second Law of Motion

$$\text{Force} = \text{mass } (m) \times \text{acceleration } (g)$$

$$F = m \times g \qquad \text{(equation (2))}$$

Equations (1) and (2) can be combined to obtain another simple relationship:

$$F = \frac{G \times M \times m}{R^2} = m \times g; \quad \text{thus } g = \frac{G \times M}{R^2} \qquad \text{(equation (3))}.$$

This shows that the magnitude of the acceleration due to gravity on Earth (g) is directly proportional to the mass (M) of the Earth and inversely proportional to the square of the Earth's radius (R). Theoretically, acceleration due to gravity should be constant over the Earth. In reality, gravity varies from place to place because the Earth has the shape of a flattened sphere (like an orange or an inverted pear),

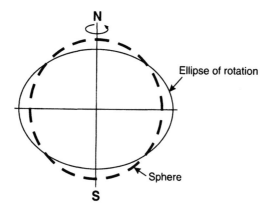

Figure 2.1 Exaggerated difference between a sphere and an ellipse of rotation (spheroid)

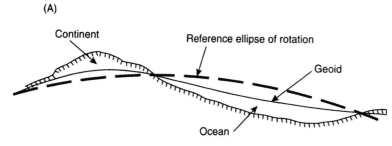

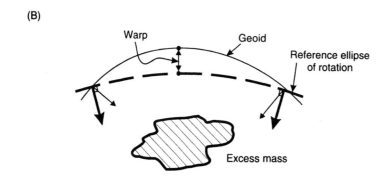

Figure 2.2 Warping of the geoid: (A) continental-scale effects, and (B) localised effects due to a subsurface excess mass

rotates, and has an irregular surface topography and variable mass distribution (especially near the surface).

The shape of the Earth is a consequence of the balance between gravitational and centrifugal accelerations causing a slight flattening to form an oblate spheroid. Mathematically it is convenient to refer to the Earth's shape as being an *ellipse of rotation* (Figure 2.1).

The sea-level surface, if undisturbed by winds or tides, is known as the *geoid* and is particularly important in gravity surveying as it is horizontal and at right angles to the direction of the acceleration due to gravity everywhere. The geoid represents a surface over which the gravitational field has equal value and is called an *equipotential surface*. The irregular distribution of mass, especially near the Earth's surface, warps the geoid so that it is not identical to the ellipse of rotation (Figure 2.2). Long-wavelength anomalies, which can be mapped using data from satellites (Wagner *et al.* 1977), relate to very deep-seated masses in the mantle (Figure 2.2A), whereas density features at shallow depths cause shorter-wavelength warps in the geoid (Figure 2.2B). Consequently, anomalies within the gravitational field can be used to determine how mass is distributed. The particular study of the gravitational field and of the form of the Earth is called *geodesy* and is used to determine exact geographical locations and to measure precise distances over the Earth's surface (*geodetic surveying*).

2.2.2 Gravity units

The first measurement of the acceleration due to gravity was made by Galileo in a famous experiment in which he dropped objects from the top of the leaning tower of Pisa. The normal value of g at the Earth's surface is 980 cm/s^2. In honour of Galileo, the c.g.s. unit of acceleration due to gravity (1 cm/s^2) is the *Gal*. Modern gravity meters (gravimeters) can measure extremely small variations in acceleration due to gravity, typically 1 part in 10^9 (equivalent to measuring the distance from the Earth to the Moon to within a metre). The sensitivity of modern instruments is about ten parts per million. Such small numbers have resulted in sub-units being used such as the milliGal (1 mGal = 10^{-3} Gals) and the microGal (1 μGal = 10^{-6} Gals). Since the introduction of SI units, acceleration due to gravity is measured in μm/s^2, which is rather cumbersome and so is referred to as the *gravity unit* (g.u.); 1 g.u. is equal to 0.1 mGal [10 g.u. = 1 mGal]. However, the gravity unit has not been universally accepted and 'mGal' and 'μGal' are still widely used.

2.2.3 Variation of gravity with latitude

The value of acceleration due to gravity varies over the surface of the Earth for a number of reasons, one of which is the Earth's shape. As the polar radius (6357 km) is 21 km shorter than the equatorial radius (6378 km) the points at the poles are closer to the Earth's centre of mass (so smaller value of R) and, therefore, the value of gravity at the poles is greater (by about 0.7%) than that at the equator (Figure 2.3) (see equation (3) under Box 2.2). Furthermore, as the Earth rotates once per sidereal day around its north–south axis, there is a centrifugal acceleration acting which is greatest where the rotational velocity is largest, namely at the equator (1674 km/h; 1047 miles/h) and decreases to zero at the poles (Figure 2.3). The centrifugal

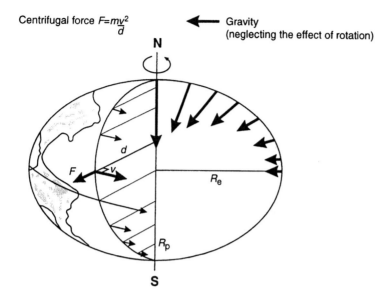

Centrifugal force $F = m\dfrac{v^2}{d}$

← Gravity
(neglecting the effect of rotation)

Figure 2.3 Centrifugal acceleration and the variation of gravity with latitude ϕ (not to scale)

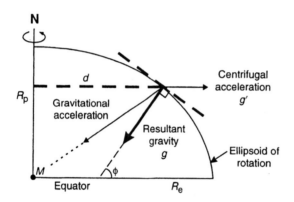

Figure 2.4 Resultant of centrifugal acceleration (g') and the acceleration due to gravity (g) (not to scale); the geographic (geodetic) latitude is given by ϕ. After Robinson and Coruh (1988)

acceleration, which is equal to the rotational velocity (ω) squared times the distance to the rotational axis (d), serves to decrease the value of the gravitational acceleration. It is exactly the same mechanism as that which keeps water in a bucket when it is being whirled in a vertical plane.

The value of gravity measured is the resultant of that acting in a line with the Earth's centre of mass with the centrifugal acceleration (Figure 2.4). The resultant acts at right-angles to the ellipsoid of rotation so that a plumb line, for example, hangs vertically at all locations at sea level. The angle ϕ in Figure 2.4 defines the geodetic (ordinary or geographic) latitude. The resultant gravity at the poles is 5186 mGal (51 860 g.u.) greater than at the equator and varies systematically with latitude in between, as deduced by Clairaut in 1743.

Subsequent calculations in the early twentieth century, based on Clairaut's theory, led to the development of a formula from which it was possible to calculate the theoretical acceleration due to gravity (g_ϕ) at a given geographic latitude (ϕ) relative to that at sea level (g_0). Parameters α and β are constants which depend on the amount of flattening of the spheroid and on the speed of rotation of the Earth.

Box 2.3 General form of the International Gravity Formula

$$g_\phi = g_0(1 + \alpha \sin^2 \phi - \beta \sin^2 2\phi)$$

In 1930 the International Union of Geodesy and Geophysics adopted the form of the *International Gravity Formula* (Nettleton 1971; p. 20) shown in Box 2.3. This became the standard for gravity work. However, refined calculations using more powerful computers and better values for Earth parameters resulted in a new formula – known as the *Geodetic Reference System 1967* (GRS67) – becoming the standard (Woollard 1975) (Box 2.4). If gravity surveys using the 1930 gravity formula are to be compared with those using the 1967 formula, then the third formula (Kearey and Brooks 1991) in Box 2.4 should be used to compensate for the differences between them. Otherwise, discrepancies due to the differences in the equations may be interpreted wrongly as being due to geological causes.

Box 2.4 Standard formulae for the theoretical value of g at a given latitude ϕ

$g_\phi(1930)$
$$= 9.78049\,(1 + 0.0052884 \sin^2 \phi - 0.0000059 \sin^2 2\phi)\,\mathrm{m/s^2}$$

$$g_\phi(1967) = 9.78031846\,(1 + 0.005278895 \sin^2 \phi$$
$$+ 0.000023462 \sin^4 \phi)\,\mathrm{m/s^2}$$

$$g_\phi(1967) - g_\phi(1930) = (-172 + 136 \sin^2 \phi)\,\mu\mathrm{m/s^2}\ (\mathrm{g.u.})$$

2.2.4 Geological factors affecting density

Gravity surveying is sensitive to variations in rock density, so an appreciation of the factors that affect density will aid the interpretation of gravity data. Ranges of bulk densities for a selection of different material types are listed in Table 2.2 and shown graphically in Figure 2.5.

It should be emphasised that in gravity surveys, the determination of densities is based on rocks that are accessible either at the surface, where they may be weathered and/or dehydrated, or from boreholes, where they may have suffered from stress relaxation and be far more

Table 2.2   Densities of common geologic materials (data from Telford *et al.* 1990)

Material type	Density range (Mg/m^3)	Approximate average density (Mg/m^3)
Sedimentary rocks		
Alluvium	1.96–2.00	1.98
Clay	1.63–2.60	2.21
Gravel	1.70–2.40	2.00
Loess	1.40–1.93	1.64
Silt	1.80–2.20	1.93
Soil	1.20–2.40	1.92
Sand	1.70–2.30	2.00
Sandstone	1.61–2.76	2.35
Shale	1.77–3.20	2.40
Limestone	1.93–2.90	2.55
Dolomite	2.28–2.90	2.70
Chalk	1.53–2.60	2.01
Halite	2.10–2.60	2.22
Glacier ice	0.88–0.92	0.90
Igneous rocks		
Rhyolite	2.35–2.70	2.52
Granite	2.50–2.81	2.64
Andesite	2.40–2.80	2.61
Syenite	2.60–2.95	2.77
Basalt	2.70–3.30	2.99
Gabbro	2.70–3.50	3.03
Metamorphic rocks		
Schist	2.39–2.90	2.64
Gneiss	2.59–3.00	2.80
Phylite	2.68–2.80	2.74
Slate	2.70–2.90	2.79
Granulite	2.52–2.73	2.65
Amphibolite	2.90–3.04	2.96
Eclogite	3.20–3.54	3.37

cracked than when *in situ*. Consequently, errors in the determination of densities are among the most significant in gravity surveying. This should be borne in mind when interpreting gravity anomalies so as not to over-interpret the data and go beyond what is geologically reasonable.

There are several crude 'rules of thumb' which can be used as general guides (Dampney 1977; Telford *et al.* 1990; Nettleton 1971, 1976). Sedimentary rocks tend to be the least dense (average density about 2.1 ± 0.3 Mg/m^3). Within the three fundamental rock classifications there are crude trends and associations which are outlined in the next section. Commonly, units are quoted in terms of grams per cubic centimetre (g/cm^3) but are herein referred to in the SI derived units of Mg/m^3 which are numerically equivalent.

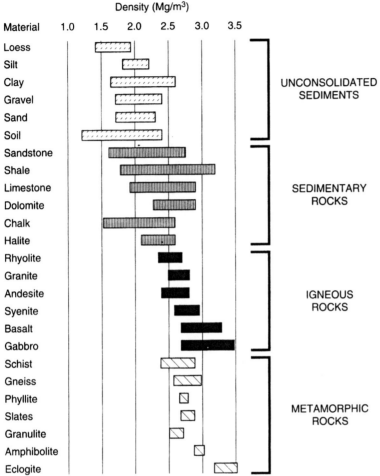

Figure 2.5 Variations in rock density for diffeent rock types. Data from Telford *et al.* (1990)

2.2.4.1 *Sedimentary rocks*

At least seven factors affect the density of sedimentary materials: composition, cementation, age and depth of burial, tectonic processes, porosity and pore-fluid type. Any or all of these may apply for a given rock mass. The degree to which each of these factors affects rock density is given in Table 2.3; but experience shows that, under normal circumstances, the density contrast between adjacent sedimentary strata is seldom greater than 0.25 Mg/m^3.

Density varies depending on the material of which the rock is made, and the degree of consolidation. Four groups of materials are listed in order of increasing density in Table 2.4. Sediments that remain buried for a long time consolidate and lithify, resulting in reduced porosity and consequently an increased density.

Table 2.3   The effect of different physical factors on density

Factor	Approximate percentage change in density
Composition	35
Cementation	10
Age and depth of burial	25
Tectonic processes	10
Porosity and pore fluids	10

Table 2.4   Approximate average densities of sedimentary rocks

Material type	Approximate average Density (Mg/m^3)
Soils and alluvium	2.0
Shales and clays	2.3
Sandstones and conglomerates	2.4
Limestone and dolomite	2.6

In sandstones and limestones, densification is achieved not by volume change but by pore spaces becoming infilled by natural cement. In shales and clays, the dominant process is that of compaction and, ultimately, recrystallisation into minerals with greater densities.

2.2.4.2   Igneous rocks

Igneous rocks tend to be denser than sedimentary rocks although there is overlap. Density increases with decreasing silica content, so basic igneous rocks are denser than acid ones. Similarly, plutonic rocks tend to be denser than their volcanic equivalents (see Table 2.5).

Table 2.5   Variation of density with silica content and crystal size for selected igneous rocks; density ranges and, in parentheses, average densities are given in Mg/m^3. Data from Telford *et al.* (1990)

Crystal size	Silica content		
	Acid	Intermediate	Basic
Fine-grained (volcanic)	Rhyolite 2.35–2.70 (2.52)	Andesite 2.4–2.8 (2.61)	Basalt 2.70–3.30 (2.99)
Coarse-grained (plutonic)	Granite 2.50–2.81 (2.64)	Syenite 2.60–2.95 (2.77)	Gabbro 2.70–3.50 (3.03)

2.2.4.3 *Metamorphic rocks*

The density of metamorphic rocks tends to increase with decreasing acidity and with increasing grade of metamorphism. For example, schists may have lower densities than their gneissose equivalents. However, variations in density within metamorphic rocks tend to be far more erratic than in either sedimentary or igneous rocks and can vary considerably over very short distances.

2.2.4.4 *Minerals and miscellaneous materials*

As the gravity survey method is dependent upon contrast in densities, it is appropriate to highlight some materials with some commercial

Table 2.6 Densities of a selection of metallic and non-metallic minerals and some miscellaneous materials. Data from Telford *et al.* (1990)

Material type	Density range (Mg/m^3)	Approximate average density (Mg/m^3)
Metallic minerals		
Oxides, carbonate		
Manganite	4.2–4.4	4.32
Chromite	4.2–4.6	4.36
Magnetite	4.9–5.2	5.12
Haematite	4.9–5.3	5.18
Cuprite	5.7–6.15	5.92
Cassiterite	6.8–7.1	6.92
Wolframite	7.1–7.5	7.32
Uraninite	8.0–9.97	'9.17
Copper	n.d.	8.7
Silver	n.d.	10.5
Gold	15.6–19.4	17.0
Sulphides		
Malachite	3.9–4.03	4.0
Stannite	4.3–4.52	4.4
Pyrrhotite	4.5–4.8	4.65
Molybdenite	4.4–4.8	4.7
Pyrite	4.9–5.2	5.0
Cobaltite	5.8–6.3	6.1
Galena	7.4–7.6	7.5
Cinnabar	8.0–8.2	8.1
Non-metallic minerals		
Gypsum	2.2–2.6	2.35
Bauxite	2.3–2.55	2.45
Kaolinite	2.2–2.63	2.53
Baryte	4.3–4.7	4.47
Miscellaneous materials		
Snow	0.05–0.88	n.d.
Petroleum	0.6–0.9	n.d.
Lignite	1.1–1.25	1.19
Anthracite	1.34–1.8	1.50

value for which the method can be used for exploration purposes. Gravity surveying becomes increasingly appropriate as an exploration tool for those ore materials with greatest densities. The densities of a selection of metallic and non-metallic minerals and of several other materials are listed in Table 2.6.

2.3 MEASUREMENT OF GRAVITY

2.3.1 Absolute gravity

Determination of the acceleration due to gravity in absolute terms requires very careful experimental procedures and is normally only undertaken under laboratory conditions. Two methods of measurement are used, namely the falling body and swinging pendulum methods. However, it is the more easily measured *relative* variations in gravity that are of interest and value to explorationists. More detailed descriptions of how absolute gravity is measured are given by Garland (1965), Nettleton (1976) and Robinson and Coruh (1988). A popular account of gravity and its possible non-Newtonian behaviour has been given by Boslough (1989); see also Parker and Zumberge (1989).

A network of gravity stations has been established worldwide where absolute values of gravity have been determined by reference to locations where absolute gravity has been measured, such as at the National Bureau of Standards at Gaithersburg, USA, the National Physical Laboratory at Teddington, England, and Universidad Nationale de Columbia, Bogata, Columbia. The network is referred to as the *International Gravity Standardisation Net 1971* (IGSN 71) (Morelli 1971) and was established in 1963 by Woollard and Rose (1963). It is thus possible to tie in any regional gravity survey to absolute values by reference to the IGSN 71 and form a primary network of gravity stations.

2.3.2 Relative gravity

In gravity exploration it is not normally necessary to determine the absolute value of gravity, but rather it is the relative variation that is measured. A base station (which can be related to the IGSN 71) is selected and a secondary network of gravity stations is established. All gravity data acquired at stations occupied during the survey are reduced relative to the base station. If there is no need for absolute values of g to be determined, the value of gravity at a local base station is arbitrarily designated as zero. Details of the data reduction procedure are given in Section 2.5.

The spacing of gravity stations is critical to the subsequent interpretation of the data. In regional surveys, stations may be located with a density of 2–3 per km^2, whereas in exploration for hydrocarbons, the station density may be increased to 8–10 per km^2. In localised surveys where high resolution of shallow features is required, gravity stations may be spaced on a grid with sides of length 5–50 m. In micro-gravity work, the station spacing can be as small as 0.5 m.

For a gravity survey to achieve an accuracy of ± 0.1 mGal, the latitudinal position of the gravimeter must be known to within ± 10 m and the elevation to within ± 10 mm. Furthermore, in conjunction with multiple gravity readings and precision data reduction, gravity data can be obtained to within $\pm 5\,\mu$Gal (Owen 1983). The most significant causes of error in gravity surveys on land are uncertainties in station elevations. At sea, water depths are measured easily by using high-precision echo sounders. Positions are determined increasingly by satellite navigation; and in particular, the advent of the Global Positioning System (GPS) (Bullock 1988), with its compact hardware and fast response time, is resulting in GPS position-fixing becoming more precise. This is particularly true with reference to airborne gravity measurements.

2.4 GRAVITY METERS

No single instrument is capable of meeting all the requirements of every survey, so there are a variety of devices which serve different purposes. In 1749, Pierre Bouguer found that gravity could be measured using a swinging pendulum. By the nineteenth century, the pendulum was in common use to measure relative variations in gravity. The principle of operation is simple. Gravity is inversely proportional to the square of the period of oscillation (T) and directly proportional to the length of the pendulum (L) (Box 2.5). If the same pendulum is swung under identical conditions at two locations where the values of accelerations due to gravity are g_1 and g_2, then the ratio of the two values of g is the same as the ratio of the two corresponding periods of oscillation T_1 and T_2.

Box 2.5 Acceleration due to gravity from pendulum measurements

Gravity = constant × pendulum length/period^2 $g = 4\pi^2 L/T^2$

$$\frac{(\text{Period}_1)^2}{(\text{Period}_2)^2} = \frac{\text{gravity}_2}{\text{gravity}_1}\qquad \frac{T_2^2}{T_1^2} = \frac{g_2}{g_1}$$

Further, the size of the difference in acceleration due to gravity (δg) between the two locations is (to the first order) equal to the product of gravity and twice the difference in periods $(T_2 - T_1)$ divided by the first period (Box 2.6). This method is accurate to about 1 mGal if the periods are measured over at least half an hour. Portable systems were used in exploration for hydrocarbons in the 1930s.

Box 2.6   Differences in gravitational acceleration

$$\text{Gravity difference} = -2 \times \text{gravity} \times \frac{\text{difference in periods}}{\text{period}_1};$$

$$\delta g = -2g \frac{(T_2 - T_1)}{T_1}$$

Another method of determining relative gravity is that of the torsion balance (Figure 2.6). English physicist Henry Cavendish devised this system to measure the gravitational constant in 1791. The method was developed for geodetic purposes in 1880 by a Hungarian physicist, Baron Roland von Eötvös. After further modification it was used in exploration from 1915 to the late 1940s. The method, which measures variations in only the horizontal component of gravity due to terrain and not vertical gravity, is capable of very great sensitivity (to 0.001 mGal) but is awkward and very slow to use in the field. The method is described in more detail by Telford *et al.* (1990).

Since about the early 1930s, variations in relative gravity have been measured using gravity meters (gravimeters), firstly stable (static) and

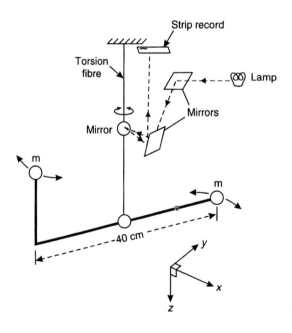

Figure 2.6 Schematic of a torsion balance

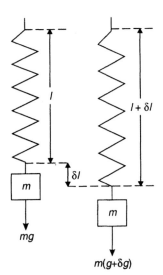

Figure 2.7 Extension (δl) of a spring due to additional gravitational pull (δg)

more recently unstable (astatic) types. Practical aspects of how to use such instruments have been detailed by Milsom (1989, Ch. 2). Gravimeters are sophisticated spring balances from which a constant mass is suspended (Figure 2.7). The weight of the mass is the product of the mass and the acceleration due to gravity. The greater the weight acting on the spring, the more the spring is stretched. The amount of extension (δl) of the spring is proportional to the extending force, i.e. the excess weight of the mass (δg). (Remember that weight equals mass times acceleration due to gravity.) The constant of proportionality is the elastic spring constant κ. This relationship is known as Hooke's Law (Box 2.7).

As the mass is constant, variations in weight are caused by changes in gravity (δg). By measuring the extension of the spring (δl), differences in gravity can then be determined. As the variations in g are very small (1 part in 10^8) the extension of any spring will also be extremely tiny. For a spring 30 cm long, changes in length of the order of 3×10^{-8} m (30 nanometres) have to be measured. Such small distances are even smaller than the wavelength of light (380–780 nm). Consequently, gravimeters use some form of system to amplify the movement so that it can be measured accurately.

Box 2.7   Hooke's Law

$$\text{Extension to spring} = \text{mass} \times \frac{\text{change in gravity}}{\text{spring constant}} \qquad \delta l = \frac{m\delta g}{\kappa}$$

$$\text{Change in gravity} = \text{constant} \times \text{extension/mass} \qquad \delta g = \kappa \delta l / m$$

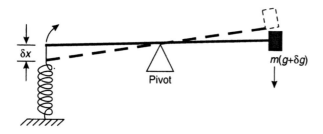

Figure 2.8   Basic principle of operation of a stable gravimeter

2.4.1   Stable (static) gravimeters

Stable gravimeters (Figure 2.8), which were developed in the 1930s,
are less sensitive than their more modern cousins, the unstable
gravimeters, which have largely superseded them. The stable
gravimeter consists of a mass at the end of a beam which is pivoted on
a fulcrum and balanced by a tensioned spring at the other end.
Changes in gravity affect the weight of the mass which is counter-
balanced by the restoring action of the spring. Different configurations of stable gravimeters are shown in Figure 2.9 and are discussed
in more detail by Garland (1965), Nettleton (1976), Telford *et al.*
(1990), and Parasnis (1986). A brief description of three stable
gravimeters is given below.

2.4.1.1   *Askania*

A beam with a mass at one end is pivoted on a main spring
S (Figure 2.9A). Changes in gravity cause the beam to tilt, so producing a deflection in a beam of light which is reflected off a mirror placed
on the mass. A photoelectric cell, the output of which is displayed on
a galvanometer, measures the displacement of the light beam. An
auxiliary spring (AS) is retensioned using a micrometer to restore the
mass to its rest position, which is indicted when the galvanometer
reading is returned to zero (nulled).

2.4.1.2   *Boliden*

The Boliden gravimeter uses the principle that the capacitance of
a parallel-plate capacitor changes with the separation of the plates
(Figure 2.9B). The mass has the form of a bobbin with a plate at each
end and is suspended by two springs between two other capacitor
plates. With a change in gravity, the mass moves relative to the fixed
plates, changing the capacitance between the upper plates; this movement can be detected easily using a tuned circuit. The lower plates are
connected to a d.c. supply which supports the bobbin mass by
electrostatic repulsion. With a change in gravity and the consequent
displacement of the bobbin relative to the fixed plates, the original or

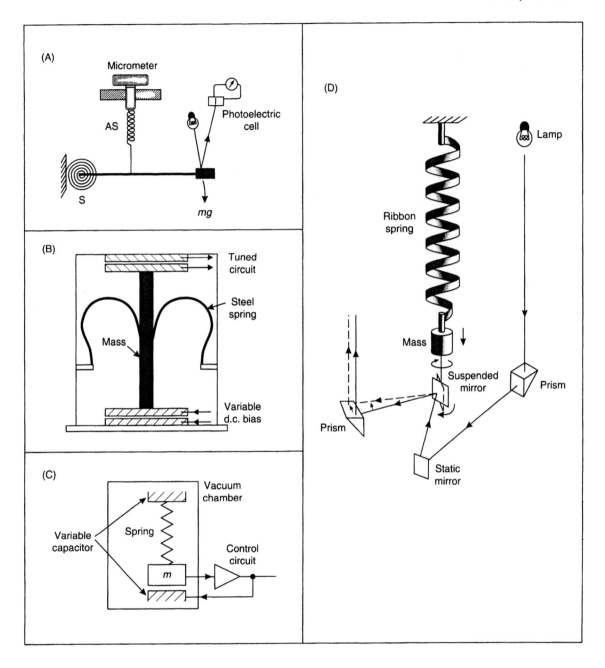

a reference position can be obtained by changing the direct voltage between the lower pair of plates. The overall sensitivity is about 1 g.u. (0.1 mGal). A modern version has been produced by Scintrex (Model CG-3) which operates on a similar principle (see Figure 2.9C). Any displacement of the mass due to a change in gravity is detected by

Figure 2.9 Types of stable gravimeter: (A) Askania; (B) Boliden; (C) Scintrex CG-3; and (D) Gulf (Hoyt). After Garland (1965), Telford *et al.* (1976), Robinson and Coruh (1988), by permission

a capacitor transducer and activates a feedback circuit. The mass is returned to its null position by the application of a direct feedback voltage (which is proportional to the change in gravity) to the plates of the capacitor which changes the electrostatic force between the plates and the mass (Robinson and Coruh 1988).

2.4.1.3   Gulf (Hoyt)

The Gulf gravimeter comprises a coiled helical ribbon spring which rotates as it changes length (Figure 2.9D). The rotation of the free end of the spring is much larger than the change in length and so is more easily measured. The range of measurement is quite small, being only 300 g.u. (30 mGal), although this can be overcome to some extent by retensioning the spring, and the accuracy of measurement is to within 0.2–0.5 g.u. (0.02–0.05 mGal).

2.4.2   Unstable (astatic) gravimeters

Since the 1930s, unstable gravimeters have been used far more extensively than their stable counterparts. In a stable device, once the system has been disturbed it will return to its original position, whereas an unstable device will move further away from its original position.

For example, if a pencil lying flat on a table is lifted at one end and then allowed to drop, the pencil will return to being flat on the table. However, if the pencil starts by being balanced on its end, once disturbed, it will fall over; i.e. it becomes unstable, rather than returning to its rest position. The main point of the instability is to exaggerate any movement, so making it easier to measure, and it is this principle on which the unstable gravimeter is based.

Various models of gravimeter use different devices to achieve the instability. The principle of an astatic gravimeter is shown in Figure 2.10. An almost horizontal beam hinged at one end supports a mass at the other. The beam is attached to a main spring which is connected at its upper end to a support above the hinge. The spring attempts to pull the beam up anticlockwise by its turning moment, which is equal to the restoring force in the spring multiplied by the perpendicular distance from the hinge (d). This turning moment is balanced by the gravitational turning moment which attempts to rotate the beam in a clockwise manner about the hinge and is equal to the weight of the mass (mg) times the length of the beam (l) multiplied by the cosine of the angle of the beam from the horizontal (θ) (i.e. $mgl\cos\theta$). If gravity changes, the beam will move in response but will be maintained in its new position because the main spring is a 'zero-length' spring. One virtue of such a spring is that it is pretensioned during manufacture so that the tension in the spring is proportional to its length. This means that if all forces were removed from the

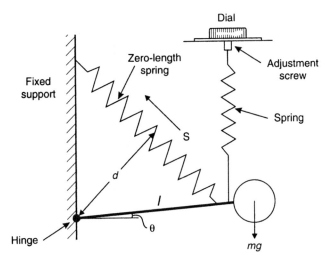

Figure 2.10 Principle of operation of an astatic gravimeter

spring it would collapse to zero length, something which is impossible in practice. Another virtue of the zero-length spring is that it results in an instrument which is linear and very responsive over a wide range of gravity values. Astatic gravimeters do not measure the movement of the mass in terms of changes in gravity but require the displaced mass to be restored to a null position by the use of a micrometer. The micrometer reading is multiplied by an instrumental calibration factor to give values of gravity, normally to an accuracy within 0.1 g.u. (0.01 mGal) and in some specialist devices to within 0.01 g.u. (0.001 mGal = 1 μGal).

2.4.2.1 *Thyssen*

Although obsolete, this gravimeter demonstrates the instability concept extremely well and is included for this reason. An extra mass is placed above a balanced beam (Figure 2.11) so producing the instability condition. If gravity increases, the beam tilts to the right and the

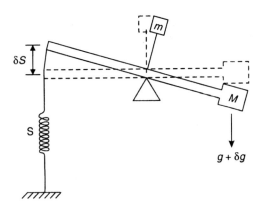

Figure 2.11 Schematic of a Thyssen gravimeter

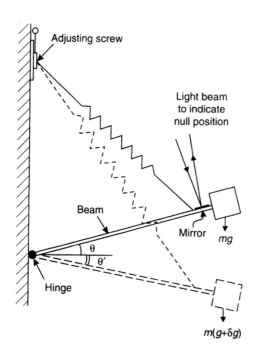

Figure 2.12   Schematic of a LaCoste–Romberg gravimeter. After Kearey and Brooks (1991), by permission

movement of the extra mass enhances the clockwise rotation about the pivot, and conversely for a reduction in gravity. When used, this type of gravimeter had a sensitivity of about 2.5 g.u. (0.25 mGal).

2.4.2.2   LaCoste–Romberg

This device is a development of LaCoste's long-period seismograph (LaCoste 1934) and is illustrated in Figure 2.12. The spring is made of metal with a high thermal conductivity but cannot be insulated totally to eradicate thermal effects and so has to be housed permanently in an enclosed container in which a stable temperature is maintained to within 0.002°C by a thermostat element. The null point is obtained by the observer viewing a scale through an eyepiece onto which a beam of light is reflected from the beam when it is in its rest position. In order to restore the position of the beam, the operator rotates a micrometer gauge on the outer casing which turns a screw which adjusts the beam position. The long length of the screw means that the gravimeter can be used worldwide without having to undergo any resets, which is a major advantage over other makes for surveys where this is important. When this type of gravimeter was manufactured in the 1930s it weighed a massive 30 kg, but modern technology has made it possible for the weight to be reduced to only about 2 kg, excluding the battery required to maintain the heating coils. The springs can be clamped and so the gravimeter is more easily trans-portable than other makes and also less sensitive to vibration. It is possible for some models of LaCoste–Romberg gravimeters to measure to 3 μGal.

2.4.2.3 *Worden*

Unlike the LaCoste–Romberg gravimeter, the Worden is made entirely of quartz glass springs, rods and fibres (Figure 2.13). The quartz construction makes it much easier to reduce thermal effects. Indeed, the whole assembly is housed in a glass vacuum flask and some models have an electrical thermostat. As the spring cannot be clamped, the Worden gravimeter is sensitive to vibration and has to be transported extremely carefully. The range of the instrument is about 20 000 g.u. (2000 mGal) with an accuracy to within 0.1–0.2 g.u. (0.01–0.02 mGal). However, quartz gravimeters such as the Worden can be quite difficult for inexperienced operators to read and a real-istic accuracy may be more like 1 g.u. (0.1 mGal). The Worden gravimeter has two auxiliary springs, one for coarse and the other for fine adjustments.

Figure 2.13 Cross-section through a Worden gravimeter (Texas Instruments Ltd). From Dunning (1970), by permission

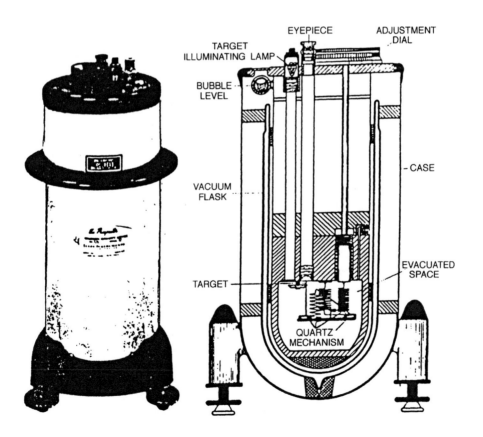

2.4.2.4   *Sodin*

This is very comparable to the Worden and operates in similar ways. Further details are given by Robinson and Coruh (1988).

2.4.2.5   *Vibrating string*

If a mass is suspended on a fibre which is forced to oscillate by an a.c. circuit, then the frequency of vibration, which can be measured electronically, will vary with changes in gravity. For a fibre of length L, and mass per unit length m_s, from which a mass M is suspended, by measuring the frequency of vibration (f), gravity can be determined (Box 2.8). However, the technology is not sufficiently developed to provide the same resolution and accuracy as other gravimeters, but it does give the impression that even more compact and lightweight gravimeters may be forthcoming in the near future. There is potential for use in airborne survey systems.

Box 2.8   Determination of g using a vibrating string

$$\text{Gravity} = \frac{4 \times \text{string length}^2 \times \text{frequency}^2 \times \text{string mass}}{\text{suspended mass}};$$

$$g = \frac{4L^2 f^2 m_s}{M}$$

2.5   CORRECTIONS TO GRAVITY OBSERVATIONS

Gravimeters do not give direct measurements of gravity. Rather, a meter reading is taken which is then multiplied by an instrumental calibration factor to produce a value of observed gravity (g_{obs}). Before the results of the survey can be interpreted in geological terms, these raw gravity data have to be corrected to a common datum, such as sea level (geoid), in order to remove the effects of features that are only of indirect geological interest. The correction process is known as *gravity data reduction* or *reduction to the geoid*. The difference between the value of observed gravity (g_{obs}) and that determined either from the International Gravity Formula/Geodetic Reference System 67 for the same location, or relative to a local base station, is known as the *gravity anomaly*. The various corrections that can be applied are listed in Table 2.7 with the sections of this book in which each one is discussed.

Table 2.7 Corrections to gravity data

Correction	Book sections
Instrument drift	2.5.1
Earth tides	2.5.2
Eötvös	2.5.7
Latitude	2.2.3 and 2.5.3
Elevation	
Free-air correction	2.5.4
Bouguer correction	2.5.5
Terrain	2.5.6
Isostatic	2.5.8

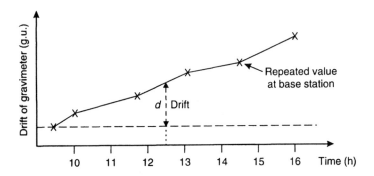

Figure 2.14 An instrumental drift curve

2.5.1 Instrumental drift

Gravimeter readings change (drift) with time as a result of elastic creep in the springs, producing an apparent change in gravity at a given station. The instrumental drift can be determined simply by repeating measurements at the same stations at different times of the day, typically every 1–2 hours. The differences between successive measurements at the same station are plotted to produce a *drift curve* (Figure 2.14). Observed gravity values from intervening stations can be corrected by subtracting the amount of drift from the observed gravity value. For example, in Figure 2.14 the value of gravity measured at an outlying station at 12.30 hours should be reduced by the amount of drift *d*. The range of drift of gravimeters is from a small fraction of one g.u. to about ten g.u. per hour. If the rate of drift is found to be irregular, return the instrument to the supplier – it is probably faulty!

2.5.2 Tides

Just as the water in the oceans responds to the gravitational pull of the Moon, and to a lesser extent of the Sun, so too does the solid earth.

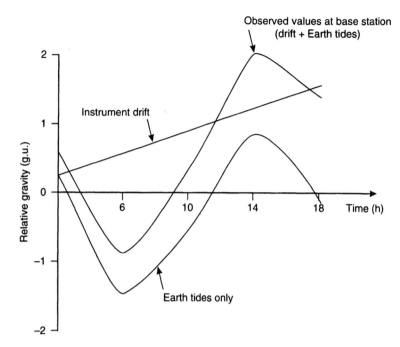

Figure 2.15 Graph of the effects of Earth tides and instrumental drift on the acceleration due to gravity

Earth tides give rise to a change in gravity of up to three g.u. with a minimum period of about 12 hours. Repeated measurements at the same stations permit estimation of the necessary corrections for tidal effects over short intervals, in addition to determination of the instrumental drift for a gravimeter (Figure 2.15). Alternatively, recourse can be made to published tide tables which are published periodically (e.g. *Tidal Gravity Corrections for 1991*, European Association of Exploration Geophysicists, The Hague).

2.5.3   Latitude

The latitude correction is usually made by subtracting the theoretical gravity calculated using the International Gravity Formula (g_ϕ) (Section 2.2.3) from the observed value (g_{obs}). For small-scale surveys which extend over a total latitude range of less than one degree, and not tied into the absolute gravity network, a simpler correction for latitude can be made. A local base station is selected for which the horizontal gravity gradient (δg_L) can be determined at a given degree of latitude (ϕ) by the expression in Box 2.9.

Box 2.9   Local latitude correction

$$\delta g_L = -8.108 \sin 2\phi \quad \text{g.u. per km N}$$

Note that the correction is negative with distance northwards in the northern hemisphere or with distance southwards in the southern hemisphere. This is to compensate for the increase in the gravity field from the equator towards the poles. For a latitude of 51°N, the local latitude correction is about 8 g.u./km. For gravity surveys conducted with an accuracy of ±0.1 g.u., the latitudinal position of the gravity station needs to be known to within ±10 m, which is well within the capability of modern position-fixing.

2.5.4 Free-air correction

The basis of this correction is that it makes allowance for the reduction in magnitude of gravity with height above the geoid (see Figure 2.16 and Box 2.10), irrespective of the nature of the rock below. It is analogous to measuring gravity in the basket of a hot-air balloon in flight – hence the term *free-air correction*. The free-air correction is the difference between gravity measured at sea level and at an elevation of *h* metres with no rock in between. A value of 3.086 g.u./m

(A)

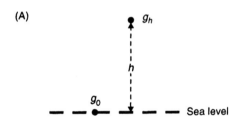

(B)

(C)

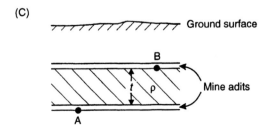

Figure 2.16 Schematic showing (A) the free-air correction, (B) the Bouguer correction, and (C) the Bouguer correction for measurements made underground

is accepted for most practical applications and is positive at elevations above sea level, and negative below. The free-air correction term varies slightly with latitude from 3.083 g.u./m at the equator to 3.088 g.u./m at the poles. With the normal measuring precision of modern gravimeters being around 0.1 g.u., elevations must be known to within 3–5 cm.

Box 2.10   Free-air correction (see also Figure 2.16)

Taking the Earth to be a sphere (rather than an oblate spheroid) with its mass concentrated at its centre of mass, then the value of gravity at sea level is:

$$g_0 = GM/R^2.$$

The value of gravity at a station at an elevation of h metres above sea level is:

$$g_h = GM/(R+h)^2 = \frac{GM}{R^2}\left(\frac{1-2h}{R}\cdots\right).$$

The difference in gravity between sea level and at h metres is the free-air correction:

$$\delta g_\mathrm{F} = g_0 - g_h = \frac{2g_0 h}{R}.$$

With $g_0 = 9\,817\,855$ g.u., $R = 6\,371\,000$ M, and with h in metres,

$$\delta g_\mathrm{F} = 3.082h \text{ g.u.}$$

Taking into account that the Earth is an oblate spheroid, rather than a sphere, the normally accepted value of the free-air correction is:

$$\boldsymbol{\delta g_\mathrm{F} = 3.086\,h \text{ g.u.}}$$

The reduction in g with increasing height above the ground is important in airborne gravimetry. Anomalies detected by helicopter-mounted gravimeters will have decreased amplitudes and lengthened wavelengths compared with those obtained from land-based surveys. To compare land gravity survey data with airborne, it is necessary to correct for the free-air attenuation of the gravity anomaly by using *upward continuation*, which is discussed in Section 2.6.

The quantity calculated by applying both the latitude and the free-air corrections is called the *free-air anomaly* and is commonly used to display corrected gravity data for oceans and continental shelves (see, e.g., Talwani *et al.* 1965).

2.5.5 Bouguer correction

Whereas the free-air correction compensates for the reduction in that part of gravity due only to increased distance from the centre of mass, the Bouguer correction (δg_B) is used to account for the rock mass between the measuring station and sea level (Figure 2.16).

The Bouguer correction calculates the extra gravitational pull exerted by a rock slab of thickness h metres and mean density ρ (Mg/m^3) which results in measurements of gravity (g_{obs}) being overestimated by an amount equal to $0.4192\rho h$ g.u. (Box 2.11). The Bouguer correction should be subtracted from the observed gravity value for stations above sea level. For an average rock density of 2.65 Mg/m^3, the Bouguer correction amounts to 1.12 g.u./m. For marine surveys, the Bouguer correction is slightly different in that the low density of sea water is effectively replaced by an equivalent thickness of rock of a specified density (Box 2.11).

Box 2.11   Bouguer correction

Bouguer correction $(\delta g_B) = 2\pi G\rho h = \beta\rho h$ (g.u.), where:

$$\beta = 2\pi G = 0.4192 \text{ g.u. m}^2 \text{ Mg}^{-1}$$
$$G = 6.67 \times 10^{-8} \text{ m}^3 \text{ Mg}^{-1}\text{s}^{-2}.$$

Density (ρ) is in Mg m^{-3} and height (h) is in metres.

For marine surveys, the Bouguer correction is given by:

$$\delta g_B = \beta(\rho_r - \rho_w)h_w \text{ (g.u.)}.$$

where ρ_r and ρ_w are the densities of rock and sea water respectively, and h_w is the water depth in metres.

A further development of this correction has to be made for gravity measurements made underground (Figure 2.16C). In this case, the Bouguer correction has to allow for the extra gravitational pull ($=0.4191\rho t$ g.u.) on Station A caused by the slab of thickness t metres between the two stations A and B, whereas the value of gravity at Station A is underestimated by the equal but upward attraction of the same slab. The difference in gravity between the two stations is *twice* the normal Bouguer correction ($=0.8384\,\rho t$ g.u.). Allowances also have to be made for underground machinery, mine layout and the

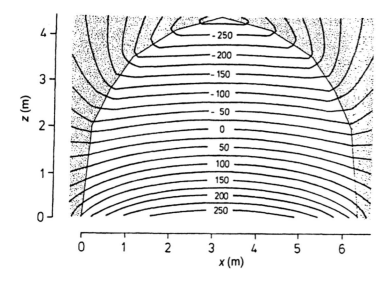

Figure 2.17   Micro-isogals for a typical gallery in a deep coal mine; the contour interval is $25\,\mu\text{Gal}$ and the density of the host rock is $2.65\,\text{Mg/m}^3$. From Casten and Gram (1989), by permission

variable density of local rocks, and this is sometimes referred to as the *Gallery correction* (Figure 2.17).

The Bouguer correction on land has to be modified by allowances for terrain roughness (see Section 2.5.6) in areas where there is a marked topographic change over a short distance, such as an escarpment of cliff. In such a situation the approximation of a semi-infinite horizontal slab of rock no longer holds true and more detailed calculations are necessary (Parasnis 1986; p. 72).

The free-air and Bouguer corrections are commonly combined into one *elevation correction* (δg_E) to simplify data handling (Box 2.12). It should be noted that in some cases, the resulting gravity anomaly may be misleading and the combined calculation should be used judiciously. For a density of $2.60\,\text{Mg/m}^3$, the total elevation correction is 2 g.u./m, which requires elevations to be known to an accuracy within 5 cm if gravity readings are to be made to within 0.1 g.u.

Box 2.12   Elevation correction

Elevation correction (δg_E) = (Free-air − Bouguer) corrections:

$$\delta g_\text{E} = \delta g_\text{F} - \delta g_\text{B}.$$

Substituting in the terms $\delta g_\text{F} = 3.086h$ and $\delta g_\text{B} = 0.4192\rho h$:

$$\delta g_\text{E} = (3.086 - 0.4192\rho)h \text{ (g.u.)}$$

where ρ is the average rock density in Mg/m^3.

One of the main problems with the Bouguer correction is knowing which density to use. For example, a difference of $0.1\,Mg/m^3$ in density for a gravity measurement made at an elevation of 250 m will result in a discrepancy of more than 10 g.u. in the Bouguer correction. In many cases, it may be possible to obtain an estimate of rock densities from appropriate surface samples, or from borehole samples, if available. Caution should be used in the latter case as rock-core samples will relax mechanically, producing many cracks, and expanding slightly in response to the reduced pressure at the surface, giving rise to an underestimate of *in situ* density.

Nettleton (1939, 1940) found a very simple way of determining the appropriateness of the chosen density using a graphical method. Corrected gravity data should show no correlation with topography as all such effects should have been removed through the data reduction process. If a range of densities is chosen and the resulting elevation corrections computed along one gravity profile, the density that shows least correlation with topography is taken as the 'correct' one (Figure 2.18). It is known, however, that this method becomes less accurate if there is any topographic expression due to dipping beds with a significant density contrast to those above and below. Examples of where this might occur are in association with an inclined dense igneous intrusion, or in a marked structural feature with significant variations in density.

A generalised Nettleton method has been proposed by Rimbert *et al.* (1987) which allows for density to vary over a geographic area. The topographic data of an area are smoothed and reduced to produce a surface that lies just below the low topographic points on the survey, and which they refer to as the 'regional topography'. The density between the 'regional' and the actual topography is considered as constant (ρ_0) for a fixed radius around each station. The density below the 'regional' topography is taken as uniform throughout the area of the survey. The variations in density above the 'regional' topography are accounted for statistically but can be plotted in map form to demonstrate the areal variation in density which can be correlated independently with maps of the known geology. Rimbert *et al.* (1987) have achieved accuracies to within 3–4 g.u. with this method in an area in the south of France in which the observed gravity anomalies ranged from 40 to 150 g.u.

Another method to calculate density is to plot a graph of the latitude-corrected gravity data from an area without marked gravity anomalies, against station elevation (Reeves and MacLeod 1986). The resulting straight-line graph yields a gradient that is numerically equal to the elevation correction. Using the result in Box 2.16 and rearranging the equation to make density the subject of the equation, it is possible to solve and find a value for the density. Reeves and MacLeod used this method on data for a survey in Belgium, and produced a graphical elevation correction of 2.35 g.u./m which gave

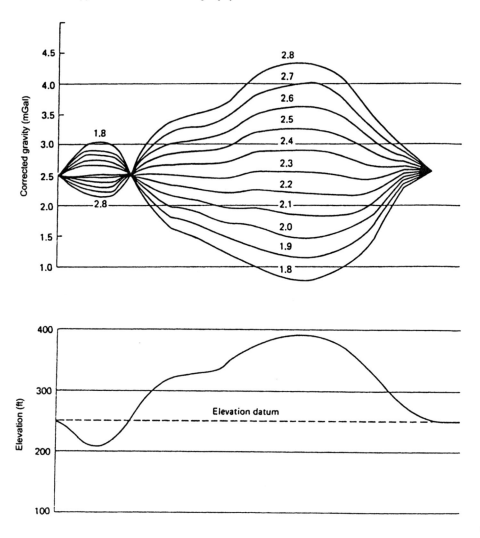

a density estimate of 1.74 Mg/m³, compared with 1.6 Mg/m³ determined using the Nettleton method.

The danger with a graphical method – even if linear regression statistics are used to determine the gradient of the best straight-line fit through the data – is that a small difference in the graphical gradient can result in an unacceptably large density discrepancy, and this makes the method rather insensitive to small changes in density. For example, the data presented by Reeves and MacLeod, rather than showing a single trend with elevation, indicate two: data up to an elevation of 215 m yield a density of 1.43 Mg/m³ and the remaining data up to 280 m yield a density of 1.79 Mg/m³, a difference of 0.36 Mg/m³. Densities should be determined to better than 0.1 Mg/m³ if possible.

Figure 2.18 The Nettleton method for determining the density of near-surface rock formations with the topography as shown. The most appropriate density is about 2.3 Mg/m³ (least correlation with the topography). Corrections are referred to a height of 76 m. From Dobrin (1976), by permission

In underground gravity surveys, a similar method is employed (Hussein 1983). The vertical gravity gradient (i.e. δg_E) is obtained by measuring g at two or more elevations separated by only 1.5–3 m at the same location within an underground chamber. The density can then be calculated as described above. As the measurements of g are strictly controlled, the error in density can be minimised routinely to within 0.08 Mg/m^3 (Casten and Gram 1989).

Rock densities for depths below which it is not possible to sample can also be estimated using the relationship of density with P-wave velocities as described, for example, by Nafe and Drake (1963), Woollard (1950, 1959, 1975), Birch (1960, 1961), and Christensen and Fountain (1975).

2.5.6 Terrain correction

In flat countryside, the elevation correction (the combined free-air and Bouguer correction) is normally adequate to cope with slight topographic effects on the acceleration due to gravity. However, in areas where there are considerable variations in elevation, particularly close to any gravity station, a special *terrain correction* must be applied. The Bouguer correction assumes an approximation to a semi-infinite horizontal slab of rock between the measuring station and sea level. It makes no allowance for hills and valleys and this is why the terrain correction is necessary.

The effect of topography on g is illustrated in Figure 2.19. Consider a gravity station beside a hill as in Figure 2.19A. The slab of rock which comprises the hill (mass M) has its centre of mass above the plane on which the gravimeter is situated. There is a force of attraction between the two masses. If the force is resolved into horizontal and vertical components and the latter only is considered, then it can be seen that the measurement of g at the gravity station will be underestimated by an amount δg. Conversely, if the gravity station is adjacent to a valley, as indicated in Figure 2.19B, then the valley represents a mass deficiency which can be represented by a negative mass $(-M)$. The lack of mass results in the measurement of g to be underestimated by an amount δg. Consequently, a gravity measurement made next to either a hill or a valley requires a correction to be added to it to make allowance for the variable distribution of mass. The correction effectively removes the effects of the topography to fulfil the Bouguer approximation of a semi-infinite rock slab.

Physical computation of the terrain correction is extremely laborious as it has to be carried out for each and every station in an entire survey. A special transparent template, known as a *Hammer chart* after its originator Sigmund Hammer (1939), consists of a series of segmented concentric rings (Figure 2.20). This is superimposed over a topographic map and the average elevation of each segment of the

(A)

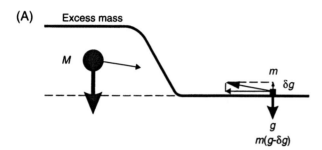

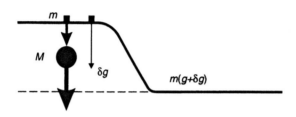

Figure 2.19 The effects of a hill and a valley on the measurement of gravity, illustrating the need for terrain corrections

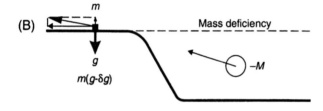

(B)

Figure 2.20 Hammer terrain correction chart with inner rings A–C shown expanded for clarity. After Dobrin (1976) and Milsom (1989), by permission

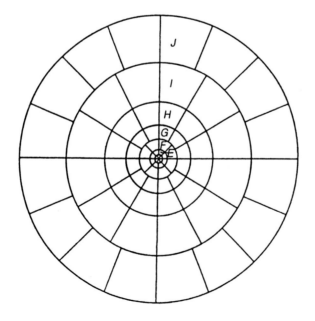

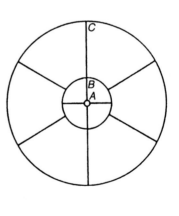

Table 2.8 Terrain corrections for Hammer zones B to M. From Milsom (1989), by permission

Zone	B	C	D	E	F	G
No. of compartments:	4	6	6	8	8	12
Correction (g.u.):			Heights (meters)			
0.01	0.5	1.9	3.3	7.6	11.5	24.9
0.02	0.7	2.6	4.7	10.7	16.3	35.1
0.03	0.8	3.2	5.8	13.1	19.9	43.1
0.04	1.0	3.8	6.7	15.2	23.0	49.8
0.05	1.1	4.2	7.5	17.0	25.7	55.6
0.06	1.2	4.6	8.2	18.6	28.2	60.9
0.07	1.3	5.0	8.9	20.1	30.4	65.8
0.08	1.4	5.4	9.5	21.5	32.6	70.4
0.09	1.5	5.7	10.1	22.9	34.5	74.7
0.10	1.6	6.0	10.6	24.1	36.4	78.7
0.20	2.4	8.7	15.1	34.2	51.6	111.6
0.30	3.2	10.9	18.6	42,1	63.3	136.9
0.40	3.9	12.9	21.7	48.8	73.2	158.3
0.50	4.6	14.7	24.4	54.8.	82.0	177.4
0.60	5.3	16.5	26.9	60.2	90.0	194.7
0.70	6.1	18.2	29.3	65.3	97.3	210.7
0.80	6.9	19.9	31.5	70.1	104.2	225.6
0.90	7.8	21.6	33.7	74.7	110.8	239.8
1.00	8.7	23.4	35.7	79.1	117.0	253.2

	H	I	J	K	L	M
	12	12	16	16	16	16
0.01	32	42	72	88	101	125
0.02	46	60	101	124	148	182
0.03	56	74	125	153	186	225
0.04	65	85	144	176	213	262
0.05	73	95	161	197	239	291
0.06	80	104	176	216	261	319
0.07	86	112	191	233	282	346
0.08	92	120	204	249	303	370
0.09	96	127	216	264	322	391
0.10	103	134	228	278	338	413
0.20	146	190	322	394	479	586
0.30	179	233	396	483	587	717
0.40	206	269	457	557	679	828
0.50	231	301	511	624	759	926
0.60	253	330	561	683	832	1015
0.70	274	357	606	738	899	1097
0.80	293	382	648	790	962	1173
0.90	311	405	688	838	1020	1244
1.00	328	427	726	884	1076	1312

Note: These tables list the exact height differences which, assuming a density of $2000\,kg/m^3$ will produce the tabulated terrain effects. Thus, a height difference of 32 m between gravity station and average topographic level in one compartment of zone E would be associated with a terrain effect of 0.20, or possibly 0.19, g.u. Almost all commercial gravity meters have sensitivities of 0.1 g.u. but an additional decimal place is necessary if large 'rounding off' errors are to be avoided in summing the contributions from all the compartments. The inner radius of zone B is 2 m. Zone outer radii are: B: 16.6 m. C: 53.3 m. D: 170 m, E: 390 m, F: 895 m, G: 1530 m, H: 2.61 km, I: 4.47 km. J: 6.65 km, K: 9.9 km, L: 14.7 km, M: 21.9 km

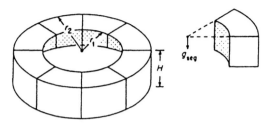

Figure 2.21 Segmented cylindrical ring used to compute terrain corrections. From Robinson and Coruh (1988), by permission

chart is estimated. The outer radius of the furthest zone is 21.9 km beyond which the effect of topography on g is negligible. In some cases, the range of the chart is extended to 50 km. Special tables (Table 2.8) are used to derive the terrain correction in g.u. for each segment, and then all the values are summed to produce a statistically weighted terrain correction for a given gravity station, each ring of the Hammer chart having a different weighting. For example, for a mean height difference of 34.7 m between the gravity station and a segment in zone D, a terrain correction of 0.95 g.u. can be determined from Table 2.8.

The terrain correction method works on the principle of determining the value of g at the centre of an annulus of inner and outer radii r_1 and r_2 (Figure 2.21) using the equation given in Box 2.13. Although topographic maps can be digitised and the terrain corrections calculated by computer (Bott 1959; Kane 1962) for the outer rings of the Hammer chart, it is still necessary to compute the terrain corrections for the innermost rings manually.

Box 2.13   Gravity of a Hammer chart segment (see Figure 2.21)

Gravity of a Hammer chart segment (δg_{seg}):

$$\delta g_{seg} = \frac{2\pi \rho G}{N} [r_2 - r_1 + (r_1^2 + z^2)^{1/2} - (r_2^2 + z^2)^{1/2}] \text{ (g.u.)}$$

where N is the number of segments in the ring, z is the modulus of the difference in elevation between the gravity station and mean elevation of the segment, and ρ is the Bouguer correction density (Mg/m^3).

There is no theoretical reason as to why the terrain correction needs to be calculated on the basis of a circular division of the terrain. It is computationally better for a regular grid to be used, in which case digitised topographic data can be used quite readily. Ketelaar (1976) has suggested a method in which the topography is represented by square prisms of side length D with an upper surface sloping at an angle α. The terrain correction due to each prism can be calculated using the expression in Box 2.14. An example of computer analysis of terrain corrections for micro-gravity surveys has been given by Blizkovsky (1979) in which he also considers the gravitational effects of walls, vertical shafts and horizontal corridors. In micro-gravity surveys, an area of radius 2 m centred on the measurement station should be flat. For terrain effects of less than 1 μGal, height variations of less than 0.3 m in Hammer zone B and up to 1.3 m in zone C can be tolerated.

Box 2.14 Terrain correction for a square prism

Terrain correction due to a square prism of side length D:

$$\delta g_{\text{prism}(i,j)} = G\rho D (1 - \cos \alpha) K(i,j)$$

where $K(i,j)$ is the matrix of prism coordinates within the grid.

The calculation of terrain corrections is labour-intensive, time-consuming and adds considerably to the total cost of the survey. It is only undertaken when dictated by the roughness of the local topography. In built-up areas, care has to be taken in estimating the likely effect of neighbouring buildings. Modern, thin-walled buildings may contribute only a small effect (of the order of 1–5 μGal), but older thick-walled constructions may give rise to a terrain correction effect of 10–30 μGal. In these cases, it is preferable to position the gravimeter more than 2.5 m away from such a building in order to keep the effect to less than 5 μGal. Otherwise, specific calculations of the terrain corrections of the buildings concerned becomes necessary. An alternative approach is to measure the gravity effect around a comparable building, assuming one exists, away from the survey area and to carry out an empirical adjustment of the actual survey data. Budgetary constraints may preclude detailed data reduction, in which case the effectiveness of the micro-gravity survey may be jeopardised. Such considerations need to be made at the design stage of the survey. In some cases, the effects of adjacent buildings are assumed to be negligible and are ignored unjustifiably as they are thought to be too difficult to determine. Yet the magnitude of the anomalies being sought may be of the same order of magnitude as the terrain correction. There is scope for collaboration between geophysicists and members of the construction industry!

2.5.7 Eötvös correction

For a gravimeter mounted on a vehicle, such as a ship or a helicopter, the measured gravitational acceleration is affected by the vertical component of the Coriolis acceleration which is a function of the speed and the direction in which the vehicle is travelling. To compensate for this, gravity data are adjusted by applying the Eötvös correction, named after the Hungarian geophysicist Baron von Eötvös who described this effect in the late 1880s.

There are two components to this correction. The first is the outward-acting centrifugal acceleration associated with the movement of the vehicle travelling over the curved surface of the Earth, and the second is the *change* in centrifugal acceleration resulting from the movement of the vehicle relative to the Earth's rotational axis. In the second case, an object that is stationary on the Earth's surface is travelling at the speed of the Earth's surface at that point as it rotates around the rotational axis in an east–west direction. If that same object is then moved at *x* km/h towards the east, its speed relative to the rotational velocity is increased by the same amount. Conversely, if it travels at a speed of *y* km/h in a westerly direction, its relative speed is slowed by the same amount. Any movement of a gravimeter which involves a component in an east–west direction will have a significant effect on the measurement of gravity. For shipborne gravimeters the Eötvös correction can be of the order of 350 g.u. For airborne gravity measurements, where speeds over 90 km/h (about 50 knots) are common, the Eötvös correction can be as high as 4000 g.u. The expression governing the Eötvös correction is given in Box 2.15, with a fuller mathematical explanation in Box 2.16.

Box 2.15   Eötvös correction

The Eötvös correction is given by:

$$\delta g_{EC} = 75.08\, V \cos\phi \sin\alpha + 0.0416\, V^2 \;(\text{g.u.})$$

or

$$\delta g_{EC} = 40.40\, V' \cos\phi \sin\alpha + 0.012\,11\, V'^2 \;(\text{g.u.})$$

where ϕ is the degree of geographic latitude, α is the azimuth in degrees, and V and V' are the speeds of the vehicle in knots and kilometres per hour respectively.

The error in the Eötvös correction $[d(\delta g_{EC})]$ in g.u. due to errors in speed (dV) and azimuth $(d\alpha)$ is:

$$d(\delta g_{EC}) = (0.705\, V' \cos\phi \cos\alpha)\,d\alpha$$
$$+ (40.40 \cos\phi \sin\alpha + 0.024\,22\, V')\,dV'.$$

At a latitude of 25°, a change of 0.1 knot (0.2 km/h) with a half a degree shift off course will result in over 7 g.u. difference in

Eötvös correction while on an easterly course ($\alpha = 90°$), or 3 g.u. on a northerly course ($\alpha = 0°$), assuming a speed of 10 km/h. Consequently, for an airborne survey to be accurate to within 10 g.u., extremely tight controls need to be kept on nagivation and on the general movement (roll, pitch, yaw, etc.) of the helicopter or plane; such accuracies are now thought to be routinely achievable (Hammer, 1982, 1984). From the last equation in Box 2.15, it can be seen that there is the greatest sensitivity to errors in speed in an east–west direction, and to errors in azimuth on a north–south course. Recent tests of airborne gravity have been discussed by Halpenny and Darbha (1995).

Box 2.16 Derivation of the Eötvös correction equation
(see Figure 2.22)

In general, the centrifugal acceleration a_1 is (velocity)^2/d. The total east–west speed of the vehicle is the linear speed of rotation of the Earth (v) plus the *east–west* component of the speed of the vehicle (V_E). Thus the centrifugal acceleration $a_1 = (v + V_E)^2/d$.

Centrifugal acceleration (a_2) along the radius vector, due simply to the movement of the vehicle in a *north–south* direction, is $a_2 = V_N^2/R$.

However, it is the *change* in acceleration that is required, so the centrifugal acceleration acting on a static body (a_3) needs to be removed: $a_3 = v^2/d$.

The total change in acceleration acting in a vertical sense is:

$$\delta g_E = a_1 \cos\phi + a_2 - a_3 \cos\phi \quad \text{(equation (4))}.$$

We note that:

$$d = R \cos\phi$$
$$v = \omega R \cos\phi$$
$$V = (V_N^2 + V_E^2)^{1/2} \quad \text{(from Pythagoras' theorem)}$$
$$V_E = V \sin\alpha, \text{ where } \alpha \text{ is the bearing to True North.}$$

Substituting into equation (4) we obtain:

$$\delta g_E = \frac{(v + V_E)^2}{R \cos\phi} \cos\phi + \frac{V_N^2}{R} - \frac{v^2 \cos\phi}{R \cos\phi}.$$

Simplifying, this becomes

$$\delta g_E = [(v + V_E)^2 + V_N^2 - v^2]/R$$

continued

continued

which reduces to

$$\delta g_E = [2vV_E + (V_E^2 + V_N^2)]/R.$$

Rewriting this in terms of ω, R, ϕ, and α using the above expressions, this becomes:

$$\delta g_E = 2\omega V \cos \phi \sin \alpha + V^2 R \quad \text{(equation (5))}.$$

Given the following values:

$$\omega = 7.2921 \times 10^{-5} \text{ radians/s}$$
$$R = 6.371 \times 10^8 / \text{cm}$$
$$1\,\text{knot} = 51.479\,\text{cm/s}$$
$$1\,\text{Gal} = 10^4 \text{ g.u.}$$

then equation (5) can be rewritten (in terms of g.u.) as:

$$\delta g_E = 2(7.2921 \times 10^{-5} \times 51.479 \times 10^4) V \cos \phi \sin \alpha$$
$$+ (51.479V)^2 \times 10^4 / 6.371 \times 10^8$$

Finally:

$$\delta g_E = 75.08\,V \cos \phi \sin \alpha + 0.0416\,V^2 \text{ (g.u.)}.$$

2.5.8   Isostatic correction

If there were no lateral variations in density in the Earth's crust, the fully reduced gravity data, after application of all the corrections so far outlined, would be the same. However, where there are lateral variations, a gravity anomaly results which is known as the *Bouguer anomaly* (discussed in more detail in Section 2.5.10). The average Bouguer anomaly in oceanic areas is generally positive, while over mountainous regions it is usually negative. These effects indicate that the rock beneath the oceans is more dense than normal while that beneath the mountains is less dense.

Two hypotheses were proved in the 1850s to account for this large-scale systematic variation in density (Figure 2.23). The geodesist G.B. Airy (1855) proposed that while mountain chains had deep roots, beneath the oceans the crust, which was assumed to have constant density everywhere, was thin. In contrast, an English Archdeacon J.H. Pratt (1859) thought that the crust extended to a uniform depth below sea level but that density varied inversely with the height of the topography.

(A)

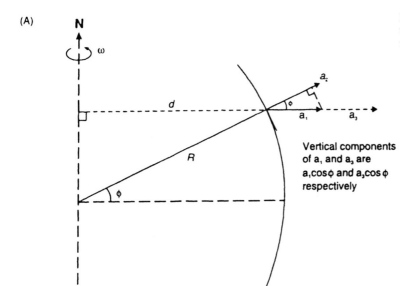

Vertical components
of a_1 and a_3 are
$a_1\cos\phi$ and $a_3\cos\phi$
respectively

Figure 2.22 Schematic illustrating the components which contribute to the Eötvös correction (see box 2.16)

(B)

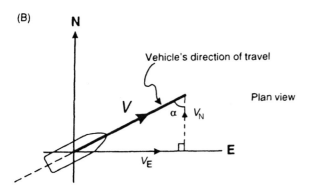

Plan view

(A)

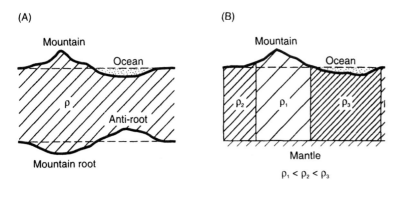

Figure 2.23 (A) Airy and (B) Pratt's models for isostacy

Airy's isostatic model is preferred geologically and seismologically, whereas Pratt's model is easier to use to calculate the isostatic (e.g. Rimbert *et al.* 1987), but the results are similar. Pratt's model has been developed by Heiskanen (1938) who suggested that density changes laterally with variable thickness of crust and that density increases gradually with depth. The aim of the isostatic correction is that effects on *g* of the large-scale changes in density should be removed, thereby isolating the Bouguer anomaly due to lateral variations in density in the upper crust (Hayford and Bowie 1912). The isostatic correction is discussed in more detail by Garland (1965), and the implications for isostatic rebound due to crustal loading and the viscosity of the mantle are discussed by Sharma (1986).

2.5.9   Miscellaneous factors

In Sections 2.5.1–2.5.8, calculable corrections to gravity data have been discussed. Gravimeters are sensitive not only to these factors but also to several others which tend to be erratic, temporal and difficult to quantify so that they may constitute gravitational noise. Such factors are: meteorological loading produced by atmospheric pressure changes; inertial acceleration caused by seismic and microseismic waves, including the effect of wind pressure on the gravimeter, and vibration from traffic and industrial machinery; and electrical noise from the gravimeter itself. Changes in atmospheric pressure can be corrected for by using -0.03 g.u. per millibar. Modern gravimeters used in micro-gravity surveys can filter out most noise above 10 Hz, and reject the microseismic noise between 0.1 and 2 Hz, so that standard deviations on gravity readings can be as small as 0.04 g.u. ($4\,\mu$Gal) (Thimus and van Ruymbeke, 1988).

2.5.10   Bouguer anomaly

The main end-product of gravity data reduction is the *Bouguer anomaly*, which should correlate only with lateral variations in density of the upper crust and which are of most interest to applied geophysicists and geologists. The Bouguer anomaly is the difference between the observed gravity value (g_{obs}), adjusted by the algebraic sum of all the necessary corrections (Σcorr; see Table 2.7 and Box 2.17), and that at some base station (g_{base}). The variation of the Bouguer anomaly should reflect the lateral variation in density such that a high-density feature in a lower-density medium should give rise to a positive Bouguer anomaly. Conversely, a low-density feature in a higher-density medium should result in a negative Bouguer anomaly.

Box 2.17   Bouguer anomaly

The Bouguer anomaly (Δg_B) is the difference between the observed value (g_{obs}), duly corrected, and a value at a given base station (g_{base}), such that:

$$\Delta g_B = g_{obs} + \Sigma(\text{corr}) - g_{base}$$

with

$$\Sigma(\text{corr}) = \delta g_L + (\delta g_F - \delta g_B) + \delta g_{TC} \pm \delta g_{EC} \pm \delta g_{IC} - \delta g_D$$

where the suffices refer to the following corrections:

L   = latitude; F = free-air; B = Bouguer;
TC = terrain correction; EC = Eötvös correction;
IC = isostatic correction; and D = drift (including Earth tides).

2.6   INTERPRETATION METHODS

There are two approaches to the interpretation of Bouguer anomaly data. One is *direct* where the original data are analysed to produce an interpretation. The other is *indirect*, where models are constructed to

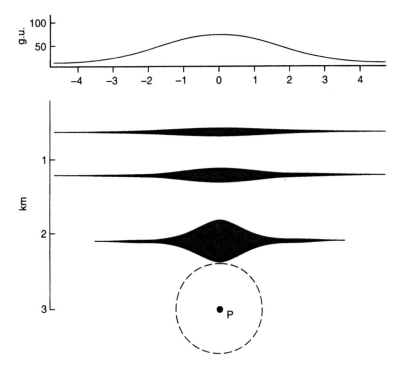

Figure 2.24   Ambiguity in geological models, all of which produce the gravity anomaly shown at the top. The lens-shaped bodies have a gravity anomaly identical to that of a sphere at P of radius 600 m and density contrast 1.0 Mg/m³. The thickness of the bodies is exaggerated by a factor of 3. From Griffiths and King (1981), by permission

compute synthetic gravity anomalies which are compared in turn with the observed Bouguer anomaly. The model producing the best fit, however, will not be unique as several alternative models may be found which also produce an equivalent fit (Figure 2.24). It is because of this type of ambiguity, which has already been discussed in Section 1.2 (see also Figure 1.1), that different geophysical methods are used together to constrain the geologic model.

2.6.1   Regionals and residuals

Bouguer anomaly maps are rather like topographic maps with highs and lows, linear features and areas where the contours (*isogals*) are closely packed and others where they are further apart. There may be a gentle trend in the gravity data, reflecting a long-wavelength gravity anomaly attributable to deep-seated crustal features; this is known as a *regional anomaly*. Shorter-wavelength anomalies arising from shallower geological features are superimposed on the regional anomaly, and it is these anomalies that are often to be isolated for further analysis. Separation of the regional from the Bouguer anomaly will leave a *residual anomaly* (Figure 2.25).

There are a number of different methods with varying degrees of complexity and effectiveness by which residual anomalies can be isolated (Nettleton, 1954). These range from curve-sketching, which is purely subjective, through to computer-based analytical methods. Graphical methods include sketching in estimated regional trends by eye on a profile (Figure 2.25A) or calculating the residual from estimated isogals on a map. Figure 2.25B illustrates how the residual is calculated. The 5.0 mGal isogal, which has been highlighted, intersects several estimated regional isogals. At points A and B, the difference (i.e. the residual) between the 5.0 mGal line and those it crosses are respectively +0.2 and +0.4 mGal and contours are drawn of the same residual value.

An example of the quantitative analytical method consists in fitting a low-order polynomial expression to the Bouguer anomaly data and then subtracting the calculated values from those observed to produce residual values, which are then plotted in map form. A more sophisticated method is the application of Fourier analysis by which a power spectrum is obtained for the Bouguer anomaly (Spector and Grant 1970; Syberg 1972). This highlights the different wavelengths of anomaly present and so allows a form of filtering to be undertaken to remove the unwanted anomalies (e.g. Granser *et al.* 1989). Dobrin (1976), Grant and West (1965) and Telford *et al.* (1990) discuss the various techniques in more detail.

Although the analytical methods appear more rigorous and thorough, there are occasions when the manual interpretation can take into account known variations in local geology more readily than an automated system.

(A)

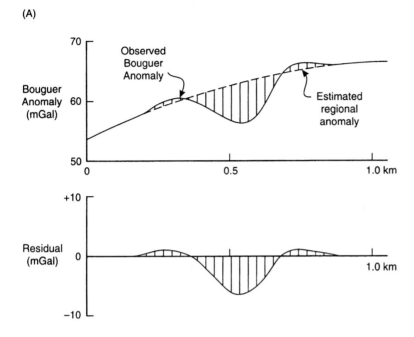

(B)

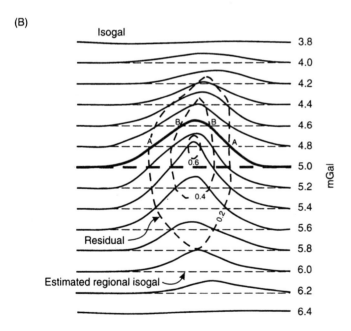

Figure 2.25 (A) Removal of a residual gravity anomaly from a regional profile, and (B) how a residual gravity map is constructed (see text for an explanation). After Dobrin (1976), by permission

2.6.2 Anomalies due to different geometric forms

Certain geologic structures can be approximated to models with known geometric forms (Nettleton, 1942). For example, a buried cavity may be represented by a sphere, a salt dome by a vertical cylinder, a basic igneous dyke by an inclined sheet or prism, etc. Another factor to be considered is whether the target to be modelled should be considered in two or three dimensions. If g is computed across a profile over a buried sphere, then that profile should hold true for any direction across the sphere. However, if the profile is across a buried horizontal cylinder, then the profile along the long-axis of the cylinder will be quite different from that across it. Also, if the strike length of the feature is greater than 20 times any other dimension, then it may be considered a two-dimensional body. Where this does not hold true, any profile will also sense the effects of the third dimension ('edge effects') and thus will not be modelled accurately if considered only in two dimensions.

Several common geometrical forms are illustrated in Figure 2.26, with their associated gravity profiles and the types of geologic features they approximate. The equations used to calculate the maximum anomaly for each geometric feature are given in Box 2.18. No attempt is made here to explain the derivations of these formulae, all of which are discussed much more fully by Dobrin (1976), Telford *et al.* (1990) and Parasnis (1986). The equations in Box 2.18 are intended only as guides to estimate the maximum values of the associated gravity anomalies. The use of the half-width $(x_{1/2})$ is discussed in more detail in Section 2.6.3 and 2.6.4.

The range of geometric forms given above is by no means complete. Details of other forms and their interpretations, such as by the use of characteristic curves, are given by Grant and West (1965) and Telford *et al.* (1990).

Calculation of gravity anomalies using the above methods should be regarded as a first step in the interpretation process. There are other, more sophisticated, and commonly computerised methods of gravity anomaly analysis. However, it is worth noting that for more complicated geological features of irregular shape which do not approximate to any of the geometrical forms, two other broad approaches can be adopted. The first is the use of graphical methods, and the second is an analytical approach. In the graphical methods, a template, which is divided into segments, is superimposed on an irregular cross-section of the geological feature to be modelled. The gravity at a point on the surface can be calculated by summing the effects of all the individual segments covering the cross-section of the feature.

Graphical methods can also be used for three-dimensional bodies. In this case, the appropriate template is superimposed on contours of the geological feature in the horizontal plane, thereby dividing it into

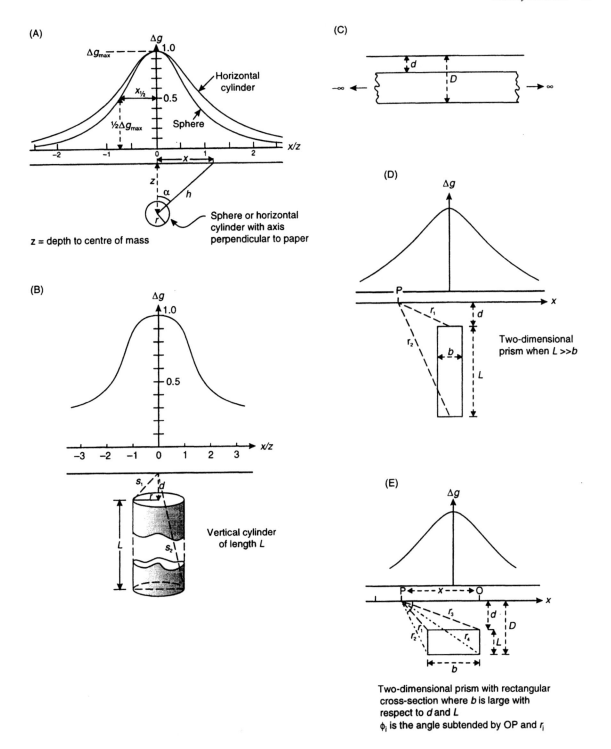

Figure 2.26 Representative gravity anomalies over given geometric forms: (A) a sphere or horizontal cylinder with its long axis perpendicular to the paper; (B) a vertical cylinder; (C) a semi-infinite horizontal slab (a Bouguer plate when $d = 0$); (D) a vertical rectangular prism; and (E) a horizontal rectangular prism

Box 2.18   Gravity anomalies associated with geometric forms
(see Figure 2.26)

Models	Maximum gravity anomaly	Notes
Sphere	$\Delta g_{max} = (4/3)\pi G \delta \rho r^3/z^2$	$z = 1.305 x_{1/2} \,(m)$
Horizontal cylinder	$\Delta g_{max} = 2\pi G \delta \rho r^2/z$	$z = x_{1/2}$
Vertical cylinder	$\Delta g_{max} = 2\pi G \delta \rho \, (s_1 - d)$	If $L \to$ infinity
	$\Delta g_{max} = 2\pi G \delta \rho r$	If $d = 0$
	$\Delta g_{max} = 2\pi G \delta \rho \, (L + s_1 - s_2)$	If L finite
		$z = x_{1/2}\sqrt{3}$
Buried slab (Bouguer plate)	$\Delta g_{max} = 2\pi G \delta \rho \, L$	For $L = 1000$ m and $\delta \rho = 0.1 \,Mg/m^3$, $\Delta g_{max} = 42$ g.u.
Infinite slab	$\Delta g_{max} = 2\pi G \delta \rho \, (D - d)$	
Horizontal rectangular prism	$\Delta g_P = 2G\delta\rho \left[x \ln\left(\dfrac{r_1 r_4}{r_2 r_3}\right) \right.$ $\left. + b \ln\left(\dfrac{r_2}{r_1}\right) + D(\phi_2 - \phi_4) - d(\phi_1 - \phi_3) \right]$	
Vertical rectangular prism	$\Delta g_{max} = 2G\delta\rho \,[b \ln(d/L)]$	$L \gg b$
Step	$\Delta g_{max} = 2G\delta\rho \,[x \ln(r_4/r_3)$ $+ \pi(D - d) - D\phi_4 + d\phi_3]$	

All distances are in metres unless stated otherwise; Δg_{max} in mGal and $\delta\rho$ in Mg/m^3, and the factor $2\pi G = 0.042$.

a pile of horizontal slabs each with a thickness equal to the contour interval.

Most computer-based analytical methods (e.g. Bott 1960) are based on the premise proposed by Talwani *et al.* (1959), that a cross-section of a two-dimensional body can be approximated by representing it by a multisided polygon (Figure 2.27A). This was developed by Talwani and Ewing (1960) for three-dimensional bodies (Figure 2.27B) which are approximated by a stack of polygonal laminae. The gravity effect of each lamina is computed and summed to give a total gravity anomaly. Enhancements of these methods have largely been centred on improving the ease of use of the software on computers that have dramatically increased in power and efficiency, and on the portability of software from mainframe machines to personal microcomputers (Busby 1987).

A development of the three-dimensional approach is to consider a geological body as a stack of cubic blocks of uniform size, each

(A)

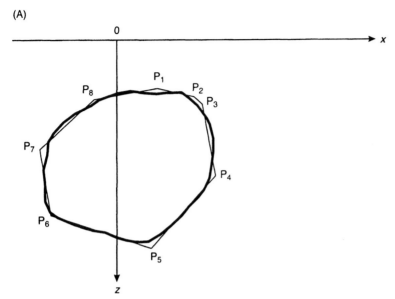

Figure 2.27 (A) Polygonal representation of an irregular vertical section
of a two-dimensional geological feature. (B) Representation of an irregular
three-dimensional geological feature
by polygonal laminae

(B)

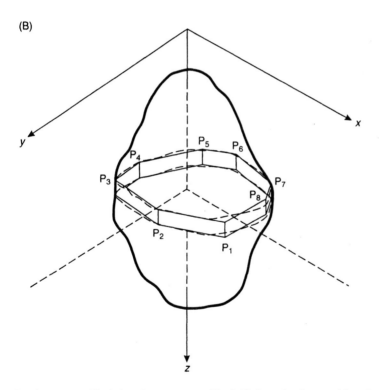

having a specified density contrast. Each little cube is considered as
a point mass and thus the total gravity anomaly for the entire body is
obtained by summing the constituent gravity components for each
mini-cube. The resultant gravity anomaly is compared with that

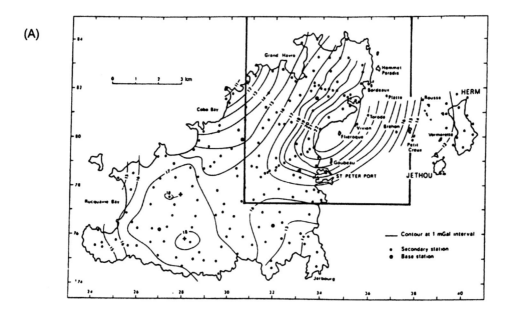

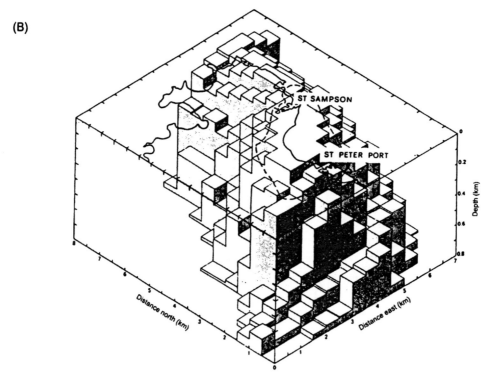

Figure 2.28 (A) Bouguer anomaly map for Guernsey, Herm and Jethou, Channel Islands, and (B) a three-dimensional model of the underlying St Peter Port Gabbro (density contrast 0.27 Mg/m^3, vertical exaggeration 5:1). The coastline and the outline of the gabbro outcrop are indicated. The gabbro is thus interpreted as a laccolith approximately 4 km in diameter and 0.8 km thick. From Briden *et al.* (1982), by permission

observed and, if necessary, the model is adjusted by trial-and-error or by automatic iterative methods (e.g. non-linear optimisation (Al-Chalabi 1972)) until the differences between the computed and observed anomalies are reduced to an acceptable, statistically defined level. Better resolution is obtained by reducing the size and increasing the number of individual cubes within the model. By having a regular cube size, the computation is eased considerably. An example of the application of this technique is given in Figure 2.28, where gravity data from Guernsey, Channel Islands, have revealed that a gabbro body, which outcrops to the north-east of the island near St Peter Port, has the form of a laccolith 0.8 km thick and about 4 km in diameter (Briden *et al.* 1982).

2.6.3 Depth determinations

Of major importance in the interpretation of any gravity data is the determination of depth to the centre of mass and/or to the top of the body causing the anomaly. The maximum depth at which the top of any particular geological body can be situated is known as the *limiting depth*. Methods of obtaining this information depend on which interpretational technique and model are being used. Let us consider various *direct* or *forward* methods where the actual gravity anomaly data are used to derive depths and also estimates of anomalous masses of the features causing the anomalies.

The commonest rules of thumb concern the use of the half-width of the anomaly; that is, the half-width $(x_{1/2})$ of the anomaly where the amplitude is half the maximum value (see Figure 2.26). Some workers define the half-width as the entire width of the anomaly at half peak amplitude and the form of the depth and mass determination equations will differ accordingly. Whichever formulae are used, care should be taken when calculating the limiting depth. The causative body has finite size and its mass is not concentrated at its centre of mass, and thus any estimate of depth will be overestimated. Also, the method will only give an approximation of depth in cases where all the constituent components have the same sense of density contrast (i.e. all negative or all positive). These formulae will also not be effective for compact mineral bodies. Formulae for a selection of given geometric forms are given in Box 2.19A and an example of one calculation is given in Box 2.19B.

Several basic 'rules', known as the *Smith Rules* after their originator (Smith 1959, 1960), have become established in the calculation of limiting depths. Two rules (1 and 2 in Box 2.20) use a *gradient–amplitude ratio* method. Consider any geological body that gives an isolated gravity anomaly (Figure 2.30) entirely of either sign with a maximum gravity (Δg_{max}) that varies along the line of the profile and thus has a horizontal gradient which reaches a maximum value at $\Delta g'_{max}$. The Smith Rules describe the various relationships between the limiting depth d to the top of any geological body and the maximum gravity (Δg_{max}) and its horizontal gradient ($\Delta g'_{max}$) as listed in Box 2.20.

Box 2.19A   Depth estimates for given geometric forms

Form	Formula	Notes
Sphere	$z = 1.305x_{1/2}$	z is depth to centre of mass
	$d = z - r$	d is depth to top of sphere of radius r
		$r^3 = \lvert \Delta g_{max}\rvert z^2/(0.028\delta\rho)$ from Box 2.18
Horizontal cylinder	$z = x_{1/2}$	z is depth to cylinder axis
	$d = z - r$	d is depth to top of cylinder of radius r
		$r^2 = \lvert \Delta g_{max}\rvert z/(0.042\delta\rho)$ from Box 2.18
Vertical cylinder	$z = 1.732x_{1/2}$	z is depth to top end of cylinder (overestimates z)
Thin dipping sheet	$z \approx 0.7x_{1/2}$	z is depth to top of sheet
	$z \approx x_{1/2}$	When $z \approx$ dip length of sheet
		When $z \gg$ dip length of sheet
		When length of sheet is very large or sheet dips at less than 60°, no solution is possible
Thick prism	$z = 0.67x_{1/2}$	z is depth to prism top = prism width, and depth to prism base is twice width
	$z = 0.33x_{1/2}$	When depth to prism base is 10 times prism width
		In both cases, estimates of z are unreliable

Box 2.19B   Example of calculation for a sphere

An air-filled cavity in rock of density 2.5 Mg/m^3 can be modelled by a sphere of radius r and depth to centre of mass, z (m). The resultant gravity anomaly is shown in Figure 2.29. Given $\Delta g_{max} = 0.048 \text{ mGal}$, $x_{1/2} = 2.2 \text{ m}$, and $\delta\rho = 2.5 \text{ Mg/m}^3$:

$$z = 1.305 \times 2.2 \text{ m} = 2.87 \text{ m}.$$

Radius of sphere $= r$:

$$r^3 = 0.048 \times (2.87)^2/(0.0286 \times 2.5) = 5.53 \text{ m}^3.$$

_____ *continued* _____

—— *continued* ——

So $r = 1.77$ m. Depth to top of sphere $d = 2.87 - 1.77 = 1.10$ m. *An air-filled cavity of this size so close to the surface could constitute a hazard*

Box 2.20 Smith Rules

(1) Where the entire anomaly has been isolated:

$$d \leqslant C. \, \Delta g_{max}/\Delta g'_{max}$$

where $C = 0.65$ for a 2-D body and $C = 0.86$ for a 3-D body.

(2) When only part of an anomaly is isolated, for any point x:

$$d \leqslant K \, \Delta g_x/\Delta g'_x$$

where $K = 1.00$ for a 2-D body and $K = 1.50$ for a 3-D body.

(3) For a maximum density contrast $\delta \rho_{max}$ and a maximum value of the second horizontal gradient $(\Delta g'_{max})$ (that is, the rate of change of $\Delta g'$ with x):

$$d \leqslant 5.4 \, G \, \delta \rho_{max}/\Delta g''_{max}.$$

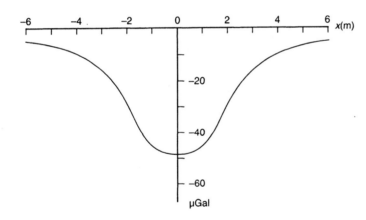

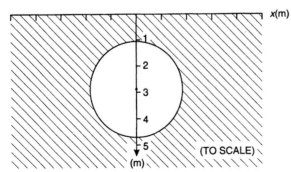

Figure 2.29 Gravity anomaly over an air-filled cavity of radius 1.77 m and 2.87 m depth to centre, in rock of density 2.5 Mg/m³

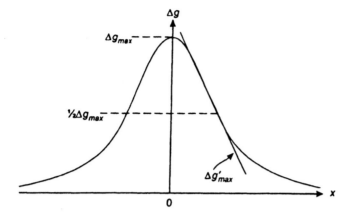

Figure 2.30 Limiting depth calculations: half-width method and gradient–amplitude ratio method (see text for details)

The third Smith Rule adopts a second-derivative method which uses the rate at which the gravity gradient changes along the profile. It is thought that second-derivative methods produce more accurate estimates of limiting depths. Second derivatives are discussed in more detail in Section 2.6.5.

2.6.4   Mass determination

Anomalous mass is the difference in mass between a geological feature and the host rock. There are two basic methods of calculating either an excess mass due to a high-density body or a mass deficiency caused by a body with a lower density.

The first method uses a rule of thumb based on the gravity anomaly half-width ($x_{1/2}$) and an assumption that the geological feature approximates to a given geometric form, such as a sphere (Box 2.21). The anomalous mass can be calculated by subtracting the mass due to a sphere (density times volume) from the mass estimated using gravity data. In the example below, the actual mass of an air-filled cavity is negligible, so the mass deficiency calculated is the mass of the missing rock.

Box 2.21   Mass of a sphere

$$\text{Total mass } M \approx 255 \, \Delta g_{\text{max}} (x_{1/2})^2 \text{ tonnes}$$

where Δg_{max} is in mGal and $x_{1/2}$ in metres.

Example
For an air-filled cavity described in Box 2.19B, the total mass deficiency of the sphere is equal to the mass of the rock that

continued

— *continued* —

would have been in the cavity, times its density ($2.5\,\mathrm{Mg/m^3}$):

$$\mathrm{Mass} = \mathrm{density} \times \mathrm{volume} = 2.5 \times (4/3)\pi\,1.77^3 = 58\ \mathrm{tonnes}.$$

Using the gravity data:

$$\mathrm{Mass} \approx 255 \times 0.048 \times 2.2^2 = 59\ \mathrm{tonnes}.$$

The second method is based on Gauss's Theorem in potential theory (Grant and West 1965) and is particularly important for two reasons. First, the total anomalous mass of a geological feature can be calculated from the associated gravity anomaly without any assumptions being necessary about the body's shape or size. Secondly, the total anomalous mass can be very important in the determination of tonnage of ore minerals (Hammer, 1945). For this method to work effectively, it is important that the regional gravity field be removed and that the entire residual anomaly be isolated clearly. The survey area is divided into a series of rings each of which is further divided into segments of area δA. The gravity effect of each segment is determined and the total for each ring is obtained and summed together (Box 2.22). Having determined the excess mass, it is then a simple matter to calculate the actual mass (M) if the densities of the host rock (ρ_0) and the anomalous body (ρ_1) are known.

Box 2.22

(1) Total anomalous mass (M_E):

$$M_E = 23.9\Sigma(\Delta g\delta A)\ \mathrm{tonnes}$$

where Δg is in mGal and δA in metres.

(2) Actual mass of a geological body (M):

$$M = M_E\frac{\rho_1}{(\rho_1 - \rho_0)}\ \mathrm{tonnes}\ (\rho_1 > \rho_0).$$

Parasnis (1966) gives an example where the total anomalous mass of the Udden sulphide orebody in northern Sweden was calculated to be 568 820 tonnes. Assuming the densities of the ore and host rock to be 3.38 and 2.70 $\mathrm{Mg/m^3}$ respectively, the actual mass of the ore was found to be 2.83 million tonnes, a value consistent with drillhole estimates.

2.6.5 Second derivatives

2.6.5.1 *Second vertical derivative (SVD) maps*

One of the problems inherent within the interpretation of Bouguer anomaly maps is that it is difficult to resolve the effects of shallow structures from those due to deeper seated ones. The removal of the effect of the regional field from the Bouguer anomaly data results in an indeterminate and non-unique set of residuals. It is possible to separate the probable effects of shallow and deeper structures by using second vertical derivatives.

The gravity field (g) which is measured by gravimeters varies with height; that is, there is a vertical gradient $(\delta g/\delta z = g')$. Over a non-uniform earth in which density varies laterally, the vertical gradient changes and the *rate* of change $(\delta g'/\delta z)$ is thus the second vertical derivative of the gravity field $(\delta^2 g/\delta z^2$. This quantity is very sensitive to the effects of shallow features (and to the effects of noise and topography).

As an illustration of how the gravity effects of shallow and deep structures can be separated, consider two equal point masses (m) at two different depths, say at depths of 1 unit and 4 units. The value of g for a point mass at a depth z is simply equal to the product of the gravitational constant (G) and the mass divided by the depth z squared, so $g = Gm/z^2$). If this is differentiated twice with respect to z, it becomes $g'' = 6Gm/z^4$. This tells us that the second derivative of the two masses, g'', is inversely proportional to z^4. Hence the ratio of the two derivatives will be, for $z_1 = 1$ and $z_4 = 4$, $g_1''/g_4'' = 256$.

It is possible to compute and plot maps of the second vertical derivative of Bouguer anomaly data. The zero-contour should indicate the edges of local geological features. The contours have units where $10^{-6}\,\mathrm{mGal/cm^2} \equiv 10^{-9}\,\mathrm{cm^{-1}\,s^{-2}} \equiv 1\,\mathrm{E\,cm^{-1}}$. (E stands for an Eötvös unit $= 10^{-6}\,\mathrm{mGal/cm}$, which is a measure of gravitational gradient.)

It should be emphasised that it is not possible to undertake any quantitative analyses of SVD maps except to produce characteristic profiles over known geometric forms. The main advantage of SVD maps is to highlight and clarify features spatially, as can be seen from Figures 2.31 and 2.32. In the first of these, the Bouguer anomaly map appears to have a consistent trend in the direction of the gravity gradient (increasingly positive to the east) with isogals aligned in a NW–SE direction. There is no obvious major feature evident on the Bouguer anomaly map. In contrast, the SVD map shows a major ENE–WSW linear feature with three closures, and it has picked out the outline of the Cement field in Oklahoma extremely well. Figure 2.32 illustrates the case when a single Bouguer anomaly is really the envelope of several smaller anomalies. In this atypical and rather extreme case, several deep boreholes were drilled on the large minimum indicated on both the Bouguer and the residual anomaly maps;

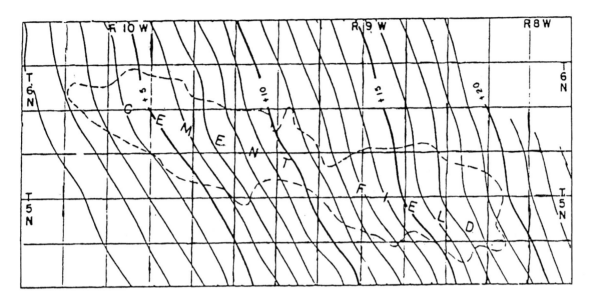

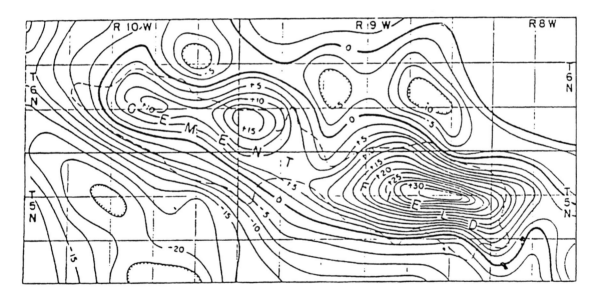

Figure 2.31 Observed Bouguer anomaly (contour interval 1 mGal) and second vertical derivative (contour interval 2.5×10^{-15} c.g.s.u.) maps over a Cement field in Oklahoma. From Elkins (1951), by permission

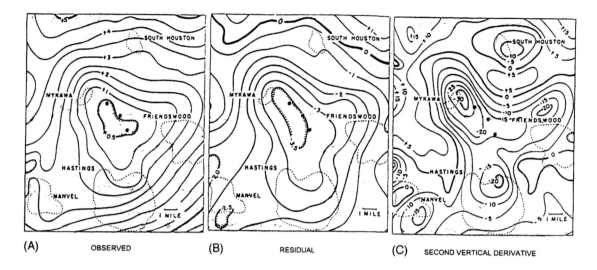

(A)   OBSERVED   (B)   RESIDUAL   (C)   SECOND VERTICAL DERIVATIVE

they were found to be dry, having missed the appropriate target, presumed to be a single salt dome. In contrast, the SVD map highlights three salt domes accurately.

Unfortunately, SVD also amplifies noise and so can produce many second-derivative anomalies that are not related to geology. Consequently, in some cases, SVD analyses provide no real advantage over the Bouguer anomaly map. An example of where extraneous anomalies shroud the geologically related features is given in Figure 2.33. Although it is possible to see on the SVD map the two main features present on the Bouguer anomaly map of the J-jaure titaniferous iron-ore region in Sweden, the SVD map also has a number of small maxima and minima that are of no structural interest. To resolve which anomalies are of geological importance, it is necessary to go back to the original gravity map and to any other source of geological information. It may even be prudent to refer back to the raw observations and corrections.

2.6.5.2 *Downward and upward continuation*

The effect on gravity of a geological mass at considerable depth is far less than if it were close to the surface (see Figure 2.24). the *principle of continuation* is the mathematical projection of potential field data (gravity or magnetic) from one datum vertically upwards or downwards to another datum. Effectively, the continuation process simulates the residual Bouguer anomaly at levels below or above sea level as if the gravity data had been obtained at those levels.

Upward continuation is relatively straightforward as the projection is usually into free space. Upward continuation serves to filter out the shorter-wavelength anomalies and reduce their amplitudes and decrease noise.

Figure 2.32 (A) Observed gravity, (B) residual gravity (contour interval 0.5 mGal), and (C) second vertical derivative (contour interval 5×10^{-15} c.g.s.u.) maps for Mykawa, Texas Gulf Coast. ● indicate dry boreholes. From Elkins (1951), by permission

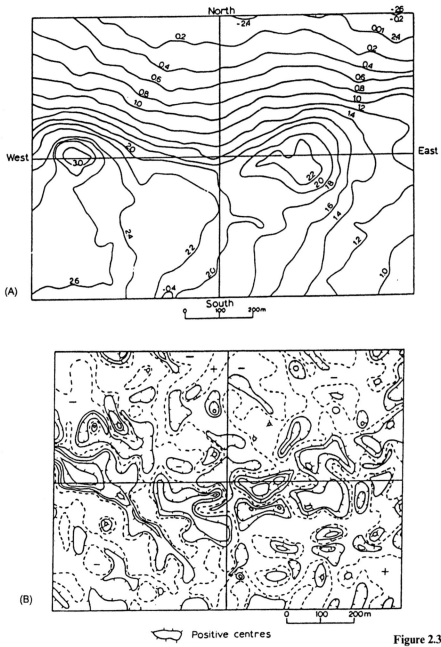

Positive centres

Negative centres

Figure 2.33 An example where the Bouguer anomaly map (A) (contour interval 2 mGal) exhibits more detail than the corresponding second vertical derivative map (B) for the J-jaure titaniferous iron-ore region in Sweden. Contours: 0 (dashed), ±0.1, ±0.2, ±0.4 in units of 0.0025 mGal/m²). From Parasnis (1966), by permission

Downward continuation is far more problematical as there is an inherent uncertainty in the position and size of the geological features as represented by the Bouguer gravity data. Furthermore, downward continuation aims to reduce each anomaly's wavelength and increase its amplitude. This mathematical amplification will also work on noise within the data and the resultant information may prove unsuitable for further analysis.

Continuation also forms a method of gravitational stripping (Hammer, 1963) where the gravity effects of upper layers are necessarily removed to reveal the anomalies due to deeper seated geological structures (e.g. Hermes 1986; Abdoh *et al.* 1990). The method uses the concept of an equivalent stratum (Grant and West, 1965). The Bouguer gravity field is continued downwards to a level that corresponds to a previously identified interface, such as from seismic reflection surveys, and an equivalent topographic surface is constructed at that level. This equivalent stratum should account for any residual anomalies at the surface arising from the interface. Continuation is discussed in much more detail by Grant and West (1965) and by Telford *et al.* (1990).

Figure 2.34 Comparison of airborne and upward continued land Bouguer gravity data with those obtained by airborne gravity surveying. From Hammer (1984), by permission

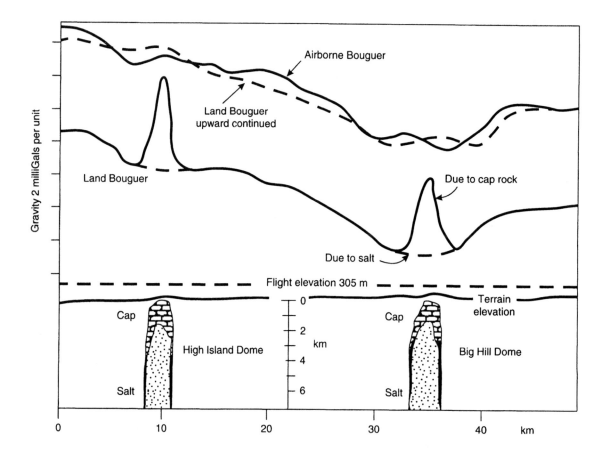

Upward continuation is used in comparisons of ground-based gravity anomalies with airborne data. It is usual to continue ground data upwards rather than to work downwards so as not to amplify noise. An example of such a comparison is shown in Figure 2.34 (Hammer 1982, 1984). Two gravity minima associated with low-density salt have shorter-wavelength maxima superimposed which are due to the high-density cap rocks. These maxima attenuate with increased elevation and the agreement between the upwardly continued land Bouguer data and the airborne is better than 5 g.u. except in the immediate vicinity of the cap rocks.

One of the major considerations in the interpretation of particularly regional gravity data is the amount of computer processing required. Considerable effort has been expended in developing computer-based methods of data enhancement. For example, image processing of data on computer-compatible tapes (CCTs) permits considerable manipulation of the data for display purposes to aid analysis and interpretation. Processes include edge enhancement to highlight lineaments (e.g. Thurston and Brown 1994), amplitude displays and spectral modelling (Figure 2.35). It is usually only economically viable to undertake such sophisticated processing on very large data sets.

Figure 2.35 Structural analysis, based on lineations from a series of colour and greyscale shaped-relief images of geophysical data can provide a basis for reassessment of regional structure, mineralisation potential and fracture patterns. This image is of observed regional Bouguer gravity data, over an area of 200 km × 140 km of the Southern Uplands. Scotland (C: Carlisle; E: Edinburgh). The data have been reduced to Ordnance Datum using a density of 2.7 Mg/m^3, interpolated to a square grid of mesh size 0.5 km and displayed as a greyscale shaded-relief image. Sun illumination azimuth and inclination are NE and 45 ° respectively. A series of NE trending features paralled to the regional strike and the Southern Uplands fault have been suppressed by the NE illumination, whereas subtle NW trending features linked to development of the Permian basins are enhanced and seen to be more extensive. For comparison, see Figure 3.48. Image courtesy of Regional Geophysics Group, British Geological Survey

2.6.6   Sedimentary basin or granite pluton?

It is very important in the interpretation of gravity data for hydrocarbon exploration to be able to distinguish between a sedimentary basin (a good possible hydrocarbon prospect) and a granitic pluton (no prospect for hydrocarbons), as both can produce negative gravity anomalies of comparable magnitude.

For example, Arkell (1933) interpreted a minimum in an initial Bouguer gravity survey in the Moray Firth, north-east Scotland, as being due to a granite pluton. It was only after further geological work (Collette, 1958) and gravity work (Sunderland, 1972) that it was realised that the minimum was due to a sedimentary basin. Simultaneously, the Institute of Geological Sciences undertook seismic reflection surveys and initiated some shallow drilling. It was not until 1978 that the Beatrice Field was discovered (McQuillin *et al.* 1984). Had the 1933 interpretation been different, the history of the development of the North Sea as a major hydrocarbon province might have been very different.

In 1962, Bott proposed a set of criteria to distinguish between a sedimentary basin and a granite boss as interpretations of gravity minima. His argument was based on the second vertical derivative of the gravity anomaly due to a semi-infinite two-dimensional horizontal slab with a sloping edge. He found that the ratio of the moduli of the maximum and minimum second vertical derivative ($|g''_{max}|/|g''_{min}|$) provides a means of distinguishing between the two geological structures, as outlined in Box 2.23 and illustrated in Figure 2.36. McCann and Till (1974) have described how the method can be computerised, and the application of Fourier analysis to Bott's method. Some authors calculate the second *horizontal* derivative ($\delta^2 g/\delta x^2$) (e.g. Kearey and Brooks 1991; figure 6.19) which responds in exactly the same way as the *vertical* derivative except that the maxima and minima are reversed, as are the criteria in Box 2.23. In order for the method to work, the gravity anomaly attributed to the appropriate geological feature (sedimentary basin or granitic pluton) needs to be clearly isolated from adjacent anomalies due to other features. The method is not applicable, however, in cases where extensive tectonic activity has deformed either a sedimentary basin by basin-shortening or a granitic pluton by complex faulting, thereby changing the gradients of the flanks of both types of model.

Box 2.23   Bott criteria (see also Figure 2.36)

---

(1) For a sedimentary basin:

$$|g''_{max}|/|g''_{min}| > 1.0.$$

Basin sides slope *inwards*.

continued

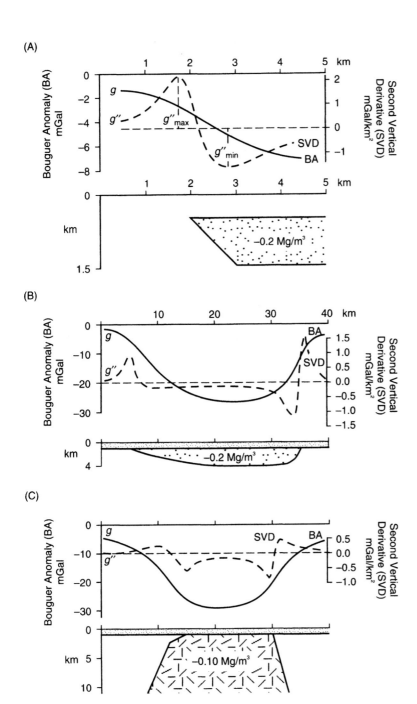

Figure 2.36 Bott criteria to distinguish between the Bouguer gravity profile over (A) a horizontal prism, (B) a sedimentary basin, and (C) a granitic pluton. After Bott (1962), by permission

---

continued

(2) For a granite pluton:

$$|g''_{max}|/|g''_{min}| \leqslant 1.0.$$

Granite pluton sides slope *outwards*.

The vertical variation of density of sediments with depth in a sedimentary basin can be represented in a number of ways. Moving away from Bott's uniform density model, consideration of the variation in density in terms of exponential and hyperbolic density contrast has been given by Rao *et al.* (1993) and Rao *et al.* (1994), for example.

2.7 APPLICATIONS AND CASE HISTORIES

In this section, a limited number of case histories are described to illustrate the diversity of applications to which the gravity method can be put. Other geophysical methods are discussed as appropriate, where they have been used in conjunction with, or to contrast with, the gravity results. These other methods are explained in their respective chapters.

2.7.1 Exploration of salt domes

2.7.1.1 *Mors salt dome, Denmark (waste disposal)*

An original interpretation of the Bouguer anomaly (Figure 2.37) over the Mors salt dome in northern Jutland was made in 1974, five years before any seismic results were known (Sharma, 1986). The investigation was connected to a feasibility study for the safe disposal of radioactive waste in the salt dome, but the methodology is identical had the study been for hydrocarbons.

The salt dome was approximated by a sphere. The values of $\Delta g_{max} \approx 16\,\text{mGal}$ and the half-width $\approx 3.7\,\text{km}$ were obtained from profiles across the feature (Figure 2.38A) used to determine the depth to the centre of mass ($z = 4.8\,\text{km}$). In order to calculate the depth to the top of the sphere, an estimate of the density contrast of the salt with the surrounding material had to be made. For a density contrast ($\delta\rho$) of $-0.25\,\text{Mg/m}^3$, this gave the radius of the sphere as 3.8 km and thus depth to the top of the sphere is about 1 km (4.8 km minus 3.8 km); with $\delta\rho = -0.2\,\text{Mg/m}^3$, the radius is 4.1 km and depth to the top is 0.7 km (4.8 km -4.1 km). This was later found to be in good agreement with the seismic results (Figure 2.38B). If the salt dome approximated to a vertical cylinder of length 5300 m, depth to top 700 m, and radius 4400 m, and density contrast $-0.2\,\text{Mg/m}^3$, the expected value of Δg_{max} is around 19 mGal, compared with an observed value of 16–18 mGal; but this is still close enough to be a reasonable approximation to the actual shape.

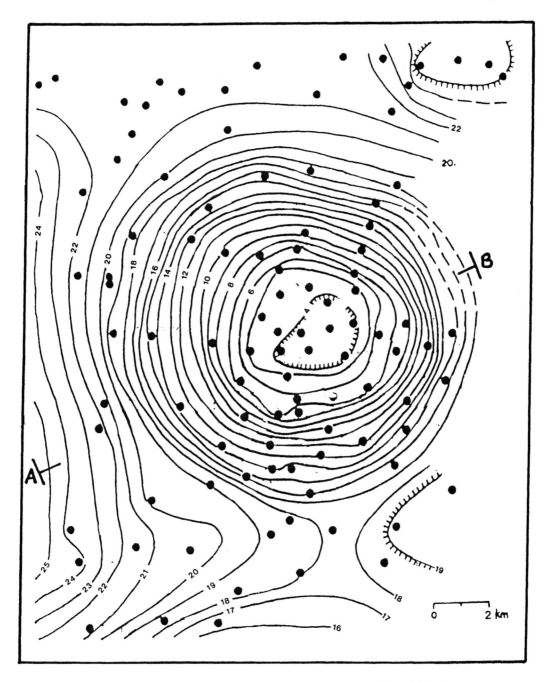

Figure 2.37 Bouguer anomaly map of the Mors salt dome, Jutland, Denmark. Solid dots represent observation points. Contour interval = 1 mGal. From Saxov (1956) and Sharma (1986), by permission

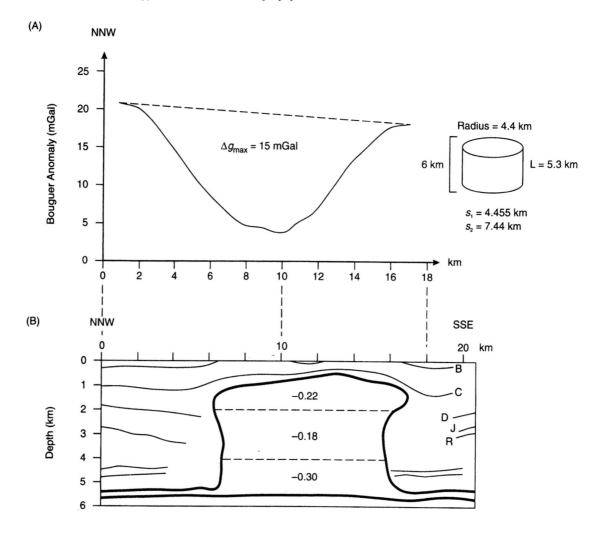

Uncertainty in the density contrast is the biggest problem in interpretation. Subsequent drilling into the salt dome and the use of the seismic profiles undertaken radially across the salt dome enabled the geological model to be enhanced. It was found that the density contrast within the salt dome could be divided into three sections with slightly different density contrasts (Figure 2.38B). The seismic sections did not give a very good image of the top of the salt dome, which is mushroom-shaped and so causes the slight discrepancies between the sphere and vertical cylinder approximations. The combined use of seismic data and gravity modelling has resulted in a much more realistic geological model.

However, it should always be remembered that any geophysical model is only a crude approximation of what can be a very complex geological structure. Consider how well (or otherwise) a vertical cylinder

Figure 2.38 (A) Bouguer anomaly profile and (B) corresponding seismic section across profile A–B in Figure 2.37. After Kreitz (1982), LaFehr (1982) and Sharma (1986), by permission

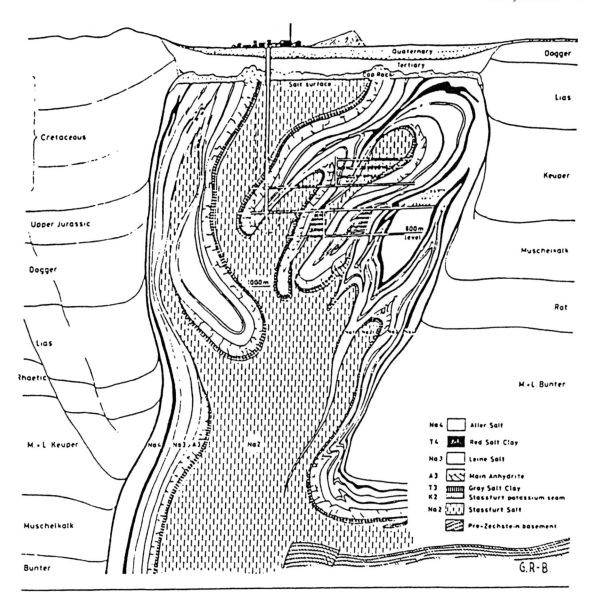

Quaternary
Tertiary
Dogger
Lias
Salt surface
Cap Rock
Cretaceous
Keuper
Upper Jurassic
Muschelkalk
Dogger
800 m
level
1000 m
Lias
Rot
Rhaetic
M.L. Bunter
M.L Keuper
Na4   Na3,A3   Na2
Muschelkalk
Bunter
G.R-B.

Na4		Aller Salt
T4		Red Salt Clay
Na3		Leine Salt
A3		Main Anhydrite
T3		Gray Salt Clay
K2		Stassfurt potassium seam
Na2		Stassfurt Salt
		Pre-Zechstein basement

model would represent the salt dome illustrated in Figure 2.39. It is clear that much fine geological detail would not be resolvable unambiguously by the interpretation of geophysical data (Sorgenfrei 1971; Richter-Bernburg 1982).

2.7.1.2 *Salt domes in NW Germany (hydrocarbons)*

Another example of gravity anomalies over salt domes is given in Figure 2.40, which is taken from part of a survey over north-west Germany (Hermes, 1986). The intention was to derive the gravity field

Figure 2.39 Schematic cross-section through a salt dome structure in north-west Germany, illustrating known complexities in contrast to the usual assumed geophysical model of a vertical cylinder with a uniform density distribution. From Sorgenfrei (1971) and Richter-Bernburg (1982), by permission

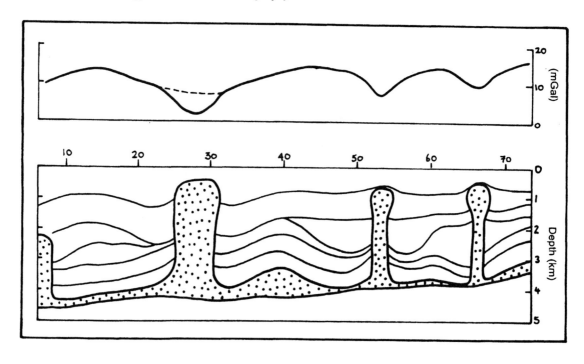

due to the pre-Zechstein and remove it, akin to gravity stripping, to be able to investigate what lies beneath it. Seismic reflection profiling has been singularly unsuccessful both in penetrating through this layer and in producing much useful information. The amplitudes of the Bouguer anomaly minima are clearly associated with the size of the salt domes, the smallest having the lowest amplitude (≈ 5 mGal). The largest minimum is a compound anomaly comprising a minimum associated with the syncline through which the low-density salt has risen. This emphasises the fact that the Bouguer anomaly map of this area is strongly influenced by the structures within the post-Zechstein and by the salt domes and walls. To determine the gravity effect of the pre-Zechstein it is thus essential that these effects be removed. Gravity stripping appears to succeed where filtering methods have been less successful.

Figure 2.40 A typical Bouguer anomaly profile compared with the corresponding sub-surface geology associated with the Zechstein in northern Germany. From Hermes (1986), by permission

2.7.2   Mineral exploration

Gravity surveys fulfil two roles in exploration for minerals: (1) for search and discovery of the ore body, and (2) as a secondary tool to delimit the ore body and to determine the tonnage of ore.

2.7.2.1   *Discovery of the Faro lead–zinc deposit, Yukon*

An integrated airborne and land geophysical exploration programme, of which gravity surveying was an integral part, led to the discovery of the Faro lead–zinc deposit in the Yukon, northern Canada (Brock 1973). Gravity was found to be the best geophysical method to delimit the ore body (Figure 2.41). It was also used to obtain an estimate of the tonnage (44.7 million tonnes), which compared very well with a tonnage proven by drilling of 46.7 million tonnes (Tanner and Gibb 1979). In contrast, vertical magnetic mapping provided an anomaly with too shallow gradients to be used, as Figure 2.41 shows.

2.7.2.2 Pyramid ore body, North West Territories

The Pyramid lead–zinc ore body, at Pine Point in the North West Territories, Canada, was discovered using the 'induced polarisation' (IP) method (Seigel *et al.* 1968). For further details of IP, see Chapter 9. Gravity was used to optimise development drilling since the gravity anomalies (Figure 2.42) correlated extremely well with the distribution of the mineralisation within the ore body. Additionally, the gravity data were used successfully to estimate total ore tonnage. Electrical resistivity (see Chapter 7) produced low-amplitude anomalies with a broad correlation with the position of the ore body, and a TURAM electromagnetic survey (see Chapters 10 and 11) was singularly unsuccessful and produced virtually no anomaly at all. The induced polarisation chargeability produced a spectacular anomaly.

2.7.2.3 Sourton Tors, Dartmoor, SW England

This is an example of where gravity did not work at all in association with a known occurrence of mineralisation (Figure 2.43), whereas electrical, electromagnetic and magnetic methods all produced significant anomalies (Beer and Fenning 1976; Reynolds 1988). The reason for the failure of the gravity method in this case is twofold:

- The scale of mineralisation, which is a stockwork of mineralised veins, was of the order of only a few metres wide.
- The sensitivity of the gravimeter was insufficient to resolve the small density contrast between the sulphide mineralisation and the surrounding rocks.

Had a gravimeter capable of measuring anomalies of the order of tens of μGal, and the station interval been small enough, then the zone of mineralisation may have been detectable. At the time of the survey (1969), such sensitive gravimeters were not as widely available as they are today.

2.7.3 Glacier thickness determination

For a regional gravity survey to be complete in areas such as Antarctica and Greenland, measurements have to be made over ice

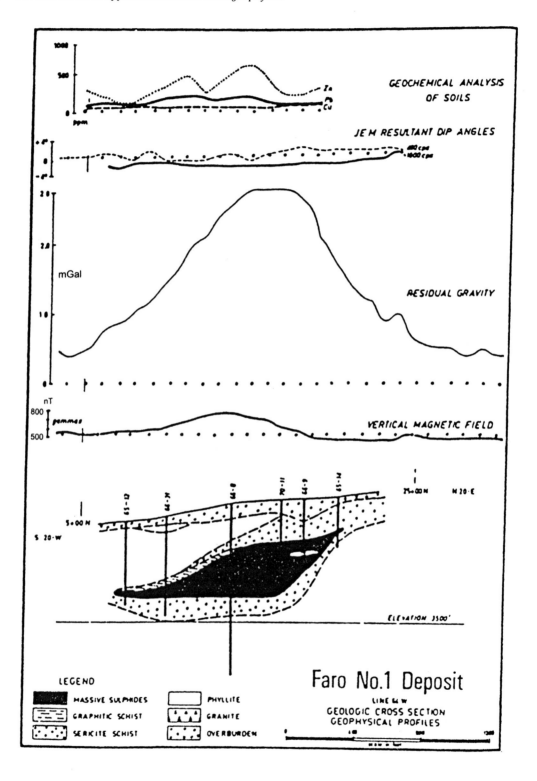

LEGEND

MASSIVE SULPHIDES

GRAPHITIC SCHIST

SERICITE SCHIST

PHYLLITE

GRANITE

OVERBURDEN

Faro No.1 Deposit

LINE 64 W

GEOLOGIC CROSS SECTION
GEOPHYSICAL PROFILES

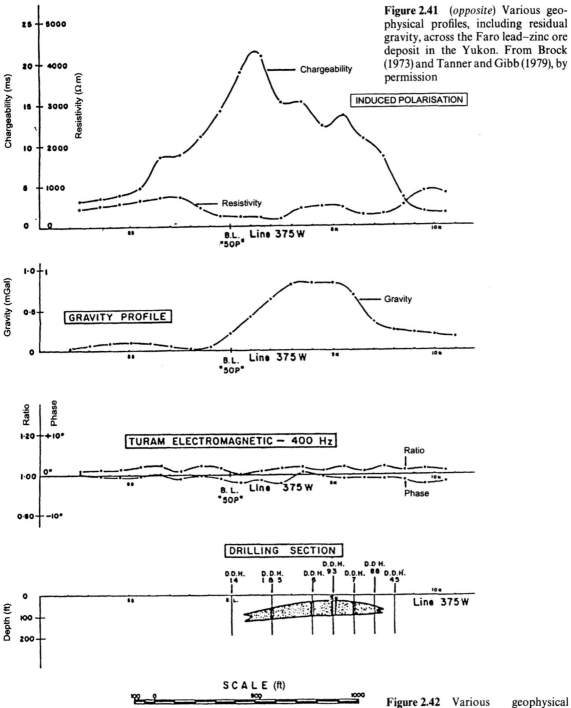

Figure 2.41 (*opposite*) Various geophysical profiles, including residual gravity, across the Faro lead–zinc ore deposit in the Yukon. From Brock (1973) and Tanner and Gibb (1979), by permission

Figure 2.42 Various geophysical profiles including residual gravity, across Pyramid no. 1 ore body. From Seigel *et al.* (1968), by permission

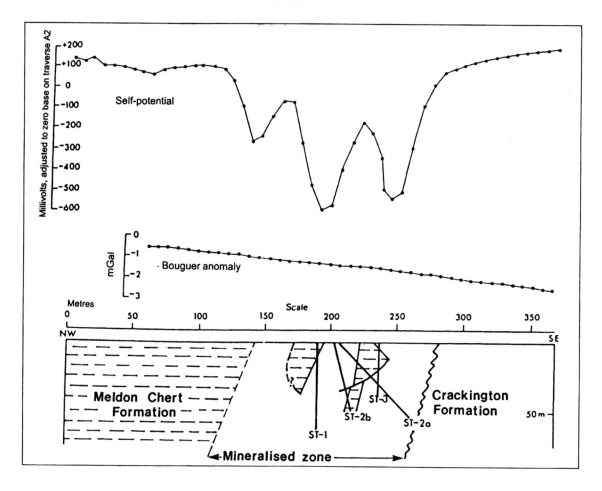

Figure 2.43 A Bouguer gravity profile across mineralised zones in chert at Sourton Tors, north-west Dartmoor, showing no discernible anomalies. From Beer and Fenning (1976), by permission

sheets and valley glaciers. Very often, these areas have incomplete information about the depth or volume of ice. The large density difference between ice (0.92 Mg/m^3) and the assumed average rock density (2.67 Mg/m^3) means that easily measured gravity anomalies can be observed and the bottom profile of the ice mass (i.e. the sub-glacial topography) can be computed.

An example of this has been given by Grant and West (1965) for the Salmon Glacier in British Columbia (Figure 2.44), in which a gravity survey was undertaken in order to ascertain the glacier's basal profile prior to excavating a road tunnel beneath it. A residual gravity anomaly minimum of almost 40 mGal was observed across the glacier, within an accuracy of ±2 mGal due to only rough estimates having been made for the terrain correction. An initial estimate of local rock densities was 2.6 Mg/m^3 and the resultant depth profile across the glacier proved to be about 10% too deep (Figure 2.44B) compared with depths obtained by drilling. Considering the approxi-

(A)

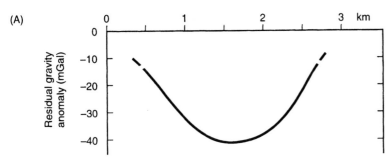

(B)

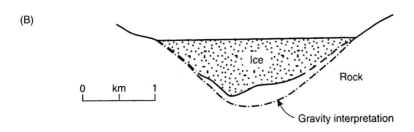

Figure 2.44 A residual gravity profile across Salmon Glacier, British Columbia, with the resulting ice thickness profile compared with that known from drilling. After Grant and West (1965), by permission

mations taken in the calculations, the agreement was considered to be fair. In addition, it was found that the average density of the adjacent rocks was slightly lower (2.55 Mg/m³). This could have indicated that there was a significant thickness of low-density glacial sediments between the sole of the glacier and bedrock. However, had more detailed modelling been undertaken on data corrected fully for terrain, the discrepancy could have been reduced.

Increasingly, ice thickness measurements are being made by seismic reflection surveying (see Chapter 5), electromagnetic VLF measurements (see Chapter 9, Section 9.6.6.3) and radio echosounding (see Chapter 9, Section 9.7.4.2). Comparisons between the various geophysical methods indicate that agreements of ice thickness within 10% can be readily obtained. Gravity measurements over large ice sheets (e.g. Renner *et al.* 1985; Herrod and Garrett 1986) can be considerably less accurate than standard surveys for three reasons:

- The largest errors are due to the imprecise determination of surface elevations. Currently, heights can be determined to within 5–10 m.
- Inaccurate corrections for sub-ice topography vary by hundreds of metres, in areas without radio echosounding control. An error in estimated ice thickness/bedrock depth of 100 m can introduce an error of ±74 g.u.
- As in all gravity surveys, the estimate of the Bouguer correction density is also of critical importance. Major ice sheets obscure the

local rock in all but a few locations, and the sub-ice geology, and its associated densities, may vary significantly.

Another glaciological application of the gravity method is the use of a gravimeter to measure oceanic tidal oscillations by the vertical movements of floating ice shelves in the Antarctic (Thiel *et al.* 1960; Stephenson 1984). The resulting tidal oscillation pattern can be analysed into the various tidal components and hence relate the mechanical behaviour of the ice shelf to ocean/tidal processes. If tidal oscillations are not found at particular locations, this may indicate that the ice shelf is grounded. Independent measurements with tilt meters, strain gauges and radio echosounding should be used to confirm such a conclusion (Stephenson and Doake 1982).

2.7.4   Engineering applications

The size of engineering site investigations is normally such that very shallow (< 50 m) or small-scale (hundreds of square metres) geological problems are being targeted. Consequently, the resolution required for gravity measurements is of the order of μGals. The use of gravity is commonly to determine the extent of disturbed ground where other geophysical methods would fail to work because of particularly high levels of electrical or acoustic noise, or because of the presence of a large number of underground public utilities (Kick 1985). Additionally, gravity is used to assess the volume of anomalous ground, such as the size of underground cavities or of ice lenses in permafrost. There are often no records of where ancient quarrying or mining has been undertaken and the consequent voids may pose considerable hazards to people and to property. The increasing development of higher latitudes brings its own engineering problems, and the application of gravity surveying, amongst other applied geophysical methods, is playing a growing and important role in site investigations in these areas. Furthermore, it is also true to say that with the increased use of geophysical methods, the very geological phenomena under investigation are becoming better studied and better known.

2.7.4.1   *Detection of back-filled quarries*

Where there is sufficient density contrast between infill material and the surrounding rock, small-scale gravity surveys can be used successfully to locate backfilled quarries (Poster and Cope 1975), as the following example demonstrates.

A proposed new railway section in Newcastle-upon-Tyne, England, was routed through a built-up area which contained a number of backfilled, late-nineteenth century sandstone quarries. The design of the section of railway included a cut-and-cover tunnel, and so it was extremely important to determine the position of the old quarry faces very accurately. The nature of the loose infill material (density 1.65 M/m³) provided a strong contrast in physical properties with the local sandstone (of density 2.1 Mg/m³) and a number of geophysical methods could have been employed. However, the site was criss-crossed by a large number of underground pipes and cables, and the superficial material contained significant amounts of scrap metal, all of which made it impractical to use electrical, electromagnetic or magnetic methods. High levels of acoustic noise which were generated by large volumes of traffic during the working day, and the lack of space due to extensive building cover, precluded the use of seismic methods. Consequently, the gravity method was selected and operated between the hours of midnight to 6 a.m., thus avoiding the problems of vibrations arising from both traffic and from heavy industrial plant, and ensuring better access through the reduction in the number of parked cars.

The contrast in density between the infill material and the local sandstone (0.5 Mg/m³) produced small but detectable residual gravity anomalies of the order of 0.7 mGal. The gravity data were reduced and interpreted with the aid of data from one borehole and a preliminary archaeological investigation. A map of the residual gravity anomalies (Figure 2.45) clearly illustrates the locations of the faces of two quarries.

2.7.4.2 *Detection of massive ice in permafrost terrain*

The thawing of massive ice masses and associated ice-rich permafrost can cause severe engineering and environmental problems. The detection and identification of such ground phenomena is thus extremely important.

Kawasaki *et al.* (1983) provide an example of the use of gravity surveying to detect the presence and the volume of massive ice within an area of permafrost at Engineer Creek, near Fairbanks, Alaska, along the route of a proposed road cut. It is well known that large bodies of massive ice, such as occur within pingos, give rise to significant gravity anomalies (Mackay 1962; Rampton and Walcott 1974). Permafrost without discrete segregated ice has a density of about 1.6 Mg/m³, compared with the density of solid ice (0.88–0.9 Mg/m³) and with that of typical Alaskan soils (1.35–1.70 Mg/m³) and should give rise to detectable residual gravity anomalies if measured with a sufficiently sensitive gravimeter.

Kawasaki and colleagues demonstrated that, although massive ice can be detected by correlation with the gravity minima along the profile shown in Figure 2.46, the measurements were also sensitive to variations of density within the schist bedrock.

The gravity method is considered an excellent tool for detailed investigation of construction sites where massive ice is suspected, but

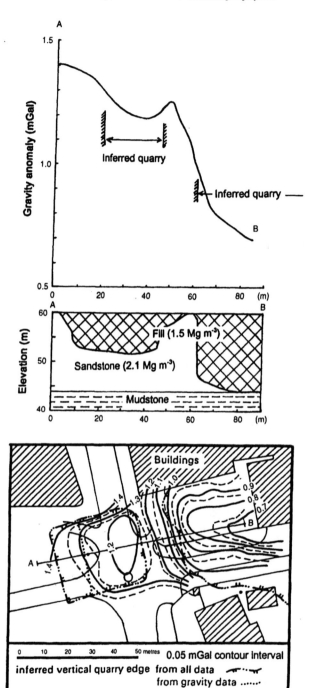

Figure 2.45 The top shows a residual gravity profile across infilled sand-stone quarries in Newcastle-upon-Tyne. The cross-section and plan are also shown. After Poster and Cope (1975), by permission

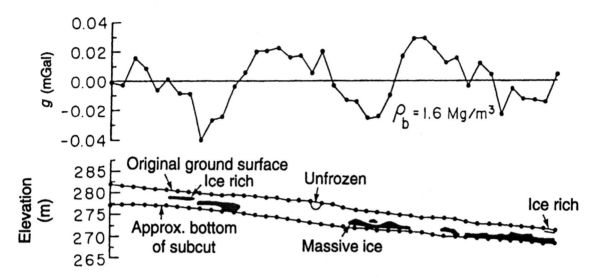

Figure 2.46 Gravity profile across massive ground ice in a road cut at Engineer Creek, near Fairbanks, Alaska. After Kawasaki *et al.* (1983), by permission

it is too slow a method for use as a reconnaissance tool over long profiles. Other geophysical methods such as electrical resistivity, electromagnetic ground conductivity and ground-penetrating radar are more efficient, particularly for reconnaissance (Osterkamp and Jurick 1980).

2.7.5 Detection of underground cavities

Hidden voids within the near-surface can become serious hazards if exposed unwittingly during excavation work, or if they become obvious by subsidence of the overlying ground (Figure 2.47). The detection of suspected cavities using gravity methods has been achieved in many engineering and hydrogeological surveys (e.g. Colley 1963). Gravimetry is increasingly of interest to archaeologists searching, for example, for ancient crypts or passages within Egyptian pyramids, such as has been achieved within the Kheops (see *First Break*, 1987 5(1): 3) with what is called 'endoscopic micro-gravity' (see also Lakshmanan 1991).

2.7.5.1 *Hidden natural cavities*

An example of the application of micro-gravimetry to the detection of underground cavities has been given by Fajklewicz (1986). Over many years he has investigated the gravity effect of both natural and man-made cavities and has helped to develop a method of detection based on calculating the vertical gradient of the gravity field. He has found that the amplitude of the gravity anomaly is commonly greater than that predicted (Figure 2.48) for reasons that are still not clear.

Figure 2.47 Catastrophic failure of the roof of an ancient flint mine in chalk in Norwich. Photo courtesy of Eastern Daily Press

A micro-gravity survey was carried out in the town of Inowrocław, Poland, where karst caverns occur to depths of around 40 m in gypsum, anhydrite, limestone and dolomite. The cavities develop towards the ground surface and have resulted in the damage and destruction of at least 40 buildings within the town. The density contrast between the cavity and the surrounding material in Figure 2.48A is $-1.8 \, \text{Mg/m}^3$ and for Figure 2.48B is $-1.0 \, \text{Mg/m}^3$, slightly lower due to the presence of rock breccia within the cavity. Fajklewicj has demonstrated that the cavity in Figure 2.48B should not have been detectable assuming that its gravity field is due entirely to a spherical cavity at the depth shown. Even the theoretical anomaly from the vertical gravity gradient is too broad to indicate the presence of a cavity, yet the observed anomaly is quite marked.

A similar approach can be taken using horizontal gravity gradients ($\Delta g/\Delta x$ or $\Delta g/\Delta y$), in which case the point at which the gravity anomaly reaches a minimum or maximum, the gradient goes through zero, and that point should lie over the centre of the body causing the anomaly (Butler 1984). An example of this (Figure 2.49) is given by Casten and Gram (1989) for a test case where gravity data were measured in a deep coal mine along an inclined drift which was known to pass at right-angles over a pump room.

Furthermore, micro-gravimetry can be used to determine the rate and extent of the development of strength-relaxation around underground excavations, as shown in Figure 2.50 (Fajklewicz 1986;

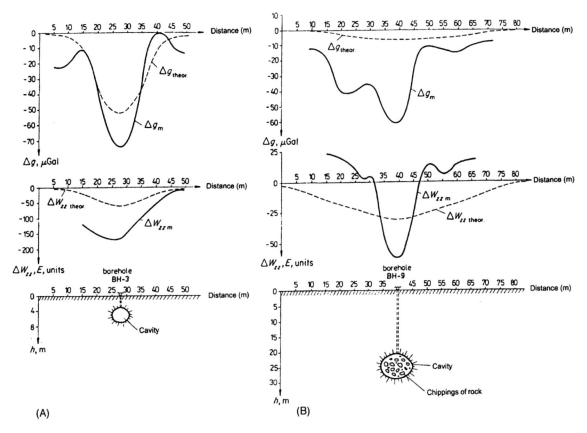

(A) (B)

Gluśko *et al.* 1981). As the rock relaxes mechanically it cracks, thereby reducing its bulk density. As the cracking continues and develops, so the changes in density as a function of time can be detected using highly sensitive micro-gravimeters, and then modelled.

2.7.5.2 *Archaeological investigations*

Blizkovsky (1979) provides an example of how a careful micro-gravity survey revealed the presence of a suspected crypt within the St 'Venceslas church, Tovacov, Czechoslovakia, which was later proven by excavation work. The dataset consisted of 262 values measured on a 1 m² or 4 m² grid to an accuracy of ±11 μGal, corrected for the gravity effect of the walls of the building (Figure 2.51). Two significant gravity minima with relative amplitudes of −60 μGal were located which indicated mass deficiencies associated with the previously unknown crypts.

2.7.6 Hydrogeological applications

Gravity methods are not used as much as electrical methods in hydrogeology but can still play an important role (Carmichael

Figure 2.48 Δg_m and $\Delta W_{zz,m}$ – micro-gravity anomalies and gravity gradient anomalies respectively–over (A) an air-filled cavity (BH-3), and (B) one partially infilled by rock fragments (BH-9). Curves labelled with suffix 'theor' represent the theoretical anomalies based on the parameters of the cavity alone. From Fajklewicz (1986), by permission

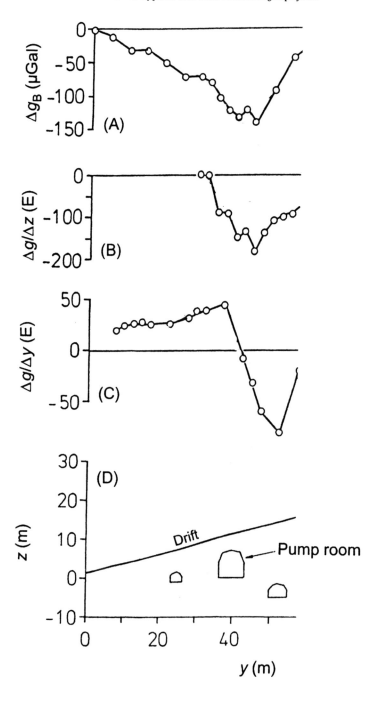

Figure 2.49 A micro-gravity survey within a deep coal mine as measured along a drift cut over a pump room and other known cavities. (A) shows observed residual gravity profiles; (B) observed and computed vertical gravity gradients; (C) observed horizontal gravity gradient, and (D) underground positions of known cavities from mine plans. From Casten and Gram (1989), by permission

and Henry 1977). Their more normal use is to detect low-density rocks that are thought to be suitable aquifers, such as alluvium in buried rock valleys (Lennox and Carlson 1967; van Overmeeren 1980).

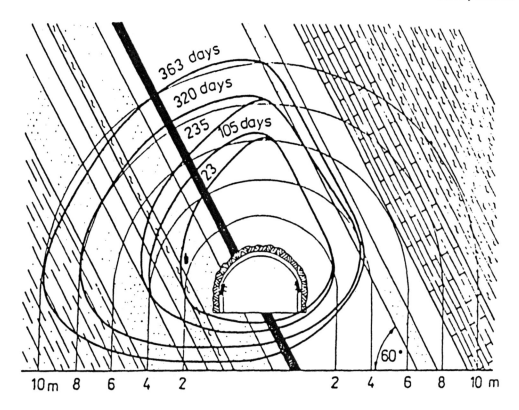

Buried valleys, which have been incised into either bedrock or glacial till and are associated with the South Saskatchewan River, have been identified by their gravity effects (Hall and Hajnal 1962). The Bouguer anomaly across the current river valley shows a minimum considerably wider than the present-day valley (Figure 2.52), thus indicating the presence of low-density material – subsequently found by drilling to be silts and sand.

Rather than interpret Bouguer anomalies, it is possible to use a gravimeter to monitor the effect of changing groundwater levels. For example, in a rock with a porosity of 33% and a specific retention of 20%, a change in groundwater level of 30 m could produce a change in g of 170 μGal. It is possible, therefore, to use a gravimeter to monitor very small changes in g at a given location. The only changes in gravity after corrections for instrument drift and Earth tides should be the amount of water in the interstices of the rock. Consequently, for an aquifer of known shape, a measured change in gravity, in conjunction with a limited number of water-level observations at a small number of wells, can be translated into an estimate of the aquifer's specific yield. Similarly, repeated gravity measurements have been used to estimate the volume of drawdown, degree of saturation of the steam zone (Allis and Hunt 1986) and the volume of

Figure 2.50 The time-dependent strength-relaxation around an underground gallery at 540 m depth as deduced from micro-gravity surveying in the Tchesnokov Colliery in the Don Basin over a period of 363 days. From Gluśko *et al.* (1981) and Fajklewicz (1986), by permission

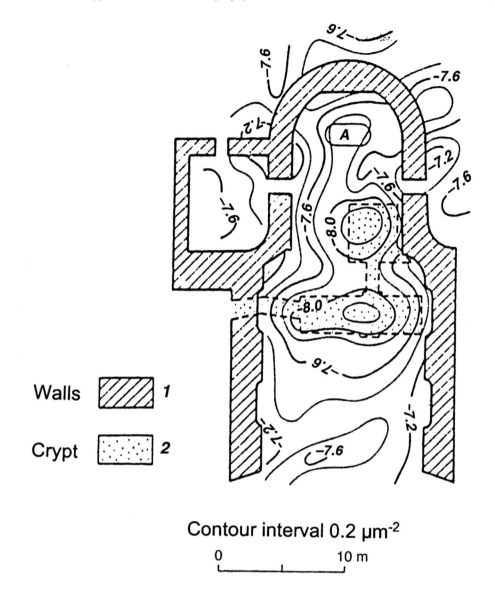

Walls [1]

Crypt [2]

Contour interval 0.2 μm⁻²

0   10 m

recharge of the Wairakei geothermal field, North Island, New Zealand (Hunt 1977).

2.7.7   Volcanic hazards

With the advent of highly accurate surveying equipment and methods, and the availability of very sensitive gravimeters, it is possible to monitor small changes in the elevations of the flanks of active volcanoes – ultimately with a view to predicting the next eruption. Such studies are often accompanied by seismic monitoring

Figure 2.51 Micro-gravity map of St 'Venceslas Church, Tovacov, Czechoslovakia showing marked anomalies over previously unknown crypts. From Blizkovskt (1979), by permission

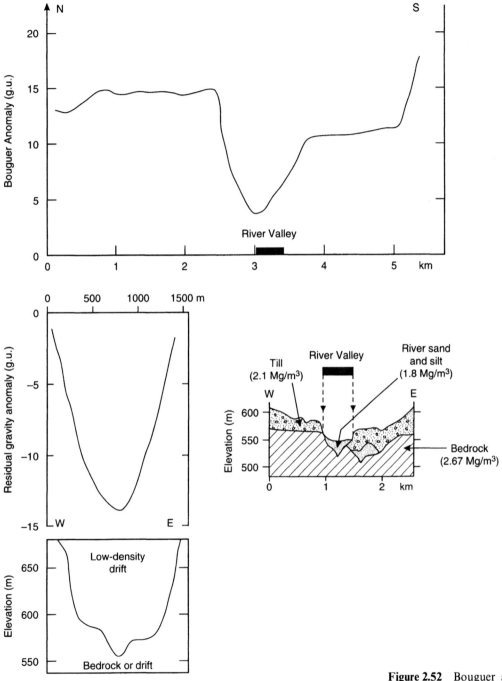

Figure 2.52 Bouguer anomaly over the South Saskatchewan River Valley, and the corresponding geological cross-section across the gravity minimum. After Hall and Hajnal (1962), by permission

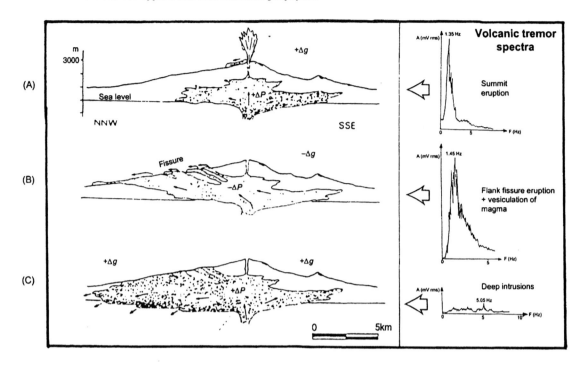

(e.g. Cosentino *et al.* (1989). Sanderson *et al.* (1983) carried out such a gravity monitoring and levelling programme on Mount Etna, Sicily, during the period August 1980 to August 1981, during which time a flank eruption took place (17–23 March 1981) from which a lava flow resulted which narrowly missed the town of Randazzo. A series of schematic diagrams is shown in Figure 2.53 to illustrate the three clear stages of the fissure eruption.

Changes in gravity in association with elevation increases were interpreted as the injection of new magma at depth during the intrusion of a new dyke at about 1.5 km depth (Figure 2.53A). Gravity decreases were observed when eruptions took place because of the reduction in material (Figure 2.53B). Where increases in gravity were observed without an increase in elevation, this was interpreted as being due to the density of the magma increasing by the enforced emplacement of new material at depth (Figure 2.53C). The magnitude of the gravity changes ($\approx 2\text{–}25\,\mu\,\text{Gal}$) coupled with the known variations in elevation ($< \approx 20\,\text{cm}$) provide a means of determining where within the volcano's plumbing the intrusions of new material and/or density changes are taking place.

Rymer and Brown (1987, 1989) have summarised the micro-gravity effects resulting from vertical movements of the magma/rock interface, vesiculation cycles with the magma column and radial changes in the dimensions of the magma body for Poás volcano in Costa Rica

Figure 2.53 Sketches of the stages of fissure eruptions on Mt Etna, Sicily, with gravity trends and typical volcanic tremor spectra. From Sanderson *et al.* (1983) and Cosentino *et al.* (1989), by permission

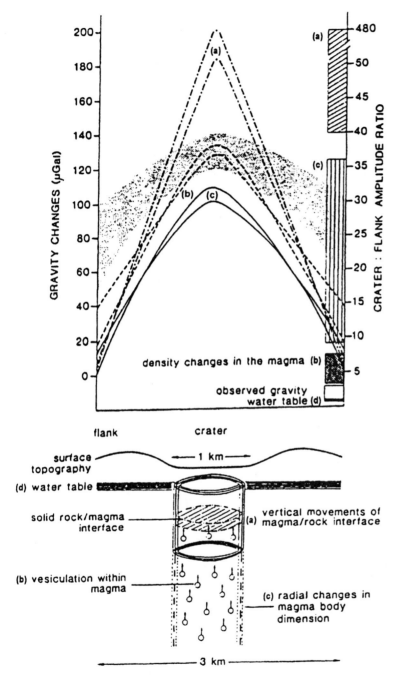

Figure 2.54 The top shows a schematic of the various gravity effects produced at the flank and summit of Poás volcano, Costa Rica (the observed range of gravity effects is shaded) and the ratio of the two (shown alongside in the vertical bar) as caused by different geological processes within the volcano (lower diagram). The processes are: (a) vertical movements of the magma/rock interface; (b) vesiculation cycles within the magma column; (c) radial changes in dimension of the magma column, and (d) variations in the level of the water table. From Rymer and Brown (1987), by permission

(Figure 2.54). The individual internal processes in this particular volcano can be clearly differentiated by using the ratio of the gravity effects measured at the volcano flank and summit. However, not all sub-terrainian activity is associated with seismic signatures. Indeed,

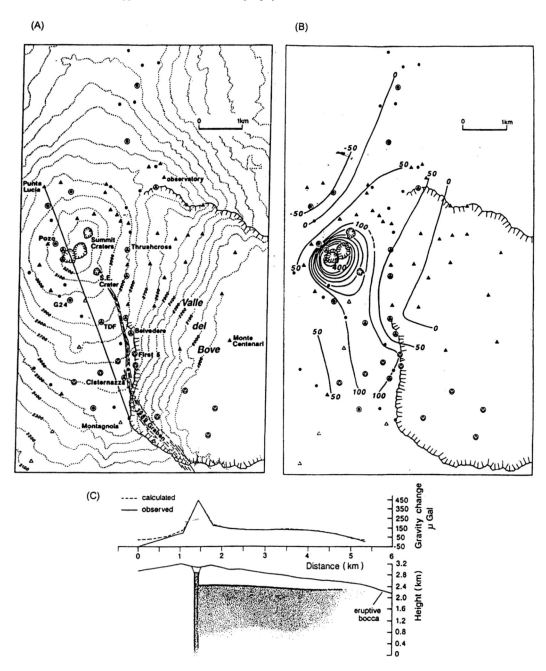

Figure 2.55　Maps showing (A) the locations of micro-gravity and ground deformation monitoring stations and (B) a contoured micro-gravity map of the summit area of Mt Etna, Sicily. The contour interval is 50 μGal. (C) Cross-section through the summit area of the Mt Etna along the profile indicated in (A). The best-fitting model for the observed gravity changes involved a dyke 4 m wide and a feeder pipe 50 m in diameter filling with magma at some time between the two sets of observations in June 1990 and June 1991. From Rymer (1993) and Rymer *et al.* (1993), by permission

Rymer (1993) reported on an increase in gravity at Mt Etna between June 1990 and June 1991 with no corresponding seismic activity. Large increases in gravity were observed around the summit craters and along an elongated zone following the line of a fracture formed during a previous eruption in 1989 (Figure 2.55). Surface elevation changes surveyed between 1990 and 1991 were only of the order of less than 3 cm. The gravity changes were an order of magnitude larger than would have been expected on the basis of elevation changes alone, which suggested that there must have been some sub-surface increase in mass. This was calculated to be of the order of 10^7 Mg and was caused by the intrusion of magma into fractures left by the 1989 eruption. The magma migrated passively into the pre-existing fractures so there was no corresponding seismic activity. Consequently, the micro-gravity measurements, in conjunction with elevations surveys, produced the only evidence of an impending eruption. The eruption of Mt Etna lasted for 16 months from 1991 to 1993, during which time lava poured out of the vent at a rate of $10\,m^3/s$, making this the largest eruption there for 300 years (Rymer 1993; Rymer *et al.* 1993).

Micro-gravity monitoring coupled with the distinctive patterns of the frequency of seismic activity (volcanic tremor spectra) are beginning to provide a very comprehensive model for volcanic eruptions, such as those at Mt Etna, and their associated processes. Many other volcanoes now have active monitoring programmes utilising gravity, seismic, and thermal investigations. Monitoring gas emissions is also proving to be a valuable additional indicator of impending volcanic activity (e.g. Pendick 1995, on the work of S. Williams). If these data can be obtained for individual volcanoes in conjunction with thermal radiation as measured by satellite, then the probability of identifying recognisable precursors to eruptions may be enhanced significantly, leading to a better prediction of volcanic activity and mitigation of potential hazards (Rymer and Brown 1986; Eggers 1987).

Chapter 3
Geomagnetic methods

3.1 INTRODUCTION

It is thought that the Chinese first used lodestone (magnetite-rich rock) in primitive direction-finding as early as the second century BC. It was not until the twelfth century in Europe that reference was made to the use of a magnetic compass for navigation. The first scientific analysis of the Earth's magnetic field and associated phenomena was published by the English physicist William Gilbert in 1600 in his book *De Magnete*. Measurements of variations in the Earth's magnetic field were made in Sweden to locate iron ore deposits as early as 1640. In 1870, Thalén and Tiberg developed instruments to measure various components of the Earth's magnetic field accurately and quickly for routine prospecting.

In 1915, Adolf Schmidt made a balance magnetometer which enabled more widespread magnetic surveys to be undertaken. As with many geophysical methods, advances in technology were made during the Second World War which enabled more efficient, reliable and accurate measurements to be made thereafter. In the 1960s, optical absorption magnetometers were developed which provided the means for extremely rapid magnetic measurements with very high sensitivity, ideally suited to airborne magnetic surveys. Since the early 1970s, magnetic gradiometers have been used which measure not only the total Earth's magnetic field intensity but also the magnetic gradient between sensors. This provides extra information of sufficient resolution which can be invaluable in delimiting geological targets.

Geomagnetic methods can be used in a wide variety of applications (Table 3.1) and range from small-scale investigations to locate pipes and cables in the very near surface, and engineering site investigations, through to large-scale regional geological mapping to determine gross structure, such as in hydrocarbon exploration. Commonly in the larger exploration investigations, both magnetic and gravity methods are used to complement each other. Used together prior to seismic surveys, they can provide more information about the sub-surface, particularly the basement rocks, than either technique on its own. Subsequent seismic reflection surveys are then used to provide more detailed imaging of the

Table 3.1   Applications of geomagnetic surveys

Locating
- Pipes, cables and metallic objects
- Buried military ordnance (shells, bombs, etc.)
- Buried metal drums of contaminated or toxic waste
- Concealed mineshafts and adits

Mapping
- Archaeological remains
- Concealed basic igneous dykes
- Metalliferous mineral lodes
- Geological boundaries between magnetically contrasting lithologies, including faults
- Large-scale geological structures

sub-surface, which is of more value to hydrocarbon exploration. The range of magnetic measurements which can now be made is extremely large, especially in the area of palaeomagnetism which will not be dealt with here. Palaeomagnetism is discussed in detail by Tarling (1983), for example.

3.2   BASIC CONCEPTS AND UNITS OF GEOMAGNETISM

3.2.1   Flux density, field strength and permeability

Around a bar magnet, a magnetic flux exists, as indicated by the flux lines in Figure 3.1, and converges near the ends of the magnet, which are known as the magnetic poles. If such a bar magnet is suspended in free air, the magnet will align itself with the Earth's magnetic field with one pole (the positive north-seeking) pointing towards the Earth's north pole and the other (the negative south-seeking) towards the south magnetic pole. Magnetic poles always exist in pairs of opposite sense to form a *dipole*. When one pole is sufficiently far removed from the other so that it no longer affects it, the single pole is referred to as a *monopole*.

If two magnetic poles of strength m_1 and m_2 are separated by a distance r, a force exists between them (Box 3.1). If the poles are of the same sort, the force will push the poles apart, and if they are of opposite polarity, the force is attractive and will draw the poles towards each other. Note the similarity of the form of the expression in Box 3.1 with that for the force of gravitational attraction in Box 2.1; both gravity and magnetism are *potential fields* and can be described by comparable potential field theory.

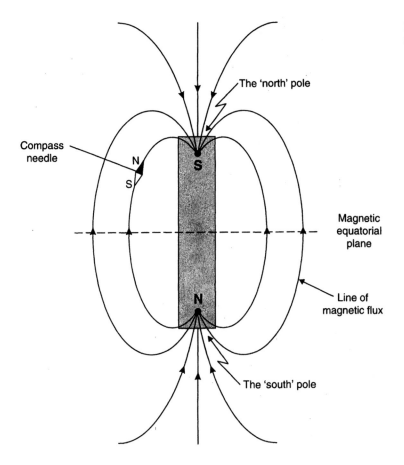

Figure 3.1 Lines of magnetic flux around a bar magnet

Box 3.1   Force between two magnetic poles

$$F = \frac{m_1 m_2}{4\pi \mu r^2}$$

where μ is the magnetic permeability of the medium separating the poles; m_1 and m_2 are pole strengths and r the distance between them.

The closeness of the flux lines shown in Figure 3.1, the flux per unit area, is the *flux density* **B** (and is measured in weber/m^2 = teslas). **B**, which is also called the 'magnetic induction', is a vector quantity. (The former c.g.s. units of flux density were gauss, equivalent to 10^{-4} T.) The units of teslas are too large to be practical in geophysical work, so a sub-unit called the nanotesla (nT = 10^{-9} T) is used instead, where 1 nT is numerically equivalent to 1 gamma in c.g.s. units (1 nT is equivalent to 10^{-5} gauss).

The magnetic field can also be defined in terms of a force field which is produced by electric currents. This *magnetising field strength H* is defined, following Biot–Savart's Law, as being the field strength at the centre of a loop of wire of radius *r* through which a current *I* is flowing such that $H = I/2r$. Consequently the units of the magnetising field strength *H* are amperes per metre (A/m).

The ratio of the flux density *B* to the magnetising field strength *H* is a constant called the *absolute magnetic permeability* (μ). Practically, the magnetic permeability of water and air can be taken to be equal to the *magnetic permeability of free space* (a vacuum), denoted μ_0 which has the value $4\pi \times 10^{-7}$ Wb A^{-1} m^{-1}. For any medium other than a vacuum, the ratio of the permeabilities of a medium to that of free space is equal to the *relative permeability* μ_r, such that $\mu_r = \mu/\mu_0$ and, as it is a ratio, it has no units.

3.2.2 Susceptibility

It is possible to express the relationship between *B* and *H* in terms of a geologically diagnostic parameter, the *magnetic susceptibility* κ (see Box 3.2 and Section 3.3.1). Susceptibility is in essence a measure of how susceptible a material is to becoming magnetised. For a vacuum, $\mu_r = 1$ and $\kappa = 0$. Although susceptibility has no units, to rationalise its numerical value to be compatible with the SI or rationalised system of units, the value in c.g.s. equivalent units (e.g. unrationalised units such as e.m.u. – electromagnetic units) should be multiplied by 4π. Some materials have negative susceptibilities (see Section 3.3).

Box 3.2   Relationship between magnetic flux density *B*, magnetising force *H*, and susceptibility κ

Given:

$$B = \mu H$$

[units: μ(Wb/Am) . *H*(A/m) = Wb/m^2 = teslas]

Since $\mu = \mu_r \mu_0$:

$$B = \mu_r \mu_0 H.$$

Rearranging to introduce $k = \mu_r - 1$:

$$B = \mu_0 H + \mu_0(\mu_r - 1)H$$
$$= \mu_0 H + \mu_0 kH = \mu_0 H + \mu_0 J.$$

Hence:

$$B = \mu_0 H(1 + k) \quad \text{and} \quad J = kH.$$

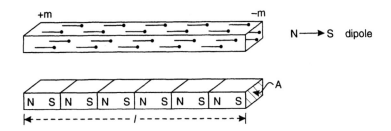

Figure 3.2 Schematic of a uniformly magnetised bar magnet as a collection of aligned dipoles producing a pole strength of $\pm m$ and as a series of minor bar magnets

3.2.3 Intensity of magnetisation

From the last expressions given in Box 3.2, it is clear that for a vacuum, $B = \mu_0 H$ (as $k = 0$). The penultimate expression in Box 3.2 indicates that in a medium other than a vacuum, an extra magnetising field strength of kH, called the *intensity of magnetisation J*, is induced by the *H*.

Another way of visualising the intensity of magnetisation is to examine a bar magnet of length *l* and cross-sectional area *A* which is uniformly magnetised in the direction of the long axis. The bar magnet can be thought of as consisting of a series of much smaller bar magnets or dipoles all aligned parallel to the long axis of the whole bar magnet (Figure 3.2). The magnetic intensities due to all the individual north and south poles will cancel out except at the end faces of the whole magnet, thus giving the whole magnet an overall magnetisation. The surface concentration of free poles, or pole strength *m* per unit area, is a measure of the intensity of magnetisation *J* (Box 3.3). The stronger the magnetisation, the greater will be the concentration of free poles. Furthermore, if a body of volume *V* is

Box 3.3   Intensity of magnetisation, *J* (amps/metre)

$$J = m/A$$

where *m* is the pole strength (amp.metre) and *A* is the cross-sectional area of the bar magnet (metre^2).
In terms of the *magnetic moment, M* (amp.metre^2):

$$J = M/V = m.l/V$$

where *l* is the length of the dipole, *V* is the volume of the magnetised body, and $M = m.l$.

The intensity of the induced magnetisation, J_i in rock with susceptibility κ, caused by the Earth's magnetic field *F* (tesla) in the sense of the flux density, i.e. the *B*-field, is given by:

$$J_i = k.F/\mu_0$$

where μ_0 is the permeability of free space, and $F = \mu_0 H$, with *H* being the magnetising force.

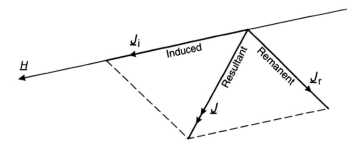

Figure 3.3 Vectorial summation of induced and remanent intensities of magnetisation

magnetised uniformly with intensity J, then that body is said to have a magnetic moment M which is defined as the product of the magnetic pole strength m and the length l separating the poles (Box 3.3). The intensity of magnetisation, which is thus the magnetic moment per unit volume, is of fundamental importance in describing the magnetic state of any rock mass.

3.2.4 Induced and remanent magnetisation

So far the discussion has centred upon a magnetisation that is induced by an applied field H where the induced intensity of magnetisation is denoted by J_i. In many cases, in the absence of an applied field (H), there is still a measurable intensity of magnetisation which is sustained by the internal field strength due to permanently magnetic particles. The intensity of this *permanent* or *remanent magnetisation* is denoted by J_r.

A rock mass containing magnetic minerals will have an induced as well as a remanent magnetisation. These magnetisations may have different directions and magnitudes of intensity (Figure 3.3). The magnitude and orientation of the resultant J dictate both the amplitude and shape of a magnetic anomaly, respectively. Consequently, interpretation of magnetic data is complicated by having greater degrees of freedom of the magnetic parameters and physical properties compared with gravity, which is largely dependent upon only rock density.

3.2.5 Diamagnetism, paramagnetism and ferromagnetism

All atoms have a magnetic moment as a result of the orbital motion of electrons around the nucleus and the spin of the electrons. According to quantum theory, two electrons can exist in the same electron shell (or state) as long as they spin in opposite directions. The magnetic moments of two such electrons, called *paired electrons*, will cancel out. In the majority of substances, when there is no external applied magnetic field, the spin magnetic moments of adjacent atoms are distributed randomly so there is no overall magnetisation. In a *diamagnetic* material, such as halite, all the electron shells are complete and so there are no unpaired electrons. When an external magnetic field is applied, a magnetisation is induced. The electrons orbit in such a way so as to produce a magnetic field which opposes the applied field, giving rise to a weak, negative susceptibility.

Unpaired electrons in incomplete electron shells produce unbalanced spin magnetic moments and weak magnetic interactions between atoms in *paramagnetic* materials such as fayerite, amphiboles, pyroxenes, olivines, garnets and biotite. In an external applied field, the magnetic moments align themselves into the same direction, although this process is retarded by thermal agitation. The result is a weak positive susceptibility but one which decreases inversely with the absolute temperature according to the Curie–Weiss Law. Paramagnetism is generally at least an order of magnitude stronger than diamagnetism.

In *ferromagnetic* materials, the susceptibility is large but is dependent upon temperature and the strength of the applied magnetic field. The spin moments of unpaired electrons are coupled magnetically due to the very strong interaction between adjacent atoms and overlap of electron orbits. A small grain in which magnetic coupling occurs forms what is called a single *magnetic domain* and has dimensions of the order of one micron. This gives rise to a strong 'spontaneous magnetisation' which can exist even when there is no external applied field. The magnetic coupling can be such that the magnetic moments are aligned either parallel or antiparallel (Figure 3.4).

Truly ferromagnetic materials occur only rarely in nature but include substances such as cobalt, nickel and iron, all of which have parallel alignment of moments. Ferromagnetism disappears when the temperature of the material is raised above the *Curie temperature* T_C as inter-atomic magnetical coupling is severely restricted and the material thereafter exhibits paramagnetic behaviour. In *antiferromagnetic* materials, for example hematite, the moments are aligned in

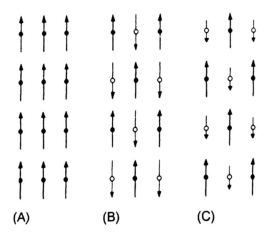

(A)   (B)   (C)

Figure 3.4 Schematic of magnetic moments in (A) ferromagnetic, (B) antiferromagnetic, and (C) ferrimagnetic crystals. After Nagata (1961), by permission

an antiparallel manner. Although the magnetic fields of the oppositely orientated dipoles cancel each other out, crystal lattice defects result in a net residual moment or *parasitic (anti)-ferromagnetism*. In ferrimagnetic materials, of which magnetite, titanomagnetite and ilmenite are prime examples, the sub-lattices are unequal and antiparallel. This results in a net magnetisation. Spontaneous magnetisation and large susceptibilities are characteristics of ferrimagnetic materials, such as in the case of pyrrhotite. Although the temperature dependence of ferrimagnetic behaviour is complex, ferrimagnetism disappears at temperatures above the Curie point. The majority of naturally occurring magnetic minerals exhibit either ferrimagnetic or imperfectly antiferromagnetic characteristics.

The Curie temperature varies with different minerals and will be different for whole rocks depending upon the composition of magnetic minerals present. In a granite rhyolite, for example, the Curie temperature for titanomagnetite is between 463 and 580°C, whereas for ilmenite–hematite series it is in the range 130–220°C. Oxidation of the iron–titanium oxides generally causes a rise in the Curie temperature. When low-temperature oxidation occurs, i.e. at temperatures lower than 300°C, in addition to increases in the Curie temperature, the intensity of magnetisation decreases. In order of increasing oxidation and decreasing intensity of magnetisation, titanomagnetite ($T_C = 100$–200°C) alters to titanomaghemite (150–450°C) then to magnetite (550–580°C) and ultimately to hematite (650–680°C) (Petersen 1990). Hematite has the lowest intensity of magnetisation. The alteration of magnetic minerals is important to remember when it comes to the interpretation of magnetic anomalies. Rocks which should display large susceptibilities and greatest intensities of magnetisation may exhibit much weaker magnetic properties owing to geochemical alteration of the magnetic minerals.

For a multidomain material in a field-free space ($H = 0$), the spontaneous magnetisation of the magnetic domains within a crystal is related to the crystal axes (Figure 3.5). The magnetisation directions of all domains cancel each other out so there is no net magnetisation intensity ($J = 0$). On increasing the applied magnetic field (H), the domain walls can move easily and reversibly should H be reduced at this point. As H increases, so the various domains reorientate themselves parallel to the applied field, but in discrete steps called *Barkhausen jumps*, which are permanent. When there is no further increase in magnetisation intensity with increasing applied field strength, all the domains are orientated parallel to the applied field direction and the material is said to be magnetically *saturated*. On reducing H to zero following saturation, only some of the magnetic domains are able to return to their former orientation, which results in a remanent magnetisation J_r.

If the magnetic permeability (μ) of a medium is independent of the magnetising force (H), the material is said to be linear in its behaviour.

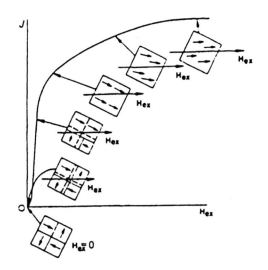

Figure 3.5 Process of magnetisation of a ferromagnetic substance according to domain theory. From Sharma (1986), by permission

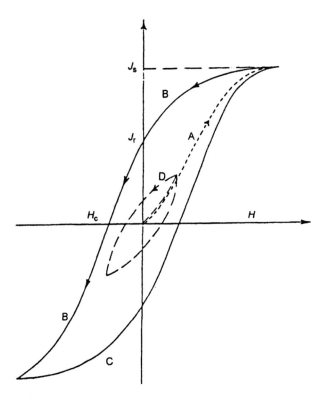

Figure 3.6 Hysteresis loop illustrating a cycle of magnetisation (curves A, B, C) of a ferromagnetic material. Small loop (D) shows the magnetisation cycle without saturation. From Sharma (1986), by permission

However, if a ferromagnetic or ferrimagnetic material, such as magnetite or pyrrhotite, with grains larger than 10 microns is placed in an increasing applied magnetic field, its magnetic intensity J increases to the point of saturation following a hysteresis loop (Figure 3.6). The physical processes by which this happens have already been described in terms of domain theory. After reaching saturation, the applied field H is reduced to zero at which point the intensity of magnetisation is that attributed to the remanent magnetisation. To eliminate this magnetisation, a negative field, $-H_c$, the *coercive force*, has to be applied. The *coercivity*, H_c, is an indication as to the 'hardness' or permanence of the magnetisation. Consequently, larger magnetic grains, which thus contain more magnetic domains, are easier to magnetise (and therefore have a higher susceptibility), than fine grains which are magnetically hard as indicated by a relatively high coercivity and low susceptibility. On increasing the applied magnetic field to full saturation, the hysteresis loop is completed. It follows that, for minerals that exhibit a nonlinear behaviour, no unique value of susceptibility exists. Values cited for such materials are usually for weak values of H and prior to saturation ever having been reached. For much more detailed discussion of rock magnetism, see the monographs by Nagata (1961), Stacey and Banerjee (1973) and O'Reilly (1984).

3.3 MAGNETIC PROPERTIES OF ROCKS

3.3.1 Susceptibility of rocks and minerals

Magnetic susceptibility is an extremely important property of rocks and is to magnetic exploration methods what density is to gravity surveys. Rocks that have a significant concentration of ferro- and/or ferri-magnetic minerals tend to have the highest susceptibilities. Consequently, basic and ultrabasic rocks have the highest susceptibilities, acid igneous and metamorphic rocks have intermediate to low values, and sedimentary rocks have very small susceptibilities in general (Table 3.2 and Figure 3.7). In this compilation of data, specific details of rock types are not available and so the values cited should be taken only as a guide. Metamorphic rocks are dependent upon their parent material and metapsammites are likely to have different susceptibilities compared with metapelites, for example.

Whole rock susceptibilities can vary considerably owing to a number of factors in addition to mineralogical composition. Susceptibilities depend upon the alignment and shape of the magnetic grains dispersed throughout the rock. If there is a marked orientation of particles, such as in some sedimentary and metamorphic rocks, a strong physical anisotropy may exist. The variation of magnetic properties as a function of orientation and shape of mineral grains is known as the *magnetic fabric*. Magnetic fabric analysis provides a very sensitive indication as to the physical composition of a rock or sediment, which in turn can be important in interpreting physical processes affecting that rock. For example, it is possible to correlate magnetic fabric variation in estuarine sediments with sonograph

Table 3.2 Susceptibilities of rocks and minerals (rationalised SI units)

Mineral or rock type	Susceptibility*
Sedimentary	
Dolomite (pure)	− 12.5 to + 44
Dolomite (impure)	20 000
Limestone	10 to 25 000
Sandstone	0 to 21 000
Shales	60 to 18 600
Average for various	**0 to 360**
Metamorphic	
Schist	315 to 3000
Slate	0 to 38 000
Gneiss	125 to 25 000
Serpentenite	3100 to 75 000
Average for various	**0 to 73 000**
Igneous	
Granite	10 to 65
Granite (m)	20 to 50 000
Rhyolite	250 to 37 700
Pegmatite	3000 to 75 000
Gabbro	800 to 76 000
Basalts	500 to 182 000
Oceanic basalts	300 to 36 000
Peridotite	95 500 to 196 000
Average for acid igneous	**40 to 82 000**
Average for basic igneous	**550 to 122 000**
Minerals	
Ice (d)	− 9
Rocksalt (d)	− 10
Gypsum (d)	− 13
Quartz (d)	− 15
Graphite (d)	− 80 to − 200
Chalcopyrite	400
Pyrite (o)	50 to 5000
Hematite (o)	420 to 38 000
Pyrrhotite (o)	1250 to 6.3×10^6
Ilmenite (o)	314 000 to 3.8×10^6
Magnetite (o)	70 000 to 2×10^7

(d) = diamagnetic material; (o) = ore; (m) = with magnetic
*$\kappa \times 10^6$ rationalised SI units; to convert to the unrationalised c.g.s. units, divide by 4π
Data from Parasnis (1986), Sharma (1986), Telford *et al.* (1990)

images of the estuary floor. In conjunction with Thematic Mapper images obtained from low-flying aircraft and simultaneous water sampling from boats, it is possible to establish a detailed model of estuarine sediment dynamic processes, as has been achieved for Plymouth Sound in south-west England (Fitzpatrick 1991). For further details of the magnetic fabric method, see the discussions by Lowrie (1990) and Tarling (1983), for example.

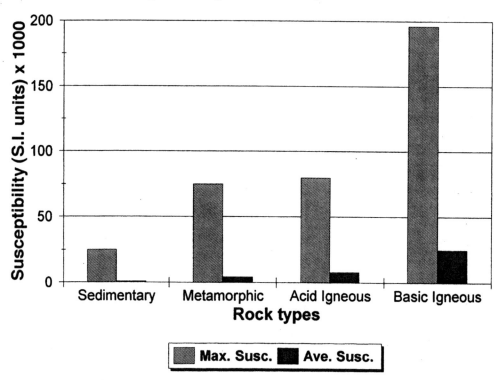

Figure 3.7 Susceptibilities of major rock types

For magnetic ore bodies with extremely high susceptibilities ($\kappa \geqslant 10^6$ SI), the measured susceptibility – more correctly referred to as the *apparent susceptibility* κ_a – can be reduced substantially by a shape demagnetisation effect (see Box 3.4). This involves a *demagnetisation factor* N_α which depends on a direction α. For a sphere, $N_\alpha = 1/3$ in all directions. In the case of a thin sheetlike body with a high true susceptibility ($\kappa \sim 10^6$ SI): $N_\alpha \sim 1$ in the transverse direction, giving a susceptibility $\kappa_a \approx 0.5K$; and $N_\alpha \sim 0$ in the longitudinal direction, so that $\kappa_a \approx \kappa$. Demagnetisation factors are discussed further by Parasnis (1986) and Sharma (1986).

Box 3.4   Apparent susceptibility κ_a and the demagnetisation factor N_α

$$\kappa_a = \kappa/(1 + N_\alpha \kappa)$$

Susceptibilities can be measured either in the field using a hand-held susceptibility meter such as the kappameter, or on samples returned to a laboratory where they can be analysed more accurately.

3.3.2 Remanent magnetisation and Königsberger ratios

In addition to the induced magnetisation, many rocks and minerals exhibit a permanent or *natural remanent magnetisation* (NRM) of intensity J_r when the applied field H is zero. The various processes by which rocks can acquire a remanent magnetisation are listed in Table 3.3 and discussed in more detail by Merrill (1990), Sharma (1986) and Tarling (1983).

Table 3.3 Types of remanent magnetisation (RM). After Merrill (1990), by permission

Type of RM	Process
Natural (NRM)	Acquired by a rock or mineral under natural conditions
Thermal (TRM)	Acquired by a material during cooling from a temperature greater than the Curie temperature to room temperature (e.g. molten lava cooling after a volcanic eruption)
Isothermal (IRM)	Acquired over a short time (of the order of seconds) in a strong magnetic field at a constant temperature (e.g. such as by a lightning strike)
Chemical (CRM)	Also crystallisation RM; acquired at the time of nucleation and growth or crystallisation of fine magnetic grains far below the Curie point in an ambient field
Thermal–chemical (TCRM)	Acquired during chemical alteration and cooling
Detrital (DRM)	Also depositional RM; acquired by the settling out of previously magnetised particles to form ultimately consolidated sediments which then have a weak net magnetisation, but prior to any chemical alteration through diagenetic processes
Post-depositional (PDRM)	Acquired by a sediment by physical processes acting upon it after deposition (e.g. bioturbation and compaction)
Viscous VMR	Acquired after a lengthy exposure to an ambient field with all other factors being constant (e.g. chemistry and temperature)
Anhysteretic (ARM)	Acquired when a peak amplitude of an alternating magnetic field is decreased from a large value to zero in the presence of a weak but constant magnetic field

Primary remanent magnetisations are acquired by the cooling and solidification of an igneous rock from above the Curie temperature (of the constituent magnetic minerals) to normal surface temperature (TRM) or by detrital remanent magnetisation (DRM). Secondary remanent magnetisations, such as chemical, viscous or post-depositional remanent magnetisations, may be acquired later on in the rock's history. This is especially true of igneous rocks which have later undergone one or more periods of metamorphism, particularly thermal metamorphism.

The intensity of the remanent magnetisation J_r may swamp that of the induced magnetisation J_i, particularly in igneous and thermally metamorphosed rocks. The ratio of the two intensities (J_r/J_i) is called the *Königsberger ratio*, Q, which can be expressed in terms of the Earth's magnetic field at a given locality and the susceptibility of the rocks (Box 3.5). Just as susceptibility can vary within a single rock type, so too can the Königsberger ratio. However, similar rock types have characteristic values of Q, some of which are listed in Table 3.4. Nagata (1961) has made four broad generalizations on the basis of Q:

- $Q \sim 1$ for slowly crystallised igneous and thermally metamorphosed rocks in continental areas;
- $Q \sim 10$ for volcanic rocks;

Table 3.4 Examples of values of the Königsberger ratio

Rock type	Location	Q
Basalt	Mihare volcano, Japan	99–118
Oceanic basalts	Northeast Pacific	15–105
Oceanic basalts	Mid-Atlantic Ridge	1–160
Sea-mount basalts	North Pacific	8–57
Cainozoic basalts	Victoria, Australia	5
Early tertiary basalts	Disko, West Greenland	1–39
Tholeiite dykes	England	0.6–1.6
Dolerite sills	North England	2–3.5
Dolerite	Sutherland, Scotland	0.48–0.51
Quartz dolerite	Whin Sill, England	2–2.9
Gabbro	Småland, Sweden	9.5
Gabbro	Minnesota, USA	1–8
Gabbro	Cuillin Hills, Scotland	29
Andesite	Taga, Japan	4.9
Granite	Madagascar	0.3–10
Granite plutons	California, USA	0.2–0.9
Granodiorite	Nevada, USA	0.1–0.2
Diabase	Astano Ticino, Switzerland	1.5
Diabase dykes	Canadian Shield	0.2–4
Magnetite ore	Sweden	1–10
Magnetite ore	South India	1–5

- $Q \sim 30$–50 for many rapidly quenched basaltic rocks;
- $Q < 1$ in sedimentary and metamorphic rocks, except when iron ore is involved.

Box 3.5   Königsberger ratio, Q

$$Q = J_r/\kappa(F/\mu_0)$$

where J_r is the intensity of remanent (NRM) magnetisation, κ is the susceptibility, μ_0 is the permeability of free space and F is the magnitude of the Earth's magnetic field (in tesla) at a given location in the same sense as the B-field (flux density).

It is also very important to consider that not only may J_r exceed J_i, but the direction of remanent magnetisation may be quite different from that of the ambient induced field at a location. Consequently, the resultant magnetisation (i.e. the vectorial sum of the remanent and induced magnetisations) will give rise to characteristic magnetic anomalies (refer back to Figure 3.3) when reversely magnetised rocks are present.

3.4   THE EARTH'S MAGNETIC FIELD

3.4.1   Components of the Earth's magnetic field

The geomagnetic field at or near the surface of the Earth originates largely from within and around the Earth's core. Currents external to the Earth in the ionosphere and magnetosphere associated with the Van Allen radiation belts (Figure 3.8), currents induced in the Earth by external field variations and the permanent (remanent) and steady-state induced magnetisations of crustal rocks, also contribute to the overall geomagnetic field. The magnetosphere is vital for the survival of life on Earth as it forms the primary force field which protects the planet from harmful radiation from the Sun. The various components of the geomagnetic field affect exploration surveys in a variety of ways which will be discussed in turn.

3.4.1.1   *The main dipole field*

The main component of the geomagnetic field is called the *dipolar field* as it behaves, to a first-order approximation, like a dipolar electromagnet located at the centre of the Earth but inclined at $11.5°$ to the rotational axis (Figure 3.9).

The *geomagnetic poles*, the positions on the Earth's surface through which the axis of the best-fitting dipole passes – which are located in Hayes Peninsula in northern Greenland and near the Russian Vostok research station in Greater Antarctica – are not the same as the

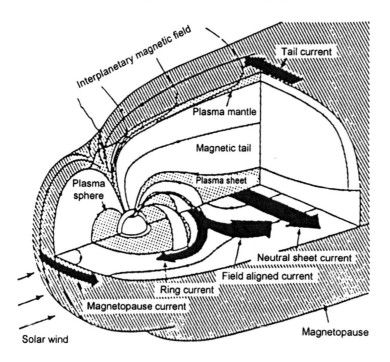

Figure 3.8 The geomagnetic field showing the magnetosphere, magnetopause and Van Allen radiation belts. From James (1990), by permission

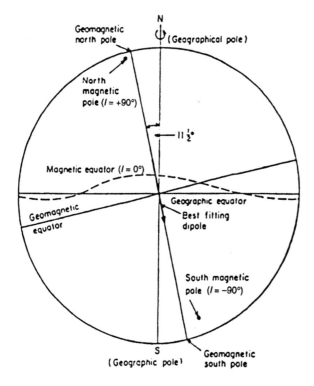

Figure 3.9 The field due to an inclined geocentric dipole. From McElhinny (1973), by permission

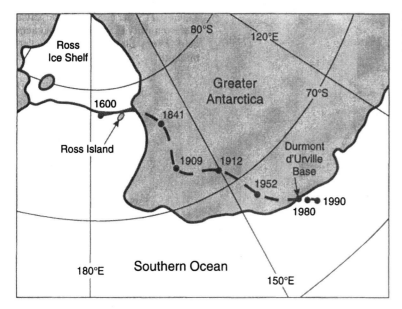

Figure 3.10 Location of the south magnetic pole and its drift since the year 1600

magnetic or *dip poles*. These are located where the magnetic field is directed vertically. The north magnetic pole is currently just north of Bathurst Island in the Canadian Arctic Archipelago and was discovered on 1 June 1831 by James Clark Ross and his uncle Sir John Ross. The south magnetic pole is currently about 150 km offshore from the French Research Station Durmont d'Urville on the Adélie Coast of Greater Antarctica. James Ross came extremely close to locating the south magnetic pole in 1841 but as it was then inland (Figure 3.10) he was thwarted by icebergs and the land ice. It was only on 15 January 1909 that Alistair Mackay, Edgeworth David and Douglas Mawson reached the south magnetic pole after an epic sledge journey from their base on Ross Island.

The geomagnetic field is produced by electric currents induced within the conductive liquid outer core as a result of slow convective movements within it (Figure 3.11). It is for this reason that the analogy of the Earth's field to that induced by an electromagnet is preferred to that of a permanently magnetised bar magnet. The liquid core behaves as a geodynamo but the precise nature of the processes involved has yet to be resolved. Models to explain the disposition of the magnetic field must also account for the slow but progressive change in field intensity and westward drift in direction known as the *secular variation*. Furthermore, the model must also explain how the Earth's magnetic field goes through reversals of magnetic polarity. The study of how the Earth's magnetic field has changed through geological time is known as *palaeomagnetism*. The use of magnetic reversals to provide global chronometric calibration of geological events is known as *magnetostratigraphy* (Tauxe, 1990).

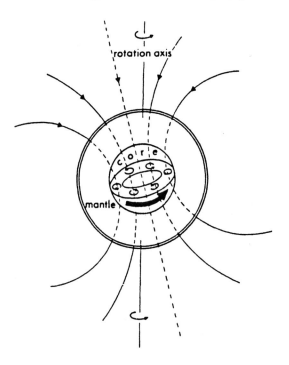

Figure 3.11 Schematic of the cause of the majority of the Earth's magnetic field. From Sharma (1986), by permission

The geomagnetic field can be described in terms of the declination, *D*, inclination, *I*, and the total force vector *F* (Figure 3.12). A freely suspended magnetised needle will align itself along the *F* vector so that at the magnetic (dip) north, the inclination is 90°; i.e. the needle will point vertically downwards. At the south magnetic (dip) pole, the needle will point vertically upwards. At the magnetic equator, the needle will lie horizontally (Figure 3.13). Furthermore, the vertical component of the magnetic intensity of the Earth's magnetic field varies with latitude, from a minimum of around 30 000 nT at the magnetic equator to 60 000 nT at the magnetic poles.

3.4.1.2 *The non-dipolar field*

While the single dipole field approximates to the Earth's observed magnetic field, there is a significant difference between them, which is known as the *non-dipole field*. The total intensity for the non-dipole field is shown in Figure 3.14, from which several large-scale features can be seen with dimensions of the order of several thousand kilometres and with amplitudes up to 20 000 nT, about one-third of the Earth's total field. Using the method of spherical harmonic analysis, it can be demonstrated that the non-dipole field and the associated large-scale features can be represented by a fictitious set of 8–12 small dipoles radially located close to the liquid core. These dipoles serve to simulate the eddy currents associated with the processes within the liquid core.

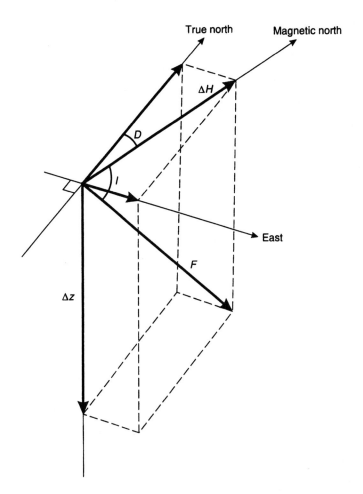

Figure 3.12 Elements of the magnetic field: inclination I, declination D, and total magnetic force F

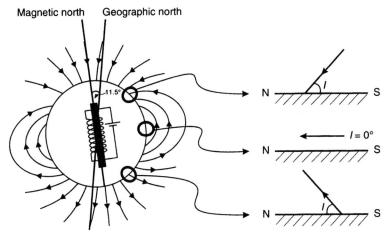

Figure 3.13 Variation of inclination with latitude

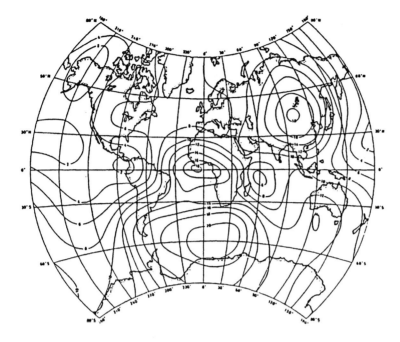

Figure 3.14 Variation in the intensity of the non-dipole field for epoch 1980. From Sharma (1986), by permission

A further use of the spherical harmonic analysis is that it provides a means whereby the spatial distribution and intensity of the total magnetic field can be calculated for the whole globe. The total field, which is calculated every five years, is called the *International Geomagnetic Reference Field* (IGRF) and the year of calculation is known as the *epoch*. It has to be recalculated regularly because of the secular variation (see, for example, IAGA (1987) and Peddie (1982)). Consequently, it is possible to obtain a theoretical value for the field strength of the Earth's magnetic field for any location on Earth (Figure 3.15). It can be seen from this figure that instead of the anticipated two maxima consistent with a truly dipolar field, there are in fact four maxima. The significance of the IGRF in processing magnetic data is discussed in Section 3.6.3.

Data used in the computation of revisions of the International Geomagnetic Reference Field have been obtained by satellite (e.g. during 1965–71, Polar Orbiting Geophysical Observatory series, POGO; October 1979 to June 1980, MAGSAT). However, at satellite orbit ranges, perturbations in the earth's magnetic field caused by magnetic materials in the crust are not resolvable. Surface or airborne measurements can detect considerable high-amplitude small-scale features within the crust down to a depth of 25–30 km where the Curie isotherm is reached. These features may be caused by induction due to the Earth's field or remanent magnetisation or a mixture of both.

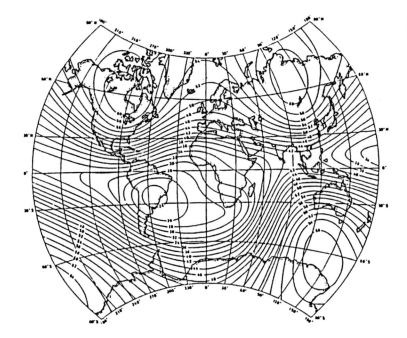

Figure 3.15 Total field intensity derived using the IGRF epoch 1980. From Sharma (1986), by permission

3.4.2 Time variable field

Observations of the Earth's magnetic field have been made for over four centuries at London and Paris. From these data, it is clear that the geomagnetic and magnetic pole positions drift with time, known as the secular variation in the magnetic field (Figure 3.16). In addition, the intensity of the main magnetic field is decreasing at about 5% per century. These rates of change, although very significant on a geological time scale, do not affect data acquisition on a typical exploration survey unless it covers large geographical areas and takes many months to complete, or if such surveys are being used to compare with historical data.

The Earth's magnetic field changes over a daily period, the *diurnal variations*. These are caused by changes in the strength and direction of currents in the ionosphere. On a magnetically 'quiet' (Q) day, the changes are smooth and are on average around 50 nT but with maximum amplitudes up to 200 nT at the geomagnetic equator. The changes are least during the night when the background is almost constant, and decrease in amplitude from dawn to midday whereupon they increase to the daily maximum about mid-late afternoon before settling down to the night-time value.

Magnetically disturbed (D) days are marked by a rapid onset of fluctuations of the order of hundreds of nanoteslas followed by slower but still erratic fluctuations with decreasing amplitude. These

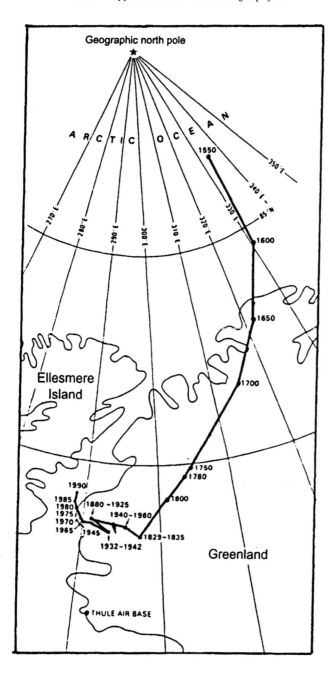

Figure 3.16 Drift of the north magnetic pole position from 1550 to 1990. From James (1990), by permission

disturbances, which are called *magnetic storms*, may persist for several hours or even days. Such frenetic magnetic activity is caused by sunspot and solar activity resulting in solar-charged particles entering the ionosphere. This may happen on fine sunny days and not necessarily in stormy weather. Magnetic observatories around the world provide an advance warning service to advise of the probability of magnetic storm activity. In severe storms, all magnetic surveying has to stop as it is not practicable to correct for such extreme fluctuations. In minor disturbances, if a continuous-reading base station magnetometer is used, the diurnal variations can be corrected. In aeromagnetic surveys, it is necessary to specify contractually what constitutes a magnetic storm. Survey data adversely affected by magnetic disturbances may have to be reflown and this obviously has cost implications. Practical details of how to correct for diurnal variations are given in Section 3.6.2.

3.5   MAGNETIC INSTRUMENTS

The earliest known device which responded to the Earth's magnetic field was a magnetised spoon used by Chinese geomancers (diviners) in the first century AD. Compass needles were introduced for navigation around the year 1000 in China and in Europe about 200 years later. The first accurate measurement of the inclination of the Earth's field was made at Radcliffe in London in 1576 by Robert Norman. He described his instruments and collected data in his book *The Newe Attractive* (1581), which was the first book ever to be devoted to geomagnetism.

Magnetometers used specifically in geophysical exploration can be classified into three groups: the torsion (and balance), fluxgate and resonance types, of which the last two have now completely superseded the first. Torsion magnetometers are still in use at 75% of geomagnetic observatories, particularly for the measurement of declination. Magnetometers measure horizontal and/or vertical components of the magnetic field (F_h and F_z respectively) or the total field F_t (see Figure 3.12).

3.5.1   Torsion and balance magnetometers

Historically the first to be devised (1640), these comprise in essence a magnetic needle suspended on a wire (torsion type) or balanced on a pivot. In the Earth's magnetic field the magnet adopts an equilibrium position. If the device is taken to another location where the Earth's magnetic field is different from that at the base station, or if the magnetic field changes at the base station, the magnet will align itself to the new field and the deflection from the rest position is taken as a measure of the Earth's magnetic field. The Swedish mine compass, Hotchkiss superdip, and Thalén-Tiberg magnetometer are all early examples of this type of device. In 1915, Adolf Schmidt devised his variometer in which a magnetic beam was asymmetrically balanced on an agate knife edge (Figure 3.17) and zeroed at a base station.

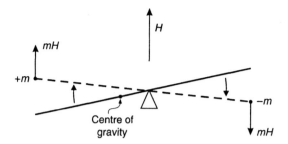

Figure 3.17 Basic principle of operation of a torsion or balance-type magnetometer

Deflections from the rest position at other locations were then read using a collimating telescope. To be used it had to be orientated at right-angles to the magnetic meridian so as to remove the horizontal component of the Earth's field. The device was calibrated using Helmholtz coils so that the magnitude of the deflection was a measure of the vertical component of the field strength.

A development of the Schmidt variometer was the compensation variometer. This measured the force required to restore the beam to the rest position. In exploration work, the greatest precision with a balance magnetometer was only 10 nT at best. For further details of these devices, see the descriptions by Telford *et al.* (1990).

3.5.2   Fluxgate magnetometers

The fluxgate magnetometer was developed during the Second World War to detect submarines. It consists of two parallel cores made out of high-permeability ferromagnetic material. Primary coils are wound around these cores in series but in opposite directions (Figure 3.18). Secondary coils are also wound around the cores but in the opposite sense to the respective primary coil. A current alternating at 50–

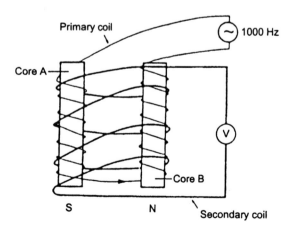

Figure 3.18 Basic operating principle of the fluxgate magnetometer

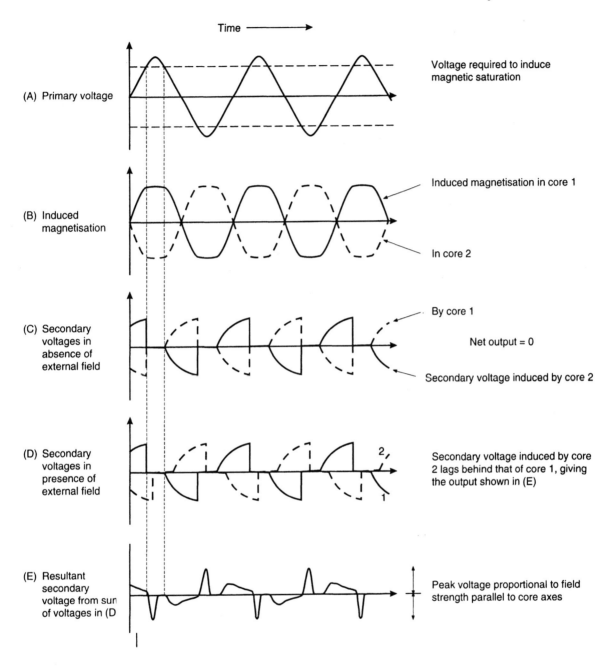

Time ⟶

(A) Primary voltage

Voltage required to induce magnetic saturation

(B) Induced magnetisation

Induced magnetisation in core 1

In core 2

(C) Secondary voltages in absence of external field

By core 1

Net output = 0

Secondary voltage induced by core 2

(D) Secondary voltages in presence of external field

Secondary voltage induced by core 2 lags behind that of core 1, giving the output shown in (E)

(E) Resultant secondary voltage from sum of voltages in (D

Peak voltage proportional to field strength parallel to core axes

Figure 3.19 Response characteristics of primary and secondary circuits in a fluxgate magnetometer

1000 Hz (Figure 3.19A) is passed through the primary coils which drives each core through a *B–H* hysteresis loop (cf. Figure 3.6) to saturation at every half-cycle (Figure 3.19B) in the absence of an external field, so inducing a magnetic field in each core. The generated alternating magnetic field induces an in-phase voltage within the secondary coils. This voltage reaches its maximum when the rate of change of the magnetic field is fastest (Figure 3.19C). As the coils are wound in opposing directions around the two cores, the secondary voltages are in phase but have opposite polarity (Figure 3.19C) so that the sum of the two voltages is at all times zero. However, when the cores are placed in the Earth's magnetic field, a component of that field will be parallel to the orientation of the cores. Consequently, the core whose primary field is reinforced by the ambient external field will reach saturation earlier than the other core whose magnetic field is opposed by the external field. This has the effect of shifting the phases of the secondary voltages (Figure 3.19D) so that the sum of the two secondary voltages is now non-zero (Figure 3.19E). The peak amplitude of the pulsed output of the combined secondary coils is proportional to the magnitude of the external field component (Primdahl 1979).

The fluxgate magnetometer can be used to measure specific magnetic components with the same attitude as the sensor cores. As the fluxgate magnetometer is relatively insensitive to magnetic field gradients, it has the advantage that it can be used in areas where very steep gradients would militate against the use of resonance-type devices which are affected. Some portable fluxgate magnetometers suffer from temperature effects owing to inadequate thermal insulation, which can reduce the resolution to only ± 10 to 20 nT, this being inadequate for ground exploration surveys. They are used quite widely in airborne surveys where better thermal insulation can be ensured and additional devices can be used to aid the consistent orientation of the sensor cores. In such cases, an accuracy to within ± 1 nT can be achieved. In addition, fluxgate instruments can provide a continuous output which is another advantage for airborne applications. Fluxgate magnetometers can also be used in down-hole logging applications in mineral exploration.

3.5.3 Resonance magnetometers

There are two main types of resonance magnetometer: the *proton free-precession magnetometer*, which is the best known, and the *alkali vapour magnetometer*. Both types monitor the precession of atomic particles in an ambient magnetic field to provide an absolute measure of the total magnetic field, *F*.

The proton magnetometer has a sensor which consists of a bottle containing a proton-rich liquid, usually water or kerosene, around which a coil is wrapped, connected to the measuring apparatus (Figure 3.20). Each proton has a magnetic moment *M* and, as it is always in motion, it also possesses an angular momentum *G*, rather like a spinning top. In an ambient magnetic field such as the Earth's (*F*), the majority of the protons align themselves parallel with this field with the remainder orientated antiparallel (Figure 3.21A). Consequently, the volume of proton-rich liquid

(A)

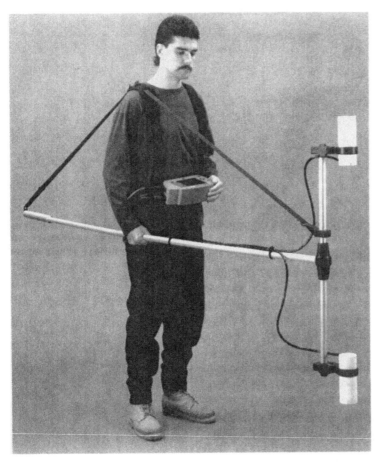

(B)

Figure 3.20 (A) A nuclear proton precession magnetometer in use, and (B) a caesium vapour magnetometer. Courtesy of Geometrics

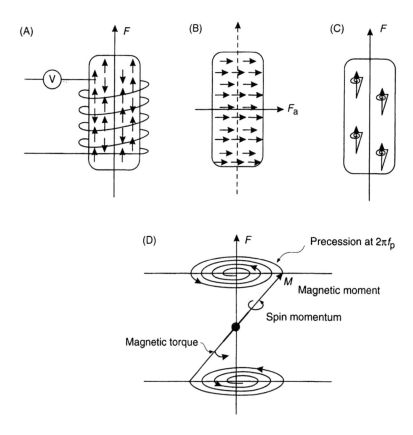

acquires a net magnetic moment in the direction of the ambient field (*F*).

A current is applied to the coil surrounding the liquid, generating a magnetic field about 50 to 100 times stronger than, and at right-angles to, the Earth's field. The protons align themselves to the new magnetic direction (Figure 3.21B). When the applied field is switched off, the protons precess around the pre-existent ambient field *F* (Figure 3.21C) at the *Larmor precession frequency* (f_p) which is proportional to the magnetic field strength *F* (Box 3.6). As protons are charged particles, as they precess they induce an alternating voltage at the same frequency as f_p into the coil surrounding the sensor bottle. Interaction between adjacent protons causes the precession to decay within 2–3 seconds, which is sufficient time to measure the precession frequency. To obtain a value of *F* to within ± 0.1 nT, frequency must be measured to within ± 0.004 Hz, which is quite easily achieved. This resolution is equivalent to 1 part in 10^6, which is 100 times less sensitive than in gravity measurements.

Figure 3.21   Basic operating principles of a proton magnetometer. After Kearey and Brooks (1991), by permission

Box 3.6   Magnetic field strength, *F*, and precession frequency, f_P

$$F = 2\pi f_P / \Phi_P$$

where Φ_P is the gyromagnetic ratio of the proton, which is the ratio of the magnetic moment and spin angular momentum (see Figure 3.21); and

$$\Phi_P = 0.26753 \text{ Hz/nT and } 2\pi/\Phi_P = 23.4859 \text{ nT/Hz}.$$

Thus:

$$F = 23.4859 f_P.$$

For example, for $F = 50\,000$ nT, $f_P = 2128.94$ Hz.

One of the limiting factors of the proton magnetometer is that its accuracy is reduced in areas of high magnetic gradient. As the sensor bottle is of the order of 15 cm long, a strong field gradient of 500 nT/m, for example, means that there is a 75 nT difference in field strength between the top and bottom of the sensor and the rate of damping is increased. The accuracy of measurement of the precession frequency is thus reduced. As a guide, if the gradient is 400 nT/m, the precision is at best 1 nT; for 200 nT/m it is 0.5 nT.

As the precession frequency is only a function of field strength, there is no need to orientate the field sensor. Modern proton magnetometers give a direct readout of the field strength in nanoteslas and data can be automatically output into a datalogger for subsequent downloading into a computer.

Proton magnetometers are used extensively not only in land surveys but also at sea and in airborne investigations. In marine surveys, the magnetometer sensor bottle, which is located in a sealed unit called a 'fish', is deployed two or three ship's lengths astern so as to be sufficiently removed from magnetic interference from the ship. In the case of aircraft, two techniques are used. One is to tow the sensor bottle at least 30 m below and behind the aircraft in what is called a 'bird', or place it in a non-magnetic boom called a 'stinger' on the nose, on the tail fin or on a wingtip of the aircraft. In the fixed mode, special magnetic compensation measures can be taken to annul the magnetisation of the aircraft; the excellence of the compensation is called the *figure of merit* (FOM) rating. The fitting of active compensation systems in modern aircraft has improved FOM values and reduced the time taken for compensation, and so helped to improve the cost-effectiveness of airborne surveys. In addition to ground, marine and airborne applications, proton magnetometers can be deployed down boreholes, and can be particularly useful in mineral exploration programmes.

A limitation on proton magnetometers, particularly in airborne surveys, is the rate at which measurements can be made. As the proton precession and measurement take a finite time (of the order of a second or longer), continuous readings are not possible and this can be restricting in some situations.

One manufacturer (GEM Systems Inc.) has produced a modified precession instrument that utilises the Overhauser Effect. An electron-rich fluid containing free radicals is added to a standard hydrogen-rich liquid. The combination increases the polarisation by a factor of 5000 in comparison with standard liquids. Overhauser proton precession uses a radio-frequency (RF) magnetic field and so needs only minimal power, in contrast with high-power direct current fields used in traditional proton precession magnetometers. Polarisation and magnetisation can occur simultaneously and thus rapid sampling of the total field strength (two readings per second) can be achieved.

The second type of resonance magnetometer is the *alkali vapour magnetometer* or *optical absorption magnetometer*, which utilises the optical pumping technique (Bloom 1962). The principle on which this method is based is illustrated in Figure 3.22. Under normal conditions of temperature and pressure, electrons exist at certain energy states (A and B) around the nucleus of the atom. According to quantum physics, it is only possible to transfer an electron from a lower energy state (A) to one with higher energy (B) in discrete jumps. If a vapour of an element such as rubidium or caesium is illuminated by a light whose filament is made of the same element, the light emitted is at the correct wavelength for incident photons to be absorbed by the vapour and the low-energy state electrons excited up to higher levels. If the incident light is circularly polarised, only electrons in the A_1 orbit will be excited or 'optically pumped' up to the B orbit. At this point, the excess photons will be transmitted through the excited vapour and will be detected by the photocell as an increase in light intensity.

A small alternating current is passed through a coil at a frequency of between 90 and 300 kHz to induce a magnetic field around the

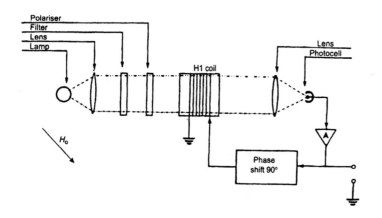

Figure 3.22 Optical pumping – principle of the alkali vapour magnetometer. From James (1990), by permission

alkali vapour cell. The frequency is tuned until it is locked into the Larmor frequency for the alkali vapour concerned. This small magnetic field energises some electrons back into their vacant A ground states. The consequence of this is that the light intensity at the photocell diminishes as photons are absorbed by the alkali vapour in the cell until saturation is reached again. Photons will continue to be absorbed until all the available electrons have been excited to the B state, when the light at the photocell will again be at its most intense. Consequently, the cycled optical pumping produces a light at the photocell that flickers at the Larmor precession frequency, which can easily be measured. As the precession frequency is dependent upon the ambient field strength of the Earth (see Box 3.6), the total field strength can be determined from a measurement of the precession frequency. The factor $2\pi/\Phi_P$ is approximately equal to 0.2141 and 0.1429 nT/Hz for rubidium and sodium respectively, which give corresponding precession frequencies of 233.5 and 350 kHz in a field of 50 000 nT. As long as the light beam axis is not parallel or antiparallel to the Earth's magnetic field (when no signals would be produced), the precession frequency can be measured with sufficient accuracy so that the magnetic field strength can be determined to within ± 0.01 nT. The measurement time is extremely small, and so alkali vapour magnetometers can be used as virtually continuous reading instruments, which makes them ideally suited to airborne surveys.

For land-based archaeological or environmental geophysical applications, self-oscillating split-beam caesium vapour magnetometers have been developed, largely from military ordnance detection instruments (e.g. the Geometrics G-822L magnetometer). Sampling rates of up to 10 readings per second are possible with a sensitivity of 0.1 nT.

3.5.4 Cryogenic (SQUID) magnetometers

The most sensitive magnetometer available is the cryogenic magnetometer which operates using processes associated with superconductivity, details of which have been given by Goree and Fuller (1976). These magnetometers are perhaps better known as *SQUID* (Superconducting QUantum Interference Device) magnetometers. Used extensively in palaeomagnetic laboratories, the SQUID magnetometer has also been developed for use in aeromagnetic surveying since the early 1980s, particularly as a gradiometer. SQUIDs can have a measurement sensitivity of 10^{-5} nT/m; this means that two sensors need only be placed 25 cm or less apart, thus making it possible to have the entire sensor system in a very small space. This has great advantages in mounting the equipment in aircraft, in borehole probes and in submarine devices where space is at a premium. Measurement accuracy of the total field strength is within ± 0.01 nT.

Technical difficulties over the use of liquid helium, which has to be maintained at a temperature of 4.2 K for superconductivity to occur, limit the widespread deployment of SQUID magnetometers. They are rarely, if ever, used in surface magnetic measurements.

3.5.5   Gradiometers

A gradiometer measures the difference in the total magnetic field strength between two identical magnetometers separated by a small distance. In airborne work, typical separations between sensors is 2 m to 5 m for stingers (Figure 3.23) and up to 30 m for birds. In ground instruments, a separation of 0.5 m is common. The magnetic field gradient is expressed in units of nT/m and taken to apply at the mid-point between the sensors. A major advantage of gradiometers is that because they take differential measurements, no correction for diurnal variation is necessary as both sensors will be equally affected. As gradiometers measure the vertical magnetic gradient, noise effects from long-wavelength features are suppressed and anomalies from shallow sources are emphasised. For detailed high-resolution surveys exploring for mineral targets, magnetic gradiometry is the preferred method (Hood 1981).

Fluxgate and resonance-type magnetometers are commonly used in ground surveys. Where continuous-reading devices are required, such as when automatic datalogging is being used, fluxgate gradiometers are preferable (e.g. Sowerbutts and Mason 1984). A detailed although dated comparison of the different types of gradiometers and magnetometers has been made by Hood *et al.* (1979).

Self-oscillating split-beam caesium vapour gradiometers have also been developed for small-scale land-based surveys, such as Geomet-

Figure 3.23 Eurocopter AS315 about to lift a 3-axis magnetic gradiometer system. Courtesy of Aerodat Inc., Canada

rics Inc.'s G-858 instrument. These instruments can sample at up to 10 readings per second with a sensitivity of 0.05 nT. An additional feature of this particular instrument is that it can be connected to a Differential Global Positioning System for the simultaneous logging of position and magnetic data. The system was first tested at Stanford University, USA, in March 1993. As with any geophysical instruments, it is always best to check with equipment manufacturers for the latest technical specifications.

3.6 MAGNETIC SURVEYING

3.6.1 Field survey procedures

As with every geophysical survey, the keeping of detailed and accurate field notes cannot be emphasised too strongly, even if dataloggers or memory-magnetometers are used. Orderly record-keeping permits more efficient and accurate data processing. Practical details of how magnetic surveys should be undertaken have been given by Milsom (1989).

In the vast majority of cases where the magnetic targets have a substantial strike length, survey profiles should, where possible, be conducted across strike with tie-lines along strike. In cases such as some archeological and engineering site investigations where the targets are more equidimensional, such as the ruins of a Roman villa or a brick-lined mineshaft, north–south and east–west orientations are commonly used.

In ground-based surveys, it is important to establish a local base station in an area away from suspected magnetic targets or magnetic noise and where the local field gradient is relatively flat. A base station should be quick and easy to relocate and re-occupy. The precise approach to the survey will depend on the type of equipment. If a manual push-button proton magnetometer is deployed, the exact time of occupation of each station is needed and at least three readings of the total field strength should be recorded. Each of the three values should be within ± 1 or 2 nanoteslas; an average of these three readings is then calculated. As the survey progresses, the base station must be re-occupied every half or three-quarters of an hour in order to compile a diurnal variation curve for later correction (see next section). Next to each data entry, where required, should be any comments about the terrain or other factors that may be considered to be important or relevant to subsequent data processing and interpretation.

If a continuous-reading base-station magnetometer is used to measure the diurnal variation, it is still worth returning to base every 2–3 hours, just in case the base magnetometer fails.

When dataloggers or memory magnetometers are used, regular checks on the recording of data are vital. It is all very well occupying hundreds of stations and taking perhaps several thousand measurements only to find that the logger is not functioning or the memory has been corrupted.

One golden rule is always to check your data as they are collected and at the end of each survey day. This serves two purposes. First, it provides a data quality check and allows the operator to alter the survey in response to the magnetic values measured. For example, if a 50 m station interval was selected at the outset and the field values indicate a rapid change over a much shorter interval, the separation between stations must be reduced in order to collect enough data to image the magnetic anomaly. Secondly, it provides a check on the consistency of the data. Odd anomalous values may indicate something of geologic interest which may need to be followed up, or may highlight human error. In either case, the next day's survey can take this into account and measurements can be made to check out the oddball values.

In the case of aeromagnetic or ship-borne surveys, the specifications are often agreed contractually before an investigation begins. Even so, there are guidelines as to what constitutes an adequate line separation or flying height, orientation of survey line, and so on. As an example, Reid (1980) compiled a set of criteria based on avoidance of spatial aliasing (Tables 3.5 and 3.6). For example, if a mean flying height over magnetic basement (h) of 500 m is used with a flight line spacing (δx) of 2 km, then $h/\delta x = 0.5/2 = 0.25$, which would indicate that 21% aliasing would occur in measurements of the total field and

Table 3.5   Degree of aliasing (Reid 1980)

$h/\delta x$	F_T	F_G
0.25	21	79
0.5	4.3	39
1	0.19	5
2	0.0003	0.03
4	0	0

F_T and F_G are the aliased power fraction (per cent) expected from surveys of total field and vertical gradient respectively

Table 3.6   Maximum line spacings (Reid 1980)

Survey type	Intended use	δx_{max}
Total field	Contour map	$2h$
Total field	Computation of gradient, etc., maps	h
Vertical gradient	Gradient contour maps	h
Total field	Modelling of single anomalies	$h/2$

as much as 79% if the vertical magnetic gradient were being measured (Table 3.5). Neither value is acceptable. The larger the value of $h/\delta x$, the less aliasing will occur; a survey is considered reasonably designed if $h/\delta x \geqslant 0.5$, depending on survey type. At this minimum value, about 5% maximum aliasing is acceptable if contouring is to be undertaken on total field data. Other maximum flight line spacings are given in Table 3.6. There have been examples of commercially flown surveys in Egypt, for instance, where the flying height was 400 m above ground, but with 1 km thickness of sedimentary cover over magnetic basement (hence $h = 1.4$ km) and the line spacing was 21 km. This gives a value of $h/\delta x \approx 0.07$, resulting in more than 65% aliasing. The contour maps of these surveys, which covered thousands of square kilometers, are virtually useless. To have been contoured adequately, the flight line spacing for data acquired at a flying height of 1.4 km above basement should have been no more than 2.8 km; i.e. from Table 3.6, $\delta x \leqslant 2h$.

The choice of survey parameters will also affect the success (or otherwise) of subsequent computer contouring. These guidelines can also apply to surface surveys using a proton magnetometer where $h = 3$ m (1 m of overburden over a magnetic target, plus 2 m for the magnetometer sensor pole) for example. In this case, an acceptable maximum line spacing would be 6 m if a contour map of the total field were required. Survey design parameters and automated contouring have been described briefly in Chapter 1 and in detail by Hood *et al.* (1979) and by Reid (1980).

3.6.2 Noise and corrections

All magnetic data sets contain elements of noise and will require some form of correction to the raw data to remove all contributions to the observed magnetic field other than those caused by sub-surface magnetic sources. In ground magnetometer surveys, it is always advisable to keep any magnetic objects (keys, penknives, some wrist-watches, etc.), which may cause *magnetic noise*, away from the sensor. Geological hammers put next to the sensor bottle of a proton magnetometer will have a significant effect, as demonstrated by students trying to simulate a magnetic storm so that they could abandon the survey and retire to the nearest hostelry! It is also essential to keep the sensor away from obviously magnetic objects such as cars, metal sheds, power lines, metal pipes, electrified railway lines, walls made of mafic rocks, etc.

The most significant correction is for the *diurnal variation* in the Earth's magnetic field. Base station readings taken over the period of a survey facilitate the compilation of the diurnal 'drift' as illustrated in Figure 3.24. Measurements of the total field made at other stations can easily be adjusted by the variation in the diurnal curve. For example, at point A in Figure 3.24, the ambient field has increased by

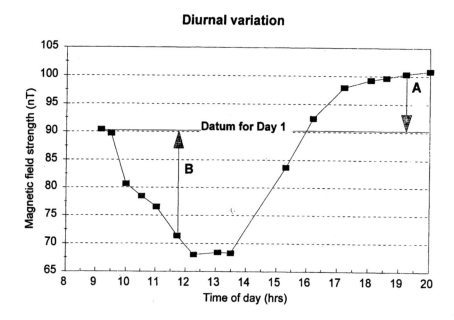

Figure 3.24 Diurnal drift curve measured using a proton magnetometer

10 nT and thus the value measured at A should be reduced by 10 nT. Similarly, at B, the ambient field has fallen by 19 nT and so the value at B should be increased by 19 nT. Further details of diurnal corrections have been given by Milsom (1989). Gradiometer data do not need to be adjusted as both sensors are affected simultaneously and the gradient remains the same between them.

In airborne and shipborne surveys, it is obviously not possible to return to a base station frequently. By designing the survey so that the track lines intersect (Figure 3.25), the dataset can be appropriately corrected. Some surveys use profiles and tie-lines at the same spacing to give a regular grid. Other surveys have tie-lines at 10 times the inter-profile line spacing. In addition to checking on diurnal variations, tie-lines also serve as a useful control on navigational and measurement accuracy. A flow chart depicting the reduction of aeromagnetic data is given in Figure 3.26. In regional surveys, a further tie is to a local Geomagnetic Observatory, if there is one within 150 km, at which all the magnetic components are measured and which can provide diurnal variations. It would then have to be demonstrated that the curve obtained from the observatory applied in the survey area.

The degree of data processing is dependent upon the resolution required in the final dataset. For a broad reconnaissance survey, a coarser survey with lower resolution, say several nanoteslas, may be

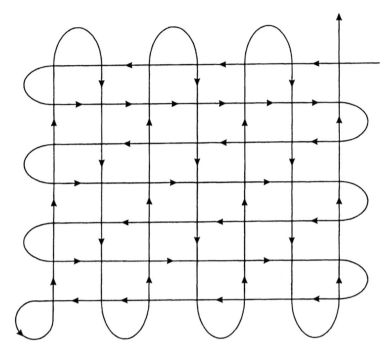

Figure 3.25 Tracks of a shipborne or airborne magnetic survey. Some surveys, rather than having equal spacings between all tracks, have tie-lines at 10 times the normal interline spacing

all that is required. In detailed surveys, however, an accuracy to within 0.1 nT will need a finer survey grid, more accurate position-fixing and diurnal drift corrections.

Rarely, a *terrain correction* may need to be applied when the ground over which a survey is conducted is both magnetic and topographically rough. Unlike the gravity case where terrain corrections, though laborious, are relatively easy to calculate, corrections for terrain for magnetic data are extremely complex. If the rough terrain is made up largely of low-susceptibility sedimentary rocks, there will be little or no distortion of the Earth's magnetic field. However, if the rocks have a significant susceptibility, a terrain factor may have to be applied. Anomalous readings as large as 700 nT have been reported by Gupta and Fitzpatrick (1971) for a 10 m high ridge of material with susceptibility $\kappa \approx 0.01$ (SI). Furthermore, magnetic readings taken in a gulley whose walls are made of basic igneous rocks will be anomalous owing to the magnetic effects of the rocks above the magnetic sensor. Considerable computational effort (see Sharma, 1986, appendix C) then has to be applied to correct the data so that they are interpretable. Similar geometric effects can also occur in radiometric surveys.

Another way of correcting for the effect of topography, or of reducing the data to a different reference plane, is by *upward continuation*. This permits data acquired at a lower level (e.g. on the ground) to be processed so that they can be compared with airborne surveys. The effect of this is to lessen the effects of short-wavelength high-

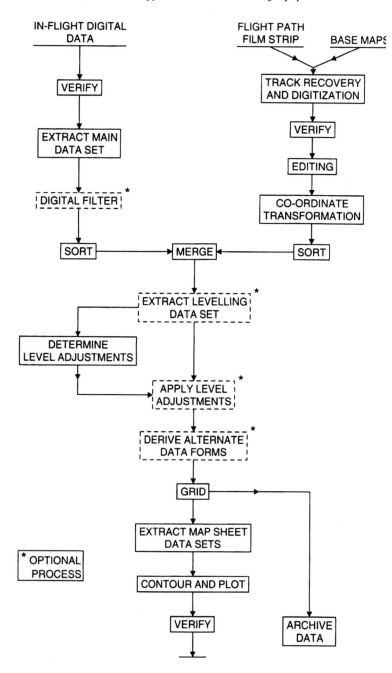

Figure 3.26 Flow chart of reduction processes on aeromagnetic data. From Hood *et al.* (1979), by permission

amplitude features as the magnetic force is indirectly proportional to the square of the distance between source and sensor (see Box 3.1). The rate of change of the field with elevation (akin to the gravitational free-air correction) is between 1.5 and 3.5 nT/100 m, with the maximum gradient being at the poles. In areas where the range of elevations of the ground surface is large, such as in the Rockies or Andes, and Himalayas, maintaining a constant terrain clearance is not practicable (or safe!). Indeed, in areas of extreme altitude, airborne surveys are not possible as the ground can be higher than the flying ceiling of available aircraft owing to the rarified atmosphere. Different flying heights can be specified for particular areas within a survey region and the various datasets processed to the same flying height or alternative datum.

In some regions, metal pipes can become inductively magnetised by currents in the atmosphere (Campbell 1986) or are cathodically protected by passing a large direct current (1–7 amps) through them to reduce internal corrosion. The presence of such pipes can contribute a significant anomaly on high-resolution aeromagnetic data (Gay 1986). In hydrocarbon exploration over sedimentary basins up to 6 km depth, basement faulting, which affects basement-controlled traps and reservoirs, can be identified from their respective magnetic anomalies which can have amplitudes of only several nanoteslas. There is then a problem over differentiating between a geologically significant fault and a cathodically protected pipe as their respective magnetic anomalies may appear to be similar.

Iron pipes have a permanent magnetisation acquired at the foundry at the time of manufacture. Being magnetisable, they also acquire a magnetisation induced by the Earth's magnetic field. Current injected into the pipes will also generate a magnetic field according to the right-hand rule. This states that a conductor carrying electrical current, indicated by the thumb on the right hand, will generate a magnetic field in the direction indicated by the coiled fingers of the right hand (Figure 3.27A). The injection of current at specific points along a pipe has one marked effect (Figures 3.27B and C). The polarity of the magnetic anomaly will be reversed either side of the injection point as the current is flowing in opposite directions away from it and into the ground. The point at which the polarity reverses (Figure 3.27B) indicates the position of current injection. The figure shows the magnitudes and senses of the western and eastern anomaly peaks for each survey line. Lines 9–13 have a positive western peak and corresponding negative eastern peak while lines 14–17 south of the injection point have a negative western peak and positive eastern peak (Figure 3.27C). The magnitude of the anomaly will also decrease away from the injection point. The magnitudes determined for survey lines south of the injection point are slightly larger than those for northern lines as the flying height was 40 m lower to the south. Mean flying height was 257 m. These characteristics are diagnostic of cathodically protected pipes and can be used to tell them apart from linear fault structures whose polarity does not switch along its length, although its magnitude may vary along the strike of the fault. Having identified a linear anomaly as being due to a cathodically protected pipe, the anomaly can be filtered out of the dataset.

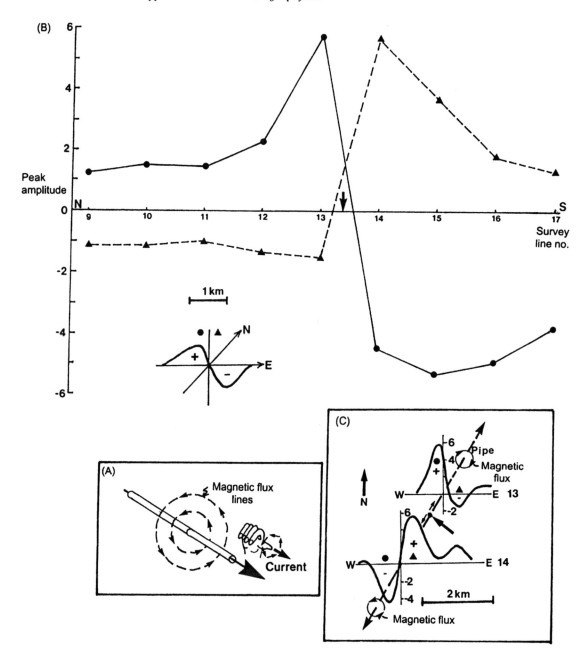

3.6.3   Data reduction

In order to produce a magnetic anomaly map of a region, the data
have to be corrected to take into account the effect of latitude and, to
a lesser extent, longitude. As the Earth's magnetic field strength varies
from 25 000 nT at the magnetic equator to 69 000 nT at the poles, the

increase in magnitude with latitude needs to be taken into account. Survey data at any given location can be corrected by subtracting the theoretical field value F_{th}, obtained from the International Geomagnetic Reference Field, from the measured value, F_{obs}. This works well in areas where the IGRF is tied-in at or near to Geomagnetic Observatories, but in many places the IGRF is too crude. Instead, it is better to use a local correction which can be considered to vary linearly over the magnetic survey area. Regional latitudinal (ϕ) and longitudinal (θ) gradients can be determined for areas concerned and tied to a base value (F_0), for example, at the south-east corner of the survey area. Gradients northwards ($\delta F/\delta\phi$) and westwards ($\delta F/\delta\theta$) are expressed in nT/km and can easily be calculated for any location within the survey area. For example, in Great Britain, gradients of 2.13 nT/km north and 0.26 nT/km west are used. Consequently, the anomalous value of the total field (δF) can be calculated arithmetically, as demonstrated by the example in Box 3.7.

Another method of calculating the anomalous field δF is to determine statistically the trend of a regional field to isolate the higher-frequency anomalies, which are then residualised in the same way that gravity residuals are calculated. The regional field is subtracted from the observed field to produce a residual field (δF) (Figure 3.28A). If the survey is so laterally restricted, as in the case of small-scale archaeological, engineering or detailed mineral prospecting surveys (e.g. $< 500 \, m \times 500 \, m$ in area), the use of regional gradients is not practicable. Instead, profile data can be referred to a local base station (F_b) which is remote from any suspected magnetic sources. In this case, the anomalous field δF is obtained by subtracting the base value (F_b) from every diurnally corrected observed value F_{obs} along the profile ($\delta F = F_{obs} - F_b$), as illustrated in Figure 3.28B.

Box 3.7 Anomalous total field strength δF

$$\delta F = F_{obs} - (F_0 + \delta F/\delta\phi + \delta F/\delta\theta) \quad (nT)$$

where F_{obs} is the measured value of F at a point within the survey area with coordinates (x, y); F_0 is the value at a reference point ($x = 0$; $y = 0$); $\delta F/\delta\phi$ and $\delta F/\delta\theta$ are the latitudinal and longitudinal gradients respectively (in units of nT/km).

Example: For a station 15 km north and 18 km west of a reference station at which $F_0 = 49\,500$ nT with gradients $\delta F/\delta\phi = 2.13$ nT/km north, and $\delta F/\delta\theta = 0.26$ nT/km west, and the actual observed field $F_{obs} = 50\,248$ nT, the magnetic anomaly δF is:

$$\delta F = 50\,248 - (49\,500 + 2.13 \times 15 + 0.26 \times 18) \, nT$$
$$= 711 \, nT$$

Figure 3.27 (*opposite*) Variation in magnetic anomalies associated with a cathodically protected pipe in north-east Oklahoma, USA. (A) Right-hand rule applied to a pipe through which a current is passed to show direction of magnetic flux lines. (B) Peak amplitudes for western and eastern parts of profile anomalies, of which two for lines 13 and 14 are shown in (C). Note the relative amplitudes of the anomalies on lines 13 and 14 either side of the current injection point (arrowed). Data from Gay (1986), by permission

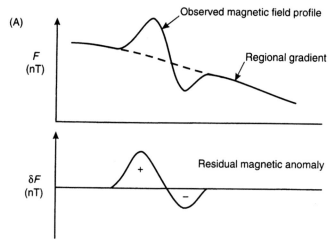

(A)

F (nT)

Observed magnetic field profile

Regional gradient

δ*F* (nT)

Residual magnetic anomaly

+

−

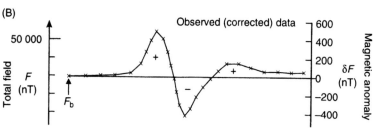

(B)

Total field

F (nT)

50 000

*F*_b

Observed (corrected) data

+

−

+

Magnetic anomaly

δ*F* (nT)

600
400
200
0
−200
−400

Figure 3.28 Magnetic residuals (A) Subtraction of a regional field gradient, and (B) reference to a local arbitrary base station

3.7   QUALITATIVE INTERPRETATION

Once magnetic data have been fully corrected and reduced to their final form, they are usually displayed either as profiles (Section 3.7.1) or as maps (Section 3.7.2) and the interpretation procedures are different for the two cases. However, it must always be borne in mind that, although the techniques used are similar to those for gravity surveys, there are two important complications. First, the Earth's magnetic field is dipolar, which means that a single anomaly can have the form of a positive peak only, a negative peak only or a doublet consisting of both positive and negative peaks. Secondly, the single largest unknown is whether there is any remanent magnetisation and, if there is, its intensity and direction (J_r) need to be ascertained. It must also be remembered that many geophysical interpretations may fit the observed data and that a given interpretation may not be unique (see Chapter 1, and Figure 1.1 in particular). For this reason, it is always useful to use other geophysical methods in the same area to help constrain the interpretations. If some geological information already exists for the area, then this should be used to help with the geophysical interpretations. However, a word of warning: be careful that some geological information may itself be an interpretation and should not be considered to be the only solution. The geophysics may disprove the geological hypothesis!

Figure 3.29 The magnetic field generated by a magnetised body inclined at 60° parallel to the Earth's field (A) would produce the magnetic anomaly profile from points A–D shown in (B). See text for details

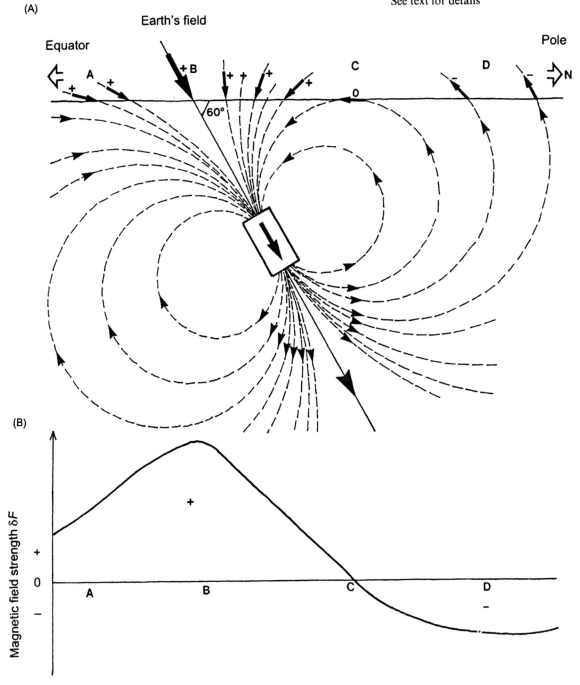

A magnetisable body will have a magnetisation induced within it by the Earth's magnetic field (Figure 3.29A). As magnetisation is a vector quantity, the anomaly shape will depend on the summation of the vectors of the Earth's field F (with intensity J) and the induced field (J_i) from the sub-surface body and from any remanent magnetisation (J_r). It can be seen from Figure 3.29 (A and B) how any magnetic anomaly is produced. The maximum occurs when the induced field is parallel to the Earth's field; the anomaly goes negative when the induced field vector is orientated upwards as at D. As the magnetic data have been corrected to remove the effect of the Earth's magnetic

Figure 3.30 Two magnetic anomalies arising from buried magnetised bodies (see text for details)

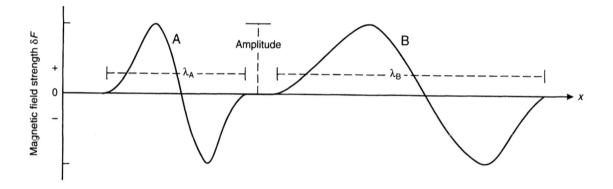

Table 3.7   Guidelines to qualitative interpretation of magnetic profiles and maps

Applies to:	Magnetic character	Possible cause
Segments of a profile and areas of maps	Magnetically quiet Magnetically noisy	Low κ rocks near surface Moderate–high κ rocks near surface
Anomaly	Wavelength $\pm$ amplitude	Short $\Rightarrow$ near-surface feature Long $\Rightarrow$ deep-seated feature Indicative of intensity of magnetisation
Profile*	Anomaly structure[†] and shape	Indicates possible dip and dip direction Induced magnetisation indicated by negative to north and positive to south in northern hemisphere and vice versa in southern hemisphere; if the guideline does not hold, it implies significant remanent magnetisation present
Profile and maps	Magnetic gradient	Possible contrast in κ and/or magnetisation direction
Maps	Linearity in anomaly	Indicates possible strike of magnetic feature
Maps	Dislocation of contours	Lateral offset by fault
Maps	Broadening of contour interval	Downthrow of magnetic rocks

* Can be determined from maps also; [†] Structure = composition of anomaly, i.e. positive peak only, negative peak only or doublet of positive and negative peaks; κ = magnetic susceptibility

field, the anomaly will be due to the vectors associated with the source body. The sign convention is positive downwards, negative upwards. Where there is no remanent magnetisation, a magnetised body in the northern hemisphere will always have a negative anomaly on its northern side and a positive anomaly on its southern side. The opposite is true for the southern hemisphere.

For the two anomalies shown in Figure 3.30, anomaly A has a short wavelength compared with anomaly B, indicating that the magnetic body causing anomaly A is shallower than the body causing B. As the amplitude of anomaly B is identical to that of anomaly A, despite the causative body being deeper, this must suggest that the magnetisation of body B is much greater than for body A, as amplitude decreases with increasing separation of the sensor from the magnetised object.

Some general guidelines for the qualitative interpretation of magnetic profiles and maps are listed in Table 3.7, and an example in Figure 3.31. The list of guidelines should be used like a menu from which various combinations of parameters apply to specific anomalies. For instance, a short-wavelength high-amplitude doublet anomaly with negative to the north, and positive to the south, with an

Figure 3.31 Magnetic anomaly map associated with a fault-bounded sedimentary basin with upthrown horst block to south-east – Inner Moray Firth, Scotland. After McQuillin *et al.* (1984), by permission

elongation in the east–west direction in the mid-latitudes of the northern hemisphere, suggests a near-surface, moderate-to-high susceptibility magnetic feature with induced magnetisation in a sheet-like body with an east–west strike and northerly dip.

3.7.1 Profiles

The simplest interpretation technique is to identify zones with different magnetic characteristics. Segments of the profile with little variation are termed magnetically 'quiet' and are associated with rocks with low susceptibilities. Segments showing considerable variation are called magnetically 'noisy' and indicate magnetic sources in the sub-surface. The relative amplitudes of any magnetic anomalies (both

Figure 3.32 Magnetic zonation of a proton magnetiometer profile across Sourton Common, north Dartmoor, England

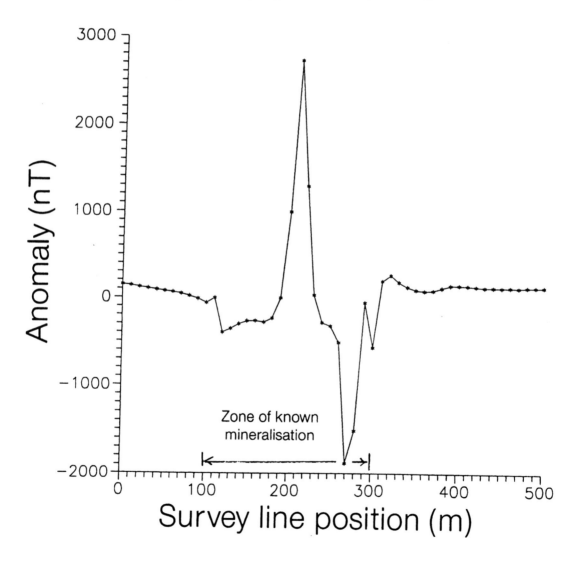

positive and negative) and the local magnetic gradients can all help to provide an indication of the sub-surface.

Figure 3.32 illustrates the differences between noisy and quiet zones in a profile over Sourton Tors, north Dartmoor, England, where the magnetically noisy segments indicate metalliferous mineralisation. The negative occurring as the northern part of the doublet indicates that there is little remanent magnetisation and that the anomaly is due largely to induction ($J_i \gg J_r$).

Figure 3.33 shows a profile measured at Kildonnan, Isle of Arran, Scotland, across a beach section over a series of vertical dolerite dykes. Two of the three dykes (A and B) give rise to large positive anomalies, while a third dyke (C) produces a broader low. All the dykes have virtually identical geochemistry and petrology and hence

Figure 3.33 Proton magnetometer profile across three vertical dolerite dykes at Kildonnan, Isle of Arran, Scotland, which have been intruded during a period when the Earth's magnetic polarity changed

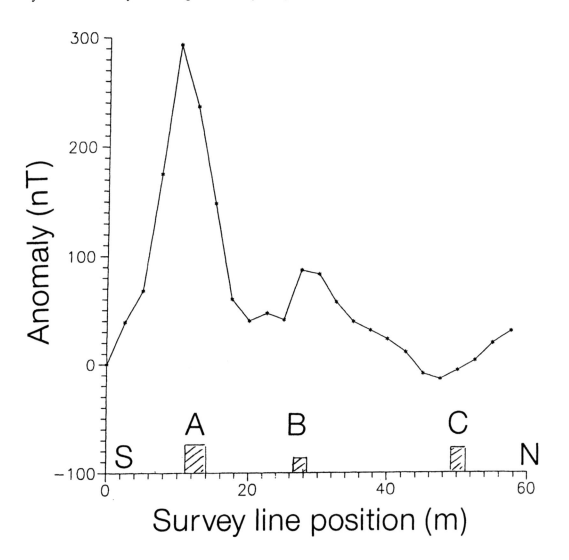

similar susceptibilities. The difference in magnetic character associated with dyke C compared with dykes A and B is attributed to
there having been a magnetic reversal between the intrusion of dyke
C relative to the other two. It is well known that in the British Tertiary
Igneous Province, which includes the western isles of Scotland,
a phase of doleritic intrusion occurred which lasted 10 million years
and straddled a magnetic reversal (Dagley *et al.* 1978). Some dykes
are normally magnetised and others have a reversed polarity. The
different magnetic characters therefore provide a means of identifying
which dykes were associated with which phase of the intrusion
episode, whereas it is virtually impossible to make such an identification in the field on the basis of rock descriptions.

Having identified zones with magnetic sources, and possibly
having gained an indication of direction of dip (if any) of magnetic

Figure 3.34 (A) Aeromagnetic map
of the south-east part of the Shetland
Islands, Scotland (Flinn 1977); (B)
(*opposite*) zonation of (A) in terms of
magnetic characteristics

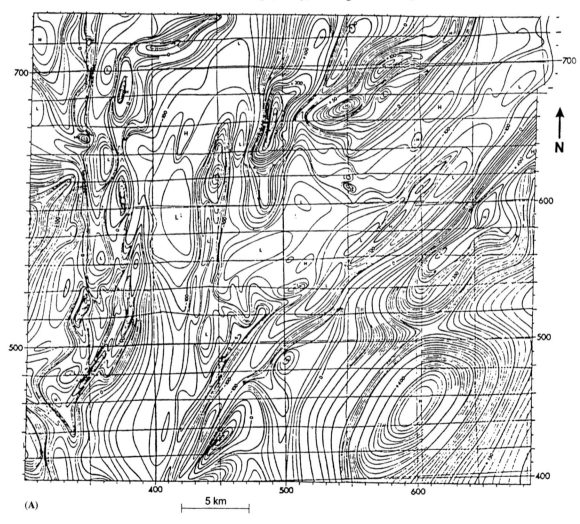

(A)   5 km

targets and relative intensities, either more detailed fieldwork can be undertaken to obtain more information and/or the existing data can be interpreted using quantitative methods (see Section 3.8).

3.7.2 Pattern analysis on aeromagnetic maps

Magnetic data acquired over a grid with no spatial aliasing are displayed as contoured maps, as three-dimensional isometric projections, or as image-processed displays (see Section 3.8.3). The various displays can be interpreted in terms of the guidelines in Table 3.7. Commonly, even such a basic level of interpretation can yield important information about the sub-surface geology very quickly. One major advantage of airborne surveys is that they can provide information about the geology in areas covered by water.

An aeromagnetic survey was acquired over the south-eastern part of the Shetland Islands, north-east Scotland, and described in detail by Flinn (1977). The original aeromagnetic data were extracted from part of a British Geological Survey map and are shown in Figure 3.34A. The corresponding pattern analysis for the same area is shown in Figure 3.34B and compares very well with Flinn's

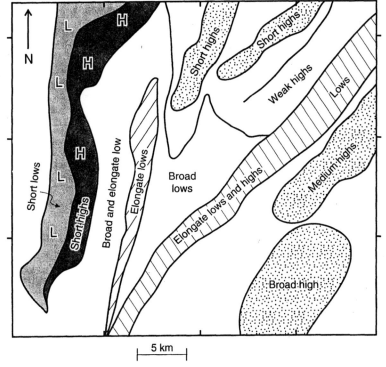

(B)

interpretation (Figure 3.35). Magnetic lows in band A are associated with psammites and migmatised psammites which have been heavily injected by pegmatite. The lows in band B correspond to gneissified semipelites and psammites intruded by granites and pegmatites. The short-wavelength but large-amplitude highs just east of band A are attributed to the magnetic pelitic schist and gneiss. In the central part of Figure 3.34A, truncation of NE–SW trending contours can be seen to be associated with the Whalsay Sound Fault. Other faults, such as the Lunning Sound Fault and the Nesting Fault, separate zones of differing magnetic character and lie parallel to the local aeromagnetic contours. A series of three localised positive highs in the south-east corner (h, i and j along the thrust beneath the Quarff Nappe;

Figure 3.35 Geological interpretation of the aeromagnetic map shown in Figure 3.34 of the south-east part of the Shetland Islands, Scotland. From Flinn (1977), by permission

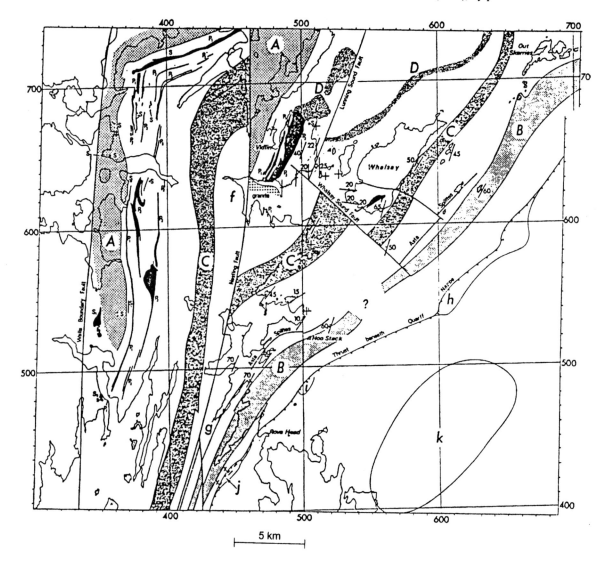

Figure 3.35) is thought to be due to phyllites with 4% hematite which in itself cannot explain the anomaly amplitudes. However, suitably orientated remanent magnetisation could account for these magnetic features. The large-amplitude anomaly (k) in the south-east corner is attributed to a large buried intrusive which is also evident on gravity data. However, this and similar anomalies in the vicinity with similar character all occur over sea and have not been sampled and so the exact nature of the rock remains unknown.

More detailed discussions on the interpretation of aeromagnetic maps have been given, for example, by Vacquier *et al.* (1951), Nettleton (1976) and Hinze (1985).

3.8 QUANTITATIVE INTERPRETATION

The essence of quantitative interpretation is to obtain information about the depth to a particular magnetic body, its shape and size, and details about its magnetisation in two possible ways. One is direct, where the field data are interpreted to yield a physical model. The other is the inverse method, where models are generated from which synthetic magnetic anomalies are generated and fitted statistically against the observed data. The degree of detail is limited by the quality and amount of available data and by the sophistication of either the manual methods or the computer software that can be used.

3.8.1 Anomalies due to different geometric forms

Just as with gravity data, magnetic data can be interpreted in terms of specific geometric forms which approximate to the shapes of sub-surface magnetised bodies. This tends to be true where profiles are to be interpreted only in terms of two dimensions. Three-dimensional models are far more complex and can be used to approximate to irregularly shaped bodies (see Section 3.8.3). Detailed mathematical treatments of the interpretation of magnetic data have been given by Grant and West (1965), Telford *et al.* (1990) and Parasnis (1986), among others.

The commonest shapes used are the sphere and the dipping sheet, both of which are assumed to be uniformly magnetised and, in the simplest cases, have no remanence. Total field anomalies (δF) for various types of model are illustrated in Figures 3.36–3.39; except where otherwise stated, the field strength is 50 000 nT, inclination $I = 60°$, declination $D = 0°$, and susceptibility $\kappa = 0.05$ (SI).

In the example of a uniformly magnetised sphere (Figure 3.36), the horizontal and vertical components are shown in addition to the total field anomaly. The anomalies associated with vertical bodies of various thicknesses are shown in Figure 3.37. The anomaly produced by a 50 m thick vertical dyke (Figure 3.37A) is both wider and has

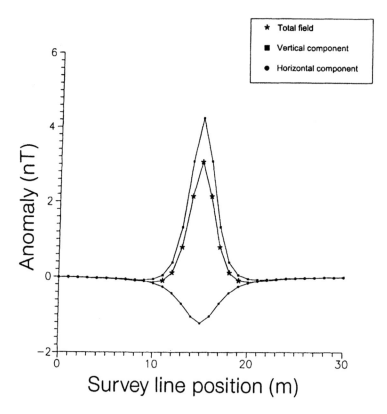

Figure 3.36 Horizontal and vertical components and the total field over a uniformly magnetised sphere with a radius of 1 m and whose centre lies at 3 m depth at position $x = 15$ m

a significantly larger amplitude (830 nT) than that of the 5 m thick dyke (peak amplitude 135 nT). Notice also that the anomaly shape changes considerably with strike direction, from being a positive–negative doublet when the dyke is striking east–west (with the negative on the northern side) to being a single symmetric positive peak when the dyke strikes north–south. In all cases, when an inductively magnetised body of regular shape is orientated north–south, its anomaly is symmetric. For a 70 m thick, 400 m long magnetised slab with its top 30 m below ground, a symmetric M-shaped anomaly is produced with the strike in a north–south direction (Figure 3.37B). When striking east–west, the positive–negative doublet is stretched to form an inflection in the middle; again the negative is on the northern side.

The effects on anomaly shape caused by changing the depth to the top of a vertical magnetised dyke are illustrated in Figure 3.38. With increasing depth, the anomaly decreases in amplitude and widens. At any given latitude, the anomaly shape will also be affected by any dip (α) of a sheet-like body, such as the inductively magnetised sheet (5 m thick) striking east–west (dip direction is towards the north) indicated in Figure 3.39. With zero dip, the body behaves like a thin horizontal slab with its northern edge off the scale of the figure.

If one end of a thick horizontal slab is sufficiently far enough away from the other, the model effectively becomes a vertical boundary separating two contrasting magnetic media (Figure 3.40A). The anomaly produced when the boundary strikes east–west has a significantly larger field strength (peaking around 1870 nT) than when in the other orientation. Furthermore, the single peak is effectively the positive peak of Figure 3.37C isolated from its negative partner.

The direction and degree of dip of a fault plane in a magnetised body also has a distinctive anomaly shape (Figure 3.40B). The negative anomaly is associated with the downthrown side. The anomaly shape is very similar to half of the M-shaped anomaly in Figure 3.37B.

Anomalies produced over a near-semicylindrical low-susceptibility body within magnetised basement, to simulate a buried rock valley

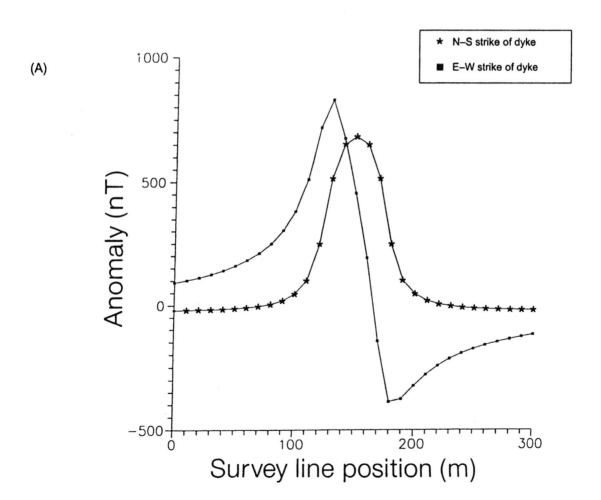

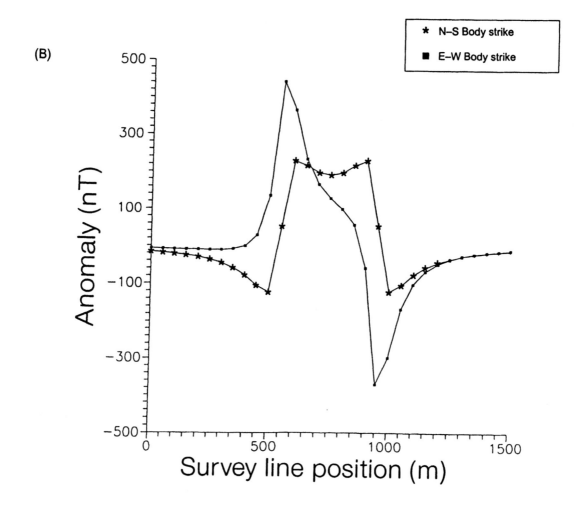

(B)

infilled by sediments in magnetic bedrock, are shown in Figure 3.41. The symmetric anomaly is obtained when the semicylinder is orientated north–south. When this body is orientated north–south, the negative anomaly is on the southern side and so can be distinguished from the anomaly over a thin vertical dyke. Furthermore, the minimum anomaly amplitude is far greater in the case of the low-susceptibility semicylinder than for a vertical dyke.

One of the largest effects on anomaly shape for a given geological structure is latitude. Anomalies in the northern hemisphere over a 5 m thick dyke dipping at 45° to the north decrease in amplitude (partly a function of the reduced field strength towards the magnetic equator) and the negative minimum becomes more pronounced (Figure 3.42A). If the same geological structure exists at different latitudes in the southern hemisphere, this trend continues (Figure 3.42B) but with the slight growth of the now northern positive anomaly. These curves

Figure 3.37 Total field anomalies over a vertical sheet-like body. (A) (*previous page*) 50 m and (B) 400 m wide. In (A), the magnetic anomaly arising from a 5 m wide body is given for comparison

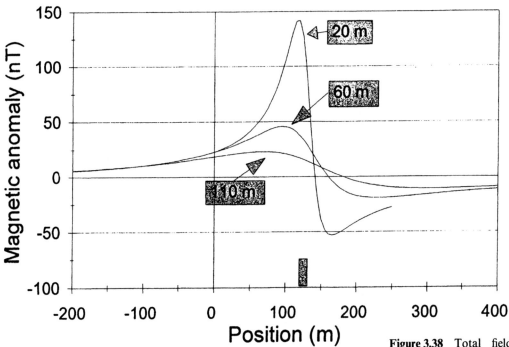

Figure 3.38 Total field anomalies over a 10 m wide vertical sheet-like body orientated east–west and buried at depths of 20 m, 60 m and 110 m; the position of the magnetised body is indicated

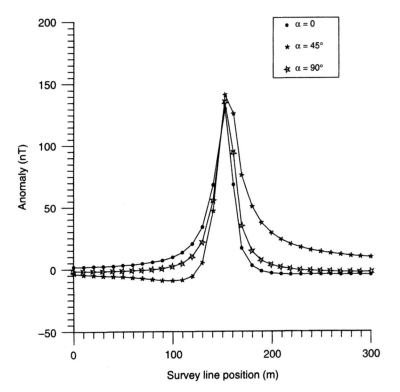

Figure 3.39 Total field anomalies over a thin dyke (5 m wide) dipping to the north at angles from $\alpha = 90°$ to $\alpha = 0°$; body strike is east–west

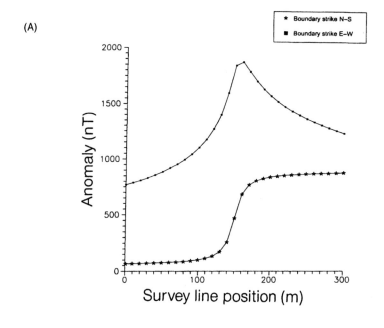

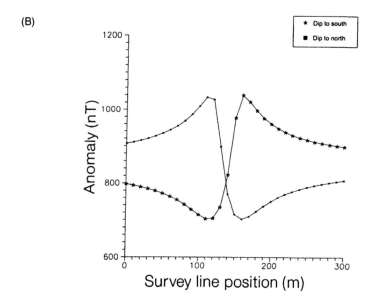

Figure 3.40 Total field anomalies over (A) a vertical contact between contrasting magnetised bodies, and (B) over a fault plane dipping at 45° to the south and to the north for a north–south fault strike direction

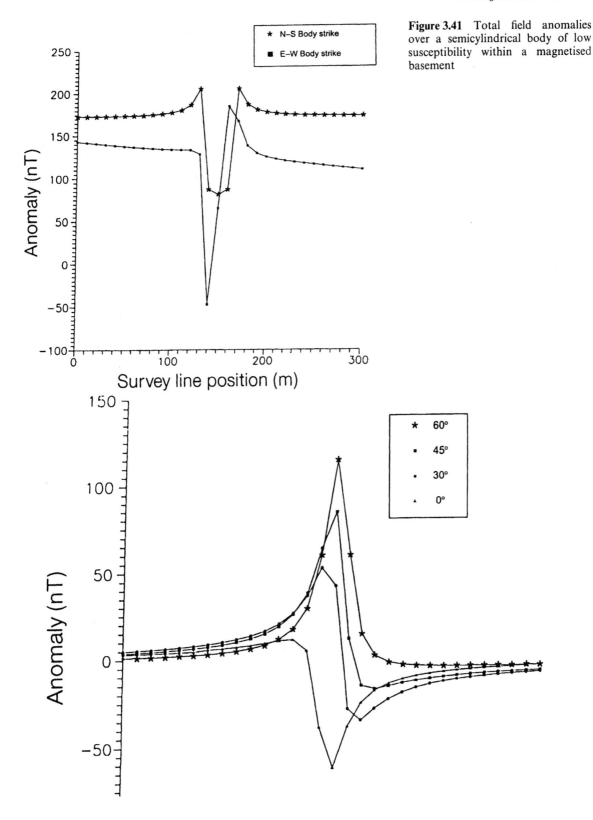

Figure 3.41 Total field anomalies over a semicylindrical body of low susceptibility within a magnetised basement

were calculated along the 0° Greenwich meridian where the field strengths in the southern hemisphere are significantly less than at equivalent latitudes in the northern hemisphere. Profiles along other lines of longitude would produce anomalies of similar shapes but with different amplitudes. It is also worth remembering that a given geological model, which produces a significant magnetic anomaly at one latitude and strike direction, may produce only a negligible anomaly at a different latitude and strike (Figure 3.42B) and so could be overlooked.

Figure 3.42 Total field anomalies over a 5 m thick dyke dipping at 45° to the north with an east–west strike direction but with different magnetic inclinations along the 0° Greenwich Meridian, with (A) (*previous page*) the northern hemisphere, and (B) (*below*) the southern hemisphere, taking into account changes in magnetic field strengh with latitude. (C) (*next page*) The total field anomaly for a vertical dyke but at two different magnetic latitudes and directions to illustrate how the magnetic anomaly over a magnetised body can become insignificant by only changing the magnetic latitude (inclination) and strike direction

(B)

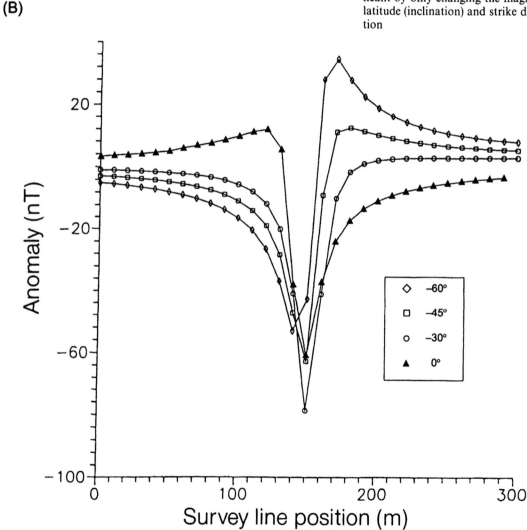

(C)

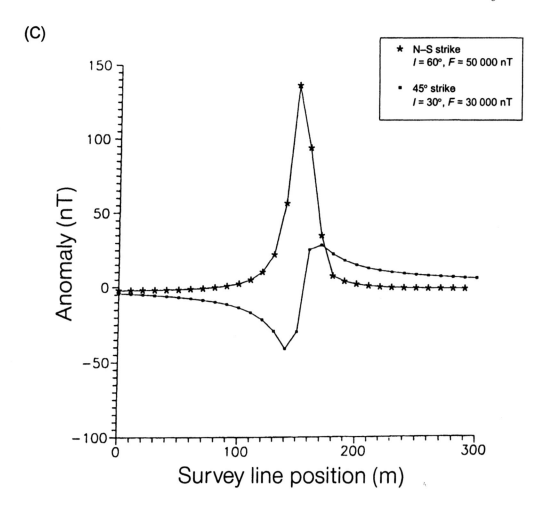

In all the above cases, it was assumed that there was no remanent magnetisation, yet if present, it can affect the shape of an anomaly significantly (Green 1960). In Figure 3.43, four profiles are illustrated, one where there is no remanent magnetisation, and three others where the intensity of remanent magnetisation is constant at 0.12 A/m (giving a strength of 150 nT). A schematic vector diagram is shown to illustrate the effects of remanence and the significance of the vertical component in high magnetic latitudes ($>45°$). When the direction of permanent magnetisation is antiparallel to the Earth's field (A), the resultant amplitude is $F'_a(\ll F)$, so the anomaly amplitude is substantially reduced. In the case B, where the remanent magnetisation is at right-angles to the Earth's field, the resultant F'_b ($<F$) has a smaller vertical component than F and so the amplitude is slightly reduced. In

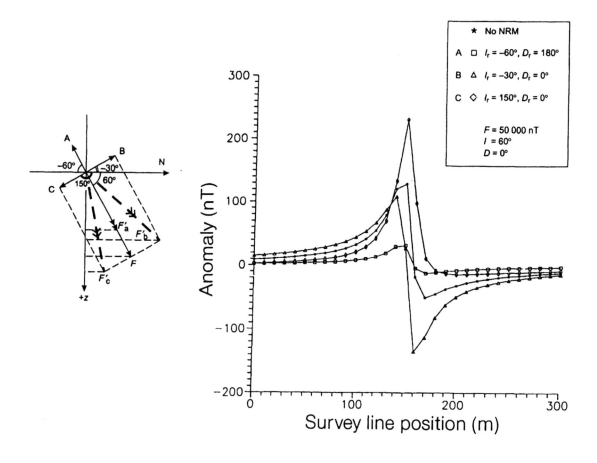

Figure 3.43 Total field anomalies over a vertical dyke striking east-west with either no remanent magnetisation (*), or remanent magnetisation of 150 nT in three directions as indicated by the schematic vector diagram

case C, where the remanent magnetisation is also at right-angles to the Earth's field but has the same positive downwards sense, F'_c, although the same magnitude as F'_b, has a larger vertical component than even the Earth's field and so the anomaly amplitude is increased substantially. In low magnetic latitudes ($<45°$), the horizontal vector component becomes more important than the vertical component when considering remanent magnetisation.

A range of different anomaly shapes demonstrating the effects of strike, latitude, dip, depth and body size has been provided above for comparison. When a field profile is to be interpreted, there is usually some background information available about the local geology (how else was the survey designed?). Consequently, many of the variables can be constrained so that a reasonable interpretation can be produced. The biggest unknown in many cases is the presence or otherwise of any remanence. Commonly it is assumed to have no effect unless found otherwise from measurements of remanence of

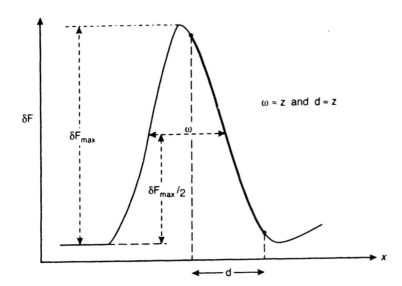

retrieved rock samples. However, to obtain a more exact interpretation it is necessary to model the observed data using computer methods (see Section 3.8.3).

3.8.2 Simple depth determinations

It is possible to obtain a very approximate estimate of depth to a magnetic body using the shape of the anomaly. By referring to either a simple sphere or horizontal cylinder, the width of the main peak at half its maximum value ($\delta F_{max}/2$) is very crudely equal to the depth to the centre of the magnetic body (Figure 3.44). In the case of a dipping sheet or prism, it is better to use a gradient method where the depth to the top of the body can be estimated. The simplest rule of thumb to determine depth is to measure the horizontal extent, d, of the approximately linear segment of the main peak (Figure 3.44). This distance is approximately equal to the depth (to within $\pm 20\%$).

A more theoretically based graphical method was devised by Peters (1949) and is known as Peters' Half-Slope method (Figure 3.45). A tangent (Line 1) is drawn to the point of maximum slope and, using a right-angled triangle construction, a line (Line 2) with half the slope of the original tangent is constructed. Two further lines with the same slope as Line 2 are then drawn where they form tangents to the anomaly (Lines 3 and 4). The horizontal distance, d, between these two tangents is a measure of the depth to the magnetic body (see Box 3.8).

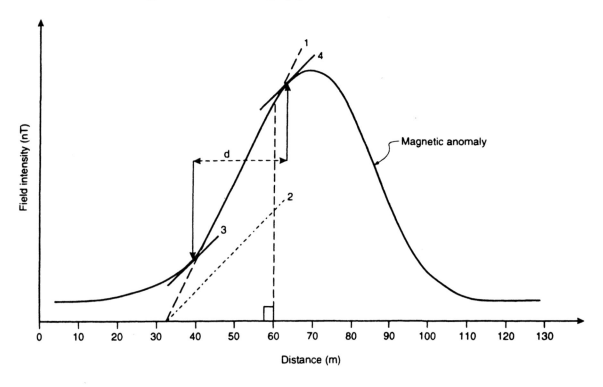

Box 3.8   Peters' Half-Slope method of depth determination
(see Figure 3.45)

Figure 3.45   Peters' Half-Slope method of determining the depth to the top of a magnetised dyke (see text for further details)

The depth (z) to the top of the magnetised body is:

$$z = (d \cos \alpha)/n$$

where d is the horizontal distance between half-slope tangents; $1.2 \leqslant n \leqslant 2$, but usually $n = 1.6$; α is the angle subtended by the normal to the strike of the magnetic anomaly and true north.

Example: If $d = 24$ m, $n = 1.6$ and $\alpha = 10°$, then $z \approx 15$ m.

Parasnis (1986) derived alternative methods of depth determination for a magnetised sheet of various thicknesses and dips using the anomaly shape as well as amplitude. Given an asymmetric anomaly (Figure 3.46) over a dipping dyke of known latitude, dip and strike directions, the position of, and the depth to, the top of the dyke can be determined from the anomaly shape. If the maximum and minimum *numerical* values are denoted δF_{max} and δF_{min} respectively (i.e. irrespective of sign), the position of the centre of the top edge of the dyke is located at the position of the station where the anomaly amplitude equals the sum of the maximum and minimum values (taking note of their respective signs, positive or negative) and which lies between the maximum and minimum values. For example, if $\delta F_{max} = 771\,nT$ and $\delta F_{min} = -230\,nT$, the position of the centre of the dyke would be located where $\delta F = 771 - 230\,nT = 541\,nT$. Expressions from which the depth to the top of the dipping sheet can be derived (see Box 3.9) relate the amplitudes of the minimum and maximum values and their respective separations. Expressions for thick dipping sheets and other models are much more complicated

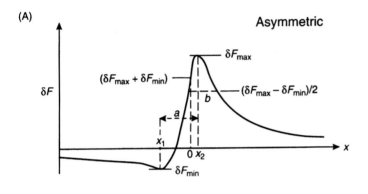

(A)

Asymmetric

(B)

Symmetric

Figure 3.46 Parasnis' method of determining the position of the centre of, and the depth to the top of, a magnetised thin sheet. (A) An asymmetric profile, and (B) a symmetric profile

to calculate and are not discussed further here. Derivations of mathematical expressions and full descriptions of methods to determine shape parameters for a variety of dipping sheet-like bodies have been given by Gay (1963), Åm (1972), Parasnis (1986) and Telford *et al.* (1990), for example. Manual calculation of parameters has been largely superseded by the ease with which computer methods can be employed.

Box 3.9   Depth estimations (see Figure 3.46 for definitions of parameters)

Thin sheet (depth to top of body z)

Asymmetric anomaly:

$$z = (-x_1 . x_2)^{1/2} (x_1 \text{ or } x_2 \text{ is negative})$$

$$\text{and } z = (a^2 - b^2)^{1/2} . |C|/2 . (1 + C^2)^{1/2}$$

where $1/C = \tan(I + i - \alpha)$; $|C|$ is the magnitude of C; $I =$ inclination of the Earth's magnetic field; $i =$ inclination of any remanent magnetisation; and $\alpha =$ angle of dip of the sheet.

Symmetric anomaly:

$$z = w/2$$

where $w =$ anomaly width at $\delta F_{max}/2$.

3.8.3   Modelling in two and three dimensions

Manual calculations of depths to the top of a particular magnetised body may provide an approximate figure provided an individual anomaly can be isolated from the background adequately. One common problem is when two magnetised bodies are so close to each other spatially that their magnetic anomalies overlap to form a complex anomaly (Figure 3.47). Sometimes such an anomaly may be misinterpreted as being due to a much thicker body with lower susceptibility than the actual cause, and so gives rise to an erroneous interpretation. Geological control may help, but having the ability to generate models quickly by computer is a great boon. 'What if...?' models can be constructed to test to see what anomaly results if different configurations of magnetised bodies are considered. A basis for many computer programs was provided by Talwani *et al.* (1959) and Talwani (1965). In essence, two-dimensional modelling requires that the body being modelled has a lateral extent at least 20 times its width so that the end effects can be ignored. For many geological features this restriction is adequate and anomalies from such bodies can be analysed successfully. In

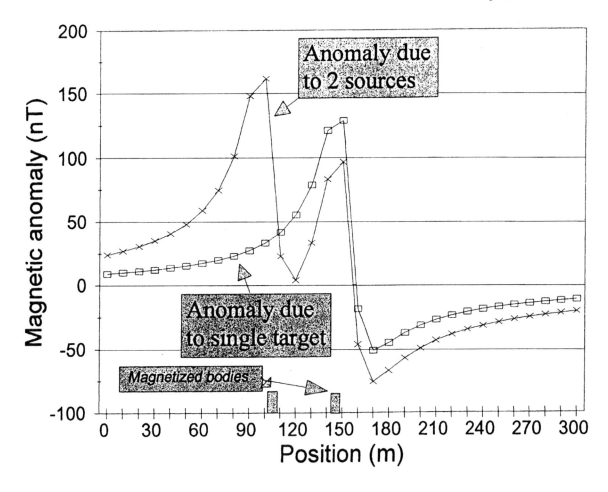

Figure 3.47 Total magnetic anomalies over two identical 5 m wide vertical dykes whose centres are separated by 45 m (their locations are indicated), and in contrast, the magnetic anomaly arising from just one of the dykes (at position 145 m), ignoring the other

many other cases, it is necessary to use more sophisticated computer methods.

With any set of profile data and the analysis of anomalies, it is absolutely essential that there be sufficient data to image the correct shape of the anomaly. Too few data leads either to incorrect evaluation of amplitude maxima or minima (consider the effects on simple depth determinations if the peak positions are not adequately defined), or in more extreme cases to severe aliasing (see Chapter 1). Limitations can be caused by too few field data values and/or by computer models producing too few point values at too large a station interval. It is therefore essential that, if computer methods are to be employed, the user is aware of what the software is actually trying to do and how it does it, and that adequate field data, both in number of values and data quality, are used.

A wide variety of computer methods exist, ranging from simple two-dimensional packages that can be run on low-cost personal computers and laptops, up to extremely sophisticated 2.5-dimensional and three-dimensional modelling packages on more powerful workstations. Two-and-a-half dimensional modelling is an extension of two-dimensional modelling but allows for end effects to be considered (e.g. Busby 1987). Some packages provide a statistical analysis on the degree of fit between the computer-generated anomaly and the observed data. Others allow successive iterations so that each time the software completes one calculation it generates better input parameters automatically for the next time round, until certain quality parameters are met. For example, a computer program may run until the sum of the least-squares errors lies within a defined limit.

One other factor that needs to be borne in mind with magnetic modelling is that some software packages compute the anomaly for each body and then sum them arithmetically at each station. This ignores the fact that adjacent magnetised bodies will influence the magnetic anomalies of the other bodies. Consequently, the computed anomalies may be over-simplifications; and even if statistical parameters of the degree of fit are met, the derived model may still be only a crude approximation. In the majority of cases involving small-scale surveys in archaeology or site investigations, these problems are not likely to be as significant.

Modelling in three dimensions permits the construction of irregularly shaped bodies made up of a stack of magnetised polygons and takes into account edge effects. An example of three-dimensional interpretation of magnetic (and gravity) data is given in Section 3.9.1.

In addition to the spatial modelling, another method of assisting interpretation is the use of spectral analysis. The magnetic field is expressed by Fourier analysis as an integral of sine and/or cosine waves each with its own amplitude and phase. In this, the amplitude or power spectrum is plotted as a function of wavelengths (from short to long), expressed in terms of wavenumber (1/wavelength). A number of methods exist to determine the power spectrum (e.g. Spector and Parker 1979; García-Abdeslem and Ness 1994). They attempt to isolate the regional field so that the residual anomalies can be identified. Once the frequency of the residuals has been determined, the dataset can be filtered to remove the regional field, thus facilitating the production of anomaly maps on which potential target features may be highlighted. It is also possible to produce susceptibility maps from this type of analysis. From the form of the Fourier spectrum, estimates of depth to the magnetised bodies can be made (Hahn *et al.* 1976). The power spectrum yields one or more straight-line segments, each of which corresponds to a different magnetised body. A recent example of such an analysis of aeromagnetic data associated with the Worcester Graben in England has been described by Ates and Kearey (1995). Further analysis of magnetic data in the wavenumber domain has been discussed by Xia and Sprowl (1992), for example.

3.8.4   Recent developments

There have been two important developments in the display and processing of magnetic data. Where a large dataset exists for a regional aeromagnetic survey, say, the data have been displayed conventionally as a contoured anomaly map. With the advent since the 1970s of image processing using computers, it is possible to digitise and process data recorded previously in analogue form. Instead of contouring, the area covered is divided into rectangular cells each of which is assigned a total field strength value. These cells are manipulated and displayed as individual pixels. The resultant display is a colour image with blue indicating negative anomalies and red the positive anomalies. Shades of colour indicate intensities, so dark red indicates a high-amplitude positive anomaly while deep blue a high-amplitude negative anomaly. Image-processing software analogous to that used to process satellite image data, such as Landsat, is used to analyse the aeromagnetic (or indeed gravity) data (Figure 3.48).

Figure 3.48   This image shows total field aeromagnetic data collected over an area of 200 km × 140 km of the Southern Uplands, Scotland (B: Bathgate; C: Carlisle; E: Edinburgh). The data were acquired in analog form along flight lines spaced approximately 2 km apart with tie lines about 10 km apart and with a mean terrain clearance of 305 m. The data were digitised, interpolated to a square grid of mesh size 5.0 km and displayed as a greyscale shaded-relief image. Sun illumination azimuth and inclination are N and 45° respectively. East–west lineations through the high-amplitude Bathgate anomaly are related to quartz-dolerite Permo-Carboniferous dykes. Numerous Teritiary dykes can be mapped trending NW–SE across the Southern Uplands. For comparison, see Figure 2.35. Image courtesy of Regional Geophysics Group, British Geological Survey

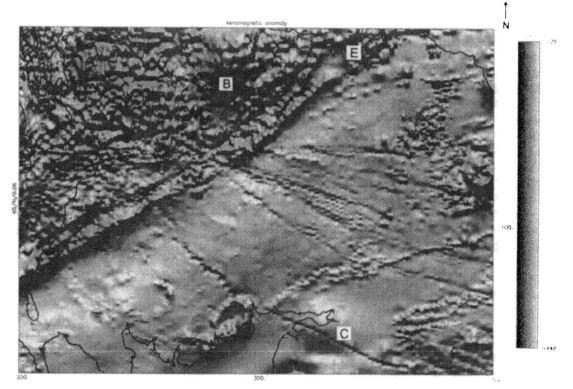

200 km x 140 km

Lineaments suggestive of faults can be identified using edge filters. Spatial filters can be applied to resolve the data into regional and residual datasets and zoom facilities allow the accurate expansion of subsets of data for high-resolution analysis of much smaller areas, thus permitting the identification of any low-amplitude anomalies. Green (1983), for example, has discussed digital image processing in detail.

Image-processing methods need not be restricted to datasets covering very large areas. The only prerequisite for image processing to be possible is that there must be a large enough dataset. Dataloggers attached to magnetometers and memory-backed magnetometers or gradiometers can also be used for ground surveys where very high spatial discrimination is required (Sowerbutts and Mason 1984). Scollar *et al.* (1986) have reviewed the various data processing and display methods used in archaeology where very high resolution is also required.

Sowerbutts (1987) mapped the largely concealed Butterton Dyke in central England using a microcomputer-based fluxgate gradiometer. More than 16 800 magnetic measurements were made in one day by one person. Data were recorded at 0.5 m intervals along 317 parallel traverses 1 m apart and the results plotted as a contour map and as an isometric display (Figure 3.49). The lines of the olivine dolerite dykes are obvious. The survey produced an extremely accurate map of the dyke, which was totally concealed in the area of the survey, and demonstrated that the dyke was not terminated by a fault, as previously thought, but died out naturally.

Calculating depths to magnetic features is still the most significant aspect of magnetic surveys. Since Peters' classic paper in 1949 there has been a steady stream of publications describing new or modified methods of depth determination. In the 1980s, one technique, known as the Euler Deconvolution method, was developed to process magnetic data and convolve them to a point source at depth (Thompson 1982). The method operates on the data directly and provides a mathematical solution without recourse to any geological constraints. This has the advantage that the Euler-derived interpretation is not constrained by any preconceived geological ideas and thus can be used critically to appraise geological, and particularly structural, interpretations.

Euler's method utilises the magnetic field strength at any point in terms of the gradient of the total magnetic field, expressed in Cartesian coordinates. Furthermore, these gradients are related to different magnetic sources by a function termed the Structural Index, N. Euler's equation is given in Box 3.10. Considering the equation asterisked, a series of seven equations can be determined from observed data by evaluating this equation in at least seven places within a window placed on the data. This provides seven equations to solve for three unknowns using a least-squares procedure. The output

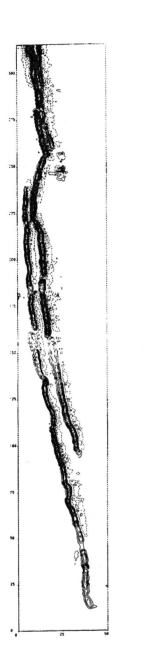

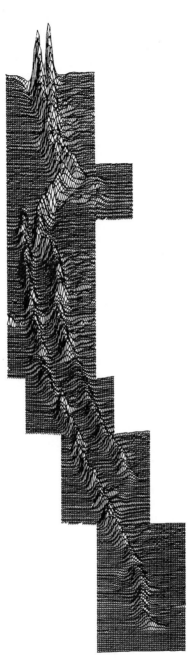

Figure 3.49 Butterton Dyke anomalies displayed as a magnetic gradient contour map with a contour interval of 20 nT/m (positives values, solid; negative values, dashed). From Sowerbutts (1987), by permission

consists of determinations of the depth to the magnetic body (z_0) producing the anomaly which can be plotted on a map as a circle, the increasing size of which indicates greater depth. Solutions that lie outside predetermined statistical bounds are rejected, thus constraining the possible number of solutions.

Box 3.10   Euler's equation

$$(x - x_0)\frac{\delta T}{\delta x} + (y - y_0)\frac{\delta T}{\delta y} - z_0 \frac{\delta T}{\delta z} = N(B - T)$$

where x_0, y_0, z_0 are the coordinates of a magnetic source whose total field intensity T and regional value B are measured at a point (x, y, z); N is the degree of homogeneity and referred to as the Structural Index.

For two dimensions (x, z), Euler's equation reduces to:

$$x_0 \frac{\delta T}{\delta t} + z_0 \frac{\delta T}{\delta z} + NB = x \frac{\delta T}{\delta x} + NT \ (*)$$

The only unknowns are x_0, z_0 and N; gradients of T with respect to x and z can be derived from the measured magnetic data; $\delta T/\delta y$ is assumed to be equal to zero.

Structural indices N:

Vertical geological contact	Edge of large tank	0– <0.5
Infinite sheet		0
Thick step		0.5
Irregular sill	Metal sheet	1
Vertical cylinder	Well/drum	2–2.25
Cylinder with unknown orientation	Drum	2.5
Horizontal cylinder	Pipeline/drum	2–2.75
Point dipole		3
Sphere	Tank	3

Reid *et al.* (1990) have used Euler's method to aid the interpretation of an area in south-central England. A simplified geological map of the survey area is given in Figure 3.50 with the aeromagnetic survey data. The Euler solutions have been plotted in Figure 3.50C which has been interpreted to form a structural map shown in Figure 3.50D. The Euler display has picked out the two north–south striking boundary faults marking the central block. Depth solutions for the Eastern Boundary Fault were within 0.3 km of those derived by other means. The magnetic anomalies over the central block yield very few

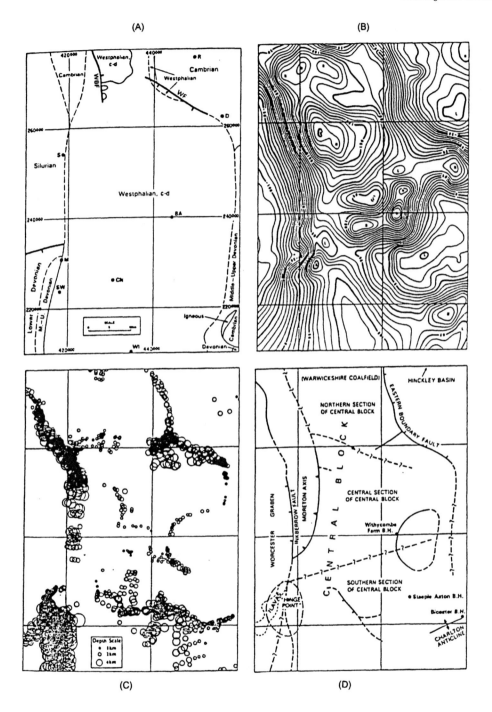

Figure 3.50   (A) Simplified geological map of the south-central part of England. (B) Aeromagnetic map of the area shown in (A). (C) Euler solutions derived from the aeromagnetic data calculated for a structural index of 1. (D) structural interpretation of the Euler solutions. From Reid *et al.* (1990), by permission

Euler solutions and this may be due to overlapping anomalies making Euler solutions statistically unacceptable and so have been rejected. Alternatively, there are insufficient data to represent the area adequately. In the south-west corner, there is a strong cluster of solutions with large depth values. This has been interpreted as being due to a magnetic body intruded at a depth of about 10 km. The main benefit of using the Euler method appears to be the delimitation of lineaments due to faults or major linear intrusions, and these features are best imaged by the Euler solutions if they are dipping, rather than vertical. The technique offers a method of independent interpretation which can be used to appraise existing geological models.

Euler's method has been incorporated into commercially available software (GRIDEPTH [Registered trademark of the Simon–Robertson Group plc] Geosoft Ltd) and has been demonstrated to be applicable to the location of near-surface utilities (pipes, drums, etc.) in addition to regional structural features. Examples of the use of the method in locating buried man-made targets have been given by Yaghoobian *et al.* (1992). They have described a detailed magnetic survey carried out at the Columbia Test Site, University of Waterloo, Canada, and corresponding Euler analysis. The results are shown in Figure 3.51. This example is perhaps too contrived to demonstrate the effectiveness in a real situation with a typical amount of background noise, but it does serve to show that the method works. Where there is a significant background noise problem, accurate depths and/or locations may be hard to achieve, especially if the measured data yield only weak signals. With careful analysis and selection of processing parameters (especially the Structural Index and GRIDEPTH window size) best-possible solutions may be obtained. It is also important that the field data be of as high quality as can be achieved and that appropriate filtering to remove obvious noise is undertaken before Euler analysis.

In addition to Euler analysis, increasing attention is being paid to the use of three-dimensional analytic signals (Nabighian 1984; Roest *et al.* 1992; MacLeod *et al.* 1993). The general principle behind the method is illustrated schematically in Figure 3.52. Given a magnetic anomaly (in the case illustrated, this arises from a square prism magnetised in an arbitrary direction), the three derivatives can be obtained from which the analytic signal is derived (Box 3.11). The amplitude of the three-dimensional analytic signal of the total magnetic field produces maxima over magnetic sources irrespective of their magnetisation direction (MacLeod *et al.* 1993; Parlowski *et al.* 1995). The three-dimensional analytic signal can be calculated readily from an accurately gridded dataset. A 3×3 filter can be applied to obtain the horizontal gradients ($\delta T/\delta x$ and $\delta T/\delta y$). A fast Fourier transform (FFT) can be used to obtain the vertical gradient ($\delta T/\delta z$). Alternatively, if a gradiometer has been used to acquire the data, the vertical gradient is a measured parameter.

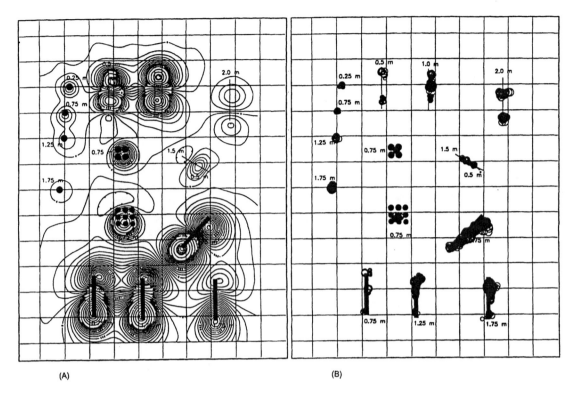

(A) (B)

From the form of the resultant analytic signal, solutions are obtained as to the location (x, y, z) of the source. These are shown graphically as circles, much the same as from Euler solutions. The distance between the inflection points of the analytic signal anomaly is directly proportional to the depth to the top of the magnetic source. It has been demonstrated that both Euler and analytic solutions can be used to produce an integrated model of the causative magnetised structure: the analytic signal can be used to delineate the magnetic contrasts and give estimates of their approximate depths, while the Euler solution yields more detailed interpretation at depth (Roest et al. 1992; Parlowski et al. 1995). It should be noted that the use of analytic signals can also be applied to gravity data or a combination of potential-field data.

Box 3.11 The amplitude of the three-dimensional analytic signal (MacLeod et al. 1993)

The amplitude of the analytic signal ($|A(x, y)|$) at any location (x, y) is given by:

$$|A(x, y)| = [(\delta T/\delta x)^2 + (\delta T/\delta y)^2 + (\delta T/\delta z)^2]^{1/2}$$

where T is the measured field at (x, y).

Figure 3.51 (A) Total magnetic field anomaly map over the Columbia Test Site, University of Waterloo, Canada, with the known depths to the various targets listed. (B) The Euler solutions calculated from (A). From Yaghoobian et al. (1992), by permission

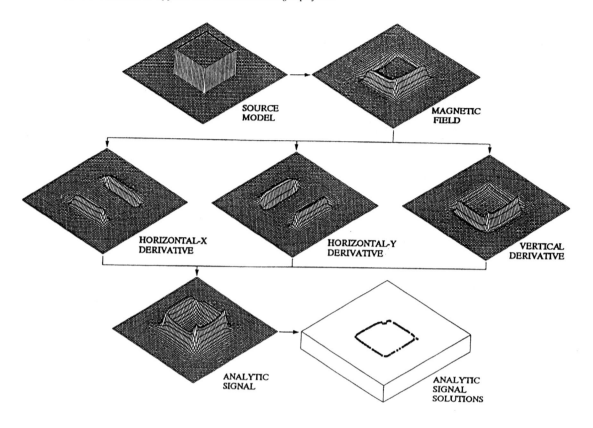

SOURCE MODEL

MAGNETIC FIELD

HORIZONTAL-X DERIVATIVE

HORIZONTAL-Y DERIVATIVE

VERTICAL DERIVATIVE

ANALYTIC SIGNAL

ANALYTIC SIGNAL SOLUTIONS

In shallow environmental applications, sources at very shallow depth (< 1 m) give rise to extremely high frequency signals which are difficult to resolve using these techniques. It is probably more cost-effective to dig with a shovel on the position of magnetic anomalies to locate the target source than to go through the analytic procedures outlined above. However, as soon as the depth is beyond easy excavatability, the value in the use of analytic solutions increases. Accuracies in depth determination are of the order of 30% of the target depth. This figure is one that may be improved upon in time with future developments and experience.

Figure 3.52 Schematic outline of the analytic signal method. Horizontal and vertical derivatives are calculated from the total field anomaly over a square prism and combined to yield the absolute value of the analytic signal. The locations of the maxima and the shape of this signal can be used to find body edges and corresponding depth estimates. From Roest *et al.* (1992), by permission

3.9   APPLICATIONS AND CASE HISTORIES

3.9.1   Regional aeromagnetic investigations

An aeromagnetic survey over the Lizard Peninsula in Cornwall, which was interpreted in conjunction with gravity data (Rollin 1986), provides an example of three-dimensional interpretation. The Lizard Complex comprises what is thought to be an ophiolite suite made up of a lower tectonic sheet of hornblende schists and metasediments and structurally overlain by peridotite and gabbro. This upper sheet shows lateral variation from harzburgite peridotite, gabbro and a dyke complex in the east to therzolite peridotite and cumulate complex to the west. Comparison of the simplified geological map with the corresponding aeromagnetic map (Figure 3.53) demon-

Figure 3.53 (A) Simplified geological map of the Lizard Peninsula, south Cornwall, and (B) the corresponding aeromagnetic map. From Rollin (1986), by permission

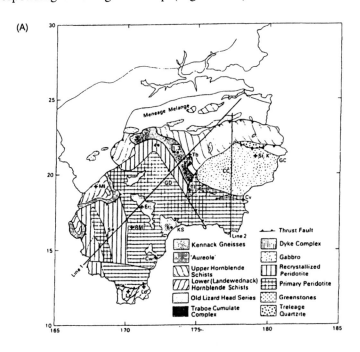

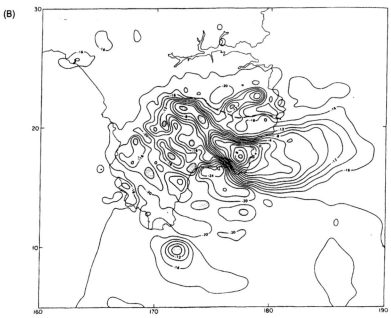

strates the correlation of the most prominent magnetic feature, a magnetic high which extends offshore to the east, onshore across the harzburgite peridotite south of Coverack north-westwards with diminishing amplitude over the Traboe Cumulate Complex. Western Lizard is characterised by small localised anomalies over the western peridotite. These features correlate with small outcrops of interlayered basic gneiss. A summary of the tectonic units and their respective schematic geophysical responses are shown in Figure 3.54. From this it can be seen that the lherzolite peridotite has a much smaller magnetic effect than the harzburgite peridotite while they both have comparable effects on the gravity field. It was found that the harzburgite peridotite was depleted particularly in TiO_2 compared with the lherzolite peridotite. Low concentrations of titanium in these rocks are indicative of complex geochemical processes which affected the formation of magnetite during serpentinisation.

When modelling these aeromagnetic anomalies, Rollin found that it was necessary to investigate the remanent magnetisation which is

Figure 3.54 Tectonics units of the Lizard and a schematic representation of their geophysical responses. From Rollin (1986), by permission

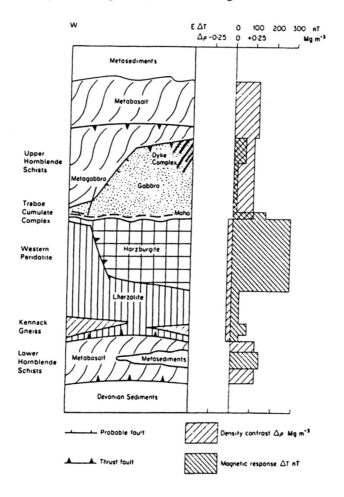

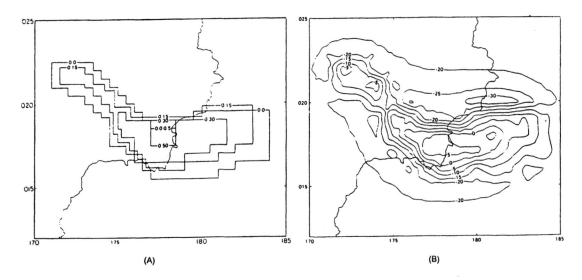

(A)   (B)

significant, as might be expected for such a suite of rocks. On the basis of field and laboratory measurements, the overall Königsberger ratio (remanent:induced magnetisation) was taken to be 0.43 with a declination and inclination of the remanent magnetisation of 286° and 75° respectively, with a resultant magnetisation up to 220 nT and a mean susceptibility of 0.0774 (SI). A simple three-dimensional model comprising stacked horizontal polygons was then constructed for the eastern area, which produced a calculated anomaly (Figure 3.55)

Figure 3.55   Three-dimensional modelling of the south-east part of the Lizard. (A) Stacked horizontal polygons, and (B) the resulting calculated magnetic anomaly. From Rollin (1986), by permission

Figure 3.56   Aeromagnetic and gravity anomaly profiles along Line 2 (Figure 3.53A) across the south-eastern part of the Lizard. From Rollin (1986), by permission

$\Delta\rho$ Mg m^{-3}	ΔT nT		
0.29	220-250	▉	Traboe Complex
-0.05	25-100	⊹	Peridotite
-0.05	75-100	▨	Gneiss-Schist
0.29	0	▥	Hornblende schist
0.15	0	⦙	Meneage
0.24	50	⦂	Gabbro

+ _ + _ +   Observed and calculated (+) gravity anomaly

● ● ●   Observed and calculated (●) magnetic anomaly

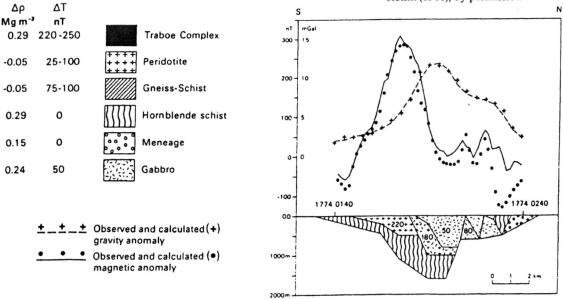

comparable to the observed aeromagnetic anomaly. The magnetic interpretation indicated that the Traboe Complex has a thickness of only 150 m while the eastern harzburgite peridotite is about 500 m thick (Figure 3.56). Independent gravity modelling gave the thickness of the eastern peridotite as 560 m, which is in reasonable accord with the magnetic interpretation.

3.9.2   Mineral exploration

In Finland, in regions of Pre-Cambrian basement comprising black graphite-schists, the location of sulphide ores is becoming increasingly difficult. The easy-to-find ores have already been identified. The more difficult ones, such as pyrrhotite-rich black schists and associated mineralisation, have been more easily eroded by glacial ice than the surrounding rocks and have become buried beneath a veneer of overburden, commonly up ·to 30 m thick. The overburden is characterised by high resistivities so the relatively conductive ore zones can be located using electrical methods. However, this is not always practicable so magnetic methods have also been used in conjunction with geochemical surveys. Furthermore, it is difficult to discriminate between the geophysical anomalies produced by economic sulphide mineralisation and those caused by the black schists and other rocks. Consequently, the identification of copper ores is particularly difficult.

The Saramäki orebody, located about 370 km north-east of Helsinki, forms a compact copper mass with some cobalt, and occurs in geological complexes comprising serpentinites, dolomites, skarns, quartzites and black schists, in a country rock of mica schist. The exploration strategy has been discussed in detail by Ketola (1979). The whole mineralised complex can be traced using aerogeophysical methods for over 240 km. The specific orebody, which was located in 1910 by drilling, is 4 km long, 200–400 m wide and extends down to about 150 m. Average copper content is 3.8% with 0.2% cobalt.

Problems with identification of the orebody can be seen by reference to Figure 3.57. The electrical resistivity (SLINGRAM) anomalies are caused by the black schists. The whole skarn zone forms a denser block (0.23 Mg/m^3 higher) compared with the country rock and so the sub-units within the mineralised zone cannot be differentiated using gravity. Susceptibility measurements made using the boreholes suggested that the mineralised zone was magnetised heterogeneously and that its upper part influenced the magnetic anomaly. In addition it was found by two-dimensional modelling that the main magnetised zones had significant remanent magnetisation. If remanence is ignored, the dip of the two main zones is inconsistent with the known dips, derived from drilling, even though the computed anomaly fits the observed one very well (Figure 3.58A). By incorporating remanence, with inclination 45° and declination 90° (compared

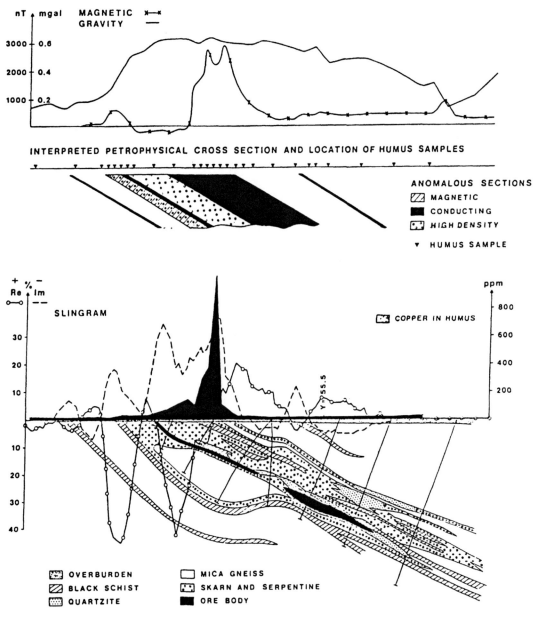

Figure 3.57 Magnetic, gravity and SLINGRAM electrical data with a geological cross-section for the Saramäki orebody, Finland. From Ketola (1979), by permission

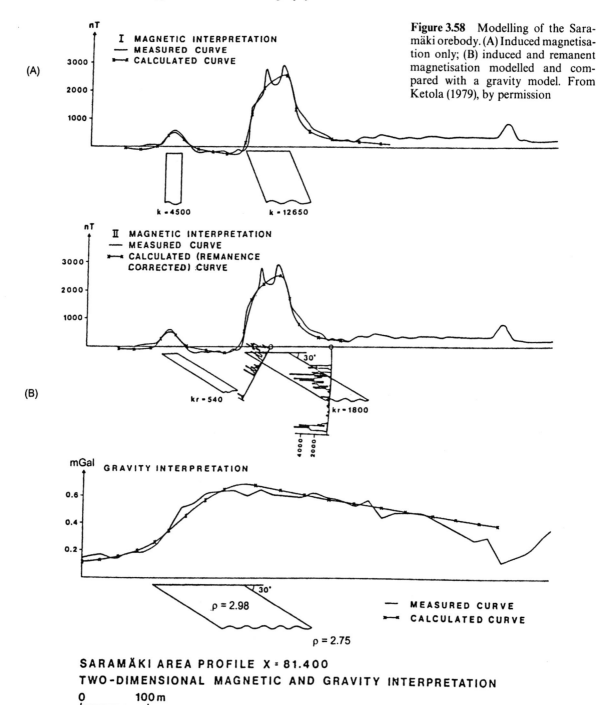

Figure 3.58 Modelling of the Saramäki orebody. (A) Induced magnetisation only; (B) induced and remanent magnetisation modelled and compared with a gravity model. From Ketola (1979), by permission

with 75° and 7°, respectively for the Earth's field), geologically compatible models were then produced which were also in accord with those used in the gravity modelling (Figure 3.58B).

Ketola concluded from this and other similar examples in Finland that geophysical or geochemical methods alone were insufficient to resolve these complex orebodies. It was necessary to use a wide variety of geophysical and geochemical techniques together, in conjunction with drilling, to provide a successful differentiation of the sub-surface geology.

Other illustrations of magnetic anomalies over mineralised bodies have been shown in Figure 1.9D, for a lode at Wheel Fanny, northwest Devon, where the course of the lode is picked out clearly by the linear magnetic anomalies. Figure 3.32 illustrates a ground magnetometer profile over sulphide mineralisation at Sourton Tors, north-west Dartmoor, Devon.

3.9.3 Engineering applications

3.9.3.1 Detection of underground pipes

Sowerbutts (1988) has provided a clear example of how high-resolution magnetic gradient surveys can identify not only the location of underground metal pipes but also the position of joints between individual sections of pipe. If joints can be located remotely, the amount of excavation required to examine and repair a pipe can be minimised, saving time, money and inconvenience.

The magnetic anomaly along a pipe consists of a series of smaller anomalies each of which corresponds to an individually cast segment of pipe which behaves like a magnetic dipole. For a pipe buried close to the surface and away from extraneous magnetic sources, a clear repetition of anomalies can be identified (Figure 3.59A). In the case illustrated, the pipe had a diameter of 0.5 m and was made of ductile iron in 6.3 m long sections and buried at a depth of 0.5 m. Magnetic highs, with gradients up to 4000 nT/m, are centred between 0.5 m and 1.0 m along from the socket end. For a pipe orientated east–west, a typical negative (north) and positive (south) anomaly doublet is produced along a profile aligned north–south in the northern hemisphere (cf. Figure 3.37A). A magnetised body orientated north–south produces a more symmetric positive-only or negative-only anomaly, and this is also seen in pipes orientated north–south (Figure 3.59B). For a smaller 76 mm diameter pipe (Figure 3.59C) this anomaly doublet effect is clearly repeated along the line of this north–south orientated pipe. In an urban environment, identifying the anomalies associated specifically with the buried pipe can be very difficult. Other magnetic materials are likely to be nearby and this can obscure or defocus the pipeline anomaly (Figure 3.59D). In the last case, the pipe was 0.15 m diameter and buried about 1.5 m down. Local metal grids

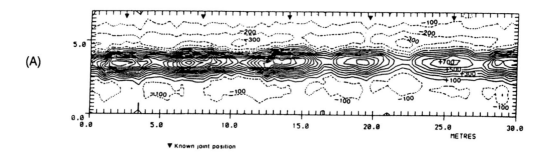

(A)

▼ Known joint position

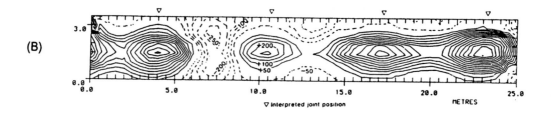

(B)

▽ Interpreted joint position

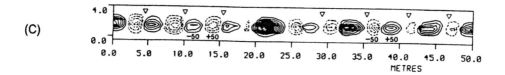

(C)

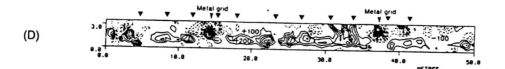

(D)

produce large magnetic anomalies. The anomaly pattern associated with this pipe is now irregular and complex, and it is not possible to identify pipe joints. Even seamless steel pipes with welded joints can be identified in areas with a quiet magnetic background.

A further example of the use of magnetometry to locate near-surface pipes has been given by Geometrics Inc., who acquired data over a test site at Stanford University, USA, using a caesium magnetometer. The data were contoured and displayed as both a map and an isometric projection which are shown with a map of the utilities in Figure 3.60. The correlation between buried objects and the corresponding magnetic anomalies is obvious.

Figure 3.59 Magnetic gradiometer anomalies over buried pipelines: (A) Ductile iron pipe, diameter 0.5 m buried at 0.5 m depth, E–W trend of pipe; contour interval 200 nT/m. (B) Cast iron pipe, N–S trend of pipe; contour interval 50 nT/m. (C) 76 mm diameter cast iron gas pipe trending N–S. (D) 0.15 m diameter pipe, buried about 1.5 m down, in an urban environment with extraneous magnetic anomalies caused by adjacent metal grids. From Sowerbutts (1988), by permission

(A)

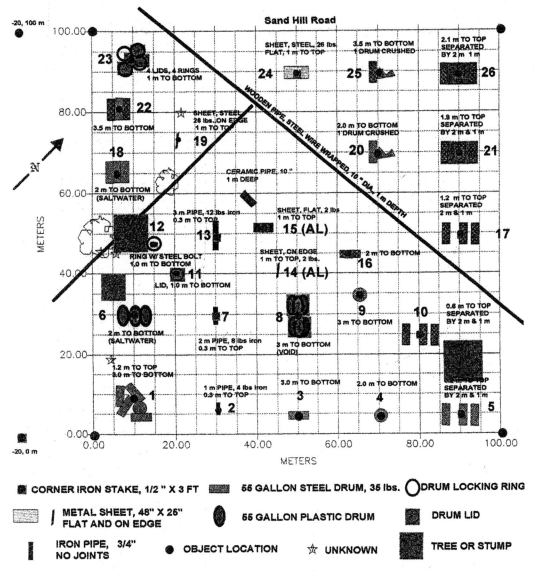

STANFORD UNIVERSITY ENVIRONMENTAL TEST SITE
GEOPHYSICAL TEST OBJECT BURIAL LOCATION
FINAL REVISION

(B)

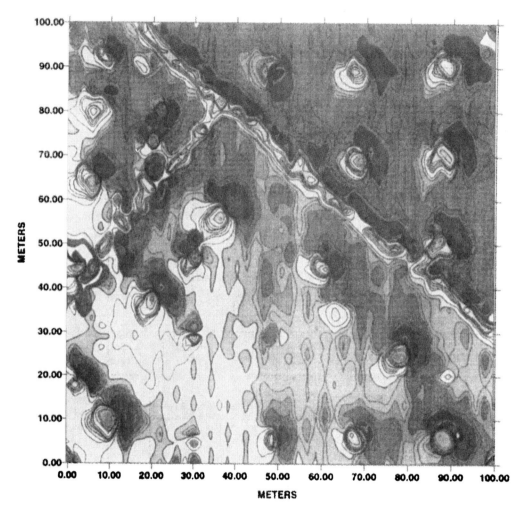

STANFORD UNIVERSITY ENVIRONMENTAL TEST SITE

G 822L CESIUM MAGNETOMETER DATA
VARIABLE CONTOUR INTERVAL
FINAL MAGNETIC CONTOUR MAP

(C)

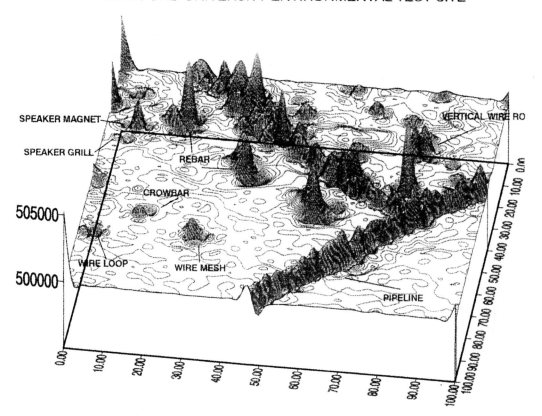

GEOMETRICS, INC

STANFORD UNIVERSITY ENVIRONMENTAL TEST SITE

G-822L CESIUM MAGNETOMETER DATA

20 GAMMA CONTOUR INTERVAL

INTERSECTING PIPELINES

CREATED WITH WINSURF BY GOLDEN

Anomalies have been produced which have been characteristic of induced magnetisation in many cases, but what has not been considered is that pipes acquire a permanent magnetisation on cooling during manufacture. It may be necessary to examine the orientation of the pipes as they cool at their respective foundries (hence determine the inclination and declination of permanent magnetisation). Having ascertained the permanent magnetisation, and knowing where these pipes have been buried, it should then be possible to determine more accurately the magnetic anomaly due to each segment of pipe.

3.9.3.2   *Detection of buried infill*

In areas where clay infills hollows in bedrock such as chalk, the slight contrast in magnetic susceptibility can still be sufficient for magnetic survey methods to be useful.

In many engineering investigations, an arbitrary spacing or grid of boreholes is used that in all too many cases is inadequate to indicate the actual variation on ground conditions. To drill holes at small enough separations may be prohibitively expensive as well as being unnecessarily intrusive. In such cases, geophysical methods can be used to image the ground variability.

McDowell (1975) has provided an example of how magnetic mapping of magnetic anomalies with only 15 nT amplitudes provided a clearer picture of a clay with flints cover over chalk in Upper Enham, Hampshire. A magnetic anomaly map (Figure 3.61A) of the site shows the possible clay infill and clearly demonstrates that the inter-borehole separation was far too large to provide a representative impression of ground conditions. Should a contractor have started excavations on the basis of ground conditions predicted by the borehole data, a claim for compensation could have justifiably been made to recompense the additional work required to cope with the unexpected variations in ground conditions. The cost of the claim probably would have exceeded that of the geophysical survey in the first place! The magnetic profile across the clay pipe, which is orientated north–south, shows a form similar to that illustrated in Figure 3.41 but with opposite polarity, as the clay with flints has the slightly higher magnetic susceptibility than the chalk.

3.9.4   Detection of buried containers

There are many situations where it is imperative that potentially harmful objects, such as bombs or drums of toxic waste, be located passively. Drilling holes to locate drums of poisonous waste, or using electromagnetic signals to detect hidden bombs, could have devastating consequences. As both types of object will produce a magnetic anomaly, it is possible to use magnetic methods to locate them without risk. The amplitudes of magnetic anomalies detectable for

Figure 3.60 (*previous pages*)(A) Map of the Stanford University, USA, environmental test site showing the details of the various buried targets. (B) Magnetic anomaly map produced using a caesium magnetometer sampling at 10 readings per metre along survey lines at 2 m spacings. (C) Isometric projection of the data in (B). Courtesy of Geometrics Inc.

(A)

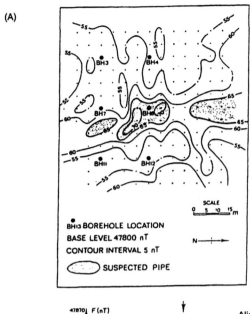

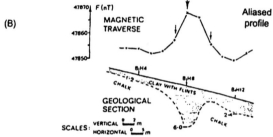

(B)

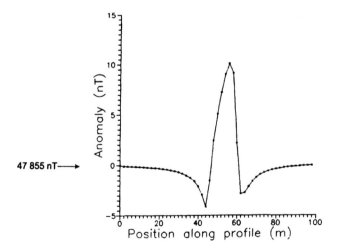

Figure 3.61 Megnetic anomalies over a clay-with-flints infill over chalk at Upper Enham, Hampshire, showing (A) how the magnetic survey resolved the variable ground conditions far better than the inadequately spaced boreholes, and (B) the shape of the profile across the clay pipe infill, which trends N–S, compared with a modelled anomaly (cf. Figure 3.41). (A) and (B) from McDowell (1975), by permission

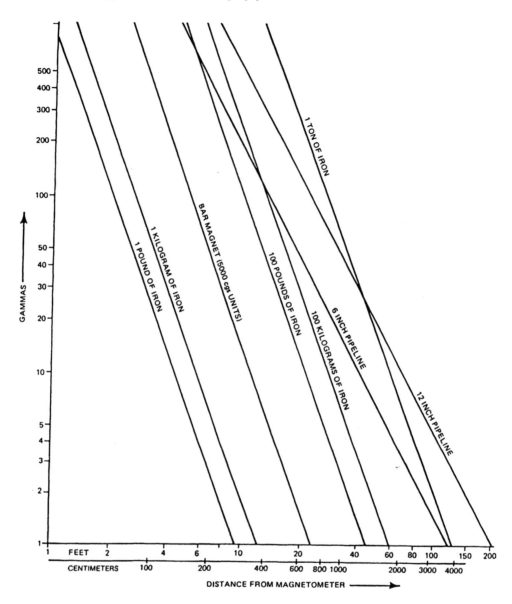

various types of ordnance of different size and depth of burial are illustrated in Figure 3.62, assuming a maximum detectability of 0.1 nT. However, in practice, the smallest anomaly likely to be discernible above background noise is likely to be around 1 nT. This means that it would technically be possible to detect a 1000 lb bomb buried at a depth of 22 m.

Metal drums also give rise to strong magnetic anomalies and, if dumped together in a random way, will produce large-amplitude but highly variable anomalies that will stand out from background noise. An example of the kind of anomaly produced is shown in Figure 3.63.

Figure 3.62 Minimum detectable anomaly amplitudes for different types of ordnance at various depths of burial. Note that the distances cited are those between the sensor and the target, not the depth below ground of the target. From Breiner (1981), by permission

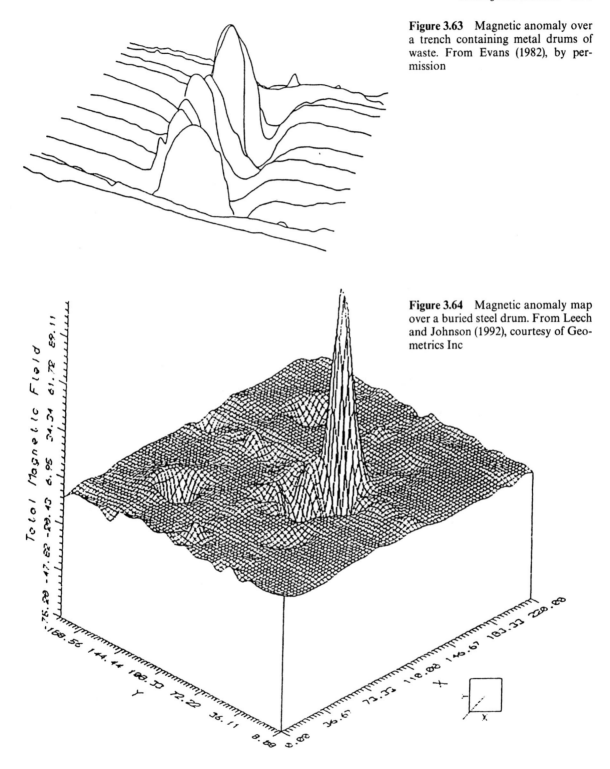

Figure 3.63 Magnetic anomaly over a trench containing metal drums of waste. From Evans (1982), by permission

Figure 3.64 Magnetic anomaly map over a buried steel drum. From Leech and Johnson (1992), courtesy of Geometrics Inc

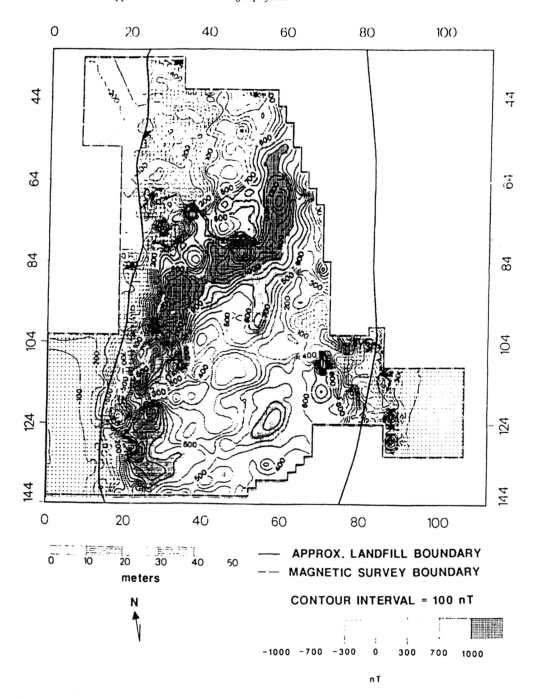

Figure 3.65 Contour map of total field magnetic intensity data measured at 1 m height, at the Thomas Farm landfill site, USA. From Roberts *et al.* (1990), by permission

To resolve the number of drums and their possible size, it may be preferable to use a gradiometer survey and datalogger so that the anomalies can be imaged with greater resolution. The magnetic anomaly over a buried steel drum is shown in Figure 3.64, from which it can be seen that local anomaly amplitudes are as high as several thousand nanoteslas, and field gradients several thousand nanoteslas per metre (Leech and Johnson 1992). The main difficulty with locating illegally buried drums of toxic waste is that they are usually dumped in ground with other rubbish which would help to mask their anomalies. However, by careful analysis of the anomaly shapes, amplitudes, orientations and overall characteristics, it should be possible to differentiate between various buried objects, if not actually identify them.

3.9.5   Landfill investigations

One of the aspects of landfills is that they are likely to contain large amounts of ferrometallic debris deposited at irregular angles. Consequently, a magnetic anomaly map produced over a former landfill will show a considerable amount of high-frequency noise from near-surface ferrometallic objects. There is perhaps a tendency to think that such noise is likely to dominate the magnetic anomaly map to produce a highly chaotic and largely unhelpful anomaly map.

One aspect of old closed landfills is that their previous tipping history may have been lost, or was never recorded. It might be useful to be able to obtain some idea as to whether a site has had different tipping sequences, such as periods of waste of a similar character being tipped and then having another type of waste with different magnetic properties. Consequently, the magnetic method lends itself to the rapid surveying of closed landfills in order to assess previous tipping histories and the zonation of waste types within a site.

An example of a magnetic survey over the Thomas Farm landfill in the USA has been given by Roberts *et al.* (1990). The site was surveyed on a $2 \times 2\,m$ grid with a sensor height at $1\,m$ above the ground. Roberts and co-workers demonstrated that, by careful data processing, the high-frequency noise could be filtered out to reveal longer-wavelength anomalies more closely associated with broad types of waste. An example of one of their maps is given in Figure 3.65. In this case, the lateral extent of the landfill was already known. The magnetic anomaly map reveals several zones with quite distinctive magnetic anomaly characters. For example, note the band of strong magnetic anomalies ($> 1000\,nT$) orientated NE–SW. To the south-east of this the magnetic anomalies are broader in wavelength and have low positive amplitudes, whereas those anomalies found north-west of this band are slightly higher frequency but predominantly negative. The actual wastes were domestic refuse in the north-west and brush, wood-cuttings and construction debris in the southern part.

Section 2
APPLIED SEISMOLOGY

Chapter 4
Applied Seismology:
introduction and principles

4.1   INTRODUCTION

The basic principle of exploration seismology is for a signal to be generated at a time that is known exactly and for the resulting seismic waves to travel through the sub-surface media and be reflected and refracted back to the surface where the returning signals are detected. The elapsed time between the source being triggered and the arrival of the various waves is then used to determine the nature of the sub-surface layers. Sophisticated recording and subsequent data processing enable detailed analyses of the seismic waveforms to be undertaken. The derived information is used to develop images of the sub-surface structure and a knowledge of the physical properties of the materials present.

Exploration seismic methods were developed out of pioneering earthquake studies in the mid-to-late nineteenth century. The first use of an artificial energy source in a seismic experiment was in 1846 by Robert Mallet, an Irish physicist, who was also the first to use the word 'seismology'. John Milne introduced the drop weight as an energy source in 1885. His ideas were further developed by August Schmidt who, in 1888, devised travel time–distance graphs for the determination of seismic velocities. In 1899, G.K. Knott explained the propagation, refraction and reflection of seismic waves at discontinuity boundaries. In 1910, Andrija Mohorovičić identified distinct phases of P and S waves on travel-time plots derived from earthquake data. He attributed them to refractions along a boundary separating material with a lower velocity above and a higher velocity at greater depth. This boundary, which separates the Earth's crust from the lower-lying mantle, is now called the 'Moho'.

Significant developments in the refraction method were made during the First World War by both the Allies and Germany, particularly by Ludger Mintrop. Research was undertaken to develop methods by which the location of heavy artillery could be achieved by studying the waves generated by the recoil of the guns on firing. This work was developed further by Mintrop who obtained the first patent for a portable seismograph in 1919 (Keppner 1991). On 4 April 1921, Mintrop founded the company Seismos Gesellschaft in order to carry out seismic refraction surveys in the search for salt domes acting as trap structures for hydrocarbons. In 1924, the Orchard Salt Dome in Texas, USA, was discovered using seismic refraction experiments undertaken by Seismos on behalf of Gulf Production Co., thus demonstrating the effectiveness of the method as an exploration tool.

The first seismic reflection survey was carried out by K.C. Karcher between 1919 and 1921 in Oklahoma, USA, based on pioneering work by Reginald Fessenden around 1913. By 1927, the seismic reflection method was being used routinely in exploration for hydrocarbons and within 10 years had become the dominant method world-wide in the exploration for oil and gas.

The use of fan shooting was also finding favour in the early 1920s due to the encouragement by L.P. Garrett who was head of seismic exploration at Gulf. Parallel to these developments, research work was also being undertaken at the US Bureau of Standards. In 1928, O. von Schmidt, from Germany, derived a method of analysis of refraction data for dipping two-layer structures to obtain the angle of dip and true velocity within the lower layer. In 1931, he published a solution to solve the dipping three-layer case. The so-called *Schmidt method* is still commonly used to determine weathered layer corrections in seismic reflection surveying.

In 1938, T. Hagiwara produced a method whereby, in addition to determining the lower layer velocity, the depths to this horizon could be determined at all shot and receiver positions along a single profile. Details of contributions made by Japanese engineering seismologists have been given by Masuda (1981).

As with just about all geophysical methods, the Second World War provided advances in technology that increased the usefulness of the various seismic methods.

In 1959, J.G. Hagedoorn published his 'Plus–Minus' method (see Section 5.4.2). In 1960, Carl Savit demonstrated that it was possible to identify gaseous hydrocarbons directly using seismic methods by identifying 'bright spots'. In 1961, L.V. Hawkins introduced the 'reciprocal method' of seismic refraction processing that has been subsequently and substantially developed by D. Palmer (1980, 1991) as the 'generalised reciprocal method' (GRM; see Section 5.4.3). Both Hagedoorn's and Palmer's methods are similar to Hagiwara's. A very good review of the seismic refraction method has also been given by Sjögren (1984).

Major developments in the seismic methods have come about by revolutions within the computing industry. Processing once thought possible only by mainframe computers is now being handled on personal computers and stand-alone workstations. With the vast increase in computer power, currently at a rate of an order of magnitude every two years, has come the ability to process data far more quickly and reliably, and this has opened up opportunities for seismic modelling. Obviously, with the degree of sophistication and specialisation that now exists in the seismic industry, it is not possible to provide anything like a comprehensive account here. There are many books available which deal extensively with exploration seis- mology, such as those by Claerbout (1976, 1985), McQuillin *et al.* (1984), Hatton *et al.* (1986), Waters (1987), Yilmaz (1987), and Dobrin and Savit (1988), among others.

There are two main seismic methods – refraction and reflection. Since the 1980s there has been a major shift towards using high- resolution seismic reflection surveying in shallow investigations (i.e.

to depths less than 200 m and especially less than 50 m). Previously, of the two seismic methods, refraction had been used principally within engineering site investigations. Neither seismic sources with suitably high frequencies, nor the data processing capability, were available or cost-effective for small-scale surveys. This is no longer so, and shallow seismic investigations are now much more common both on land and over water. Data obtained by signal-enhancement seismographs can be processed in similar ways to data acquired in large-scale hydrocarbon exploration surveys. Consequently, following a brief overview of the basic principles of applied seismology in this chapter,

Table 4.1   Derived information and applications of exploration seismology

Gross geological features:
  Depth to bedrock
  Location of faults and fracture zones
  Fault displacement
  Location and character of buried valleys
  Lithological determinations
  Stratigraphy
  Location of basic igneous dykes

Petrophyscial information:
  Elastic moduli
  Density
  Attenuation
  Porosity
  Elastic wave velocities
  Anisotropy
  Rippability

Applications:
  Engineering site investigations
  Rock competence
  Sand and gravel resources
  Detection of cavities
  Seabed integrity (for siting drilling rigs)
  Degassing or dewatering of submarine sediments
  Preconstruction site suitability for:
    new landfill sites
    major buildings
    marinas and piers
    sewage outfall pipes
    tunnel construction etc.
  Hydrogeology and groundwater exploration
  Ground particle velocities
  Forensic applications:
    location of crashed aircraft on land
    design of aircraft superstructures
    monitoring Nuclear Test Ban Treaty
    location of large bore military weapons

seismic refraction data processing and interpretation techniques are discussed in the next chapter, with seismic reflection surveying being discussed in detail in Chapter 6. These will provide a brief introduction to the shallow refraction and reflection methods (to which emphasis is given), and briefly to the processes used in the seismic industry for hydrocarbon exploration.

In addition to hydrocarbon exploration, seismic methods have a considerable number of other applications (Table 4.1), ranging from crude depth-to-bedrock determinations through to more subtle but fundamental information about the physical properties of sub-surface media, and from the obvious applications such as site suitability through to the apparently obscure uses such as in forensic investigations in aircraft crashes on land, such as the Lockerbie air disaster in Scotland in 1989. Details of some of these applications are given in Section 6.6.

4.2 SEISMIC WAVES

4.2.1 Stress and strain

When an external force F is applied across an area A of a surface of a body, forces inside the body are established in proportion to the external force. The ratio of the force to area (F/A) is known as *stress*. Stress can be resolved into two components, one at right-angles to the surface (normal or dilatational stress) and one in the plane of the surface (shear stress). The stressed body undergoes *strain*, which is the amount of deformation expressed as the ratio of the change in length (or volume) to the original length (or volume). According to *Hooke's Law*, stress and strain are linearly dependent and the body behaves *elastically* until the *yield point* is reached. Below the yield point, on relaxation of stress, the body reverts to its pre-stressed shape and size. At stresses beyond the yield point, the body behaves in a *plastic* or *ductile* manner and permanent damage results. If further stress is applied, the body is strained until it fractures.

Earthquakes occur when rocks are strained until fracture, when stress is then released. However, in exploration seismology, the amounts of stress and strain away from the immediate vicinity of a seismic source are minuscule and lie well within the elastic behaviour of natural materials. The stress/strain relationship for any material is defined by various elastic moduli, as outlined in Figure 4.1 and Box 4.1.

4.2.2 Types of seismic waves

Seismic waves, which consist of tiny packets of elastic strain energy, travel away from any seismic source at speeds determined by the

(A)

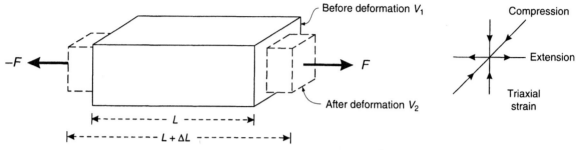

(B)

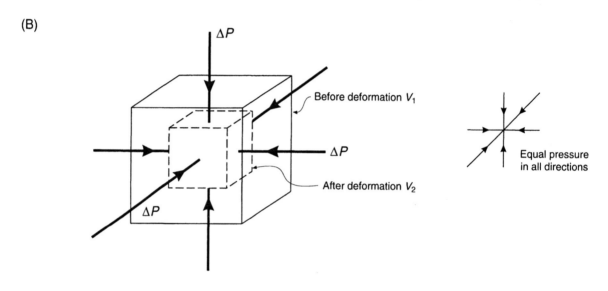

(C)   (D)

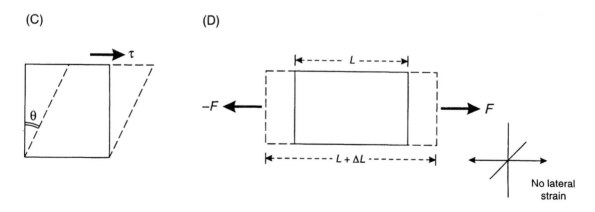

elastic moduli and the densities of the media through which they pass (Section 4.2.3). There are two main types of seismic waves: those that pass through the bulk of a medium are known as *body waves*; those confined to the interfaces between media with contrasting elastic properties, particularly the ground surface, are called *surface waves*. Other types of waves encountered in some applications are *guided waves*, which are confined to particular thin bands sandwiched between layers with higher seismic velocities by total internal reflection. Examples of these are *channel* or *seam waves*, which propagate along coal seams (Regueiro 1990a,b), and *tube waves*, which travel up and down fluid-filled boreholes.

Figure 4.1 (*opposite*) Elastic moduli. (A) Young's modulus; (b) bulk (rigidity) modulus; (C) shear modulus; and (D) axial modulus

4.2.2.1   *Body waves*

Two types of body wave can travel through an elastic medium. *P-waves*, which are the most important in exploration seismology, are also known as *longitudinal*, *primary*, *push*, or *compressional* waves. Material particles oscillate about fixed points in the direction of wave propagation (Figure 4.2A) by compressional and dilatational strain, exactly like a sound wave. The second type of wave is the *S-wave*, also known as the *transverse*, *secondary* or *shear* wave. Particle motion is at right-angles to the direction of wave propagation and occurs by pure shear strain (Figure 4.2B). When particle motion is confined to one plane only, the S-wave is said to be *plane-polarised*. The

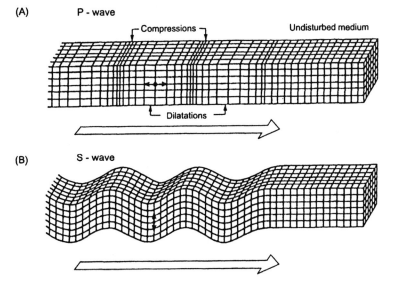

(A)   P - wave

Figure 4.2   Elastic deformations and ground particle motions associated with the passage of body waves. (A) A P-wave, and (B) an S-wave. From Bolt (1982), by permission

Box 4.1   Elastic moduli

Young's modulus

$$E = \frac{\text{Longitudinal stress } \Delta F/A}{\text{Longitudinal strain } \Delta L/L} = \frac{\sigma}{\varepsilon}$$

(in the case of triaxial strain)

Bulk modulus

$$k = \frac{\text{Volume stress } \Delta P}{\text{Volume strain } \Delta v/v}$$

(in the case of excess hydrostatic pressure)

Shear (rigidity) modulus (a Lamé constant)

$$\mu = \frac{\text{shear stress } \tau}{\text{shear strain } \varepsilon}$$

($\mu \approx 1\text{--}7 \times 10^4\,\text{MPa}$; $\mu = 0$ for fluids)

Axial modulus

$$U = \frac{\text{Longitudinal stress } \Delta F/A}{\text{Longitudinal strain } \Delta L/L} = \frac{\sigma}{\varepsilon}$$

(in the case with no lateral strain)

Relationships between Young's modulus (E), Poisson's ratio (σ), and the two Lamé constants (μ and λ)

$$E = \frac{\mu(3\lambda + 2\mu)}{(\lambda + \mu)} \quad \sigma = \frac{\lambda}{2(\lambda + \mu)} \quad k = \frac{3\lambda + 2\mu}{3}$$

and

$$\lambda = \frac{E\sigma}{(1 + \sigma)(1 - 2\sigma)}$$

Poisson's ratio ranges from 0.05 (very hard rocks) to 0.45 (for loose sediments). Elastic constants for rocks can be found in handbooks of physical constants.

identification and use of polarised shear waves in their vertical and horizontally polarised modes (SV and SH respectively) are becoming

(A)

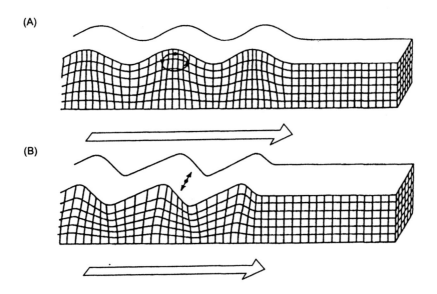

(B)

increasingly important in exploration seismology, as will be discussed later.

All the frequencies contained within body waves travel through a given material at the same velocity, subject to the consistency of the elastic moduli and density of the medium through which the waves are propagating.

Figure 4.3 Elastic deformations and ground particle motions associated with the passage of surface waves. (A) A Rayleigh wave, and (B) a Love wave. From Bolt (1982), by permission

4.2.2.2 *Surface waves*

Waves that do not penetrate deep into sub-surface media are known as surface waves, of which there are two types, *Rayleigh* and *Love waves*. Rayleigh waves travel along the free surface of the Earth with amplitudes that decrease exponentially with depth. Particle motion is in a retrograde elliptical sense in a vertical plane with respect to the surface (Figure 4.3A) and, as shear is involved, Rayleigh waves can travel only through a solid medium. Love waves occur only where a medium with a low S-wave velocity overlies a layer with a higher S-wave velocity. Particle motion is at right-angles to the direction of wave propagation but parallel to the surface. These are thus polarised shear waves (Figure 4.3B).

Surface waves have the characteristic that their waveform changes as they travel because different frequency components propagate at different rates, a phenomenon known as *wave dispersion*. The dispersion patterns are indicative of the velocity structure through which the waves travel, and thus surface waves generated by earthquakes can be used in the study of the lithosphere and asthenosphere. Surface wave dispersion has been described in detail by Grant and West (1965, pp. 95–107), by Gubbins (1990, pp. 69–80) and by Sheriff and

Geldart (1982, p. 51). Body waves are non-dispersive. In exploration seismology, Rayleigh waves manifest themselves normally as large-amplitude low-frequency waves called *ground roll* which can mask reflections on a seismic record and are thus considered to be noise. Seismic surveys can be conducted in such a way as to minimise ground roll, which can be further reduced by filtering during later data processing.

4.2.3 Seismic wave velocities

The rates at which seismic waves propagate through elastic media are dictated by the elastic moduli and densities of the rocks through which they pass (Box 4.2). As a broad generalisation, velocities increase with increasing density. Examples of P- and S-wave velocities for a range of geological materials are listed in Table 4.2. The seismic wave velocities in sedimentary rocks in particular increase both with depth of burial and age (cf. Chapter 2, Section 2.2.4.1; see Box 4.3).

Box 4.2   Seismic wave propagation velocity

Velocity of propagation V through an elastic material is:

$$V = (\text{Appropriate elastic modulus/density } \rho)^{1/2}.$$

Velocity of P-waves is:

$$V_{\mathrm{P}} = \left(\frac{k + 4\mu/3}{\rho}\right)^{1/2}.$$

Velocity of S-waves is:

$$V_{\mathrm{S}} = (\mu/\rho)^{1/2}.$$

The ratio $V_{\mathrm{P}}/V_{\mathrm{S}}$ is defined in terms of Poisson's ratio (σ) and is given by:

$$\frac{V_{\mathrm{P}}}{V_{\mathrm{S}}} = \left(\frac{1 - \sigma}{1/2 - \sigma}\right)^{1/2}. \qquad (*)$$

Note that $\mu = 0$ for a fluid, as fluids cannot support shear, and the maximum value of Poisson's ratio is 0.5; $\sigma \approx 0.05$ for very hard rocks, ≈ 0.45 for loose, unconsolidated sediments, average ≈ 0.25.

From the last equation (asterisked) in Box 4.2, it is clear that Poisson's ratio has a maximum value of 0.5 (at which value the

Table 4.2   Examples of P-wave velocities

Material	V_P (m/s)
Air	330
Water	1450–1530
Petroleum	1300–1400
Loess	300–600
Soil	100–500
Snow	350–3000
Solid glacier ice*	3000–4000
Sand (loose)	200–2000
Sand (dry, loose)	200–1000
Sand (water saturated, loose)	1500–2000
Glacial moraine	1500–2700
Sand and gravel (near surface)	400–2300
Sand and gravel (at 2 km depth)	3000–3500
Clay	1000–2500
Estuarine muds/clay	300–1800
Floodplain alluvium	1800–2200
Permafrost (Quaternary sediments)	1500–4900
Sandstone	1400–4500
Limestone (soft)	1700–4200
Limestone (hard)	2800–7000
Dolomites	2500–6500
Anhydrite	3500–5500
Rock salt	4000–5500
Gypsum	2000–3500
Shales	2000–4100
Granites	4600–6200
Basalts	5500–6500
Gabbro	6400–7000
Peridotite	7800–8400
Serpentinite	5500–6500
Gneiss	3500–7600
Marbles	3780–7000
Sulphide ores	3950–6700
Pulverised fuel ash	600–1000
Made ground (rubble etc.)	160–600
Landfill refuse	400–750
Concrete	3000–3500
Disturbed soil	180–335
Clay landfill cap (compacted)	355–380

*Strongly temperature dependent (Kohnen 1974)

denominator becomes zero). When Poisson's ratio equals 0.33, S-wave velocities are half P-wave velocities. Of the surface waves, Love waves travel at approximately the same speed as S-waves, as they are polarised S-waves, but Rayleigh waves travel slightly slower at about $0.92V_S$ (for Poisson's ratio = 0.25).

Box 4.3   Elastic wave velocity as a function of geological age and depth (after Faust 1951)

> For shales and sands, the elastic wave velocity V is given by:
>
> $$V = 1.47\,(ZT)^{1/6}\ \text{km/s}$$
>
> where Z is the depth (km) and T the geological age in millions of years.

Care has to be taken in comparing seismic velocities. Velocities can be determined from seismic data (see Sections 6.3.4 and 6.4.1 as to how this is done) and from laboratory analysis. When velocities have been determined using seismic refraction, the range of velocities obtained for a given type of material should be cited, and preferably with the standard deviation. *In situ* measurements made using refraction studies may yield velocities that are significantly different from those obtained from laboratory measurements. This occurs when the *in situ* rock is heavily jointed or fractured. The refraction velocities sample the rock and the discontinuities whereas a laboratory measurement on a solid sample cannot by virtue of the scale of the sample. Detailed measurements which may yield more representative *in situ* velocities on a fine scale are those obtained using a Schmidt hammer system. Two geophones are attached to the exposed material at a small distance apart. A hammer is used to generate a P-wave directly on to the rock at a known distance from the receivers. The velocity is obtained from the difference in travel time between the two receivers relative to their separation. No one velocity is absolute. In the case of laboratory measurements, ultrasonic transducers are used to transmit a pulse through a sample of the material in question. From the measured travel time of this pulse through the material whose length is known, a velocity can be calculated. Ultrasonic frequencies (0.5–1.5 MHz) are three to four orders of magnitude higher than typical seismic frequencies and so the velocities may not be directly comparable.

In addition to knowing the frequency of the transducers, it is also important to determine whether the samples have been measured dry or fully saturated, and if the latter, the salinity of the fluid used and the temperature at which the sample was measured. Of greater significance is the problem over mechanical relaxation of retrieved core pieces. If a sample of rock is obtained from significant depth below ground where it is normally at substantial pressure, the rock specimen expands on its return to the surface, resulting in the formation of microcracks due to the relaxation of confining pressure. These microcracks thereby increase the porosity and decrease the density of the rock. In the case of saturated samples retrieved from below the seabed, gas bubbles form within the sample as the pressure is reduced and these can reduce the acoustic velocity significantly compared with the *in situ* velocity. This may be noticeable even if samples are retrieved from less than 10 m of water depth.

In porous rocks, the nature of the material within the pores strongly influences the elastic wave velocity; water-saturated rocks have different elastic wave velocities compared with gas-saturated rocks; sandstones with interstitial clay have different propagation characteristics compared with clean sandstones, for example. Seismic velocities can be used to estimate porosity using the time-average equation in Box 4.4. If the P-wave velocities of both the pore fluid and the rock matrix are known, the porosity can be deduced. The form of this equation applies both for water-saturated and frozen rocks. In the case of permafrost, the velocity depends upon (a) the type of geological material, (b) the proportion of interstitial ice, and (c) the temperature. The importance of seismic velocities in interpretation are discussed in Chapter 6. The P-wave velocity in water is dependent upon temperature and salinity (Box 4.5) but is normally considered to be around 1500 m/s for a salinity of 35 parts per thousand at 13°C. In high-resolution surveys conducted where water masses with different temperatures and salinities occur, such as at the mouth of a river where fresh river water flows into and over salt water, the stratigraphy of the water column can become important in determining the correct P-wave velocities for use in subsequent data processing.

Box 4.4   Time-average equation to estimate rock porosity

The P-wave velocity V for a rock with fractional porosity (ϕ) is given by:

$$\frac{1}{V} = \frac{\phi}{V_f} + \frac{1-\phi}{V_m}$$

where V_f and V_m are the acoustic velocities in the pore fluid and the rock matrix respectively (Wyllie *et al.* 1958).

Typical values: $V_f = 1500\,\text{m/s}$, $V_m = 2800\,\text{m/s}$.

Box 4.5   P-wave velocity as a function of temperature and salinity in water

$$V = 1449.2 + 4.6\,T - 0.055\,T^2 + 0.0003\,T^3$$
$$+ (1.34 - 0.01\,T)(S - 35) + 0.016d$$

where S and T are the salinity (parts per thousand) and the temperature (°C); d is depth (m) (Ewing *et al.* 1948; cf. Fofonoff and Millard 1983)

In stratified media, seismic velocities exhibit anisotropy. Velocities may be up to 10–15% higher for waves propagating parallel to strata than at right-angles to them. Furthermore, some materials with strongly developed mineral fabrics can also demonstrate anisotropy, as for example in glacier ice and in highly foliated metamorphic rocks. In high-resolution surveys over sediments with marked anisotropy, significant differences in seismic character and data quality may be observed. In situations where such anisotropy is anticipated, it is essential to run test lines orientated at different azimuths in order to identify directions of shooting which either degrade the data quality or provide good resolution and depth penetration. For example, at the mouth of a fast-flowing river, lines parallel and at right-angles to water flow would be good directions to try. Once tests have been undertaken, an appropriate survey grid can then be established to ensure that the best-quality data are acquired. Towing the hydrophone streamer against the water flow produces better control and signal-to-noise ratios than across the water currents. Cross-track currents can also lead to excessive feathering of the source–receiver system (see Sections 4.6.2 and 6.3.2.2).

4.3 RAYPATH GEOMETRY IN LAYERED GROUND

4.3.1 Reflection and transmission of normally incident rays

Huygens' Principle is of critical importance to the understanding of the propagation of seismic waves through layered ground. Huygens' Principle states that every point on a wavefront can be considered to be a secondary source of spherical waves. The new wavefront is the envelope of these waves after a given time interval (Figure 4.4). If this is borne in mind, it is easier to understand how reflection, refraction and diffraction occur. Instead of always considering the wavefront of each wave, it is often easier to consider a line at right-angles ('normal') to the wavefront as a ray along which energy travels. Consequently, the propagation of seismic waves is frequently discussed in terms of rays and raypaths.

Whenever a wave impinges upon an interface across which there is a contrast in elastic properties, some of the energy is reflected back off the interface, and the remainder passes through the boundary and is refracted on entering the second medium. The relative amplitudes of the partitioned energy at such an interface into reflected and transmitted components are described by Zoeppritz–Knott equations in terms of the seismic velocities and densities of the two layers (Sheriff and Geldart 1982; see Telford *et al.* 1990, pp. 155ff, for a detailed discussion). The product of the density (ρ) and the seismic velocity (V) for each respective layer is known as the acoustic impedance (Z).

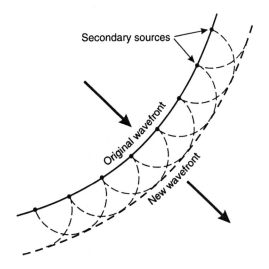

Figure 4.4 Propagation of a wave-front according to Huygens' Principle

Secondary sources

Original wavefront

New wavefront

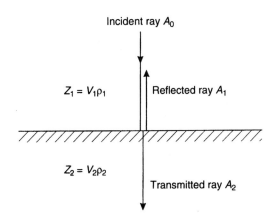

Figure 4.5 Partitioning of energy of a normally incident ray with amplitude A_0 into reflected and transmitted rays with respective amplitudes A_1 and A_2. Z_1 and Z_2 are the acoustic impedances of the two layers

Incident ray A_0

$Z_1 = V_1 \rho_1$

Reflected ray A_1

$Z_2 = V_2 \rho_2$

Transmitted ray A_2

Generally speaking, the more consolidated a rock, the greater will be its acoustic impedance. In order to propagate seismic energy most efficiently across a boundary, the acoustic impedance contrast should be small.

In the case of normal incidence, if the amplitude of incident energy is A_0 and those of the reflected and transmitted energy are respectively A_1 and A_2 (Figure 4.5), then – assuming no loss of energy along any raypath – the energy within the incident wave must equal the sum of the energy contained within the reflected and transmitted waves (i.e. $A_0 = A_1 + A_2$).

The degree of reflectivity of an interface for normal and low angles ($<20°$) of incidence is described by the *reflection coefficient* (R), which is the ratio of the amplitudes of the reflected wave (A_1) and the incident wave (A_0). Furthermore, the reflection coefficient is also a solution to Zoeppritz's equations, and is given by the ratio of the difference in acoustic impedances to the sum of the impedances (Box 4.6).

It can be seen that the magnitude of reflection coefficient lies in the range of $\leqslant \pm 1$. When a ray passes from a high-velocity medium into one with low velocity, the reflection coefficient is negative and a phase reversal occurs (π or $180°$) within the reflected ray. Typical values of R are between <0.1 (weak reflection), 0.1–0.2 (moderate reflection), >0.2 (strong reflection). In the case of an emergent ray in water impinging upon the water–air interface, such as from a submerged seismic source, the reflection coefficient is -1 and the reflected ray undergoes a phase reversal at the interface. (To calculate this as a check, the values used are: $V_{water} = 1500 \, \text{m/s}$; $\rho_w = 1.0 \, \text{Mg/m}^3$; $V_{air} = 330 \, \text{m/s}$, $\rho_a = 1.293 \, \text{kg/m}^3$ at sea level. As we are dealing here with an emergent ray from water to air, layer one is the water, and layer two is the air.)

The degree of *transmittivity* of an interface for normal and low angles ($<20°$) of incidence is described by the *transmission coefficient* (T), which is the ratio of the amplitudes of the transmitted wave (A_2) and the incident wave (A_0). Furthermore, the transmission coefficient is also a solution to Zoeppritz's equations, and is given by the ratio of twice the acoustic impedance of the first layer to the sum of the impedances (Box 4.6).

The proportions of *energy* that are reflected or transmitted are also referred to as the reflection and transmission coefficients. However, in this case, the forms of the equations are different (see Box 4.6). It should be noted that the sum of the reflected and transmitted energy must equal one.

If the reflection coefficient is ± 1, or $E_R = 1$, then all the incident energy is reflected and none is transmitted (i.e. $T = 0$ and $E_T = 0$). Conversely, if $R = 0$ and $E_R = 0$, then all the incident energy is transmitted ($T = 1$ and $E_T = 1$), suggesting that there is no contrast in acoustic impedance across the interface (i.e. $Z_1 = Z_2$). In such a situation there still may be differences in both velocity and density between the two materials. For example, if the seismic velocities and densities for layers 1 and 2 are $1800 \, \text{m/s}$, $1.6 \, \text{Mg/m}^3$ and $1600 \, \text{m/s}$, $1.8 \, \text{Mg/m}^3$, respectively, there would be no contrast in acoustic impedance as $Z_1 = Z_2 = 2880 \, \text{Mg/m}^2\text{s}$.

Box 4.6   Reflection and transmission coefficients (see Figure 4.5)

For normal and low angles ($<20°$) of incidence:

Reflection coefficient

$$R = A_1/A_0 = (Z_2 - Z_1)/(Z_2 + Z_1)$$
$$R \leqslant \pm 1.$$

———— *continued* ————

--- *continued* ---

Transmission coefficient

$$T = A_2/A_0 = 2Z_1/(Z_2 + Z_1).$$

Z_1 and Z_2 are the acoustic impedances of the first and second layers, respectively. $Z = V\rho$, where V and ρ are the seismic velocity and density of a given layer; A_0, A_1 and A_2 are the relative amplitudes of the incident, reflected and transmitted rays, respectively.

Of the incident energy, the proportions of energy reflected (E_R) and transmitted (E_T) are given by:

Reflected energy

$$E_R = (Z_2 - Z_1)^2/(Z_2 + Z_1)^2$$

Transmitted energy

$$E_T = 4Z_1Z_2/(Z_2 + Z_1)^2$$

Note that $E_R + E_T = 1$.

For the derivation of these formulae, see Telford *et al.* (1990, p. 156).

In the above discussion, it has been assumed that the reflection from an interface arises from a point. In reality, it is generated from a finite area of the reflector surface as defined by the first Fresnel zone (Figure 4.6). The second and subsequent Fresnel zones can be ignored in the case of normal incidence. Effectively, the incident wavefront has a discrete footprint on the reflector surface. The reflection coefficient for a given interface is thus the average response over the first Fresnel zone.

Furthermore, surface roughness of the interface also becomes important if the amplitude of the roughness is of the same order or greater than the quarter-wavelength of the incident wave. The rougher the surface, the more it behaves as a specular reflector returning rays at a very wide variety of angles. The amount of energy reaching the surface is therefore much reduced and the observed reflection coefficient is much less than that predicted for a given contrast in acoustic impedances. The radius (r) of the first Fresnel zone is related to the depth of the reflector below the source (h) and the wavelength of the incident wave (λ) such that $r^2 \approx \lambda h/2$ (Box 4.7). As two-way travel times are considered, the quarter-wavelength is used as opposed to classical optics where only a half-wavelength is used.

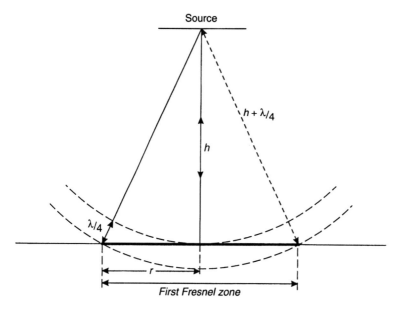

Figure 4.6 The first Fresnel zone on a reflector at a depth h below the source of the incident spherical wave

For a more detailed discussion, see McQuillin *et al.* (1984), Yilmaz (1987), Knapp (1991) and Eaton *et al.* (1991). It is clear from this that the first Fresnel zone becomes larger as a result of increasing depth and decreasing frequency (i.e. larger wavelengths). The effect that this has on horizontal resolution is considered in Chapter 6 for seismic reflection surveying and in Chapter 12 for ground penetrating radar. The determination of Fresnel zones for travel time measurements has been discussed in detail by Hubral *et al.* (1993).

Box 4.7   First Fresnel zone (see Figure 4.6)

The radius (r) of the first Fresnel zone is given by:

$$r^2 = \lambda h/2 + \lambda^2/16 \approx \lambda h/2$$

$$r \approx (\lambda h/2)^{1/2} = (V/2)(t/f)^{1/2}$$

where h is the distance between the source and the reflector, and λ is the wavelength of the incident wave of frequency f, and propagation speed V in the material and a reflector at a two-way travel time t on a seismic section.

4.3.2   Reflection and refraction of obliquely incident rays

In the case of an incident wave impinging obliquely on an interface across which a contrast in acoustic impedance exists, reflected and

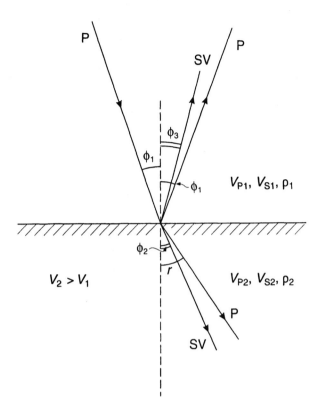

Figure 4.7 Geometry of rays associated with a P-wave (shown as a ray) incident obliquely on a plane interface, and converted vertically polarised S-waves (SV; shown as a ray). V_P and V_S are the respective P- and S-wave velocities and ρ is the density. Suffixes 1 and 2 depict the layer number

transmitted waves are generated as described in the case of normal incidence. At intermediate angles of incidence, reflected S-waves generated by conversion from the incident P-waves (Figure 4.7) may have larger amplitudes than reflected P-waves. This effect is particularly useful in the study of deep reflection events in crustal studies where very large offsets (source–receiver distances) are used.

In general, the P-wave amplitude decreases slightly as the angle of incidence increases. This is equivalent to a decrease in P-wave amplitude with increasing offset. However, there are cases where this does not occur, such as when Poisson's ratio changes markedly, perhaps as a result of gas infilling the pore space within a rock. This phenomenon has been reported by Ostrander (1984) for seismic field records obtained over gas reservoirs and can be used as an indicator of the presence of hydrocarbon gas.

When a P-wave is incident at an oblique angle on a plane surface, four types of waves are generated: reflected and transmitted P-waves and reflected and transmitted S-waves. The relative amplitudes of these various waves are described by Zoeppritz's equations (Telford *et al.* 1990). The direction of travel of the transmitted waves is changed on entry into the new medium, and this change is referred to as *refraction*. The geometry of the various reflected and refracted waves relative to the incident waves is directly analogous to light and can be described using Snell's Laws of refraction (Box 4.8). These state that the incident and refracted rays, and the normal at the point of incidence, all lie in the same plane; for any given pair of media, the ratio of the sine of the angle of incidence to the sine of the angle of refraction is a constant. In its generalised form, Snell's Law also states that for any ray at the point of incidence upon an interface, the ratio of the sine of the angle of incidence to the velocity of propagation within that medium remains a constant, which is known as the *raypath parameter*.

Box 4.8   Laws of reflection and Snell's Laws of refraction (Figure 4.7)

Snell's Laws:

$$\frac{\sin i}{V_{P1}} = \frac{\sin r}{V_{P2}} = \frac{\sin \beta_1}{V_{S1}} = \frac{\sin \beta_2}{V_{S2}} = p$$

where i and r are the angles of incidence and refraction respectively, and V_1 and V_2 are the speeds of propagation in layers 1 and 2 respectively for P- and S-waves as indicated by suffix, and where p is the raypath parameter. Conversely:

$$\frac{\sin i}{\sin r} = \frac{V_1}{V_2}.$$

In the case of critical refraction:

$$\frac{\sin i_c}{\sin 90°} = \frac{V_1}{V_2}$$

Since $\sin 90° = 1$, $\sin i_c = V_1/V_2$, where i_c is the critical angle.

Laws of reflection:

- The angle of incidence equals the angle of reflection.
- The incident, reflected and refracted rays and the normal at the point of incidence all lie in the same plane.

4.3.3   Critical refraction

When the angle of incidence reaches a particular value, known as the critical angle, the angle of refraction becomes 90°. The refracted wave travels along the upper boundary of the lower medium, whose speed of propagation is greater than that of the overlying medium (i.e. $V_2 > V_1$). The material at the interface is subject to an oscillating stress from the passage of the refracted wave, which in turn generates

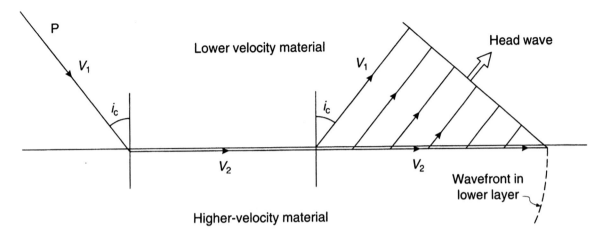

upward moving waves, known as *head waves*, which may eventually reach the surface (Figure 4.8). The orientation of the ray associated with the head wave is also inclined at the critical angle (Figure 4.8). Critical refraction is discussed further in Chapter 5.

Figure 4.8 Critical refraction at a plane boundary and the generation of a head wave

4.3.4 Diffractions

If a wave impinges upon a surface which has an edge to it, such as a faulted layer, then the wavefront bends around the end of the feature and gives rise to a diffracted wave (Figure 4.9). Similarly, boulders etc., whose dimensions are of the same order as the wavelength of the incident signal, can also give rise to diffractions. The curvature of the diffraction tails are a function of the velocity of the host medium

Figure 4.9 Diffracted wavefronts arising from a truncated reflector. The shaded area is a shadow zone where, according to ray theory, no energy should be observed. Huygens' Principle of wavefront generation explains why signals are observed in this shadow zone

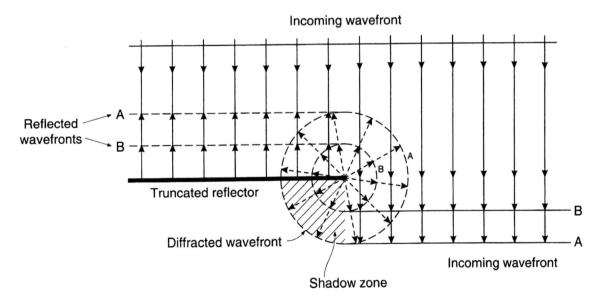

(A)

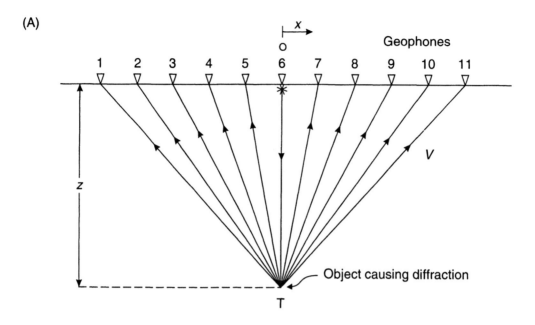

(B)

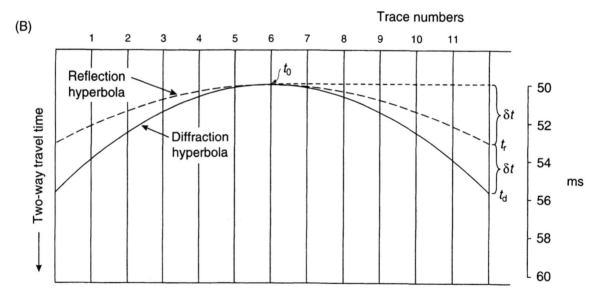

(Figure 4.10). While diffractions are usually considered as noise and attempts are made at resolving them through data processing, they can be used as an interpretational aid (see Chapter 6).

The reason that diffraction occurs is best explained by Huygens' Principle of secondary wavefronts. The object causing the diffraction acts as a secondary source of waves which spread out spherically from that point and can travel into areas where, according to ray theory, there should be no signals observed, such as the shadow zone shown in Figure 4.9. In the case of an isolated cause, such as a boulder, where the shot is located over the source of the diffraction, a hyperbolic travel time response is obtained (Figure 4.10; see also Box 4.9). For comparison, the two-way travel time for a shot–receiver pair with increasing offset (i.e. starting with both at station 6, then shot–receiver at 5 and 7, 4 and 8, 3 and 9, etc.) is given in Figure 4.10.

Whereas a diffraction from a point source in a uniform-velocity field is symmetrical, a diffraction caused by the termination of a reflector undergoes a 180° phase change on either side of a diffracting edge (Trorey 1970; see Figure 4.11).

Figure 4.10 (*opposite*) (A) Raypath geometry for diffracted rays arising from a point target situated beneath a short position; reflections arise from increasing offset between shot–receiver pairs 6 and 6 (coincident), 5 and 7, 4 and 8, etc. Depth, $z = 45$ m, in a material with $V_p = 1800$ m/s. The geophone interval is 5 m. (B) Corresponding time section to illustrate the shape and symmetry of the diffraction as compared with the reflection hyperbola

Box 4.9   Diffraction travel time calculations (see Figure 4.10)

Two-way travel time for a reflected signal, t_r, is given by:

$$t_r = (x^2 + 4z^2)^{1/2}/V \approx 2z/V + x^2/4Vz.$$

For normal incidence, the two-way travel time, t_0 is given by:

$$t_0 = 2z/V \quad \text{and} \quad \delta t = x^2/4Vz.$$

Hence

$$t_r = t_0 + \delta t.$$

Total travel time for a diffracted wave (t_d) with a source at O is given by the sum of the travel time along OT ($= z/V$) and the travel time along any oblique raypath, such as TA ($= (x^2 + z^2)^{1/2}/V$), such that:

$$t_d = z/V + (x^2 + z^2)^{1/2}/V$$
$$= 2z/V + x^2/2Vz = t_0 + 2\delta t.$$

Note that the difference between the two travel times is that t_d is delayed by an extra term δt relative to the reflected arrival.

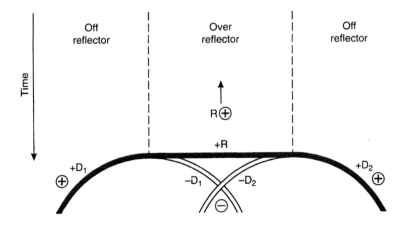

Off
reflector

Over
reflector

Off
reflector

Time

R⊕

+R

+D₁

⊕

−D₁   −D₂

⊖

+D₂

⊕

Figure 4.11 Change of polarity by 180° (from + to −) on either side of a diffracting edge. D_1 and D_2 are diffractions with their polarities indicated; R is a reflection. After Trorey (1970) and Waters (1978)

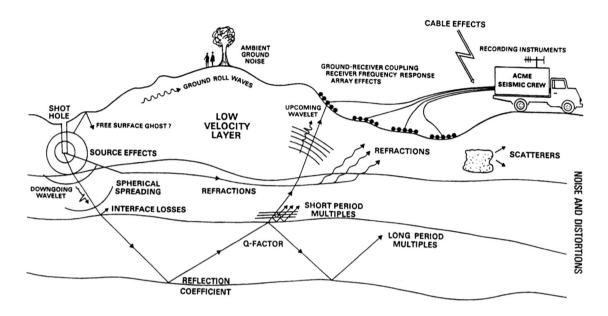

4.4 LOSS OF SEISMIC ENERGY

The loss of amplitude with distance travelled by a seismic wave occurs in three main ways: spherical divergence, intrinsic attenuation, and scattering. These are summarised in Figure 4.12. The amount of transmitted energy also decreases each time a boundary in acoustic impedance is crossed since proportion of energy is reflected (see Section 4.3.1).

4.4.1 Spherical divergence or geometrical spreading

Seismic wave energy propagates radially away from the source and decreases in amplitude with increasing distance. The loss of amplitude can be considered by reference to Figure 4.13. The total energy (E) generated at the shot instant is spread out over a spherical shell with a radius (r) that increases with time. The energy is spread out over the surface of the sphere such that the energy density (i.e. energy per unit area) is $E/4\pi r^2$. Some time later, when the shell has radius R, the energy density is $E/4\pi R^2$. As $R > r$, the energy density is now smaller. The energy thus diminishes in proportion to $1/r^2$. Amplitude, which is proportional to the square-root of the energy density, thus varies in proportion to $1/r$.

4.4.2 Intrinsic attenuation

In addition to spherical divergence, elastic energy is absorbed by the medium by being transferred into heat by the friction of individual

Figure 4.12 Phenomena causing the degradation of a seismic wave. Courtesy of BP

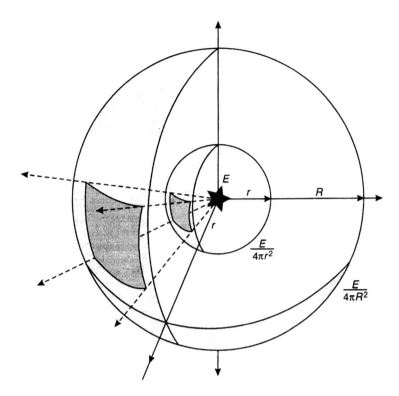

Figure 4.13 The progressive diminution of energy per unit area caused by spherical propagation from an energy source at E

particles moving against each other as the seismic wave passes through the medium, amongst other mechanisms.

The exact processes by which seismic waves are attenuated are still not clearly understood. However, it is known that this energy absorption or intrinsic attenuation decreases exponentially with distance travelled. Attenuation also varies with the type of material through which the wave passes and is characterised by the attenuation coefficient α. If both spherical divergence (the $1/r$ term) and absorption $(\exp(-\alpha r))$ are combined together, the reduction in amplitude with distance is given by the expression in Box 4.10 – in which a further coefficient is introduced, namely the *quality* (or *slowness*) factor (Q) and the *specific dissipation function* Q^{-1}. Q has been used extensively in seismological studies in plate tectonics (Jacobs 1992) and particularly around island arcs. The attenuation coefficient is physically diagnostic of different types of rock. Much attention is now being paid to methods of determining Q directly from seismic data, particularly in shallow seismic surveys. Attenuation data have been traditionally obtained from laboratory measurements using ultrasonic transducers and through VSP experiments. It is through this kind of work that the processes of attenuation are being investigated (e.g. Jones 1986; Klimentos and McCann 1990). One of the outstanding problems which is receiving much attention with respect to hydrocarbon exploration is the effect of clay on the petrophysical properties of sandstones and shales. With the increase in capability of imaging the sub-surface with finer resolution, unconsolidated sediments near the surface should also be studied in more detail.

Box 4.10   Attenuation of seismic waves

For a homogeneous material:

$$\frac{A}{A_0} = \frac{r_0}{r} \exp\{-\alpha(r-r_0)\}$$

where A and A_0 are the amplitudes at distance r and r_0 from the source, respectively; α is the attenuation coefficient which is related to the velocity of elastic waves (V) and their frequency (f) by:

$$\alpha = \pi f/QV \quad \text{and} \quad Q^{-1} = 2\alpha\lambda$$

where Q is the quality factor and λ the wavelength.

For sandstones with porosity ϕ (per cent) and a clay content C (per cent):

$$\alpha = 0.0315\,\phi + 0.241\,C - 0.132 \quad \text{(dB/cm)}$$

and:

$$Q = 179\,C^{-0.843}$$

at 1 MHz and 40 MPa (Klimentos and McCann 1990).

The attenuation coefficient is a measure of the fractional loss of energy per unit distance, and $2\pi/Q$ is the fractional loss per wavelength. Therefore the more cycles that occur, the greater will be the attenuation. From the first equation shown in Box 4.10, it is clear that the attenuation coefficient increases with increasing frequency and so low-frequency waves will be attenuated more slowly than the higher-frequency wave. The effect of this on the shape of a wave pulse is for the higher frequencies to be attenuated faster than the low frequencies, which produces a pulse lengthening with distance travelled (Figure 4.14).

4.4.3   Scattering

Scattering of incident energy is evident as an apparent attenuation that takes place by reflection, refraction and diffraction of seismic waves. There are three levels of scattering that can be described in

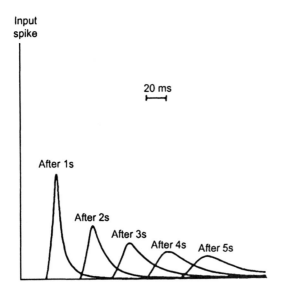

Input
spike

20 ms

After 1s

After 2s

After 3s

After 4s   After 5s

Figure 4.14 The progressive change of shape of an original spike pulse during its propagation through the ground, due to the effects of absorption

terms of the product of the wavenumber ($k = 2\pi f/V$) and scale of the heterogeneity (a).

- When $ka \ll 0.01$, the material through which the seismic waves are travelling are said to be *quasi-homogeneous*, and the scatterers are too small to be seen by the seismic waves except as an apparent anisotropy if enough cracks within the rock are aligned.
- Where the wavelength of the seismic waves is large compared with the heterogeneities ($ka < 0.1$), *Rayleigh scattering* occurs and produces an apparent attenuation. This situation is the most common.
- In areas where there is very rapid variation in seismic wave velocity and density laterally and vertically, such as over old landfill sites, scattering can be very significant and will be evident on the seismic record as a chaotic jumble of signals that appears as noise. In this case, ka lies in the range 0.1 to 10, and the energy dissipation is known as *Mie scattering*.

In high-resolution shallow seismic surveys, sediments such as boulder clay with boulders of the order of 0.5–1 m across are sufficiently large relative to the wavelength of the incident wave for scattering to be a significant cause of amplitude loss. For a survey with seismic waves of frequency 500 Hz, in a medium with P-wave velocity 2000 m/s, boulders of the size given above would yield a value of ka between 0.8 and 1.6, in which case Mie scattering would occur.

Table 4.3   Requirements of a seismic source

Technical:

1. Sufficient energy to generate a measurable signal with a good signal-to-noise ratio
2. Short-duration source pulse (with high enough frequency) for the required resolution
3. A source wave of known shape (or minimum phase)
4. Minimal source-generated noise

Operational:

5. Efficient to use, especially if multiple shots or fast repetition rates are required
6. Repeatable pulse shape
7. Safe to operate and with minimal maintenance
8. To be operated by as few people as possible
9. Reasonably priced both to buy/hire and to use

4.5   SEISMIC ENERGY SOURCES

The prime requirements of a seismic source are listed in Table 4.3. The aim of using any seismic source is to produce a large enough signal into the ground to ensure sufficient depth penetration and high enough resolution to image the sub-surface. There are a large number of different sources that can be used in a wide variety of environments, and a great deal of development has been and is being done to make seismic sources more efficient and effective. Selection of the most appropriate source type for a particular survey is very important. Overviews of different source types have been given by Miller *et al.* (1986, 1992, 1994) with particular reference to engineering and environmental applications.

Generally speaking, there are three types of seismic source: impact, impulsive and vibrator. They can be used on land, in water and down boreholes, and are summarised in Table 4.4. In selecting a source,

Table 4.4   Seismic sources

	On land	On water	
Impact:	Sledge hammer Drop-weight Accelerated weight		
Impulsive:	Dynamite Detonating cord Airgun Shotgun Borehole sparker	Airgun Gas gun Sleeve gun Water gun Steam gun	Pinger Boomer Sparker
Vibrator:	Vibroseis Vibrator plate Rayleigh wave generator	Multipulse GeoChirp	

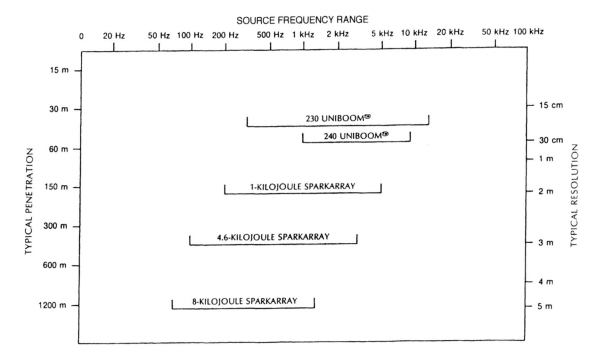

Figure 4.15   Typical penetration and resolution for Uniboom and Sparker sources as a function of their respective frequency bandwidth. Courtesy of EG & G

there is always a trade-off between depth penetration and minimum resolution, which is dependent upon one-quarter wavelength. To achieve good depth penetration requires a low-frequency source but this has a lower resolution. High-resolution shallow seismic surveys require higher-frequency sources and thus have restricted depth penetration. The broad relationship between depth penetration and frequency for several high-resolution water-borne seismic sources is given in Figure 4.15. Detailed discussions of seismic sources and their characteristics have been given by Lugg (1979), Sieck and Self (1977) and Reynolds (1990a), among others.

4.5.1   Impact devices

These are devices where the seismic signal is generated by the impact of anything from a heavy sledge hammer through to a 3 tonne block being dropped from 4 m from the back of a drop-weight truck.

4.5.1.1   *Sledge hammer*

The easiest of all sources must be the humble sledge hammer. A standard sledge hammer is struck against a heavy-duty rubber baseplate. An inertia switch on the hammer handle or a geophone beneath the baseplate triggers the seismograph when the hammer head strikes the plate. The source is easily repeatable, although to obtain similar source pulses the same person should use the hammer for each set of shots. Even with the same weight of hammer head, operators who are not the same height (or even build) swing the hammer differently, so producing source signals that are not exactly reproducible. Furthermore, if care is not taken, the more a baseplate is hit at one location, the more it tends to angle away from the operator, giving some sense of directionality to the signal generated and an increase in groundroll. The hammerer must also try to stop the hammer head from bouncing on the baseplate after the main strike as even slight bounces may produce a tail to the source pulse.

In engineering work, signal enhancement seismographs stack the signal to improve the signal-to-noise ratio. The bad news for the person wielding the hammer is that to improve the signal-to-noise ratio ten times requires the plate to be struck 100 times! The rate of improvement corresponds to the square-root of the number of shots or hits.

The advantage of this source is that it is cheap, causes minimal environmental damage and is extremely easy to use. For shallow seismic work, the hammer source may provide sufficient energy for spreads up to refraction line lengths of over 200 m, and interface depths of 30 m of more, depending on the local geological conditions.

4.5.1.2 Drop-weight devices

The depth penetration obtainable from drop-weight sources is dependent upon the momentum of the mass as it strikes the ground. For lightweight portable systems, a small weight is suspended from an electromagnet at the top of a tube 3 m long. On triggering the seismic recording system, the electromagnet releases the weight causing it to fall and strike a baseplate. The disadvantage with this system is that often the falling weight rattles against the side of the tube during its descent, triggering the recording system prematurely. Also the weight may bounce on the baseplate unless a catcher device is incorporated, giving rise to additional but unwanted signals.

Portable gantry systems can be used where a tower is constructed out of pre-assembled sections to reach up to 5 m high. An electromagnet again is used to release the weight. In this case the weight used can be something like an old gas cylinder filled with concrete. The heavier the drop-weight and the taller the gantry, the less portable the system becomes.

In the seismic industry, special weight-drop trucks (Figure 4.16A) have gantries that can drop a 3 tonne weight from 4 m height routinely with reproducible source characteristics. Often more than one shot is used so that the total source signal is made up of a number of wallops and the received signals are summed. The disadvantages are the relative expense of such a system, and the fact that the source

(A)

Figure 4.16 (A) Weight-drop truck (courtesy of BP). The mass is dropped from the gantry at the rear of the vehicle. (B) Accelerated weight-drop system from Bison Instruments Inc.

(B)

noise is often high, making it only useful in regions, such as deserts, with very low ambient noise levels. Furthermore, with some drop-weights, they fall unevenly so that one edge lands before the rest and so slaps the ground rather than providing a sharp impulse. Consequently, the source shape is often not evenly reproducible or well-constrained. However, the field effort involved in undertaking an extensive programme with a weight-drop truck can be less than that required for shooting with explosives for which pre-drilled holes in which to fire the shots are required.

4.5.1.3   *Accelerated weights*

In order to maintain a lightweight system yet not sacrifice depth penetration, accelerated weight-drop devices have been developed. These use a shorter tower and smaller weight but have a mechanical system to accelerate the weight towards the baseplate rather than allowing it to freefall. The Dinoseis uses compressed air. An alternative propulsion system is the vertical slingshot method employed in the elastic wave generator (Bison Instruments Inc.). Thick elastic slings and gravity accelerate a 250 kg hammer into a steel anvil on the ground surface (Figure 4.16B).

4.5.2   Impulsive sources

A further type of source is one that uses the detonation of a small charge to propel either lead shot, bullets, or wadding into the ground. A common make is the so-called 'Buffalo gun' that comprises a metal pipe inserted vertically up to 1 m into the ground. A shotgun shell located at the lower end is detonated via a firing rod, which protrudes to the surface, being struck by a mallet. In some countries, such as the United Kingdom, and depending on the policy of local police forces, a firearms certificate may be needed before such a device can be used. Cartridges made up with different gunpowder mixes coupled with variable amounts of wadding present will affect the amount of energy discharged with each shot.

A modification of this which does not require a hole to be predrilled is called the Betsy gun™ (Betsy Seisgun) where the entire assembly is housed in a wheelbarrow assembly that is wheeled to the required location and the shot fired on to the ground surface inside a protected chamber (Figure 4.17A). Another source consists of a 0.5 calibre machinegun mounted on the back of a pickup truck with the barrel pointing vertically downwards into a muzzle. Belts of standard ammunition are fed through the gun and fired in bursts. There are restrictions in some parts of the USA on using such weaponry as seismic sources because of the resulting sub-surface lead pollution. For sources of shear waves, mortars and anchored Howitzer artillery guns have been used when firing blanks so as to utilise the recoil to produce S-waves.

A less dramatic and perhaps safer way to generate shear waves is to place a heavy wooden beam on the ground and weight it down using the front wheels of a heavy vehicle. The end of the beam can be struck using a pivoted sledge hammer (Figure 4.17B).

(A)

Figure 4.17 (A) Betsy gun. The bags suspended from the barrel are used to hold live and spent cartridges. (B) Generation of S-waves using a pivoted sledge hammer against a wooden sleeper weighted down by a heavy vehicle

(B)

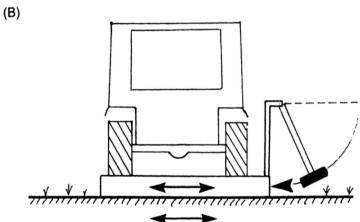

4.5.3 Explosive sources

Explosives are now used in only 40% of land seismic surveys. Once the main source in marine shooting, the high fish kill occasionally caused by the shock wave from the explosion was considered to be environmentally unsatisfactory and has led to its decline in usage. Also, explosions were not identically repeatable nor could the exact time of detonation be obtained with the accuracy required in modern seismic surveying. In land shooting, a single shot can provide enough energy into the ground and with sufficient frequency bandwidth for rapid coverage of survey lines. It is necessary for the shot to be acoustically coupled to the ground. This is best achieved by blasting in water-filled holes. Some shot firers advocate using the nearest river or stream rather than going to the expense or effort of drilling a hole and providing adequate amounts of water to fill it.

The common form of explosive is dynamite or gelignite (ammonium nitrate) which is quite safe when handled properly as it can only be detonated by very high temperatures as achieved by the shock from detonator caps. Handling gelignite causes headaches if the chemicals make contact with the skin. Anyone considering using explosives should attend appropriate safety courses and become fully conversant with the safe use of such material. Although explosives provide a very lightweight and portable high-energy source, the practical problems involved with transporting and storing explosives can be a major disadvantage. So, too, can be the problem of dealing with explosives that have failed to detonate. Liaison with local police is almost universally necessary and in many countries only licensed explosives-handlers are permitted to work as shot firers.

For very shallow work, detonator caps on their own can be used as a source. It is best to use 'zero delay' seismic detonators because other types, as used for example in quarry blasting, have a finite delay of several milliseconds between being triggered and detonating. Detonator caps need to be handled with great care as, although small, they can still blow off fingers or hands. Another form of explosive is detonating cord, such as Primacord™ (Ensign Bickford Co.), Geoflex™ and its marine equivalent, Aquaseis™ (both Imperial Chemical Co.). Explosive cord is ploughed into the ground using special bulldozer rippers and detonated as a line rather than a point source. The explosion starts at a point but ripples along the cord and may be used to trigger other explosives attached along its length. Alternatively, cord can be deployed from hovercraft or other platforms in shallow water-borne surveys.

In marine multichannel seismic surveying, various configurations of detonation of small amounts of explosives are available, such as the Maxipulse™ (Western Geophysical Co.) and Flexotir™ (Compagnie Général de Géophysique) (Figure 4.18). In the Maxipulse system, a small explosive charge (Figure 4.18A) of nitrocarbonitrate is flushed down a hose to the trigger mechanism. The charge strikes its detonator on a firing wheel as the charge is ejected and triggers a delay fuse which detonates the main charge at between 7 and 15 m depth after a period sufficient for the gun to have been towed from the vicinity. The ensuing bubble pulse train has to be removed by specialised processing.

(A)

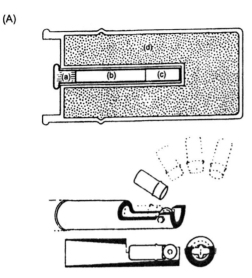

Figure 4.18   (A) The top shows a schematic cross-section of the Superseis charge used in the Maxipulse system: (a) = rim fire percussion cap; (b) = delay column; (c) = booster; and (d) = nitrocarbonitrate main charge. Underneath is shown the Maxipulse gun. A perspective view of the gun shows the charge striking the firing wheel, then successive stages of ejection. The cross-section and end views show the charge in the gun at the instant of striking the firing wheel. Courtesy of Western Geophysical; McQuillin *et al.* (1984) with kind permission from Kluwer Academic Publishers. (B) Flexotir system. After Sheriff (1991)

(B)

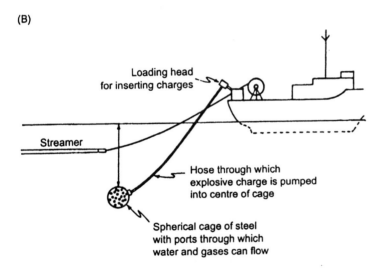

Loading head
for inserting charges

Streamer

Hose through which
explosive charge is pumped
into centre of cage

Spherical cage of steel
with ports through which
water and gases can flow

In the Flexotir system, a very small amount of charge (typically about 60 g) is pumped down a flexible hose by water under pressure into a steel cage rather like a spherical colander in which the charges are detonated (Figure 4.18B). Water expelled by the exlosion flows in and out through the holes in the cage.

4.5.4   Non-explosive sources

All the non-explosive marine sources (cf. Table 4.4) give rise to varying degrees of a bubble pulse. When the source is discharged,

(A)

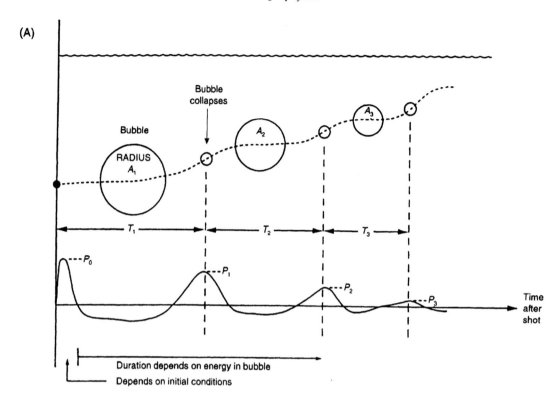

(B)

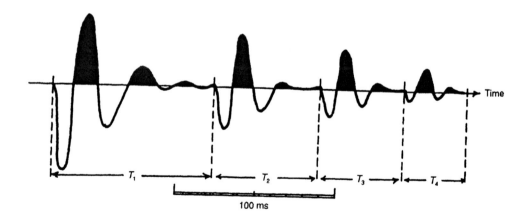

cavitation occurs when the surrounding water is displaced by the formation of a gas bubble, usually of air. Hydrostatic pressure causes the bubbles to collapse, so pressurising the gas which heats up and expands to reform smaller bubbles, which too, in turn, collapse and reform (Figure 4.19A), until such time that the hydrostatic pressure is too great and the remaining gas dissolves in the water or finally vents to the surface. The acoustic noise from the successive collapses of the bubbles produces noise which is undesirable (Figure 4.19B) and which can either be removed by later processing or is minimised using a tuned array of sources. Airguns (see next section) with different volume capacities are fired simultaneously but the respective bubble pulses occur at different times and with varying amplitudes. By summing all the individual signatures of an array, a much more acceptable overall source signature can be produced (Figure 4.20). Bubble pulses can also be reduced by using a waveshape kit within each airgun, for example. Air is bled under pressure into the expanding bubble, thereby reducing the rate at which it collapses with a consequent reduction in the overall length and severity of the bubble train. The effectiveness of a marine source can be measured in terms of the ratio of the amplitudes of the primary to first bubble signal.

4.5.4.1 Airguns and sleeve guns

An airgun is used to inject a bubble of highly compressed air into the surrounding water and is the most commonly used of all seismic sources (Figure 4.21). Airguns, although usually deployed in marine surveys, can also be used in modified form in marshlands, in a land airgun, and as a borehole source. Compressed air is fed through a control chamber (Figure 4.22A) into a lower main chamber and in so doing a shuttle is depressed, closing off the external ports. Opening a solenoid valve (Figure 4.22B) releases the sealing pressure in the upper control chamber and the shuttle moves rapidly upwards, so releasing the high-pressure air in the lower chamber to vent explosively through the ports (Figure 4.22C and D). The shape of the source pulse is dependent upon the volume of air discharged, its pressure, and the depth at which the gun is discharged. It has been found (Langhammer and Landrø 1993) that airgun signatures are also affected by the temperature of the surrounding water. The primary-to-bubble ratio and the bubble time period both increase with increasing water temperature.

Airguns range in volume capacity (lower chamber) from a few cubic inches to around 2000 cubic inches ($0.033\,\mathrm{m}^3$), and pressures from 2000 to 4000 pounds per square inch ($1400–2800\,\mathrm{Mg/m}^2$). Older airguns had 4 or 6 cylindrical holes through which air was expelled. More recently, *sleeve guns* have been developed which allow the air to escape through an annulus as a doughnut-shaped bubble

Figure 4.19 (*opposite*) (A) Archetypal marine source showing the response of the bubble from the point of forming from the firing of the source through to venting at the water surface. The respective bubble pulse periods (T) and amplitudes (P) are indicated as a function of time. Courtesy of BP. (B) The corresponding bubble pulse train

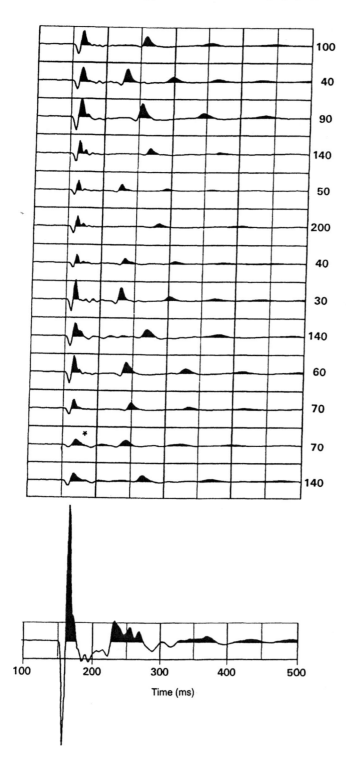

Gun volume (inches³)

100
40
90
140
50
200
40
30
140
60
70
70
140

Time (ms)

Figure 4.20 Individual airgun signatures and the simulated array signature produced by summing them. The asterisk indicates a malfunction. Courtesy of BP

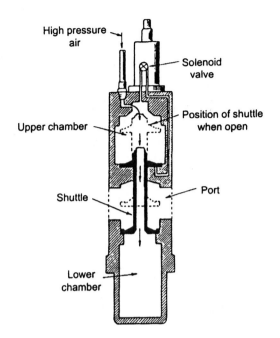

Figure 4.21 Schematic cross-section through an airgun. High-pressure air flows continuously into the upper chamber and through the shuttle into the lower chamber. Opening the solenoid valve puts high-pressure air under the upper shuttle seat, causing the shuttle to move upward, opening the lower chamber and allowing its air to discharge through the ports to form a bubble of high-pressure air in the surrounding water. The size of an air-gun is indicated by the volume in cubic inches of its lower chamber (Sheriff 1991). Courtesy of Bolt Associates

(Figure 4.23) and the effect is to reduce the bubble pulse. When the gun is fired, instead of an internal shuttle moving, an external sleeve moves.

4.5.4.2 *Water guns*

Instead of expelling compressed air into the surrounding water, which often has had deleterious effects on local fish stocks, air is used to force a piston forward which discharges a slug of water. When the piston stops, cavitation occurs behind the expelled water, resulting in an implosion which creates the seismic pulse. The firing sequence is outlined in Figure 4.24. The major advantage of a water gun is that no real bubble pulse is produced.

4.5.4.3 *Gas guns/sleeve exploders*

A mixture of butane or propane with oxygen or air is exploded under water using a gas gun. Alternatively, the mixture is detonated inside a tough rubber bag which expands to contain the explosion, and then collapses creating an implosion which is the main source pulse. The exhaust gases are vented from the sleeve, up hosing to the surface, so that there is no bubble pulse. A major trade name is Aquapulse™ (Esso Production Research) and the system is used under licence by several seismic contractors.

(A)   CHARGED

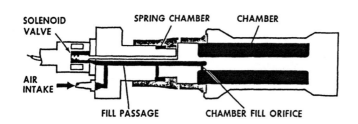

Figure 4.22   The operation of a sleeve gun is very similar to that of an airgun. Instead of a shuttle moving to release the compressed air, an outer sleeve is used instead. (A) The sleeve gun is charged ready to fire; (B) fired; (C) the lower chamber discharges its compressed air into the surrounding water, until (D) the lower chamber is exhausted of air

(B)   FIRED

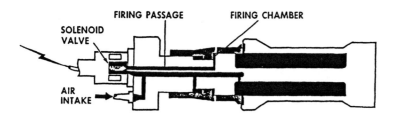

(C)   EXHAUSTING

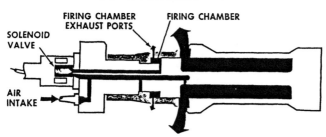

(D)   EXHAUSTED

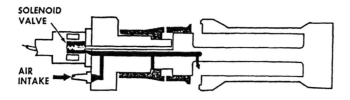

(A)

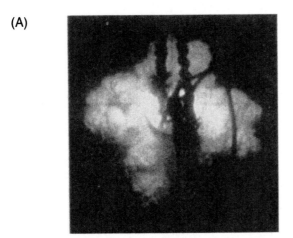

(B)

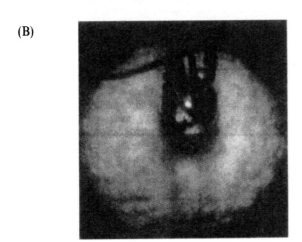

Figure 4.23 The exhaust bubbles from (A) a conventional airgun, and (B) a sleeve gun. Courtesy of Texas Instruments

4.5.4.4   *Steam gun and Starjet*

Compagnie Générale de Géophysique (CGG) has developed a steam gun under the name Vaporchoc™. Steam generated on board ship is fed into the water through a remotely controlled valve (Figure 4.25A). This releases a bubble of steam, causing a small amount of acoustic noise which precedes the main seismic pulse. This pre-emptive noise is known as the *source precursor*. On closing the valve, steam in the bubble condenses and the internal gas pressure drops to less than hydrostatic pressure, so causing the bubble to implode radiating acoustic energy with negligible bubble pulse. However, to overcome the problems over the precursor signal, CGG has developed the Vaporchoc principle in a system known as Starjet, which uses four systems each of which has four tunable guns. By varying the timing of the discharge of each of these systems, the precursors can be

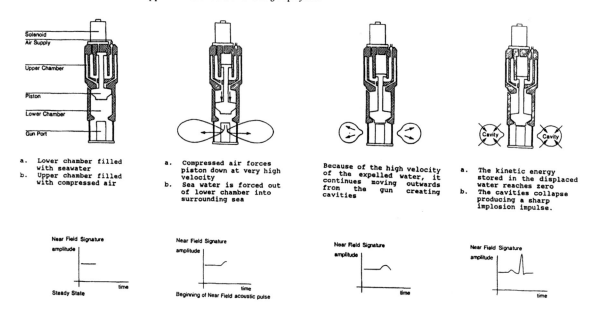

manipulated to become self-cancelling. The overall source pulse shape is far cleaner with minimum ringing, as can be seen in Figure 4.25B.

4.5.5 High-resolution water-borne sources

In the last few years an increasing amount of attention has been paid to high-resolution water-borne surveys using high-frequency sources. The types of applications for such surveys are listed in Table 4.5. There are three main types of source in this category, namely pingers, Boomers™ (EG & G) and sparkers (Figure 4.26), each with a different frequency bandwidth, resolution and depth penetration (see Table 4.6).

A *pinger* has the highest frequency of the three types and greatest resolution but least penetration. A *boomer* consists of a capacitor bank that discharges through a transducer which consists of two flat parallel plates. Eddy currents are established in the plates, forcing them suddenly apart (by magnetic repulsion) to produce a low-pressure region between the plates into which surrounding water rushes, creating an implosive pressure wave. To Boomer™ is a surface tow device which is very portable (liftable by two people) and can easily be operated out of relatively small vessels, making it ideal for inshore work. A deep-tow version can be used for deep-water applications. The energy source is deployed close to the seabed to maximise the energy transmission into the substrate.

A *sparker* consists of a capacitor bank that discharges through a submerged cage which houses a number of electrodes (3–300). An

Figure 4.24 The operation of a water gun: (A) charged ready to fire; (B) fired; (C) discharge of water from the gun creates cavities surrounding the gun; (D) the cavities collapse producing a sharp implosion impulse. Courtesy of SSL

(A)

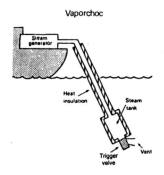

Vaporchoc

Steam generator

Heat insulation

Steam tank

Trigger valve

Vent

Figure 4.25 (A) The Vaporchoc system. (B) The use of Starjet to modify the source characteristics

TUNING THE GUNS WITHIN ONE SUBARRAY

(B)

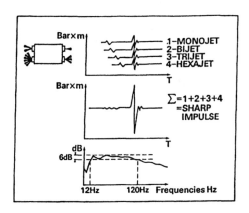

Bar×m

.1–MONOJET
2–BIJET
3–TRIJET
4–HEXAJET

T

Bar×m

Σ=1+2+3+4
=SHARP
IMPULSE

T

dB

6dB

12Hz   120Hz Frequencies Hz

REDUCING THE EFFECTS OF THE FORERUNNERS WITHIN ONE SUBARRAY

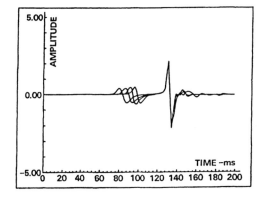

5.00

AMPLITUDE

0.00

TIME –ms

−5.00

0   20 40 60 80 100 120 140 160 180 200

Table 4.5 Principal applications of commercial water-borne site investigation using high-resolution seismic methods

---

Near/inshore marine/estuarine environments:

    Bridges, tunnels, viaducts
    Harbours, jetties, quay walls, marinas, canals
    Pipelines and sewage outfalls
    Dredging for access channels to ports/harbours

Marine:

    Hydrocarbon pipelines
    Hydrocarbon production platforms/wellheads
    Siting of drilling rigs
    Shallow gas surveys
    Sand and gravel resources
    Dredge and contaminated spoil-dump surveys

---

electric arc is created between the electrode tips and the frame, so ionising seawater to produce hydrogen bubbles that then implode, creating the pressure pulse.

Of the three types of sources, sparkers undoubtedly have the most variable source characteristics. The source pulse shape depends critically on the output power of the system and the number and configuration of the electrodes in the sparker array (Figure 4.27; Reynolds 1990a). Better pulse shapes are achieved by increasing the number of electrode tips. During a survey some electrode tips may burn back to the insulation and cease to work, thereby reducing the number of active tips and the effective output – both of which serve to degrade the pulse shape. It is important, therefore, to check on the state of the electrode tips during each day's survey and to monitor signal quality. In addition, the depth of the frame below the water surface also affects the pulse shape by virtue of the interference of the surface ghost and bubble pulse.

4.5.6 Vibrators

4.5.6.1 Vibroseis

On land, an alternative to explosive or impulsive sources is the use of a vibrating plate to generate a controlled wavetrain. A system developed by Conoco, known as Vibroseis – and described in detail by Baeten and Ziolkowski (1990) – consists of a vibrator plate mounted on the underside of a special truck. When on location the plate is set on the ground surface and the truck is jacked up so that its weight is transferred to the plate (Figure 4.28). A low-amplitude sinusoidal vibration of continuously varying frequency (between 60 and 235 Hz) is applied over a *sweep period* which lasts between 7 and 60 seconds.

(A)

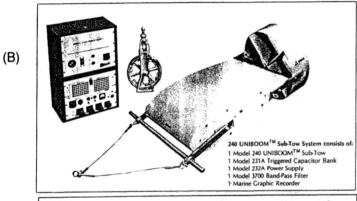

230 UNIBOOM™ System consists of:
1 Model 230-1 UNIBOOM™ Sound Source
1 Model 234 Energy Source
1 Model 265 or 262 Hydrophone
1 Model 3700 Band-Pass Filter
1 Marine Graphic Recorder

(B)

240 UNIBOOM™ Sub-Tow System consists of:
1 Model 240 UNIBOOM™ Sub-Tow
1 Model 231A Triggered Capacitor Bank
1 Model 232A Power Supply
1 Model 3700 Band-Pass Filter
1 Marine Graphic Recorder

(C)

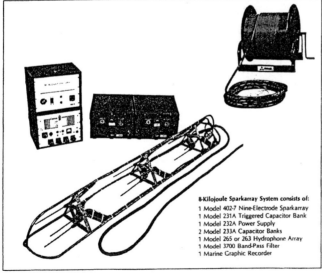

8-Kilojoule Sparkarray System consists of:
1 Model 402-7 Nine-Electrode Sparkarray
1 Model 231A Triggered Capacitor Bank
1 Model 232A Power Supply
2 Model 233A Capacitor Banks
1 Model 265 or 263 Hydrophone Array
1 Model 3700 Band-Pass Filter
1 Marine Graphic Recorder

Figure 4.26 Three high-resolution seismic sources: (A) Uniboom™ surface tow catamaran with short streamer; (B) Univoom™ sub-tow fish; and (C) a 9-tip sparker system. Courtesy of EG and G

Table 4.6   Theoretical resolution and depth penetration of three common high-frequency source types

Source	Frequency bandwidth (kHz)	Resoloution (m)	Depth of penetration (m)
Pingers	3.5–7	0.1–1	$\leqslant$ tens
Boomers	0.4–5	≈ 1	tens to 100 +
Sparkers	0.2–1.5	2–3	$\geqslant 1000$

(A)   (B)

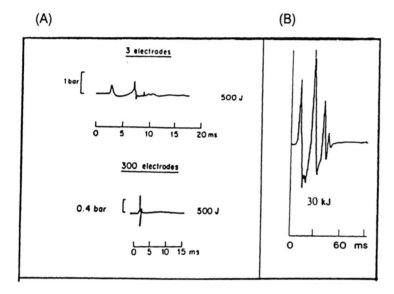

Figure 4.27 Sparker source pulse shapes with (A) 3 and 300 electrode tips with 500 J output (Lugg 1979) and (B) a far-field signature for a system with 30 kJ output (Hatton *et al* 1986)

Progressively increasing frequencies are used in *upsweeps* and progressively decreasing frequencies with time are used in *downsweeps*. Usually the change of frequency is linear, although nonlinear sweeps may be used where higher frequencies are used for longer to compensate for loss of high-frequency information through the propagation of the signal. The resulting field record is the superposition of the reflected wavetrains (Figure 4.29) and is correlated with the known source sequence (pilot sweep). At each sample point along each trace the pilot is cross-multiplied with the superimposed signals to produce a *correlogram* trace. When the pilot sweep matches its reflected wavetrain (i.e. autocorrelates) it produces a significant zero-phase wavelet (known as a *Klauder wavelet*) on the trace (Figure 4.29). The display of adjacent correlogram traces produces a seismic section or correlogram which resembles its conventional seismic counterpart obtained using a high-energy impulsive source.

   A major advantage of the Vibroseis method is that it is a rapid and easily repeatable source that is convenient to use, especially in urban areas. The cross-correlation data-reduction method enables the ex-

Figure 4.28 (A) A Vibroseis truck. The weight of the vehicle is transferred to the jack in the middle of the vehicle when operating (Courtesy of SSL) (B) Schematic of a Vibroseis truck. Courtesy of BP

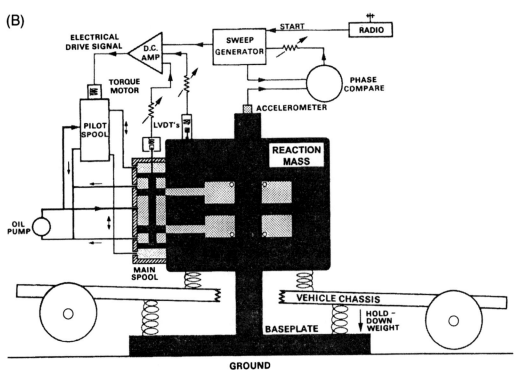

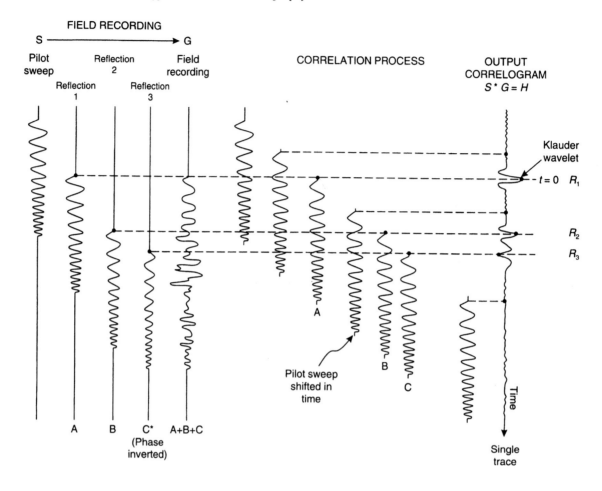

Figure 4.29 Schematic to illustrate the process of generating a Vibroseis pilot sweep, acquiring field recordings and correlating the field record with the pilot sweep to obtain the output correlogram with Klauder wavelets

traction of sensible signals even in areas with high cultural noise. Furthermore, the Vibroseis method does not damage the ground surface and thus can be used with care on macadamised roads and over public utilities.

The Vibroseis method is best deployed over firm ground such as made-up roads as it requires good ground coupling. It does not work very well on soft ground. In order to increase the energy input into the ground for greater depth penetration, a number of vibrators can be used simultaneously in a group provided they are phase-locked, i.e. they all use the same sweep at exactly the same time. Multiple sweeps can be used and summed (stacked) to improve the signal-to-noise ratio.

4.5.6.2   *Small-scale land vibrators*

On a much reduced scale, the Vibroseis principle can be used in the form of a number of small vibrator sources. One of these is the Mini-

Source sensor

Figure 4.30 Mini-Sosie land vibrator source; a signature recording sensor is mounted on the top side of the baseplate. Courtesy of Société Nationale Elf Aquitaine (SNEA)

Sosie source that consists of a pneumatic hammer impacting on a baseplate (Figure 4.30). The thumper delivers a random series of blows to the baseplate on which a detector is located which records the source sweep. The variable-frequency source wavetrain with low amplitude is transmitted into the ground. The detected signals are cross-correlated with the recorded pilot sweep to produce correlograms. The source can produce 10 pops/second and several hundred pops are summed at each shot point. The frequency range of this method is higher than for dynamite, thereby providing higher resolution (as can be seen in the comparison shown in Figure 4.31).

For a shallow hydrocarbon exploration, mining or geotechnical surveys, a number of small land vibrators were developed during the early 1990s. One of these has been described by Christensen (1992). His device consists of a 0.91 m diameter baseplate on which a 136 kg mass is located (Figure 4.32). A small engine drives a hydraulic pump which energises the vibrator over a frequency bandwidth of

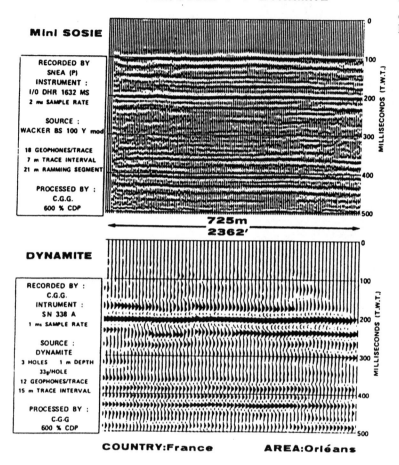

COMPARISON TEST

Mini SOSIE ◄───► DYNAMITE

Mini SOSIE

RECORDED BY
SNEA (P)
INSTRUMENT :
I/O DHR 1632 MS
2 ms SAMPLE RATE

SOURCE :
WACKER BS 100 Y mod

18 GEOPHONES/TRACE
7 m TRACE INTERVAL
21 m RAMMING SEGMENT

PROCESSED BY :
C.G.G.
600 % CDP

725m
2362'

DYNAMITE

RECORDED BY :
C.G.G.
INTRUMENT :
SN 338 A
1 ms SAMPLE RATE

SOURCE :
DYNAMITE
3 HOLES   1 m DEPTH
33g/HOLE
12 GEOPHONES/TRACE
15 m TRACE INTERVAL

PROCESSED BY :
C.G.G
600 % CDP

COUNTRY:France   AREA:Orléans

Figure 4.31 Comparison test between the use of Mini-Sosie and Conventional dynamite as a seismic source over the same ground. From McQuillin *et al.* (1984), by permission

10–550 Hz at 4.4 kN output. The peak force is 30.5 kN. By slanting the mass-actuator through 45° (Figure 4.32), both P- and S-waves can be generated. A PC-based controller drives the vibrator using open-loop amplitude and phase control. It also measures the output of the vibrator, utilising appropriate sensors, and records this information for signal correlation purposes.

It is becoming increasingly important to measure *in situ* stiffness of the ground rather than relying solely on measurements made on samples in a laboratory. One way to achieve this is to use a Rayleigh wave generator. A mass is suspended in a small gantry (Figure 4.33) that can be excited by a signal oscillator. The inertia of the mass is such that the vibrations that are in a horizontal plane are used to strain the ground by generating Rayleigh waves (Abbiss 1981; Powell and Butcher 1991).

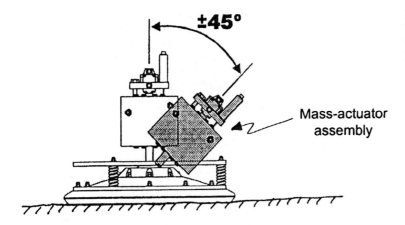

Figure 4.32 Land vibrator that can be used to generate both P- and S-waves. From Christensen (1992), by permission

4.5.6.3 Marine vibrators

In 1988, Western Geophysical introduced the Multipulse™ (Hydroacoustics Inc.) marine vibrator which operates over the frequency range 5–100 Hz and has a sweep period of 6 seconds. The advantages of this system over traditional airgun arrays are.the minimal

Figure 4.33 Field layout of a ground stiffness test using a Rayleigh wave vibrator. From Abbiss (1981), by permission

Figure 4.34 Multipulse marine vibrator showing the shipboard installation. Courtesy of Western Geophysical

disturbance to marine life due to reduced cavitation; a smaller radius of interference away from the source so that it can be operated near production platforms without affecting diving activities; and particularly, that the source characteristics are extremely well known and controlled. The vibrator plates are powered by a pneumatic-nitrogen supply and compressed air, and are deployed in a flotation gantry. The system configuration is illustrated in Figure 4.34 and sample records are shown in Figure 4.35. Marine vibrators have been discussed in more detail by Baeten *et al.* (1988).

For smaller scale surveys, a new system has been developed by GeoAcoustics Ltd, UK, called the GeoChirp. The system can be deployed in either an over-the-side mounted assembly or as a towed fish (Figure 4.36). The instrument consists of up to four source transducers with frequency bandwidths of either 2–8 kHz or 1.5–11.5 kHz. An upsweep of frequencies (the 'chirp') is transmitted in a pulse lasting either 16 or 32 ms and repeated 4 or 8 times a second.

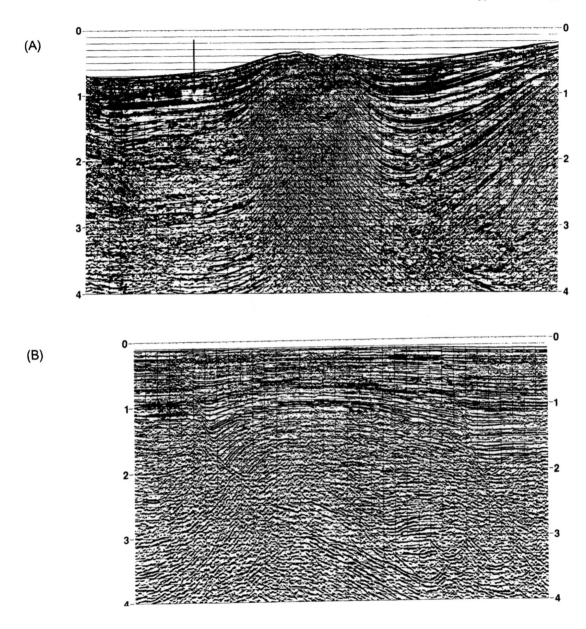

Figure 4.35 Multipulse migrated sections. (A) 5–80 Hz sweep at 6 dB/octave; the section shows a bright spot shadow at 1 s (arrowed), probably caused by shallow gas. (B) 7.5–100 Hz sweep at 6 dB/octave; the section shows well-resolved faults

Immediately behind the source mounting (which looks like the head of a tadpole) is located an 8-element streamer that is about 1 m long. As the source transducer characteristics have been measured by the manufacturer, the system is programmed with these characteristics in order to provide an autocorrelated 'de-chirped' analogue output. This can be printed directly on to a hard-copy device, such as a thermal linescan printer. Alternatively, the analogue signals can be put through a sonar enhancement processing unit that provides

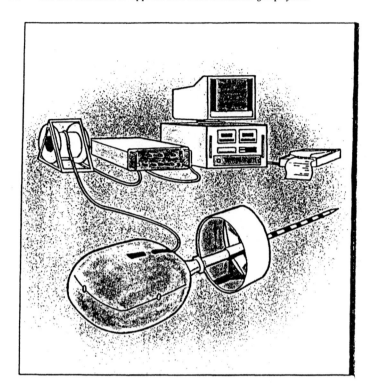

Figure 4.36 GeoChirp configuration with the transmitting transducers mounted in the tadpole head and a short streamer aft. Courtesy of GeoAcoustics Ltd

facilities for real-time filtering, gain control, etc., and then printed out as required.

4.6   DETECTION AND RECORDING OF SEISMIC WAVES

Seismic surveys would not be possible without sensors to detect the returned signals. These detectors are called *geophones* and are used on a substrate of some kind, normally the ground surface, or down boreholes. They are used to convert seismic energy into a measurable electrical voltage. A special form of geophone is the *accelerometer* which, as its name suggests, is used to measure acceleration. The water-borne equivalent of a geophone is a *hydrophone* and these, too, can be used down water-filled boreholes.

4.6.1   Geophones and accelerometers

Most geophones, which are also known as jugs (and people who implant geophones are known as 'jug-handlers' or 'juggies'), are of the 'moving-coil' type. A cylindrical coil is suspended by a leaf-spring in a magnetic field provided by a small permanent magnet which is fastened to the geophone casing (Figure 4.37). By suspending the coil from a spring, an oscillatory system is created with a resonant frequency dependent upon the mass of the spring and the stiffness of the suspension. The geophone is implanted into the ground with a spike attached to the base of the casing (Figure 4.37A) to ensure good ground coupling. Shear-wave geophones are slightly different in that they can have two spikes mounted side-by-side. Typical geophone construction is shown in Figure 4.37B. Some geophones are used as transducers to monitor vibrations in heavy engineering machinery but these work on exactly the same principle as their geological counterparts.

The passage of a seismic wave at the surface causes a physical displacement of the ground which moves the geophone case and magnet in sympathy with the ground but relative to the coil because of its inertia. This relative movement of the magnet with respect to the coil results in a small voltage being generated across the terminals of the coil in proportion to the relative velocity of the two components (above the natural frequency of the geophone). Geophones thus respond to the *rate* of movement of the ground (i.e. particle velocity), not to the *amount* of movement or displacement. The maximum sensitivity of any geophone occurs when the coil axis is parallel to the direction of the maximum ground movement. As the incident reflected and refracted P-waves are usually steeply orientated with respect to the ground surface and produce a vertical displacement of the

Figure 4.37 (A) Field geophone with spike. (B) Typical geophone construction

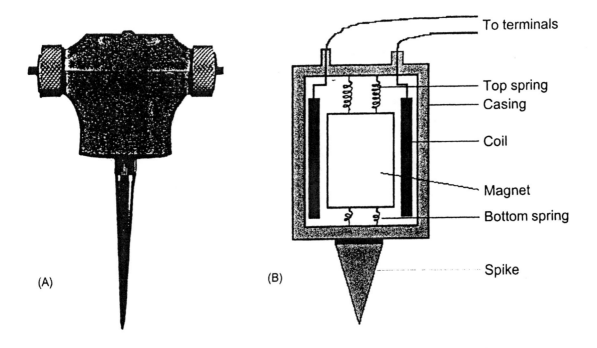

To terminals

Top spring
Casing

Coil

Magnet
Bottom spring

Spike

(A)

(B)

ground, geophones with vertical coil orientations are deployed to detect them. Similarly, as horizontally polarised shear waves tend to produce ground displacements in the plane of the surface, the coil assembly is mounted horizontally in shear-wave geophones. Deployment of geophones with a specifically orientated coil axis can filter out orthogonal responses. If either P- or S-wave geophones are planted crookedly, they are designed to stop working and give a 'dead trace'. This condition is usually looked for at the beginning of any survey and the problem rectified by replanting the misaligned geophone.

Geophones need to respond quickly to the arrival of seismic waves but should not ring as this would affect their ability to sense other seismic arrivals. Thus geophones need to be damped so that, after their initial response to ground movement, the relative movement of the coil and casing stop quickly ready for the arrival of the next event. *Critical damping* (μ_c) is the minimum amount required which will stop any oscillation of the system from occurring (Figure 4.38A). Geophones are inherently damped as the oscillatory movement of the coil is retarded by the relative movement of the coil and magnet. Current generated in the coil by the initial relative movement of the permanent magnet induces a magnetic field which interacts with that of the permanent magnet in such a way as to oppose the motion of the coil. This damping can be changed by the addition of a shunt resistance across the coil terminals. This resistance controls the amount of current in the coil; the lower the resistance the greater is the degree of damping. Most geophones are slightly underdamped, typically around 0.6–0.66 μ_c (Figure 4.38B). Geophones are designed to respond to different frequencies with the common natural frequencies being in the range 4–30 Hz for seismic refraction and deep reflection surveys. For shallow high-resolution reflection surveys, geophones with a natural frequency greater than or equal to 100 Hz should be used. Near and below the natural frequency, the geophone response is heavily attenuated and for most seismic work this helps to filter out unwanted very-low-frequency noise. Well above the natural frequency, the geophone response is usually flat (Figure 4.38B and C).

An accelerometer is a device whose output is proportional to the acceleration to which it is subjected. A geophone or seismometer whose response is below its natural frequency can be used as an accelerometer. They are used in inertial navigation systems and with shipboard and airborne gravimeter systems to record variations in acceleration which are necessary for the reduction of the gravity data (see Section 2.5.7).

Seismometers are devices used to measure the waveform of earthquake-type waves and are normally installed in specially engineered recording stations. Extremely good ground coupling is achieved by bolting the device to bedrock or to a concrete plinth which is in contact with bedrock. Multicomponent seismometers record in two horizontal but orthogonal directions and vertically. The output goes

(A)

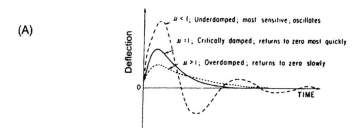

(B)

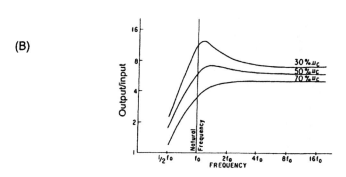

(C)

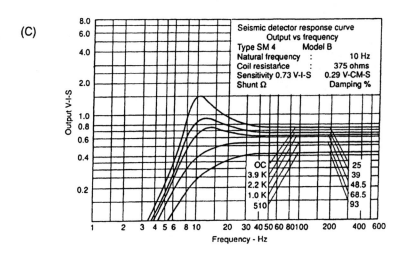

Figure 4.38 (A) Deflection as a function of time for underdamped, critically damped and overdamped modes of a geophone. (B) The amplitude response as a function of frequency for three states of underdamping relative to critical damping (μ_c). (C) Variation of response with frequency for a typical moving-coil geophone. Courtesy of Sensor Nederland BV

to three channels recording simultaneously. Further details of the operating principles of seismometers have been given by Howell (1990).

The earliest known seismometer or seismoscope was built in China in the year 132 by Heng Zhang, Imperial Historian, and represents

not only a remarkable piece of early engineering but also a beautiful work of art (Figure 4.39). The device consisted of a hollow dome in which was suspended a pillar. Eight cups in the shape of dragons' heads were arranged symmetrically around the upper periphery. In the mouth of each dragon was a delicately poised ball. Immediately below each dragon's head was an ornamental frog with an open mouth pointing upwards. When the instrument was disturbed, ostensibly by an earthquake wave, the suspended pillar swayed in the direction of motion and triggered a mechanism which knocked the ball out of the nearest dragon's head and into the mouth of the frog beneath (Wartnaby 1957). It is known that the device registered an earthquake in Gansu Province in China in 138.

Figure 4.39   The earliest known seismoscope was developed in 132 by Heng Zhang and was used successfully to indicate the occurrence of an earthquake in the year 138. Courtesy of *Stop Disasters* and the World Intellectual Property Organization

(A)

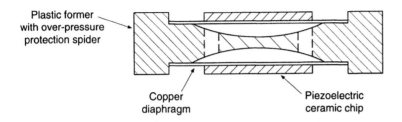

Plastic former
with over-pressure
protection spider

Copper
diaphragm

Piezoelectric
ceramic chip

(B)

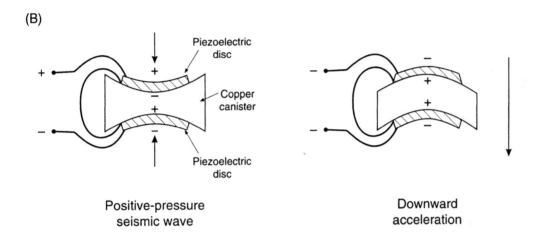

Piezoelectric
disc

Copper
canister

Piezoelectric
disc

Positive-pressure
seismic wave

Downward
acceleration

4.6.2 Hydrophones and streamers

A hydrophone is in essence a water-borne equivalent of a geophone, although the principle of operation is quite different. A hydrophone responds to variations in pressure, whereas a geophone responds to ground particle motion. Hydrophones can be used in open water, down a water-filled borehole, and in water-saturated marshy conditions.

A hydrophone consists of two piezoelectric ceramic discs (e.g. barium titanate or lead zirconate) cemented to a hollow sealed copper or brass canister (Figure 4.40A). A pressure wave effectively squeezes the canister and bends the piezoelectric disc, thus generating a voltage between the top and bottom of each disc. The two discs are polarised and connected in series so that the voltage generated by the passage of a pressure wave (i.e. seismic wave) are added but those due to acceleration will cancel (Figure 4.40B). Piezoelectric hydrophones have a high electrical impedance and thus any signals must pass through impedance-matching transformers before being transmitted through to the recording instruments.

Figure 4.40 (A) Hydrophone construction, and (B) acceleration-cancelling hydrophone. Courtesy of SSL

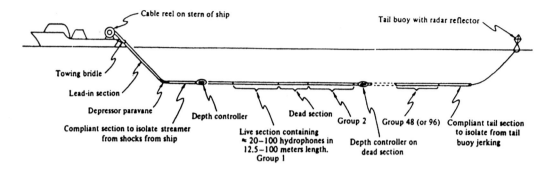

In marine work, hydrophones are deployed in a *streamer*, a marine cable up to 6 km long, which is designed to be towed continuously through the water. A streamer is made up of a number of elements (Figure 4.41). The main section comprises active or live sections (groups) in each of which 30 hydrophones are connected in parallel with impedance-matching transformers. In multiplexing cables, the signals from each hydrophone group are amplified, multiplexed, and converted into digital form within the cable, and then transmitted to the recording vessel along a single wire. The individual hydrophones are acoustically coupled to the surrounding water by being immersed in oil, which also assists in keeping the streamer neutrally buoyant. The streamer skin, which is kept from collapsing by plastic bulkheads, is made either of PVC for use in warm water, or of polyurethane for use in cold water. If the wrong skin is used – for example, if a warm-water skin is used in cold water – the streamer stiffens, which generates acoustic noise as the cable is towed through the water.

From the front end of the streamer, the first section encountered is the towing bridle which takes the strain of towing and connects the hydrophone communication cables to the ship's recording systems. At the rear end of the lead-in section is a depressor hydrovane (paravane) which is used to keep the nose of the streamer at a given level below the surface. Immediately aft of this is a compliant or stretch section which is designed to absorb and attenuate the jerks caused by uneven tow rates and/or sea conditions, and to isolate the streamer from shocks to the ship (such as ploughing through a swell). Another stretch section is located right at the end of the streamer where it is connected to the tail buoy, which is a polystyrene raft housing a radar reflector and radio antenna, and is used to isolate the streamer from jerks from uneven movement of the tail buoy.

Along the line of the streamer, depth controllers are located at specific points. These devices have servo-controlled fins which change their aspect in order to keep the streamer at a predefined depth below the surface. Pressure sensors on each controller are used to measure

Figure 4.41 Basic structure of a hydrophone streamer. From Sheriff (1991), by permission

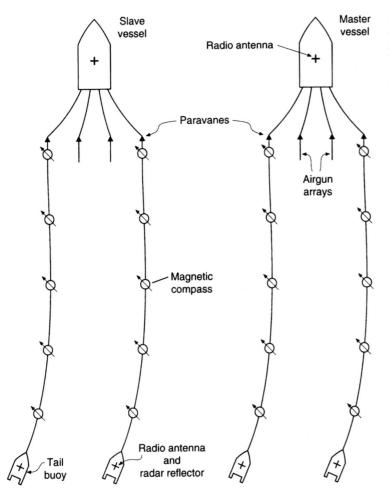

Figure 4.42 Three-dimensional multi-streamer deployment. Each streamer has a series of magnetic compasses with which it is possible to determine the position of each element of the streamer

float depth, and if the streamer deviates from the required level the fin angles are adjusted to compensate. Also, a series of compasses is located along the length of the streamer (Figure 4.42) and each transmits signals back to the ship so that the position of each active group of the streamer can be determined and plotted (see Section 6.2.2.3). If the ship is towing a long streamer across a side current which causes the streamer to drift off track (called *feathering*), the feather angle and the streamer position can still be determined (Figure 4.43). Such information is vital for subsequent data processing. Knowing the position of each towed element is vital not only for the success of the seismic survey, but also for marine safety. Seismic surveys are sometimes undertaken in busy shipping lanes, and other ships need to know where the streamers are so as to be able to avoid snagging them. When a streamer is not being used it is coiled up on a drum on the ship's deck (Figure 4.44).

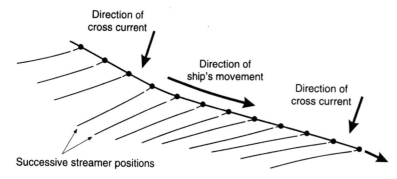

Direction of
cross current

Direction of
ship's movement

Direction of
cross current

Successive streamer positions

Figure 4.43 Schematic plot of successive streamer locations when significant feathering occurs caused by strong crosstrack currents

Knowing the precise location of the streamers deployed is all the more important when three-dimensional seismic data are being acquired. Usually, up to four streamers, each up to 6 km long, are deployed from two ships steaming parallel to each other. In addition, there may be four seismic gun arrays deployed from the same two ships. The most modern seismic vessels are capable of deploying 12–16 streamer arrays each 6 km long, and longer in special cases. In order to be able to deploy so many streamers, a specially designed ship has been built (the *Ramform Explorer*) based on the Norwegian intelligence-gathering 'stealth' ship, the *Marjatta*. The 83 m long *Ramform Explorer*, which is operated by the Norwegian Petroleum Geo-Services Group (PGS), is radically different from previous seismic vessels in that she is triangular in shape with a 40 m aft beam (Figure 4.45) and automatic streamer spooler system. The wide beam and massive deck space of this ship are sufficient to accommodate a Chinook-size helicopter operation. This allows for complete crew changes at sea, thereby increasing the ship's productive time. The simultaneous use of up to 16 streamers is for the acquisition of 3-D seismic data. The *Ramform Challenger*, sister ship of the *Ramform Explorer*, became operational in 1996.

For use in snow-covered areas, a gimballed geophone streamer has been developed. Towed land cables are not new, having been in use since the early 1970s. However, a new system was produced after field trials for an exploration survey in 1988 in Svalbard. The system used comprised a main cable with a central stress member surrounded by insulated conductors. Half-gimballed geophones were connected to the main cable by short geophone cables up to a few metres long. The survey was carried out using a tracked over-snow vehicle towing the snow streamer. Each geophone sensor was self-orientating along one horizontal axis and was enclosed in an oil-filled cylindrical metal case (20 cm long, diameter 4.5 cm with a mass of 1 kg). The weight of the sensor in its case coupled well with soft snow. For each shot, the streamer was stationary. No appreciable difference in data quality was noted between data acquired using the streamer and using standard spiked geophones implanted within the snow cover. How-

ever, the increased speed of surveying achieved using the streamer resulted in significant savings in survey time and hence costs. The snow streamer has also been used in the Antarctic. Given the increased amount of exploration work, particularly in the Arctic, both onshore and over sea-ice, the use of the snow streamer is likely to increase (Eiken *et al.* 1989).

Figure 4.44 Coiled streamer on the deck of a survey vessel. Courtesy of BP

4.6.3 Seismographs

There is a wide variety of seismographs available designed for different applications and budgets. For small-scale engineering studies, a single-channel seismograph is available for operation by one person using constant offset profiling. For general usage, 12-, 24- and 48-channel signal enhancement seismographs with 'digital instantaneous floating point' (DIFP) gain control have been produced. The advantage of signal enhancement instruments is that, instead of having to supply all the required energy in a single blast, many but smaller 'shots' can be used to build up the seismic record. By so doing, the process of summation also helps to reduce incoherent noise, thus improving the signal-to-noise ratio. To illustrate what the control panel of a modern seismograph looks like, the top panel of

a Geometrics ES-2401 seismograph is shown in Figure 4.46. Comparable seismographs are produced by Bison, and ABEM, for example.

For large-scale three-dimensional land surveys, multichannel systems capable of handling up to 600 channels have been developed for use by specialist seismic contractors, predominantly for use in hydrocarbon exploration (Bertelli *et al.* 1993). Modern digital seismographs are also used in vertical seismic profiling surveys, cross-hole investigations as well as for surface applications. DIFP removes the need for manual gain controls on a seismograph. The amplifier gains are adjusted automatically each time the record is sampled. Consequently, both large-amplitude first breaks and small-amplitude late arrivals can be recorded precisely.

An outline of the basic system architecture of Geometrics' Strata-View system is shown in Figure 4.47. The seismograph incorporates a '486 computer or utilises an external Pentium-based controller for onboard processing and analysis. Modern seismographs have preloaded software for refraction processing so that in-field solutions can be obtained very quickly. Not only does this provide rapid preliminary interpretation, but, more importantly, it highlights bad first-break picks and inconsistent modelling. With the increased field

Figure 4.45 *Ramform Explorer* in operation showing the distinctive broad stern from which up to 12 streamers can be deployed simultaneously. Photograph courtesy of PGS Exploration, Norway

Figure 4.46 (*opposite*). The display panel of an ES-2401 seismograph. Courtesy of Geometrics Inc.

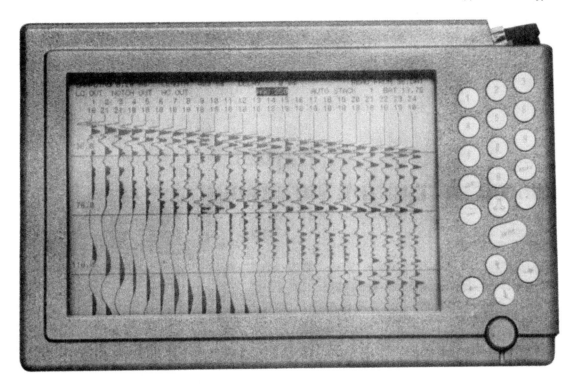

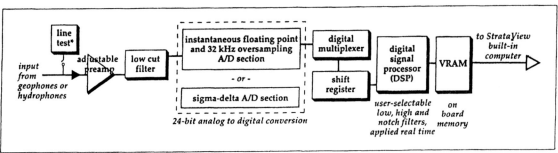

Figure 4.47 System architecture for the Geometrics StrataView seismograph. Courtesy of Geometrics Inc.

production rates, it is also possible to undertake preliminary surveys and, with the benefit of the in-field analysis, to obtain additional data to provide for detailed information where it is needed while the field crews are still deployed. By producing high-quality digital data, powerful seismic data processing techniques can be applied to the data in order to retrieve the maximum amount of information (see Chapter 6).

Chapter 5
Seismic refraction surveying

5.1 INTRODUCTION

Seismic refraction experiments can be undertaken at three distinct scales: global (using earthquake waves), crustal (using explosion seismology), and near-surface (engineering applications). For the purposes of this book, emphasis is placed on shallow investigations. Discussion of passive seismic refraction in earthquake studies can be found in other texts, such as those by Brown and Mussett (1981), Gubbins (1990) and Kearey and Vine (1990).

The major strength of the seismic refraction method is that it can be used to resolve lateral changes in the depth to the top of a refractor and the seismic velocity within it. The most commonly derived geophysical parameter is the seismic velocity of the layers present. From such values, a number of important geotechnical factors can be derived, such as assessing rock strength, determining rippability (the ease with which ground can be ripped up by an excavator), and potential fluid content. In addition to the more conventional engineering applications of foundation studies for dams and major buildings etc., seismic refraction is increasingly being used in hydro-geological investigations to determine saturated acquifer thickness, weathered fault zones etc. The location of faults, joints, and other such disturbed zones using seismic refraction is of major importance in the consideration of the suitability of potential sites for the safe disposal of particularly toxic hazardous wastes.

5.2 GENERAL PRINCIPLES OF REFRACTION SURVEYING

5.2.1 Critical refraction

The seismic refraction method is based on the principle that when a seismic wave (P- and/or S-wave) impinges upon a boundary across which there is a contrast in velocity, then the direction of travel of that wave changes on entry into the new medium. The amount of change of direction is governed by the contrast in seismic velocity across the boundary according to Snell's Law as described in Chapter 4 (see Boxes 4.8 and 5.1 and Figures 4.7 and 5.1). The critical angle for a given boundary for P-waves may be different from that for S-waves as the respective ratio in velocities between the two layers for P- and S-waves may not be the same.

Box 5.1   Snell's Law (see Figure 5.1)

Snell's Law:

$$\sin i/\sin r = V_1/V_2 \text{ for general refraction}$$
$$\sin i_c = V_1/V_2 \text{ for critical refraction}$$

where i_c is the angle at which critical refraction occurs, and V_2 is greater than V_1. V_1 and V_2 are the seismic velocities of the upper and lower layers respectively; i and r are the angles of incidence and refraction.

The refraction method is dependent upon there being an increase in velocity with depth. If, however, the lower medium has a velocity lower than that from which the wave is emerging (i.e. V_2 is less than V_1, a *velocity inversion*) then the refracted wave will bend towards the normal. This gives rise to a situation known as a *hidden layer*, which is discussed in more detail in Section 5.4.4.

(A)

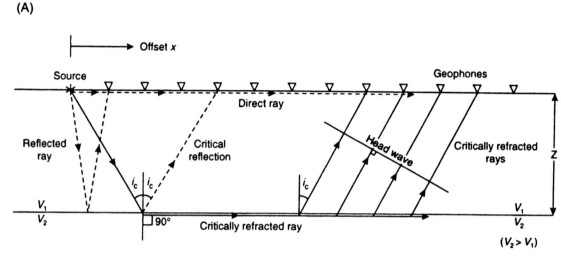

(B)

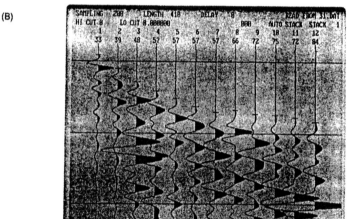

It is assumed in the refraction method that the thickness of each layer present is greater than the wavelength of the incident energy, and that each successive layer is as thick as, or is thicker than, the one lying above it. A further assumption is that the raypaths are constrained to lie in the vertical plane of the refraction profile line such that there is no seismic energy (refracted or reflected) arising from boundaries out of that plane – a phenomenon known as *side-swipe*. In some engineering applications, this factor is significant and has to be considered during data acquisition.

In a refraction survey, only the P-wave is usually considered. There are occasions when the additional use of S-waves is beneficial to the overall interpretation. For the purposes of the discussion which follows, we shall confine our considerations to P-waves only.

The basic components of a seismic refraction experiment are shown schematically in Figure 5.1A. A source, such as a sledge hammer on

(C)

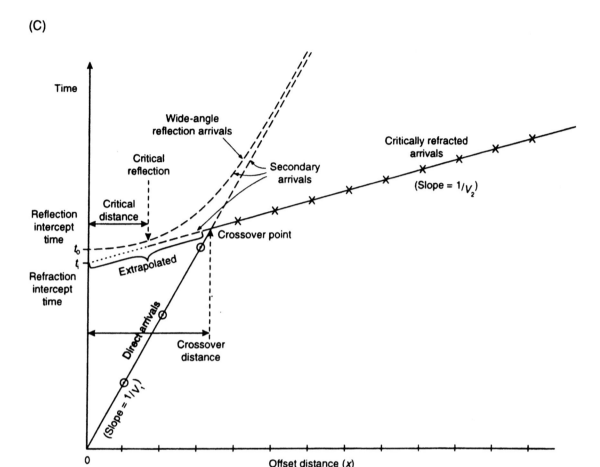

a baseplate or shotgun blanks in a buffalo gun, is used to generate the P-waves. The waves produced travel in three principal ways: directly along the top of the ground surface (direct wave); by reflection from the top of the refractor; and, of greatest importance, by critical refraction along the top of the refractor(s). The arrival of each wave is detected along a set of geophones and recorded on a seismograph, with the output of each geophone being displayed as a single trace. The onset of each arrival for each geophone is identified (Figure 5.1B) and the associated travel time is measured and plotted on a time–distance graph (Figure 5.1C).

At a distance called the *critical distance*, the reflected arrival is coincident with the first critically refracted arrival and the travel times of the two are identical. The critical distance is thus the offset at which the reflection angle equals the critical angle. This distance should not be confused with the *crossover distance*, which is the offset at which the critically refracted waves precede the direct waves. The crossover

Figure 5.1 (A) (*opposite*) Raypath diagram showing the respective paths for direct, reflected and refracted rays. (B) (*opposite*) Example of a seismic refraction record as seen on the display of an EG&G Smartseis seismograph, with first arrivals picked on each trace (indicated by the small cross line). (C) (*top*) Arrival times plotted on a time–distance graph.

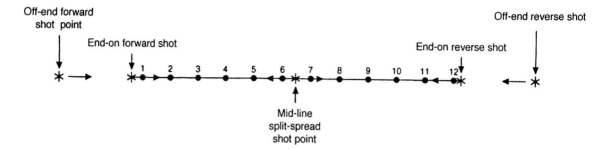

point marks the change in gradient of the time–distance graph from the slope of the direct arrivals segment to that for the refracted signals. While the travel time hyperbola associated with reflected arrivals is shown in Figure 5.1C, only direct and refracted arrival times are usually considered in the refraction analysis.

Figure 5.2 Geophone spread for a refraction survey with shot locations indicated

5.2.2   Field survey arrangements

5.2.2.1   *Land surveys*

For a seismic refraction survey on land, the basic layout is shown in Figure 5.2. A number of geophones, usually 12 or 24, are laid out along a cable with a corresponding number of takeouts along a straight line. This set of geophones constitutes a 'spread'; it should be noted that this is not a geophone 'array' (see Chapter 6). The seismic source (shot), whatever type it happens to be for a given survey, is located in one of five locations. The simplest case is for the shot to be positioned at the start and the end of the spread ('end-on' shots). A source located at a discrete distance off the end of the spread is known as an 'off-end' shot. A source positioned at a location along the spread is known as a 'split-spread' shot; usually this is either at mid-spread or at a quarter or three-quarters along a spread. Shots are usually fired into a spread from each end (end-on and off-end shots) in forward and reverse directions. The positioning of shots relative to a given spread is to achieve adequate coverage of the refractor surface and to provide adequate lateral resolution. As will be described in later sections, for each shot, appropriate travel times can be picked on each trace on the seismic record obtained. With each shot location into the same spread, additional data are acquired to provide sufficient information for detailed data reduction and analysis.

5.2.2.2   *Water-borne surveys*

Seismic refraction can also be undertaken in an aquatic environment but special systems are required. There are two main methods of detecting the critically refracted signals: one is to use a bottom-drag hydrophone cable and the other is to use sonobuoys. Each will be described briefly in turn.

A bottom-drag cable is similar to a normal reflection hydrophone streamer but is usually only 55 m long. A boat housing all the shot-firing and seismograph recording systems tows the cable on to station, when the cable is allowed to sink to the river/seabed. In some cases, divers are used to anchor the cable to the substrate as required. Shots are fired as necessary and then the cable is towed to the next station and the process repeated. By careful operation of the boat and the positioning of the bottom-tow cable, long seismic refraction lines can be acquired.

The sonobuoy method does away with hydrophone streamer cables and is not limited by streamer cable lengths. The sonobuoy consists of a flotation device containing batteries which, when in contact with seawater, are activated, causing a radio telemetering antenna to be raised and one or more hydrophones to be suspended on cables from below. When a shot is fired from the survey ship, the signals received by the hydrophones are radioed back to the ship where they are recorded. The offset between the shot and the sono- buoy is determined from the travel time of the direct wave through the water. In the case of inshore surveys, land-based surveying methods can be used to track the sonobuoy. One advantage of the sonobuoy method if used during a major marine seismic reflection survey is that the sonobuoys are expendable and sink after a finite period. The refraction data can, however, be acquired while the reflection survey is in progress.

A third method of marine seismic refraction surveying requires at least two ships, one towing the hydrophone streamer and the other firing the shots. If reverse shooting is to be used, a third ship may be required – two ships steam in opposite directions firing shots in sequence while a third ship steams along the profile track towing a hydrophone streamer.

In hydrocarbon exploration, the use of multiship surveys and the slow rate of data acquisition make this an extremely expensive operation. Similarly, water-borne refraction surveys in harbours and estuaries takes considerably longer to acquire than a comparable line-meterage of reflection profiling. Marine refraction surveys are very valuable in investigations for proposed tunnel routes under harbours for sewerage schemes, for example.

5.3 GEOMETRY OF REFRACTED RAYPATHS

The basic assumption for seismic refraction interpretation is that the layers present are horizontal or only dipping at shallow angles and are, in the first instance, planar surfaces.

5.3.1   Planar interfaces

5.3.1.1   Two-layer case

In Figure 5.3, it can be seen that the raypath taken by a signal originating from the source S travels to A where it undergoes critical refraction and travels towards and ultimately beyond position B. The headwave originating from the refractor at B travels through layer 1 where it is detected by a geophone at G. The geophone is offset from the shot by a distance x. The total travel time taken is the sum of the three component travel times, details of which are presented in Box 5.2. The time–distance graph for the two-layer case (Figure 5.1C) is used to calculate the velocities of the two layers, V_1 and V_2, from the gradients of the two straight-line segments (gradient $= 1/V$). The extrapolation of the segment from the critically refracted arrivals on to the time axis gives an intercept time t_i from which the depth to the refractor (z) can be calculated (Box 5.2, equation 8), given values of V_1 and V_2 derived from the time–distance graph.

Box 5.2   Travel time calculations for a two-layer case
(see Figures 5.1 and 5.3)

Total travel time is:

$$T_{SG} = T_{SA} + T_{AB} + T_{BG} \tag{1}$$

where:

$$T_{SA} = T_{BG} = z/(V_1 \cos i_c) \tag{2}$$

$$T_{AB} = (x - 2z \tan i_c)/V_2. \tag{3}$$

Substituting expressions (2) and (3) into (1), we obtain:

$$T_{SG} = z/(V_1 \cos i_c) + (x - 2z \tan i_c)/V_2 + z/(V_1 \cos i_c)$$

which simplifies to:

$$T_{SG} = (1/V_2)x + 2z (\cos i_c)/V_1. \tag{4}$$

This has the form of the general equation of a straight line, $y = mx + c$, where $m =$ gradient and $c =$ intercept on the y-axis on a time–distance graph. So, from equation (4), the gradient is $1/V_2$ and c is the refraction time intercept t_i (see Figure 5.1) such that $t_i = 2z (\cos i_c)/V_1$.

Remember that $\sin i_c = V_1/V_2$ (Snell's Law), and hence:

$$\cos i_c = (1 - V_1^2/V_2^2)^{1/2} \text{ (from } \sin^2 \theta + \cos^2 \theta = 1).$$

continued

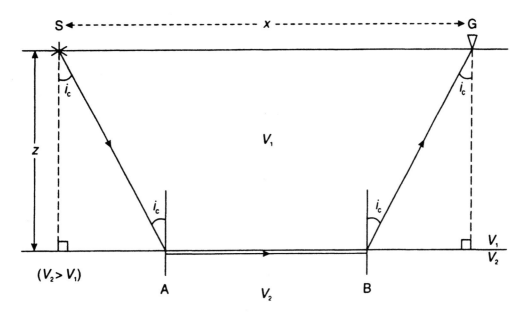

S $\leftarrow$ - - - - - - - - - - - - - - - - - - x - - - - - - - - - - - - - - - - $\rightarrow$ G

i_c

V_1

i_c

z

i_c

i_c

V_1
V_2

$(V_2 > V_1)$

A

V_2

B

Figure 5.3 Simple raypath for a two-layer structure

continued

An alternative form to equation (4) is:

$$T_{SG} = x(\sin i_c)/V_1 + 2z(\cos i_c)/V_1 \qquad (5)$$

or

$$T_{SG} = x/V_2 + t_i \qquad (6)$$

where

$$t_i = 2z(V_2^2 - V_1^2)^{1/2}/V_1 V_2 \qquad (7)$$

$$z = t_i V_1 V_2/2(V_2^2 - V_1^2)^{1/2} \qquad (8)$$

On Figure 5.1, it can be seen that the travel times for the direct and critically refracted ray are the same at the crossover distance. Consequently, solving equation (1) in Box 5.3 in terms of the crossover distance x_{cross} provides an additional means of calculating the depth to the refractor. Furthermore, equation (2) in Box 5.3 shows that the depth to the refractor is always less than half the crossover distance.

Box 5.3   The use of crossover distance to calculate refractor depth

Travel time of direct ray at the crossover distance is x_{cross}/V_1. Travel time of critically refracted ray at the crossover distance is (from equation (6) in Box 5.2) given by:

$$T = x_{cross}/V_2 + 2z(V_2^2 - V_1^2)^{1/2}/V_1 V_2.$$

continued

— continued —

Hence, at the crossover point:

$$x_{cross}/V_1 = x_{cross}/V_2 + 2z(V_2^2 - V_1^2)^{1/2}/V_1 V_2 . \qquad (1)$$

Solving for x_{cross} and reorganising gives:

Depth to the refractor:

$$z = \tfrac{1}{2} x_{cross} [(V_2 - V_1)/(V_2 + V_1)]^{1/2} \qquad (2)$$

Crossover distance:

$$x_{cross} = 2z[(V_2 + V_1)/(V_2 - V_1)]^{1/2}$$

5.3.1.2 Three-layer case

The simple raypath geometry for critical refraction to occur in a three-layer model with horizontal interfaces is shown in Figure 5.4A and its corresponding travel time–distance graph in Figure 5.4B. The expressions governing the travel time–velocity relationships are given in Box 5.4. The effect of reducing the thickness of layer 2 on the time–distance graph is to reduce or even remove completely the straight-line segment corresponding to refracted arrivals from the top of layer 2 (Lankston 1990). The signal travels from the source down to the first refractor (at A), where it is refracted into the second medium through to the second interface (at B), at which point it is then critically refracted. From there the generated head wave from the lowest refractor travels back from C through the overlying layers to arrive at the geophone at G.

Box 5.4   Travel time calculations for a three-layer case
(see Figure 5.4)

Total travel time is:

$$T_{SG} = T_{SA} + T_{AB} + T_{BC} + T_{CD} + T_{DG}$$

where:

$$T_{SA} = T_{DG} = z_1/V_1 \cos \theta_1$$
$$T_{AB} = T_{CD} = z_2/V_2 \cos \theta_c$$
$$T_{BC} = (x - 2z_1 \tan \theta_1 - 2z_2 \tan \theta_c)/V_3 .$$

Combining these gives:

$$T_{SG} = x/V_3 + (2z_2 \cos \theta_c)/V_2 + (2z_1 \cos \theta_1)/V_1 \qquad (1)$$

$$T_{SG} = x/V_3 + t_2 \qquad (2)$$

— continued —

continued

where

$$\frac{\sin \theta_1}{V_1} = \frac{\sin \theta_c}{V_2} = \frac{1}{V_3} \text{ from Snell's Law.}$$

Thicknesses of refractors are given by:

$$z_1 = t_1 V_1 V_2 / 2(V_2^2 - V_1^2)^{1/2}$$

$$z_2 = t_2 V_2 V_3 / 2(V_3^2 - V_2^2)^{1/2}$$

$$- z_1 V_2 (V_3^2 - V_1^2)^{1/2} / V_1 (V_3^2 - V_2^2)^{1/2}$$

The analysis works by determining V_1, V_2, t_1 and t_2 from the travel time graph for the top two layers, and hence the thicknesses of the first two refractors can be calculated using the equations in Box 5.4. The thicknesses of refractors are usually underestimated by about 3%, with the percentage inaccuracy increasing the larger the number of layers involved.

5.3.1.3 Multilayer case

The general expressions governing the travel time–velocity relationships for the situation where more than three horizontal planar layers are present are given in Box 5.5. The form of the equations and of the analysis of the travel time graphs follow the procedures explained for a three-layer case, but are extended to the relevant total number of layers.

Box 5.5   Travel time calculations for a multilayer case

The total travel time T_{SG} in an n-layer case is given by:

$$T_{SG} = x/V_n + \sum_{i=1}^{n-1} [(2z_i \cos \theta_i)/V_i]$$

where $\sin \theta_i = V_i/V_n$.

Note that θ_i are not critical angles except for θ_{n-1}.

In engineering applications, it is unusual to have more than three or at most four layers present; commonly only two-layer cases are used, with any variations in velocity within the first layer being taken into account by consideration of the value of the first layer velocity being an average. Commonly, in areas of alluvial fill or over glacial

(A)

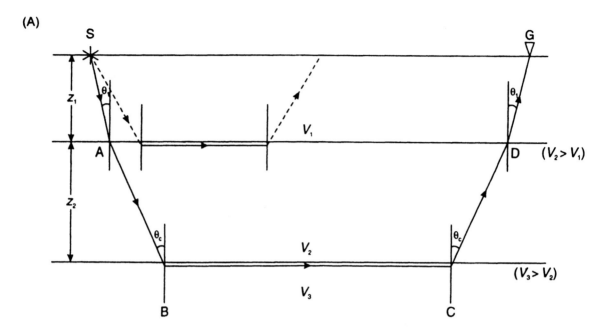

(B)

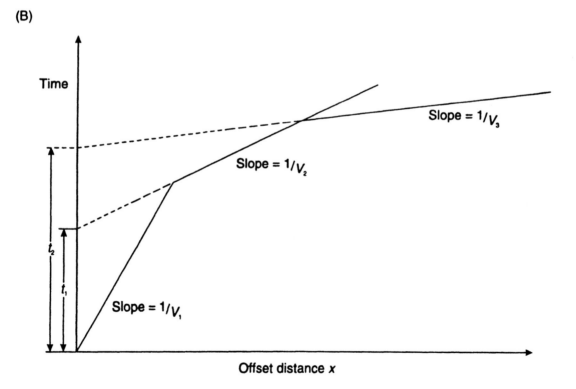

sediments, the local lateral and vertical variations in velocity can be significant. It is often advisable to undertake micro-spreads, with geophone intervals as small as 1 m, in order to measure the near-surface velocities along a spread. If it is known from desk study work that the area has Quaternary and/or Recent sediment cover, the survey design should include micro-spreads in order to provide better control on the determination of local velocities. The variation in near-surface velocities makes it quite inappropriate to quote calculated values of velocity to the nearest metre per second. Careful thought should be given to determining the statistical significance of variations in P-wave velocities determined from refraction surveys. Calculated values of velocity should be quoted to perhaps the nearest 50 m/s, for example, with an estimation of the standard deviation as calculated from a relevant statistical analysis. If due consideration is not given to careful analysis of the variation of velocities, any depth estimates arising from the refraction survey are likely to be inaccurate, possibly being as much as 30–40% of the actual refractor depth. Depths determined to a refractor should be quoted with *realistic* error bars, not 'hoped-for' values!

Where refraction surveys are carried out as part of a hydrocarbon exploration programme or for crustal research, larger numbers of refractors can be encountered. However, there still needs to be a very careful appraisal of the errors and inaccuracies involved in whatever analysis is undertaken.

5.3.1.4 *Dipping-layer case*

When a refractor lies at an angle to the horizontal, the simple geometries so far described are complicated by the *angle of dip*. It is no longer adequate to undertake only one direction (forward) of shooting; it becomes necessary to carry out both forward and reverse shooting in order to determine all the parameters required to solve for the refractor geometry. Furthermore, the refractor velocities determined in the case of dip are referred to as *apparent* velocities, as the values of velocity determined from the inverse gradients of the straight-line segments on the travel time–distance graphs are higher in the upslope direction (V_u) and lower in the downslope direction (V_d).

The raypath geometry for a dipping two-layer case is shown in Figure 5.5A with its travel time–distance graph in Figure 5.5B. Following the logic used in discussing the two-layer planar refractor in Section 5.3.1.1, a set of equations can be produced (Box 5.6) that relate velocity, layer thickness and angle of refractor dip from which the geometry of the dipping refractor can be determined. The depths (d_a and d_b) to the refractor vertically below the spread end-points can easily be calculated from the derived values of perpendicular depths (z_a and z_b) using the expression $d = z/\cos \alpha$.

Figure 5.4 (*opposite*) (A) Simple raypath diagram for refracted rays, and (B) their respective travel time–distance graphs for a three-layer case with horizontal planar interfaces

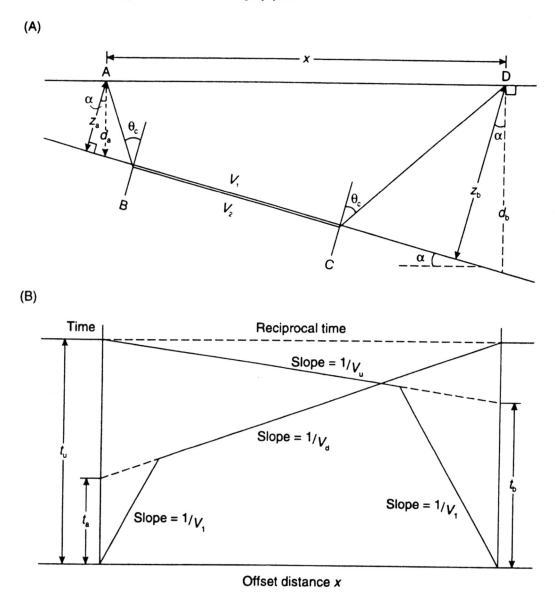

It should also be noted that the total travel times obtained from both forward and reverse shooting should be the same ($T_{AD} = T_{DA}$) and these are hence also known as 'reciprocal times'. When undertaking forward and reverse shooting, the reciprocity of the measured travel times should be checked. If the total travel times are not the same, the picking of first arrivals and the offsets of the geophones should be checked for errors.

Figure 5.5 (A) Raypath geometry over a refractor dipping at an angle α, and (B) the respective travel time-distance graph for the forward (down-dip) and reverse (up-dip) shooting directions.

Box 5.6   Travel time calculations for a dipping refractor
(see Figure 5.5)

Total travel time over a refractor dipping at an angle α is given by:

$$T_{ABCD} = (x \cos \alpha)/V_2 + [(z_a + z_b) \cos i_c]/V_1$$

where V_2 is the refractor velocity, and z_a and z_b are the distances perpendicular to the refractor.
The down-dip travel time t_d is given by:

$$t_d = x[\sin(\theta_c + \alpha)]/V_1 + t_a \qquad (1)$$

where $t_a = 2z_a(\cos \theta_c)/V_1$.

$$t_u = x[\sin(\theta_c - \alpha)]/V_1 + t_b \qquad (2)$$

where $t_b = 2z_b(\cos \theta_c)/V_1$.
Equations (1) and (2) above can be written in terms of the apparent up-dip velocity (V_u) and down-dip velocity (V_d) such that:

$$t_d = x/V_d + t_a, \qquad \text{where } V_d = V_1/\sin(\theta_c + \alpha)$$

$$t_u = x/V_u + t_b, \qquad \text{where } V_u = V_1/\sin(\theta_c - \alpha).$$

An approximate relationship between true and apparent velocities for shallow angles of dip ($\alpha < 10°$) is given by:

$$V_2 \approx (V_d + V_u)/2.$$

5.3.1.5   *The case of a step discontinuity*

So far the refractor has been assumed to be planar and continuous. There are situations where a step discontinuity may occur in the refractor, such as in an old backfilled quarry where a former quarry face has been covered over with backfill material (Figure 5.1B). In such a situation, the refractor velocity should remain the same along its length, the only perturbation arising from the distortion is raypath (Figure 5.6).

If the step discontinuity has been caused by a normal fault, for example, there may be an additional complication of the refractor velocity having different magnitudes across the fault plane; if the

(A)

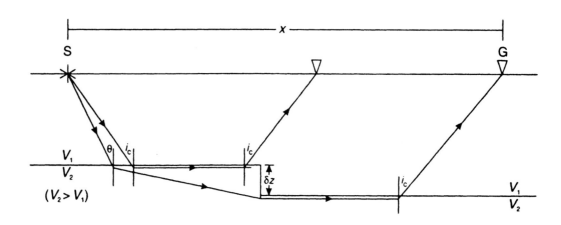

(B)

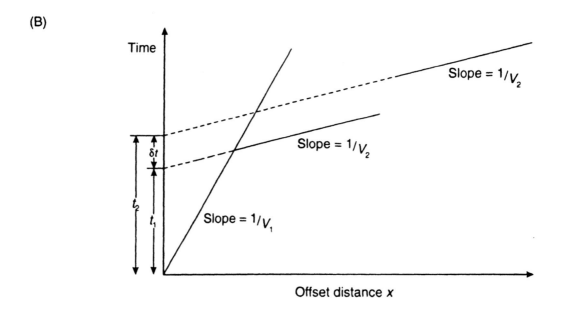

Figure 5.6 (A) Raypath geometry over a refractor with a step discontinuity but no lateral change in layer velocity, and (B) the corresponding travel time–distance graph

throw on the fault is significant, different geological materials may now be juxtaposed with consequent difference in seismic velocities. Indeed, the presence of a lateral change in refractor velocity in a two-layer case may give rise to a travel time–distance graph from a single end-on forward shot identical to that produced by three horizontal layers but with uniform layer velocities. The easiest way to discriminate between the two cases is to carry out an off-end forward shot. If the crossover point on the travel time–distance graph is shifted laterally (i.e. along the *x*-axis), this indicates that the three-layer case is correct. If, instead, the crossover point is shifted vertically on the time axis (i.e. there is a delay time) then this is indicative of a two-layer case, such as across a fault plane at which there is no topographic effect on the refractor but where there is a significant velocity contrast. This situation has been discussed in more detail by Lankston (1990).

Where the size of the step discontinuity is small with respect to the depth of the refractor, then a simple equation (Box 5.7) can be used to estimate the difference in depth to the refractor. While the delay time (δt) may be observed on a time–distance graph from a shot in a forward direction, commonly the amplitudes of the refracted signals from a reverse shot are very small and it may not be possible to obtain satisfactory data from a reverse shot.

Box 5.7 Determination of depth to a refractor with a step discontinuity (see Figure 5.7)

The step size (δz) in a discontinuity in a refractor is given by:

$$\delta z = \delta t \, V_1 V_2 / (V_2^2 - V_1^2)^{1/2}.$$

5.3.2 Irregular (non-planar) interfaces

The preceding discussion has been for planar layers that are either horizontal or dipping uniformly and with no lateral variation in velocity in the refractor or vertical variation in velocity within the surface layer. It is important whenever a refraction survey is designed and then executed that consideration be given to the type of layer case present as judged from the quality and nature of travel time–distance data obtained. In a significant number of cases the simple models are obviously not appropriate, as in the situation of a refraction survey over a buried rock valley, for instance. It becomes necessary to carry out interpretation on the basis that the refractor is no longer planar or necessarily horizontal, but irregular in form, but where the wave-length of variation in refractor elevation is greater than the depth to the refractor. When an irregular refractor is encountered, it is usually analysed on the basis of only a two-layer case, details of which are given in the next section.

5.4 INTERPRETATIONAL METHODS

Before any detailed interpretation can begin it is important to inspect the travel time–distance graphs obtained, (a) as a check on quality the of data being acquired, and (b) in order to decide which

Table 5.1   Travel time anomalies (see Figure 5.7)

(i)	Isolated spurious travel time of a first arrival, due to a mispick of the first arrival or a mis-plot of the correct travel time value
(ii)	Changes in velocity or thickness in the near-surface region
(iii)	Changes in surface topography
(iv)	Zones of different velocity within the intermediate depth range
(v)	Localised topographic features on an otherwise planar refractor
(vi)	Lateral changes in refractor velocity

Figure 5.7 (*opposite*) Travel time anomalies and their respective causes. (A) Bump and cusp in layer 1. (B) Lens with anomalous velocity (V_2') in layer 2. (C) Cusp and bump at the interface between layers 2 and 3. (D) Vertical but narrow zone with anomalous velocity (V_3') within layer 3. After Odins (1975) in Greenhalgh and Whiteley (1977), by permission

interpretational method to use – simple solutions for planar layers and for a dipping refractor, or more sophisticated analysis for the case of an irregular refractor. Checks to be considered are listed in Table 5.1, and some of the associated time anomalies ((iii)–(vi)) are illustrated in Figure 5.7. Indiscriminate smoothing of 'cusps' in travel time data can obscure what might be important features and result in serious errors in interpretation.

Several different interpretational methods have been published, falling into two approaches: delay-time and wavefront construction methods. Examples of the former are Gardner (1939, 1967), Wyrobek (1956), Barry (1967), and Palmer (1980), and those of the latter are Thornburgh (1930), Hales (1958), Hagedoorn (1959), Schenck (1967), Hill (1987), Vidale (1990), and Aldridge and Oldenburg (1992). However, two methods emerge as the most commonly used, namely Hagedoorn's 'plus–minus' method (Hagedoorn 1959) and the generalised reciprocal method (GRM) (Palmer 1980), which are described in Sections 5.4.2 and 5.4.3 respectively. The more recent publications tend to be modifications of earlier methods or new computational procedures.

5.4.1   Phantoming

The sub-surface coverages of a dipping refractor obtained by forward and reverse shooting overlap along only part of the refractor and are not totally coincident, as shown in Figure 5.8 (Lankston 1990). The apparent velocities obtained up-dip and down-dip are in fact derived from different segments of the refractor. If the interface is planar, then the use of such apparent velocities to derive layer thicknesses and dip angles is justified.

In order to increase the sub-surface coverage, and depending on the suitability of the local geology, both the shot locations and the geophone spread have to be moved along the line of the profile. If there is no lateral variation in refractor velocity, then the resultant travel time–distance graph from a headwave generated from shot at increased offset is delayed in travel time but has the same gradient as that from the end-on shot (Figure 5.9). The observed parallelism of the two travel time graphs indicates that the difference in travel time is

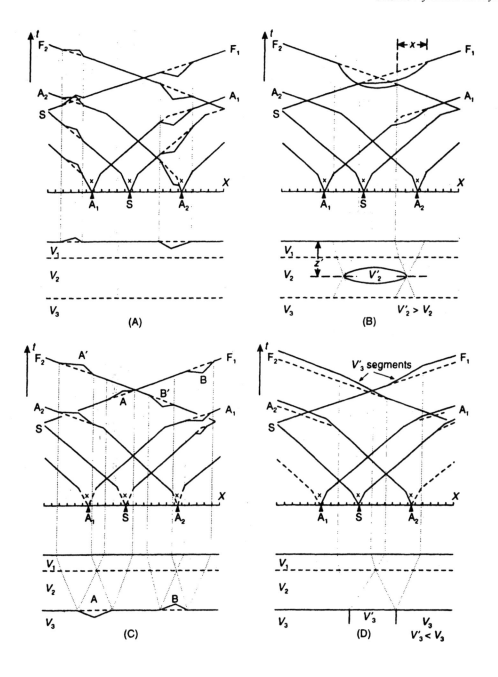

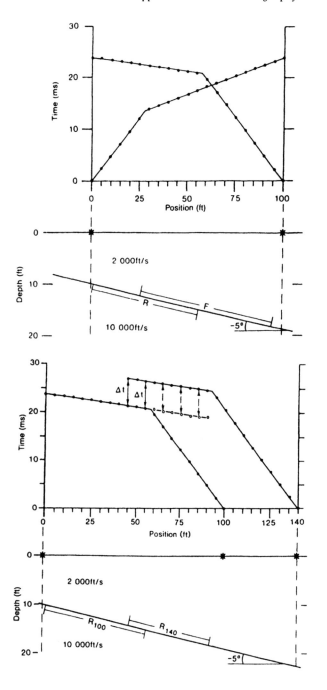

Figure 5.8 Overlapping zones of sub-surface coverage from forward and reverse shooting over a dipping refractor, and corresponding travel time–distance graphs. From Lankston (1990), by permission

Figure 5.9 Travel time–distance graphs and sub-surface coverages from two overlapping reverse direction spreads; open circles denote phantom arrivals. From Lankston (1990), by permission

a constant for each geophone location. By substracting this time difference from the headwave arrivals from the refractor in the second, offset shot, then the reduced travel times are those that would have been recorded from the first shot over the distance between the end-on shot and the crossover distance.

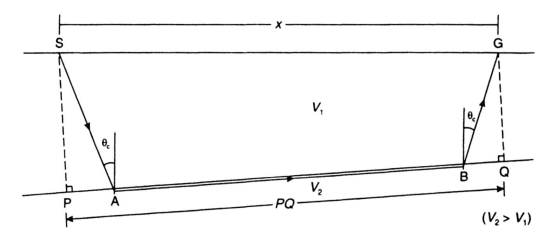

The time-shifted arrivals are known as *phantom arrivals* (Redpath 1973) and the process of time-shifting is referred to as *phantoming*. This process provides a means of obtaining real first-arrival information between the end-on shot and the original crossover distance which could not have been recorded from the first shot. The phantom arrivals, therefore, remove the necessity to extrapolate the travel time graph from beyond the crossover point back to the zero-offset point to obtain the intercept time, as indicated previously in Figure 5.1.

Figure 5.10 The principle of delay time (see text for details)

5.4.2 Hagedoorn plus–minus method

In Hagedoorn's method, it is assumed that the layers present are homogenous, that there is a large velocity contrast between the layers, and that the angle of dip of the refractor is less than 10°. The method uses intercept times and delay times in the calculation of the depth to a refractor below any geophone location. Referring to Figure 5.10, the delay time (δt) is the difference in time between (1) the time taken for a ray to travel along a critically refracted path from the source via the sub-surface media and back to the surface (T_{SG} along SABG), and (2) the time taken for a ray to travel the equivalent distance of the source–geophone offset (x) along the refractor surface (i.e. T_{PQ} along the projection PQ of the refracted raypath on to the refractor surface). The total delay time (δt) is effectively the sum of the 'shot-point delay time' (δt_s) and the 'geophone delay time' (δt_g). In cases where the assumptions on which the method is based are realistic, it is sufficiently accurate to consider the distance PQ to be approximately the same as the source–geophone offset (SG). The mathematical equations relating the various delay times and velocities are given in Box 5.8.

Box 5.8 **Determination of delay times** (see Figure 5.10)

The total delay time is given by:

$$\delta t = T_{SG} - T_{PQ}$$

and

$$T_{SG} = (SA + BG)/V_1 + AB/V_2 \quad \text{and} \quad T_{PQ} = PQ/V_2.$$

Thus:

$$
\begin{aligned}
\delta t &= (SA + BG)/V_1 - (PA + BQ)/V_2 \\
&= (SA/V_1 - PA/V_2) + (BG/V_1 - BQ/V_2) \\
&= \delta t_s + \delta t_g \approx T_{SG} - x/V_2.
\end{aligned}
$$

Alternatively:

$$T_{SG} = x/V_2 + \delta t_s + \delta t_g \qquad (1)$$

where δt_s and δt_g are the shot-point and geophone delay times.

In the case of a horizontal refractor, $\delta t = t_i$, the intercept time on the corresponding time–distance graph (see Figure 5.1 and Box 5.2, equation 6).

Hagedoorn's plus–minus method of analysing refraction data from both forward and reverse shooting provides a means of determining the delay times and hence the layer velocities and the depth to the refractor beneath any given geophone.

In Figure 5.11A, A and B are the locations of the forward and reverse shot points, and G is an arbitrary location of a geophone in between. Using equation (1) from Box 5.8, the travel time for a refracted ray from a shot point to any geophone G can be defined in terms of the respective delay times (see Box 5.9). Hagedoorn's 'plus term' (T^+) is the sum of the travel times from shot to the geophone from each end, minus the overall travel time between one shot point and the other (Box 5.9, equation 4). Hagedoorn's 'minus term' (T^-) is the difference in travel times taken by rays from each shot point to a given geophone.

T^- values can be plotted on a graph either against offset distance directly, in which case the velocity is determined from the slope which is $2/V_2$, or against $(2x - L)$ in which case the velocity is simply the inverse slope $(1/V_2)$. Values of the first layer velocity can be obtained in the conventional way from the inverse slope of the time–distance graph $(1/V_1)$. In each case, the T^- term provides

(A)

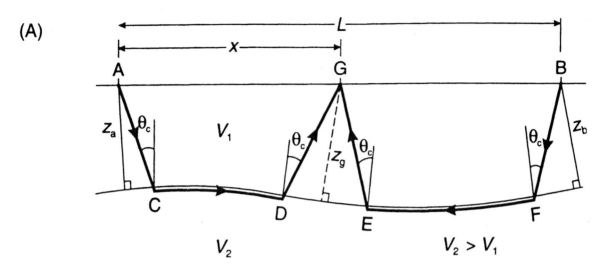

(B)

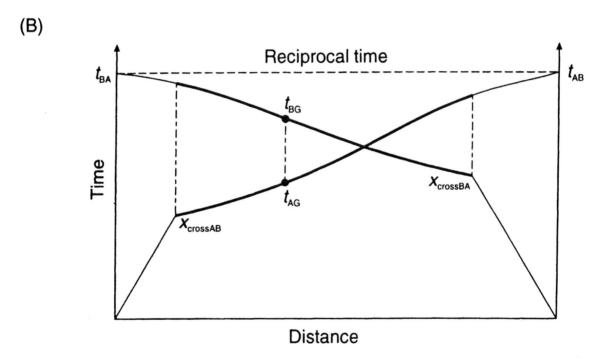

Figure 5.11   (A) Raypath geometry for forward and reverse shots over a refractor with irregular surface topography, and (B) the corresponding travel time–distance graphs. The segments between respective crossover distances are highlighted by thicker lines as it is these segments that are used in Hagedoorn's plus-minus analysis

a means of examining the lateral variation in refractor velocity with a lateral resolution equal to the geophone separation. Better accuracy of the estimate of velocity can be achieved, whichever graphical method is used, by carrying out linear regression analysis over the appropriate segments of the graph to determine the slope(s) most accurately and to provide some means of determining standard deviations.

Once values of layer velocities have been calculated, the depth (z_g) perpendicular to the refractor below each geophone can be computed using the T^- term. Calculations are made easier if velocities are expressed in km/s and time in milliseconds as depths are then given in metres. Given the assumptions made, the accuracy of depth calculation should be to within 10%.

An example of the graphical output from the plus–minus method is shown in Figure 5.12.

Box 5.9   Hagedoorn plus – minus terms (see Figure 5.11)

The travel time of a refracted ray at any geophone G is given by:

$$t_{AG} = x/V_2 + \delta t_g + \delta t_a \tag{1}$$

$$t_{BG} = (L - x)/V_2 + \delta t_g + \delta t_b \tag{2}$$

$$t_{AB} = L/V_2 + \delta t_a + \delta t_b \tag{3}$$

Hagedoorn's plus term T^+ is given by:

$$T^+ = t_{AG} + t_{BG} - t_{AB} = 2\delta t_g = 2z_g(\cos\theta_c)/V_1. \tag{4}$$

Therefore, the depth to the refractor beneath any geophone (z_g) is given by:

$$z_g = (T^+)V_1/2\cos\theta_c = (T^+)V_1 V_2/2(V_2^2 - V_1^2)^{1/2}. \tag{5}$$

Hagedoorn's minus term T^- is given by:

$$T^- = t_{AG} - t_{GB} = (2x - L)/V_2 + \delta t_a - \delta t_b. \tag{6}$$

The Hagedoorn plus–minus method is readily adapted for calculation using a programmable calculator (Cummings 1988), or a personal computer using either stand-alone proprietary software (e.g. GREMIX from Interpex Ltd; and van Overmeeren 1987) or a computer spreadsheet (e.g. Morris 1992), such as Lotus 1–2–3, Microsoft's Excel or Borland's Quattro-Pro. Seismograph manufacturers are also providing their own interpretational software with their instruments.

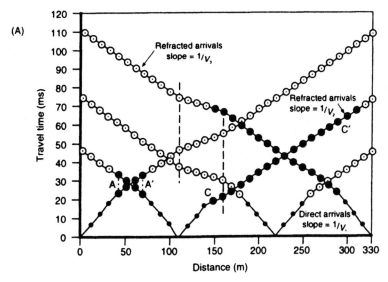

(A)

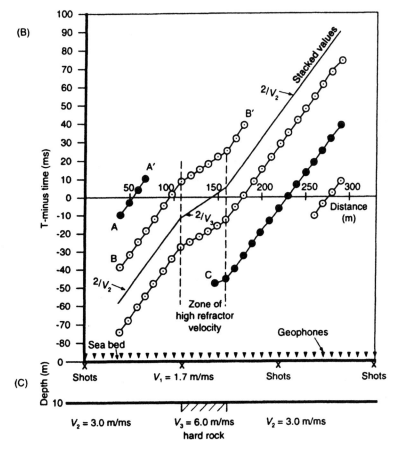

(B)

(C)

Figure 5.12 (A) Composite travel time-distance graphs; (B) Hagedoorn T^- graph; and (C) calculated depths to a refractor. After Gardener (1992), by permission

The interpretation procedure is very straightforward. Having picked the first-arrival times, the values are put into their respective spreadsheet cells for their given offset distance. By inspection of the time–distance (T–X) graphs the reciprocal times can be found and checked; the distances between end-to-end shots and hence the values of the distance L are known from the survey layout. The T–X graphs are inspected to separate out which portions of which graph are to be used in the analysis (i.e. which data give information about V_1 and which about V_2, etc.). Velocities are found using linear regression analysis on the relevant portions of the T–X graph (for V_1) and of the T^-–distance graphs (for V_2). These values are then used with the T^+ term to derive values of depth to the refractor at each geophone location. All the graphical output and the calculations can be achieved using just one software package, thus making it very convenient to carry out the analysis.

The plus–minus method assumes also that the refractor is uniform between the points of emergence on the interface (i.e. between D and F in Figure 5.11A). If this assumption does not hold but the method is used, then the non-planarity of the refractor is smoothed and an oversimplified interpretation would be obtained. This smoothing is overcome by the generalised reciprocal method which is discussed in the next section.

5.4.3   Generalised reciprocal method (GRM)

The generalised reciprocal method (Palmer 1980) is an inversion technique which uses travel-time data from both forward and reverse shots and which provides a graphical solution to resolve the geometry of sub-surface refractors. The method uses refraction migration to obtain the detailed structure of a refractor and information about any localised lateral variations within it. Refraction migration uses the offset (migration) distance which is the horizontal separation between a point on the refractor where a ray is critically refracted and that at the surface where the ray emerges.

It has been pointed out (Sjögren 1984), however, that the migration distance does not satisfactorily define narrow zones with low seismic velocities. As such zones are often the targets for refraction surveys, it is important that the interpretational method is capable of resolving them. Palmer (1991) has demonstrated that the migration distance selected to define the refractor surface is not necessarily the same as that which provides the optimum information about refractor velocities. Consequently, if the GRM is used to its full effect, it can be used successfully to define narrow low-velocity zones (see also Section 5.5.1). In order to increase the accuracy of the GRM, Zanzi (1990) has proposed that consideration should be given to nonlinear corrections in the case of an irregular refractor.

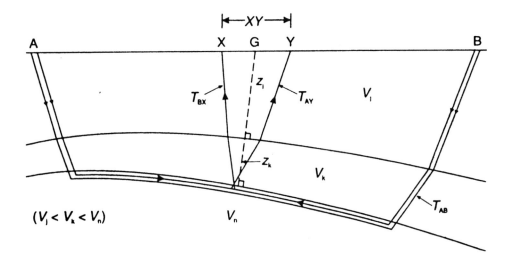

$(V_j < V_k < V_n)$

Figure 5.13 Schematic summary of parameters used in the generalised reciprocal method (GRM)

A principal difference of the GRM compared with the plus–minus method is that the critically refracted rays emerge at or very near the same point on the refractor, thereby removing the smoothing problem discussed in the previous section. They arrive at two different geophone locations separated by a distance XY (Figure 5.13).

Two functions are computed within the generalised reciprocal method: the velocity analysis function (t_v) and the time–depth function (t_G). The latter is determined with respect to a position G (Figure 5.13) at the mid-point between the points of emergence at the surface of forward and reverse rays at Y and X, respectively. Both functions are expressed in units of time and are detailed in Box 5.10. Equations (1) and (2) in that box are similar to Hagedoorn's T^- and T^+ terms, respectively, when the points X and Y are coincident (see Box 5.9). In cases where the maximum angle of dip is less than 20°, the GRM, as detailed here, provides estimates of refractor velocities to within 5%.

Both the velocity analysis and time–depth functions are evaluated and presented for a range of XY values from zero to one considerably larger than the optimum value. The increment in values of XY is usually taken as being equivalent to the geophone separation. Thus, if geophones are spaced at 5 m intervals, XY is incremented in steps of 5 m.

Box 5.10   GRM velocity analysis and time functions (Figure 5.13)

The **refractor velocity analysis function** (t_v) is given by:

$$t_v = (T_{AY} - T_{BX} + T_{AB})/2 \qquad (1)$$

where the distances AY and BX can be defined in terms of the

_____ *continued* _____

——— *continued* ———

XY and AG, such that:

$$AY = AG + XY/2 \quad \text{and} \quad BX = AB - AG + XY/2.$$

A graph of t_v plotted as a function of distance x has a slope $= 1/V_n$, where V_n is the seismic velocity in the refractor (which is the nth layer).

The **time–depth function** (t_G) is given by:

$$t_G = [T_{AY} + T_{BX} - (T_{AB} + XY/V_n)]/2. \tag{2}$$

The time–depth function, plotted with respect to position G, is related to the thicknesses (z_{jG}) of the overlying layers, such that:

$$t_G = \sum_{j=1}^{n-1} z_{jG}(V_n^2 - V_j^2)^{1/2}/V_n V_j \tag{3}$$

where z_{jG} and V_j are the perpendicular thickness below G and velocity of the jth layer, respectively.

The optimum distance XY (XY_{opt}) is related to layer thickness z_{jG} and seismic velocities V_j and V_n by:

$$XY_{opt} = 2 \sum_{j=1}^{n-1} z_{jG} \tan \theta_{jn} \tag{4}$$

where $\sin \theta_{jn} = V_j/V_n$.

Given a value of XY_{opt}, an average velocity (V') of all the layers above the refractor (layer n) is given by:

$$V' = [V_n^2 XY_{opt}/(XY_{opt} + 2t_G V_n)]^{1/2}. \tag{5}$$

At the optimum value of XY, both the forward and reverse rays are assumed to have been critically refracted at or very near the same point on the refractor. The optimum value of XY in the analysis of velocity is determined as being that which gives a time–depth function which approximates most closely to a straight line. Where there are lateral variations in refractor velocity, the optimum value of XY may differ with distance (see, for example, the case history in Section 5.5.1). In the selection of the optimum value of XY for the time–depth function, the graph which exhibits most detail should be the one chosen. In general, the optimum value of XY should be the same for both the velocity analysis and time–depth functions.

Having determined the value(s) of refractor velocity from the velocity analysis function (t_v), and hence determined the optimum value of XY, then equation (3) in Box 5.10 can be used to calculate the perpendicular depth to the refractor below each geophone position. This depth value is, however, the locus of depths centred on any given geophone position. Hence, an arc is drawn centred on a given geophone location with a radius equivalent to the calculated depth. The refractor surface is constituted by drawing the tangent to the constructed arcs, so that the true geometry of the refractor is migrated to the correct location in space. A further example of the application of this method is given in Section 5.5.1 (Lankston 1990).

5.4.4 Hidden-layer problem

A *hidden layer* or *blind zone* occurs when a layer that is present is not detected by seismic refraction. There are four causes of this problem: velocity inversion; lack of velocity contrast; the presence of a thin bed; and inappropriate spacing of geophones (Figure 5.14). In the situation where a layer with a lower velocity underlies one with a higher velocity, then the lower-lying layer may not be detected using seismic refraction methods. No critical refraction can occur in such a situation and thus no headwaves from the interface are produced. If there is little velocity contrast, then it may be extremely difficult to identify the arrivals of headwaves from the top of this zone. In addition, in the case where velocities increase with depth, but the thickness of a layer is less than one wavelength of the incident wave, then the thin bed would not be evident on the corresponding time–distance graph and would therefore effectively be hidden. Furthermore, the portion of travel time graph arising from this thin bed may not be sampled if the geophone interval is too large (Lankston 1990).

The hidden-layer problem precludes seismic refraction surveys from being undertaken where there is known to be a velocity inversion, such as where there is a layer of strongly cemented material in less consolidated material at shallow depth, such as hard-pan or duricrust. The only way seismic refraction can be undertaken in such circumstances is for each shot and geophone to be located below this hard, higher-velocity layer. This solution can lead to considerable logistical problems and reduced production rates, with a corresponding increase in cost.

5.4.5 Effects of continuous velocity change

So far it has been assumed that each layer has a uniform velocity both horizontally and vertically. In reality there are some materials where a uniform velocity is not observed. Examples of such materials are: sedimentary sequences in which grain size varies as a function of depth (fining-up and fining-down sequences), or where the degree of cementation increases with depth; some igneous bodies where fractionation has occurred, giving rise to more dense material at depth; and in polar regions over snowfields, where the degree of compaction from snow to ice increases with depth (firnification). In these situations, there is a gradual change in velocity with depth (e.g. Kirchner

(A)

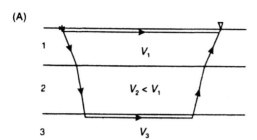

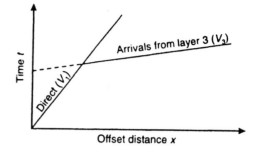

(B)

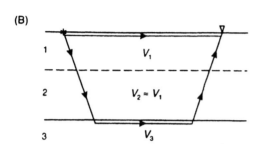

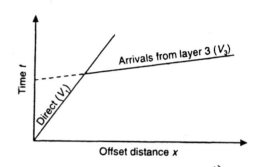

(C)

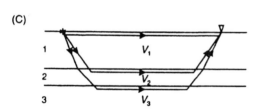

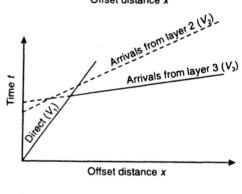

(D)

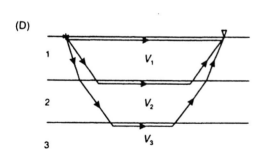

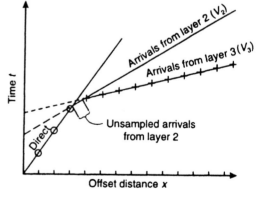

and Bentley 1979; Beaudoin *et al.* 1992; King and Jarvis 1992) and hence there is no obvious crossover point on the time–distance graph. Instead, the graph is curved, and the change from one so-called layer to another is gradational, making the usual methods of analysis inappropriate.

5.5 APPLICATIONS AND CASE HISTORIES

5.5.1 Rockhead determination for a proposed waste disposal site

Lankston (1990) has provided an example of the use of seismic refraction surveying with respect of a municipality granting a permit for a waste disposal site. The proposed location was an isolated site adjacent to a major river. Basalt, which was exposed along the side of the site farthest from the river, formed the underlying bedrock. Overlying bedrock was alluvium and windblown sand that in areas formed dunes 1–2 m high. Owing to the presence of loose sand at the surface, a sledge-hammer source was ineffective so explosives were used. The geophone interval was 50 ft (15.2 m). Each refraction spread was surveying using one long and one short offset shot at each end plus one mid-spread shot. Each successive spread overlapped with the previous one by two geophone positions.

Figure 5.14 (*opposite*) Depiction of the 'hidden layer' problem due to: (A) velocity inversion ($V_2 < V_1$); (B) lack of velocity contrast ($V_2 \approx V_1$) and (C) a thin layer (layer 2) sandwiched between layers 1 and 3. In (D) the distance between geophones is too large to permit the identification of layer 2.

Figure 5.15 Travel time–distance graphs for forward-direction shots. The first breaks for the far-offset shots (shown as crosses) for the leftmost spread arrive earlier than those from the near-offset shot (shown by squares) for this spread. This occurs when the distance between the shotpoint and the refractor decreases significantly bewteen the near- and far-offset shots. Form Lankston (1990), by permission. See text for details

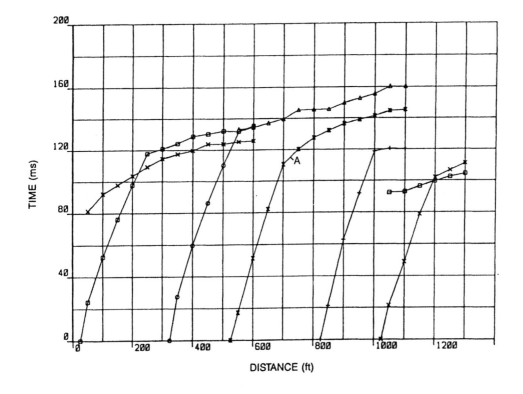

DISTANCE (ft)

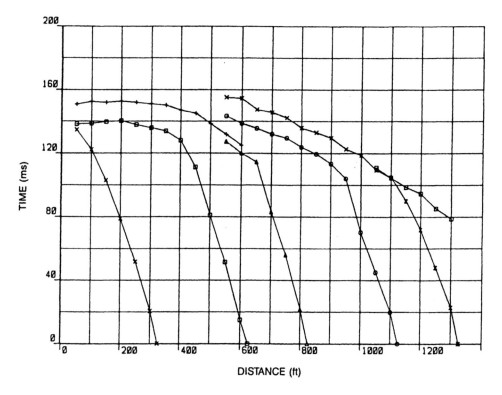

DISTANCE (ft)

The forward-direction travel time–distance graphs along one survey line are shown in Figure 5.15. The data are indicative of a two-layer case, with good parallelism evident in the high-velocity, deeper layer arrivals, except between 700 and 800 ft (213 m and 244 m) along the line. It is evident that the geophone interval is too large to provide the required resolution here. It is a good example of the occurrence of the problem illustrated in Figure 5.14D. The presence of an intermediate layer is indicated at 750 and 800 ft (indicated by the letter A in Figure 5.15).

The reverse-direction travel time graphs for the same survey line are given in Figure 5.16. However, these data do not help to resolve the problem over the presence of an intermediate layer – or to what extent the feature occurs laterally.

The data were phantomed (Figure 5.17) using the method described by Lankston and Lankston (1986). Forward data from the shot at 325 ft (99 m) and reverse data from the shot at 1125 ft (343 m) were used. Between each observed (or phantomed) point on the travel time–distance graph, travel times were interpolated in order to improve the resolution of the optimum XY value using the general reciprocal method. The resulting velocity analysis is shown in Figure 5.18. Applying Palmer's (1980) criterion of least detail, the curve for $XY = 0$ was considered to be the best for defining the refractor's velocities. Applying Palmer's (1980) maximum detail criterion to the

Figure 5.16 Travel time–distance graphs for the reverse-direction shots. From Lankston (1990), by permission

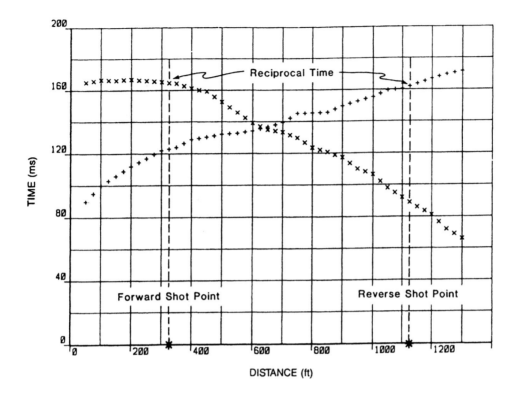

DISTANCE (ft)

time–depth data in Figure 5.19, an optimum $XY = 0$ was also chosen. In generating the time–depth graphs, the refractor velocity was required; three values were interpreted along the refractor as indicated in Figure 5.20.

With such a large inter-geophone spacing (50 ft), quantifying any change in the optimum XY value on the basis of the data acquired was not possible. However, had the layer thicknesses and velocities been known, the optimum XY value could have been predicted. By working backwards in this example from the interpreted velocities and depths, the optimum XY value was predicted to be equal to about 6 m (20 ft).

Having determined the refractor velocities and, from microspreads, the velocities of the near-surface layers, within the constraints of the actual survey, the depth of the refractor below each geophone position was determined. The uppermost layer was determined to have a velocity of 305 m/s (1000 ft/s) and a thickness of 3.1 m (10 ft) along the length of the survey line. The second layer was assigned a velocity of 550 m/s (1800 ft/s). The calculated depth to the refractor is shown graphically below the appropriate geophone as an arc whose radius is equal to the calculated depth at that point (Figure 5.21). The actual surface of the refractor is then indicated by the envelope of tangents to the circular arcs. Also indicated in Figure 5.21

Figure 5.17 Forward- and reverse-direction travel time–distance graphs after intra-line phantoming and interpolation of new arrivals at locations halfway between observed arrivals. The reciprocal time as interpreted from the forward and reverse direction experiments is indicated. From Lankston (1990), by permission

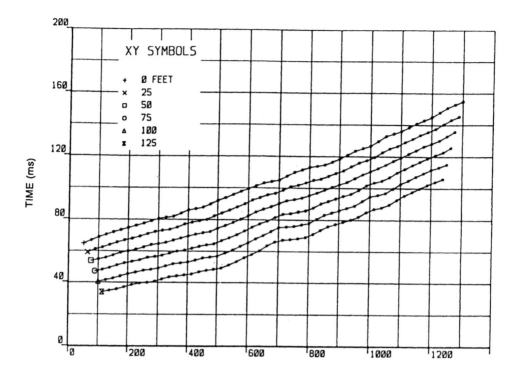

are the three values of refractor velocity. As it was known that the bedrock was basalt and that its structure was generally flat-lying, then the interpreted structure depicted in Figure 5.21 appears to be reasonable, with the three intra-refractor velocities corresponding to layers within the basalt.

Figure 5.18   Velocity analysis curves for *XY* spacings from zero to 48 m (125 ft). From Lankston (1990), by permission

Having produced the interpretation shown in Figure 5.21, this does not necessarily reflect the true geological picture. Earlier, it was suggested that there was an intermediate layer whose presence was intimated in the data in Figure 5.15 (at A). This has not been taken into account in the final model. If it had been, the depression evident in the refractor surface at 750 ft along the profile (which corresponds to the location of A) would have had a greater calculated depth to its surface.

In this example, Lankston (1990) reported that at the intersection of cross lines, the interpreted depths to the refractor agreed to within 5%. It was also stated that the final model produced was a fair reflection of the geological structure and that the survey had achieved the client's objectives.

5.5.2   Location of a buried doline

Palmer (1991) has described a seismic refraction survey over a known collapsed doline (sink hole) in Silurian metasediments in central

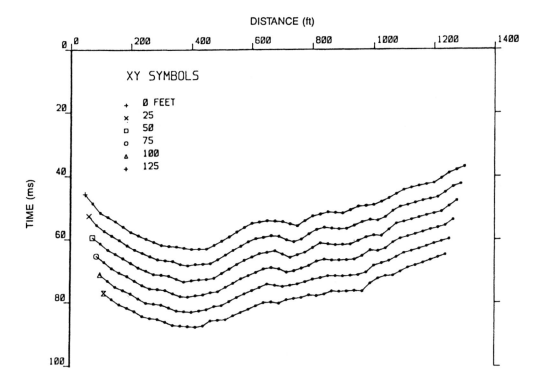

Figure 5.19 Time-depth curves for XY spacings fron zero to 48 m (125 ft). Refractor velocities used to generate these curves are indicated in Figure 5.20. From Lankston (1990), by permission

eastern Australia. In contrast to the previous example, the geophone interval was 2.5 m with a shot point interval nominally of 30 m. The travel time–distance graphs obtained are shown in Figure 5.22.

The refractor velocity analysis for shots at stations 12 and 83 are shown in Figure 5.23 for values of XY from zero to 15 m, in increments of 2.5 m. Velocity analysis for the refractor revealed values of 2750 m/s between stations 24 and 46, and 2220 m/s between stations 46 and 71. Palmer determined that the optimum value of XY was 5 m. The travel time–depth graphs over the same spread and range of XY values are shown in Figure 5.24. From these graphs the optimum value of XY was confirmed as being 5 m. A deepening of the refractor is evident between stations 40 and 52.

Evaluating equation (5) in Box 5.10 with an optimum value of XY of 5 m, a time-depth of 15 ms is obtained and a refractor velocity of 2500 m/s, with an average seismic velocity of 600 m/s for the material above the refractor. This latter value is comparable to that determined directly from travel time–distance graphs. The similarity of the near-surface velocities determined using the two approaches suggests strongly that there is no hidden-layer problem. Using these velocities, a depth section was produced (Figure 5.25).

In Figure 5.25, the arcs, representing the loci of the refractor surface beneath stations 39 and 46, all intersect in a cluster beneath station 39.

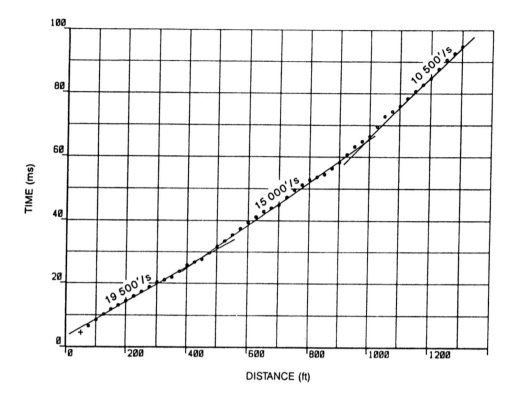

Similarly, the arcs from stations 47 and 54 cluster beneath station 53. Such clusters indicate that all arrivals associated with the collapsed doline at these locations are diffractions from the edges. The refractor surface shown in Figure 5.25 is the minimum equivalent refractor rather than a true image of the doline. A drill hole sited near station 47 intersected almost 50 m of siliceous sandstone and voids. It was concluded that, with the large vertical extent of the doline relative to its 30 m diameter, and a roughly circular plan section, most of the seismic energy probably travelled around the doline rather than underneath it. Consequently, the minimum equivalent refractor was thought to be a hemisphere with a diameter of about 30 m.

From this analysis using the GRM, it was also concluded that there was no low-velocity zone associated with the collapsed doline. If a value for XY of zero was used in the velocity analysis function, it was possible to infer a seismic velocity of about 750 m/s between stations 44 and 47. This in itself would warrant further investigation. However, as it was thought that the seismic energy travelled around the doline, the volume beneath the feature was thought not to have been adequately sampled. Nevertheless, it was concluded that any low-velocity zone, if there was one, would have been of only limited lateral extent.

Figure 5.20 Velocity analysis curve for the $XY = 0$ case; the refractor is interpreted to have two lateral velocity changes. From Lankston (1990), by permission

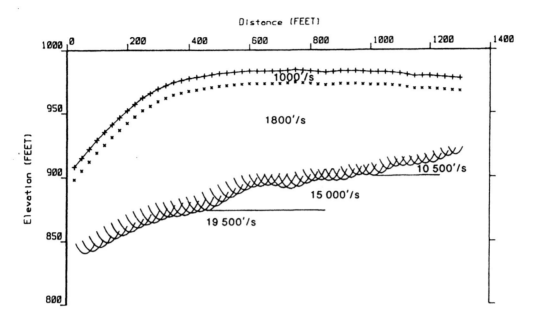

Another feature of this example is that it demonstrates the ability of the GRM to take diffractions into account without having to recognise them beforehand. In this respect, the migrated depth section (Figure 5.25) is very useful in revealing the clusters of arcs that are indicative of the diffractions. Furthermore, inclusion of the drilling results also helped to elucidate the mechanism of wave propagation in the vicinity of the collapsed doline. The combined interpretation that resulted was more realistic than if either the refraction survey or the drilling had been used in isolation.

5.5.3   Assessment of rock quality

There is an increasing requirement for geophysical surveys carried out during geotechnical investigations to provide direct information about rock quality or other geotechnical parameters. With the paucity of information to correlate geophysical results with actual rock properties, this is still difficult to achieve. Much more research needs to be done to address this. However, New (1985) has described one detailed study designed specifically to investigate this matter.

As part of the United Kingdom's radioactive waste management research programme, the Transport and Road Research Laboratory undertook a short programme of seismic velocity measurements at Carwynnan, Cornwall. The site was an experimental excavation in massive granite towards the north-western margin of the Carnmenellis Granite boss. The Cornubian batholith, of which the Carnmenellis

Figure 5.21 Final migrated section. The refractor surface is the envelope of the tangents to the suite of arcs. The arcs are mathematically circular, but they appear elliptical because of vertical exaggeration of the elevation scale. The ground surface is denoted by the continuous curve. The small crosses represent the interface between the 305 m/s (1000 ft/s) unit and the underlying 550 m/s (1800 ft/s) unit. The positions of the lateral velocity changes from Figure 5.20 are shown and have been interpreted as being due to layering with the basalt. From Lankston (1990), by permission

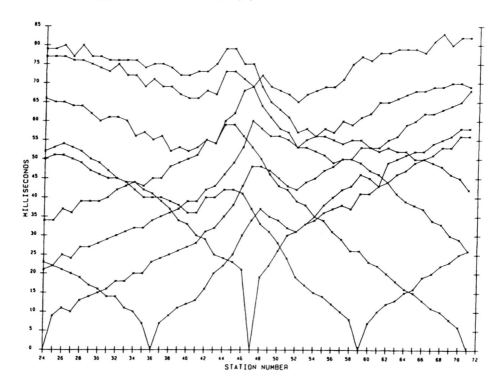

boss is part, is predominantly coarse-grained biotite–muscovite granite with phenocrysts of potassium feldspar. The rock is generally very strong and is weakly weathered. A predominant sub-vertical joint set, with a 120° strike direction and a joint spacing usually greater than 1 m, was found throughout the test site area.

The survey was undertaken within three horizontal headings at a depth of 30 m below ground level. Twenty reference points were fixed along three sides of a rectangular section of rock some 900 m² in area. Transmitting and receiving locations were numbered from 0 to 19 (Figure 5.26A). Geophones were bolted rigidly to the rock and surveyed in to an accuracy of ±0.5 cm. The seismic source used was a sledge hammer.

The travel times and waveforms of the seismic signal were recorded and analysed using Fourier processing and a representative velocity (V_R) determined for each cell of a grid calculated for the rock mass. Contoured values of V_R were displayed in map form (Figure 5.26B). It is obvious that rock affected by the excavation of the headings has a reduced seismic velocity (< 5000 m/s) in comparison with that for undisturbed rock (≈ 5500 m/s).

A site-specific rock classification system was used to derive a numerical classification value (C) for each area of wall between reference points. This parameter depended upon:

Figure 5.22 Travel time-distance graphs recorded over a collapsed doline in eastern Australia; the inter-geophone spacing is 2.5 m. From Palmer (1991), by permission

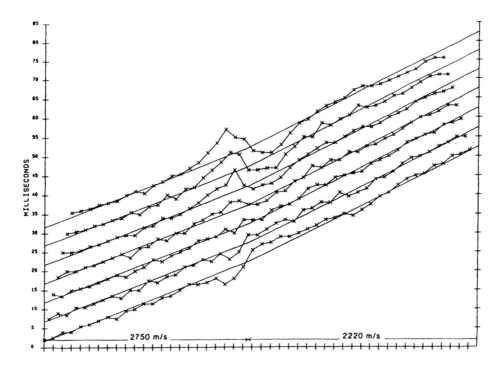

Figure 5.23 The GRM velocity analysis function computed with XY values from zero, the lower set of points, to 15 m, the upper set of points, in increments of 2.5 m. The reciprocal time has been increased by 10 ms for each XY value for clarity of display. From Palmer (1991), by permission

- the spacing of the joints;
- the condition of the joints (aperture, filling etc.);
- the general intact rock condition;
- the degree of excavation-induced blast damage;
- the density of shot holes in a given area – these were drilled to depths of between 1 m and 3 m.

The representative seismic velocity (V_R) was plotted as a function of the derived rock quality classification number (C) (Figure 5.27) for each pair of reference points.

It is clear from Figure 5.27 that there is a correlation between the seismic velocity and the rock quality. Indeed, New (1985) calculated the correlation as $V_R = 48.4 C + 4447$ (with a correlation factor $r^2 = 0.86$). The very lowest velocities were found to be associated with a damaged and weathered area of rock with several major joints at the western extremity of the survey area. The north-western and south-western walls of the test area had been affected by heavy blast damage and extensively drilled compared with the smaller heading along the south-eastern wall.

This experiment, although limited in extent, did demonstrate that classification of rock using seismic methods can satisfactorily indicate the engineering properties of a rock mass. See also papers by Sjögren *et al.* (1979) and Gardener (1992).

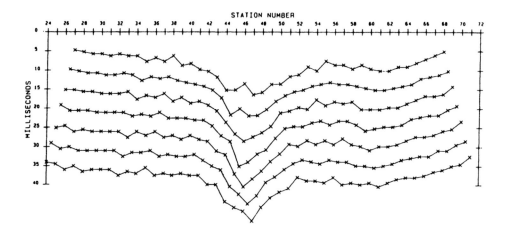

A very well-known application of seismic velocities is the determination of rippability – the ease with which the ground may be excavated using machines. In 1958, the Caterpillar Tractor Company developed the use of seismic velocities determined from surface refraction experiments to produce a practical chart (Figure 5.28) of rippability. Using this, contractors can estimate the ease (or otherwise) of excavating a given tract of ground using a mechanical device. Obviously, the chart is not exhaustive, but it provides general guidelines for practical use. Similarly, rippability is also a function of vehicle size, power and machine type. More specific charts relate geological material and rippability to machine types (Caterpillar Tractor Co. 1988). Other charts have been produced by other bulldozer manufacturers (e.g. Komatsu). However, disputes have arisen because the determination of rippability is not precise and contractors have found that ground reported to be rippable was patently not so, or their production rates were considerably slower than expected because of greater difficulty in excavating the ground. The estimation of rock rippability, including the use of seismic velocities, has been reviewed extensively by MacGregor *et al.* (1994).

Figure 5.24 The GRM travel time–depth function computed with XY values from zero, the lower set of points, to 15 m, the upper set of points, in increments of 2.5 m. The reciprocal time has been increased by 10 ms for each XY value for clarity of display. From Palmer (1991), by permission

Figure 5.25 The depth section computed with $XY = 5$ m and an average velocity in the overburden of 600 m/s; the position of the collapsed doline is evident. From Palmer (1991), by permission

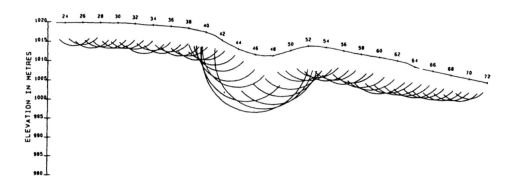

(A)

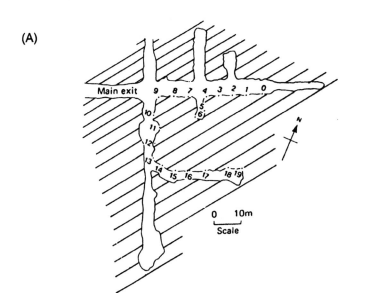

Figure 5.26 (A) Map of the three underground headings with source-receiver locations numbered. (B) Contoured map of the representative velocity (V_R in m/s), based on measurements between reference locations. From New (1985), by permission

(B)

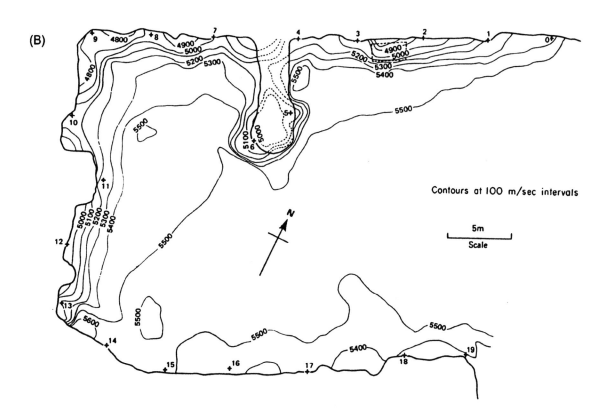

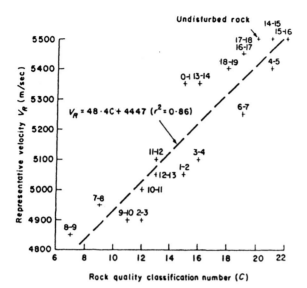

Figure 5.27 Correlation of representative velocity with rock quality classification. From New (1985), by permission

5.5.4   Landfill investigations

Refraction rather than reflection seismology has been used on landfills for three principal reasons:

- most landfills are too shallow for reflection seismology to be of much use;
- landfill material is highly attenuating and thus it is difficult to put much energy into the material and detect any significant signals;
- the cost of seismic reflection surveys is much greater than that of refraction surveys.

Even refraction surveys are seldom used over closed landfills (Reynolds and McCann 1992). One of the reasons for this is the difficulty in energy input – a hammer source may not be powerful enough and shotgun shells cannot be used where there is an active generation of methane gas! The general use of seismic methods on landfills is thus quite rare.

For selected sites where conditions permit, refraction seismology can be used to determine (1) the depth of fill; (2) P-wave velocities of the fill and substrates; (3) the location of old quarry walls that may delimit the edge of a landfill; and (4) the state of a clay cap, if it is of sufficient thickness. An example of a seismic refraction record with displaced first arrivals due to a vertical step associated with a step within an old backfilled quarry is shown in Figure 5.29.

Knight *et al.* (1978) have described the results of a seismic refraction survey carried out at a waste disposal site at Lucas Heights near

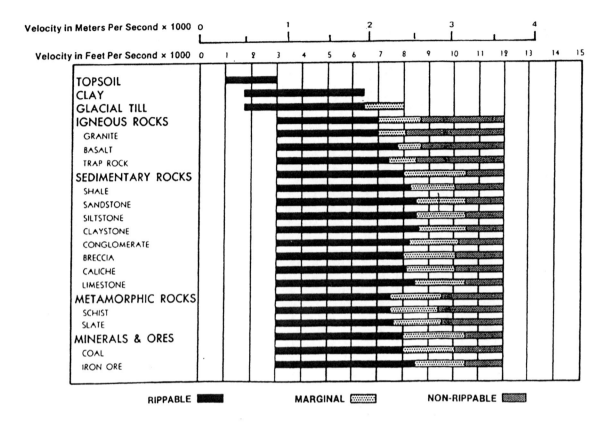

Velocity in Meters Per Second × 1000

Velocity in Feet Per Second × 1000

TOPSOIL
CLAY
GLACIAL TILL
IGNEOUS ROCKS
  GRANITE
  BASALT
  TRAP ROCK
SEDIMENTARY ROCKS
  SHALE
  SANDSTONE
  SILTSTONE
  CLAYSTONE
  CONGLOMERATE
  BRECCIA
  CALICHE
  LIMESTONE
METAMORPHIC ROCKS
  SCHIST
  SLATE
MINERALS & ORES
  COAL
  IRON ORE

RIPPABLE ▉   MARGINAL ▨   NON-RIPPABLE ▨

Figure 5.28 Typical chart of ripper performances related to seismic P-wave velocities. © Caterpillar Tractor Company, by permission

Sydney, Australia, to determine the depth of the fill material and the properties of the underlying bedrock. The fill material was made up of glass, metal and plastic, with small quantities of paper, cardboard and wood. At the base of the fill was a 'brown viscous sludge' at a depth of about 4 m below ground surface. The depth of fill was 6.5 m and it was underlain by sand containing a second water table and shale at a depth of 19 m. Their results were generally of poor quality because of the high attenuation of the seismic energy in the fill material. With a geophone interval of 3 m, the close-spaced survey indicated a P-wave velocity of 450 m/s for the fill material and 1900 m/s for the underlying sand.

Nunn (1978) described a refraction survey to determine the depth of fill at a landfill site at Brownhills, Warwickshire, UK. The fill material was colliery spoil and domestic waste. The actual site was an old lake bed over a shallow depression in Etruria Marl. The fill material was underlain by an unknown thickness of glacial drift deposits, interbedded clays, sands and gravels, which were known to vary both laterally and vertically. Nunn found P-wave velocities of 500 m/s for unsaturated fill material, and 1300 m/s for saturated fill below the water

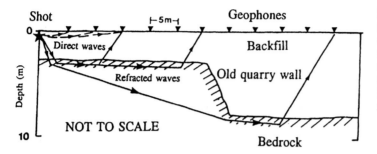

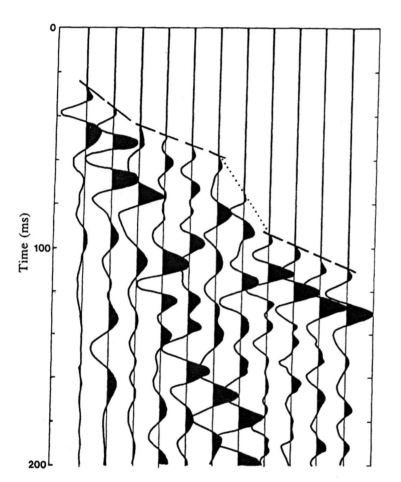

Figure 5.29 Seismic section over a buried quarry face within a landfill site backfilled with marl with the corresponding refraction record; the lines of the first breaks are indicated by the dashed and dotted lines. From Reynolds and McCann (1992), by permission

table. There was no direct evidence of the saturated drift layer in the results and thus it was thought that this layer was 'hidden'.

As far as the assessment of the geotechnical properties of a landfill is concerned, the work of Baoshan and Chopin (1983) on the measurement of shear waves in a tailings deposit at four tailings dam sites in China is of particular significance. Surface to borehole and cross-hole

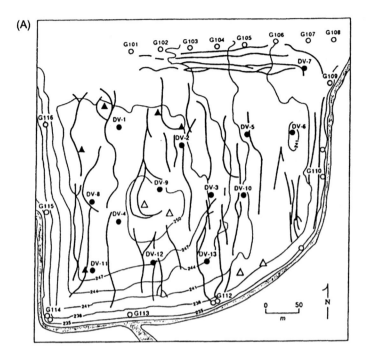

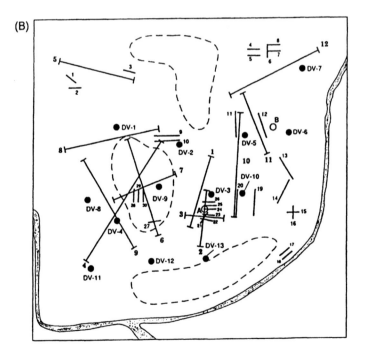

Figure 5.30 Map of the Mallard North landfill, showing (A) topography, major fissures, shallow/deep gas vents, and monitoring wells, and (B) lines of resistivity soundings, seismic refraction profiles, and azimuthal resistivity arrays marked ('A' and 'B'). Areas of cover replacement during the experiment are also shown in (B). From Carpenter *et al.* (1991), by permission

seismic measurements were made to determine the shear wave velocity (V_s). Their results showed excellent correlation between V_s and the geotechnical properties of the tailings material.

An integral part of the successful management of an enclosed landfill is the maintenance of the integrity of the compacted clay cap overlying the waste material. As long as this impermeable layer remains intact, gases are kept beneath (to vent in a controlled manner through appropriate outlets) and rain water/snow melt is kept out to run off into surface drains. However, erosion can occur into this clay cap and it can also degrade through differential settlement of the waste beneath.

Carpenter *et al.* (1991) reported on their use of both seismic refraction and electrical resistivity surveys to examine the integrity of a clay cap over a municipal landfill at Mallard North, near Chicago, USA. They demonstrated that detailed mapping of P-wave velocities could be used to identify areas where the clay cap had been fractured (giving rise to low P-wave velocities) compared with the intact clay cap (with higher P-wave velocities). Similarly, variability in electrical resistivity with azimuth around a central point indicated the orientation of fractures within the clay cap. Maps of the site and of their survey locations are shown in Figure 5.30.

Carpenter and co-workers found that average P-wave velocities determined along survey lines parallel and perpendicular to fractures were around 370 ± 20 m/s and 365 ± 10 m/s, respectively, compared with a value of 740 ± 140 m/s over unfractured clay cap. They also reported difficulty in obtaining refracted arrivals in some areas owing to the P-wave velocity in the underlying waste being lower than that for the clay cover. It is thought that where clay caps are of the order of 1.5–2 m thick, as in this case, electrical resistivity sub-surface imaging could provide a quick and reliable method of measuring the thickness non-intrusively.

Chapter 6
Seismic reflection surveying

6.1   INTRODUCTION

Seismic reflection surveying is the most widely used geophysical technique, and has been since the 1930s. Its predominant applications are hydrocarbon exploration and research into crustal structure, with depths of penetration of many kilometres now being achieved routinely. Since around 1980, the method has been applied increasingly to engineering and environmental investigations where depths of penetration are typically less than 200 m. Applications of shallow high-resolution seismic reflection surveying include mapping Quaternary deposits, buried rock valleys and shallow faults; hydrogeological studies of aquifers; shallow coal exploration; and pre-construction ground investigations for pipe, cable and sewerage tunnel schemes.

One of the principal reasons for the increased use of shallow seismic reflection surveying has been the improvement in equipment capabilities and the availability of microcomputers and portable digital engineering enhancement seismographs. Computing power is such that preliminary data processing can be accomplished while still in the field in some cases. This has facilitated real-time processing on board ships for quality assurance and control, the availability of portable data processing centres which can be housed within a container and flown into remote areas by helicopter, through to self-contained highly sophisticated seismographs which can be carried by one person. Perhaps more importantly, the growth of computer power has also provided the means whereby three-dimensional data acquisition can be undertaken much more cost-effectively. Data manipulation and management can be accomplished much faster than was possible 15 years ago. In the late 1970s, the costs of three-dimensional seismic surveying were at least double that of conventional two-dimensional acquisition, and it was only undertaken as a last resort. Now, 3-D seismics have become much more cost-effective and currently constitute in excess of 60% of the market share in the seismic industry.

As seismic reflection surveying is such an established technique, and as a very large amount of research and development has been undertaken within the hydrocarbon industry, there is a vast technical literature available. Members of the International Association of Geophysical Contractors, for example, are usually keen to respond to requests for the very latest details of proprietary techniques, new equipment for data acquisition and processing methods.

The literature in relation to shallow applications is, however, surprisingly sparse, with little attention having been paid to this growing area of work. This chapter deals particularly with shallow seismic reflection surveying but also provides an overview of the wider issues of the methods where appropriate.

6.2 REFLECTION SURVEYS

6.2.1 General considerations

The essence of the seismic reflection technique is to measure the time taken for a seismic wave to travel from a source (at a known location at or near the surface) down into the ground where it is reflected back to the surface and then detected at a receiver which is also at or near the surface at a known position. This time is known as the *two-way travel time* (TWTT). The seismic method is used to obtain important details, not only about the geometry of the structures underground, but also about the physical properties of the materials present. After all, the ultimate objective of hydrocarbon exploration is to find oil in sufficiently large quantities in places from where it can be extracted cost-effectively. In engineering applications, everything is scaled down and depths of penetration are typically of the order of 10–50 m and the sizes of mappable targets are correspondingly reduced in size.

The most important problem in seismic reflection surveying is the translation of two-way travel times (in the time domain) to depths (in the space domain). While travel times are measured, the one parameter that most affects the conversion to depth is seismic velocity. Consequently, a great deal of effort in research is devoted to improving our understanding of this parameter. Although this chapter is devoted to seismic reflection methods, analytical techniques used and interpretational problems are equally valid for 'ground penetrating radar' data (see Chapter 12).

A further difficulty with seismic reflection surveying is that the final results obtained from data processing undertaken by contractors are likely to be different. Although the basic processing routines may be exactly the same, the choice of parameters to be used in the various stages of processing is more subjective. Thus, the final output from a variety of processors may have different frequency content and signal-to-noise ratio, and even differences in structural detail. Interpreters beware! Furthermore, some data processing may be carried out that adds nothing to the technical content of the data but which focuses attention by the form of display on to specific zones within the data. This 'cosmetic' processing, which may have some benefit to the contractor in marketing terms, has no specific technical justification.

6.2.2 General reflection principles

The reflection method requires a source of seismic energy and an appropriate method of both detection (geophones or hydrophones) and recording (seismographs). Details of these three sets of components are discussed in Chapter 4. For a seismic wave to be reflected back to the surface, there has to be a sub-surface interface across which there is a contrast in acoustic impedance (Z), which is the

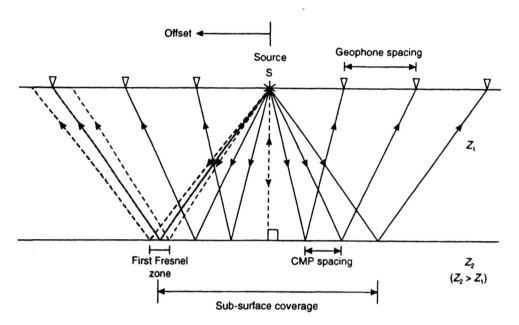

Offset

Source
S

Geophone spacing

Z_1

First Fresnel
zone

CMP spacing

Z_2
$(Z_2 > Z_1)$

Sub-surface coverage

product of the seismic velocity (V) and the density (ρ) of each layer (i.e. $Z = V_i\rho_i$ for the ith layer). The amplitude of the reflected wave is described by the *reflection coefficient* (see Section 4.3.1).

If a seismic source is discharged at a given shot point S and the reflected waves are detected at geophone locations laid out in a line each side of the shot, then the resulting raypaths are as shown in Figure 6.1. At each point of incidence on a sub-surface interface, over an area corresponding to the first Fresnel zone (see Section 4.3.1), the incident waves are reflected back. The point of reflection is half-way between the source and the detecting geophone. The spacing between reflection points on the interface is always half that of the geophone spacing. Consequently, the total sub-surface coverage of an interface is half the total geophone spread length. The distance from the source to any geophone is known as the *offset*.

If more than one shot location is used, reflections arising from the same point on the interface will be detected at different geophones (Figure 6.2A). This common point of reflection is known as the *common midpoint* (CMP). Sometimes the terms *common depth point* (CDP) and *common reflection point* (CRP) are used as being equivalent to CMP. This is true, however, only in the case of flat horizons with no lateral velocity variation. If the reflector is dipping, the points of reflection are smeared along the interface (Figure 6.2B). According to Snell's Law, the angle of reflection equals the angle of incidence, and it is for this reason that the point of reflection moves up-slope with greater shot–geophone offset. It is recommended that only the term '*common mid-point*' be used.

Figure 6.1 Schematic of reflection raypaths over a horizontal interface across which there is a contrast in acoustic impedance Z $(Z_2 > Z_1)$; the first Fresnel zone at just one location is indicated

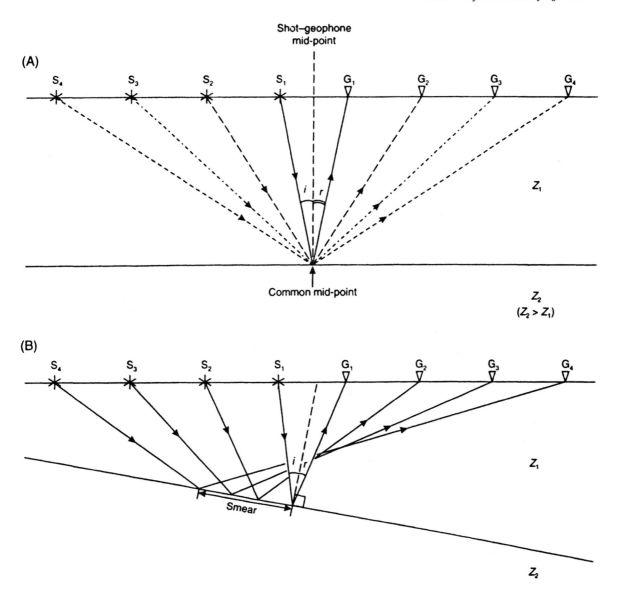

Figure 6.2 Principle of the common midpoint over (A) a horizontal and (B) a dipping interface

The number of times the same point on a reflector is sampled is known as the *fold of coverage*. If a sub-surface point is sampled only once, as in common-offset shooting, then this is termed *single-fold* or 100% coverage. If 12 different shot–geophone locations are used to sample the same point on a reflector, for example, then this is said to have *12-fold* or 1200% coverage. Folds of coverage from 6, 12, 24, 48 and 96 are frequently used in the hydrocarbon industry, with up to 1000-fold being used exceptionally; in engineering surveys, 12-fold coverage is usually considered excessive. From the number of shots

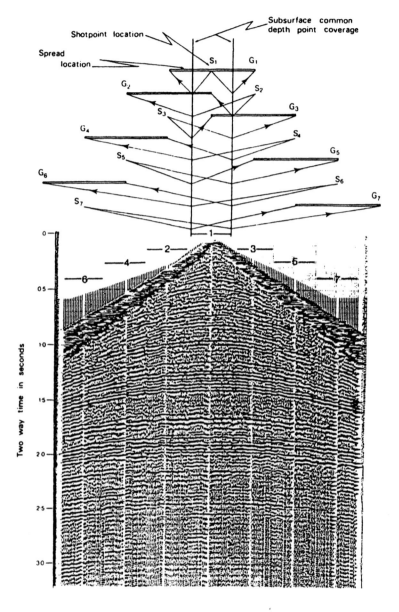

Figure 6.3 Example of a common midpoint gather

providing the fold of coverage, it is possible to gather together all the traces arising from the same CMP to produce a *common midpoint gather* (CMG) (Figure 6.3). The use of CMP gathers is described in more detail in Sections 6.3.3 and 6.3.4.

6.2.3 Two-dimensional survey methods

Survey designs vary according to the requirements of the investigation. Survey layouts for engineering investigations are considerably simpler than for large-scale hydrocarbon exploration, with corresponding differences in costs.

For an engineering shallow investigation, the layout consists of a shot of some description, and a set of 12, 24 or occasionally 48 geophones with a fundamental frequency of at least 100 Hz connected by a multicore cable (with the appropriate number of takeouts) to an engineering seismograph. The output from each geophone is displayed as one trace on the field record (seismogram).

For larger-scale surveys, more than one geophone may be connected together to form a 'group' or 'array' of geophones, the output of which is summed together and recorded in one channel. There is a variety of different group layouts: the 'in-line array', where all the geophones within the group lie along the line of the survey; the 'perpendicular array', where the geophones are aligned at right-angles to the survey line; and the 'cross array', which is a combination of the previous two. Other examples are given by Sheriff (1991, p. 16). In some cases the number of geophones in a group may be as many as several hundred laid out in a rectangular grid, in which case the set of geophones is known as a *patch*. When multiple geophones are used per recording channel, the offset distance is taken as that between the shot and the centre of the group. The distance between centres of adjacent geophone groups is known as the *group interval*.

The use of multiple sets of geophones gives a directional response to the array, designed to enhance the near-vertical upward-travelling reflected waves and to minimise any horizontally travelling coherent noise – which can be correlated from trace to trace, in contrast to random noise which has no coherency between traces. Coherent noise may be in the form of a Rayleigh wave, for example. If the individual geophones in a given group have a spacing equivalent to a half-wavelength of the Rayleigh wave, then the signals at alternate geophones will be out of phase and will be cancelled by the summation of the outputs from the geophones. An upward-travelling reflected bodywave, however, should arrive at the geophones simultaneously, and the total output is then the sum of all the in-phase signals from all the geophones within the group. This has the advantage that it also helps to filter out some random noise and has the overall effect of increasing not only the signal strength but also the signal-to-noise ratio.

Just as geophones can be laid out in groups, so too can the source shots, and for very much the same reasons. For engineering surveys, however, it is unusual to use either source or geophone arrays.

The shooting arrangements for reflection surveys on land are normally end-on shots, firing into a line of geophones, or split-spread shots, where the source is located within the line of geophones. It is usual for a reflection profile to be built up from end-on shots with split-spread shots as required. For example, if geophones have been laid out along the survey line at an interval of 2 m, the shot may be fired at the end of the line, and then every 2 m along the line, but slightly offset at right-angles from it. The fold of coverage of a reflection profile can be calculated from the expression $N/2n$, where n is the total number of geophone groups in a spread and n is the number of group intervals by which the entire spread is moved along the survey line between shots (i.e. the move-up rate). Thus for a 24-geophone spread ($N = 24$), with a move-up rate of one interval per shot interval ($n = 1$), the fold of coverage would be $24/2 \times 1 = 12$-fold. An example of the successive shot–spread layouts required to achieve a 6-fold coverage from a single 12-channel spread is given in Figure 6.4.

In water-borne surveys for hydrocarbon exploration, shots are fired using one or more tuned airgun arrays, into one or more hydrophone streamers in which a number of hydrophones are interconnected to form an active group the summed output from which feeds one channel. See Section 4.6.2 for more details of hydrophone streamers. In small-scale engineering surveys, a single airgun or higher-frequency source may be used (see Section 4.5.4) with a short hydrophone streamer, perhaps less than 10 m long, with the output from each hydrophone being summed to give one channel only. Given adequate digital data recording systems with fast enough sampling rates, very high quality records can be produced even though only single-fold coverage is achieved. The compactness of size of equipment, and the shortness of the streamer, permit seismic reflection surveys to be undertaken from small boats (< 10 m long) in small rivers and narrow canals.

In land-based shallow engineering and hydrogeological investigations, the simplest form of reflection surveying uses a single geophone offset from the source by a constant amount. This survey configuration is known as the *constant-offset* method. The source and receiver are moved progressively along the survey line in equal increments with one trace being produced from the single geophone from each shot (Figure 6.5). The seismic record is obtained by displaying successive traces side-by-side. The main advantage is that, as the incident raypath is virtually perpendicular to the ground surface, there is virtually no data processing required in order to obtain a seismic section. For short survey lines requiring only very basic information, simple and cheap seismographs are available for constant-offset surveys. For larger surveys, it is better to lay out a geophone cable with 12 or 24 takeouts and to shoot with optimum offset into one active geophone at a time, with all the rest switched off. The benefit of this approach depends upon the optimum offset chosen and the geophone interval selected for most efficient data acquisition and reflector imaging.

The method has been developed further by the Geological Survey of Canada into what is known as the *optimum-offset method*. If the source and single receiver are located at virtually the same point, groundroll and perhaps the airwave from the shot would swamp the

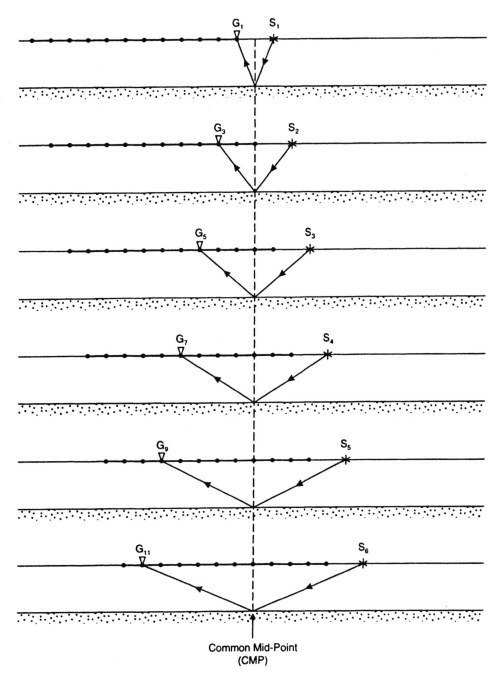

Common Mid-Point
(CMP)

geophone and no useful information would be obtained. In order to reduce the effect of the groundroll, it is necessary to offset the geophone from the source by an optimum amount (Figure 6.6). This offset distance is best determined from trial reflection shots to gauge the most appropriate offset in order to image the reflectors required.

Figure 6.4 Sequence of survey layouts to acquire 6-fold coverage; S indicates the source and G a geophone (or hydrophone)

(A)

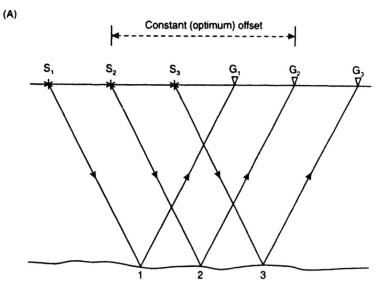

Figure 6.5 Constant-offset seismic reflection surveying using the optimum-offset window. (A) Raypaths, and (B) three adjacent traces obtained using the layout shown in (A)

(B)

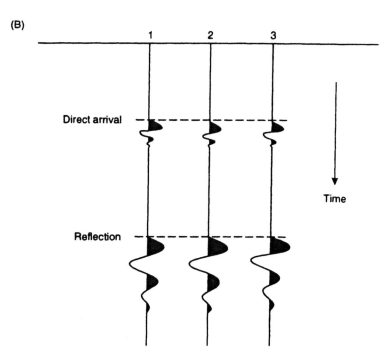

The optimum offset window is the range of offsets at which the bedrock reflection arrives before either the groundroll or airwave. The choice of incremental move-up distance between shots is dependent upon the type of target being imaged. If the reflector in question has significant dip or surface topography, then a short spacing is best

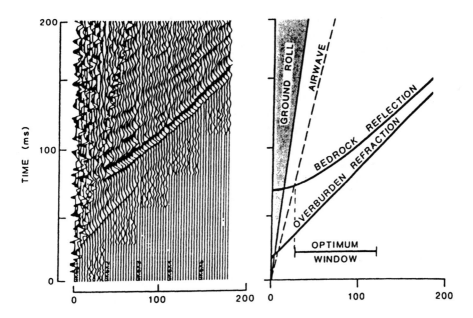

Figure 6.6 (A) Composite shallow reflection record made up of five separate 12-channel records, each of which was recorded with a different time delay between the shot instant and the start of recording. (B) Corresponding time–distance graph identifying the major seismic events on the record. The optimum window is that range of source–receiver separations that allows the target reflection to be observed without interference from other events. From Slaine *et al.* (1990), by permission

(typically 1–3 m). If the reflector targeted is reasonably planar, then the spacing can be increased, perhaps to 5 m or more. Further details of the method have been summarised by Pullan and Hunter (1990).

6.2.4 Three-dimensional surveys

Three-dimensional surveys were first undertaken in 1975. Since then there has been a rapid growth in the use of this mode of survey, particularly after 1985. Previously, it was only used over mature producing fields with good prospects for hydrocarbon recovery. With substantial improvement in three-dimensional acquisition efficiency and processing, the costs have fallen such that the method is now used increasingly as part of primary exploration surveys to aid appraisal of new fields as well as over producing fields. Three-dimensional surveys are also used to evaluate trends in reservoir quality. The proven advantages of 3-D over 2-D surveys have resulted in a massive increase in its usage; for example, of its annual expenditure on seismic methods, Shell invested 75% in 3-D (Nestvold 1992). Indeed, it is likely that 3-D seismics will become the primary survey method in hydrocarbon exploration in many areas.

While the method is obviously highly successful in the hydrocarbon industry, in engineering investigations it is used only where the high cost can be justified, such as in major investigations for nuclear power stations or for sites for the safe disposal of highly radioactive waste material deep underground. Henriet *et al.* (1992) published an account of very-high-resolution 3-D seismic reflection surveying for geotechnical investigations with a bin size of only 1 m². Two sources were used, namely an EG&G Boomer™ and a Sodera 15 in³ modified airgun firing into an array of 12 dual-channel streamers at 1 m separation; the array was orientated at right-angles to the boat track, giving a 5.5 m wide swathe. The survey imaged a clay diapir (with a reported accuracy of better than 0.25 m) over a depth range of 50 m to about 25 m below ambient ground level with a 60 m diameter.

A three-dimensional survey requires extremely careful design so as to avoid spatial aliasing. In the field, the data are acquired in subsets known as *common shot gathers*. When all the subsets are taken together, these common-shot gathers form a complete 3-D data set.

A marine 3-D survey is accomplished by shooting closely spaced parallel lines, known as *line shooting*. On land or over shallow water, the survey uses a number of receiver lines deployed parallel to each other with the shot points located along lines at right-angles in a configuration known as *swathe shooting*. In marine surveys, the direction in which the shooting vessel sails (boat track) is known as the *in-line direction*, whereas in land 3-D surveys the receiver cable is deployed along the in-line direction. The line at right-angles to the in-line direction is known as the *cross-line direction*. The line spacing in marine 3-D surveys is typically less than 50 m, and may be as small as 12.5 m. Consequently, it is essential that the locations of all shots and receivers be accurately known at all times in order to locate the area on each reflector from where the reflection originates. In 2-D surveys, traces are collated as CMP gathers; in 3-D surveys, traces are collected together as common cell gathers ('bins'). Effectively, the common point of reflection from a large number of source–receiver pairs lie within an area ('bin') on the reflector (Figure 6.7), with a typical size of 25 × 25 m. The fold of coverage is then the number of traces that pertain to a given bin.

Three independent coordinate sets need to be considered: travel time (t), and two spatial coordinates, midpoint (x_m) and offset (x_o). The same data can be described in terms of travel time (t), source coordinate (x_s) and receiver coordinate (x_r). The relationship between these sets of coordinates is given in Box 6.1 and illustrated in Figure 6.8 (Vermeer 1991).

The two methods of displaying shot–receiver geometries in Figure 6.9 are known as a *surface diagram* (shot–receiver coordinate system) and a *subsurface* or *stacking diagram* (midpoint–offset coordinate system). The design of sampling using 3-D seismics is dealt with in more detail by Vermeer (1991), for example. As the range of applications diversifies, so too does the option of source–receiver geometry in 3-D acquisition. Additional survey layouts (Sheriff and Geldart 1982) include:

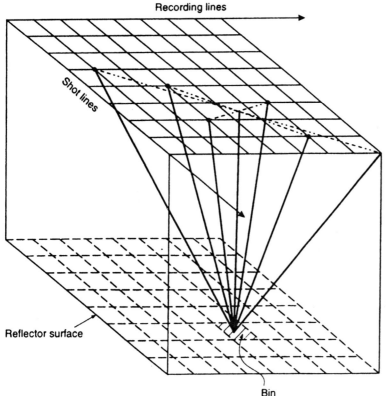

Recording lines

Shot lines

Reflector surface

Bin

Figure 6.7 Three-dimensional survey with a small number of raypaths shown for a given bin for a common cell gather. There is no requirement for the bin size to be the same as the shot–receiver grid – these are shown schematically

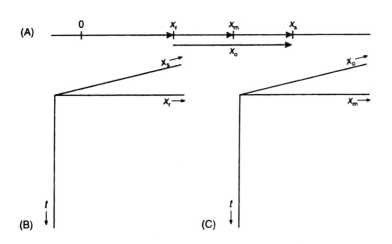

Figure 6.8 Prestack data coordinate systems: (A) the four spatial coordinates in relation to the seismic line; (B) the shot–receiver coordinate system; and (C) the midpoint–offset coordinate system, Vermeer (1991)

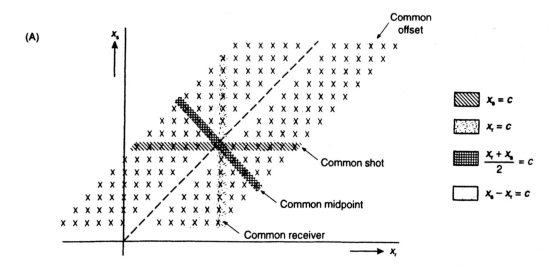

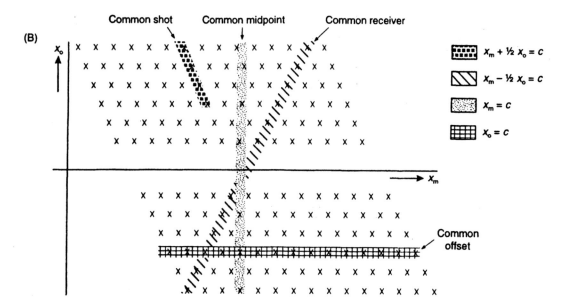

- the *wide-line layout* (Compagnie Général Geophysique), in which a single line of groups of geophones is straddled obliquely by four source points each with a different offset (Figure 6.10A);
- *zigzag* (Figure 6.10B), where the source boat zigzags a track relative to a straight streamer track (Bukovics and Nooteboom 1990);
- *block layout*, where several parallel lines of geophones are recorded simultaneously from an orthogonal line of shots (Figure 6.10C);
- and *Seisloop*, where both geophones and source points are located around the perimeter of a square (Figure 6.10D).

Figure 6.9 Descriptions of a pre-stack seismic data set in: (A) shot–receiver coordinate systems (a *surface diagram*); and (B) midpoint–offset coordinate systems (a *sub-surface diagram* or *stacking diagram*). From Vermeer (1991), by permission

(A)

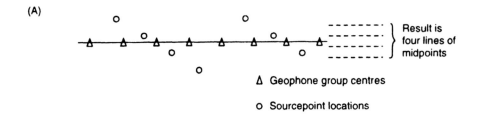

(B)

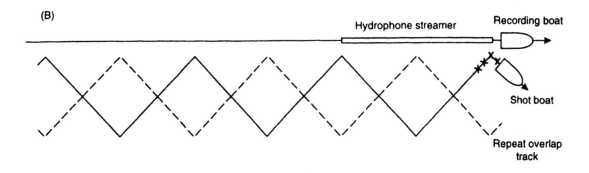

(C)

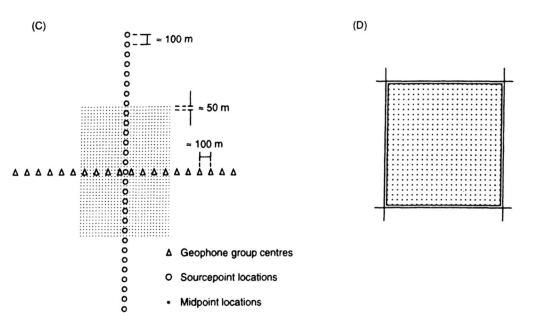

(D)

Figure 6.10 Survey layout for 3-D investigations: (A) wide-line layout; (B) zigzag (Bukovics and Nooteboom 1990); (C) block layout; and (D) Seisloop where the geophones and sources are located around the perimeter. After Sheriff (1991), by permission

Box 6.1   Spatial coordinate system for 3-D seismic surveys
(see Figure 6.8)

The distances between a coordinate reference point $(0, 0)$ and the seismic source (x_s) and receiver (x_r) produce a distance to the shot–receiver midpoint (x_m) with offset being given by x_o. Thus:

$$x_m = (x_s + x_r)/2 \quad x_s = x_m + x_o/2$$

and

$$x_o = x_s - x_r \quad x_r = x_m - x_o/2.$$

In a marine survey, a two-boat system has been devised by GECO in 1988 to provide rapid and cost-effective acquisition of 3-D seismic reflection data. It is known as Quad/Quad because it consists of four hydrophone streamers (each 3 km long with 120 hydrophone groups) and four source arrays towed behind two seismic vessels. One vessel is designated the master ship with the second acting as slave. All survey operations are controlled from the master ship. The field and shooting configurations are shown in Figure 6.11. Each source array is fired in turn at 10-second intervals; at a typical cruising speed of 5 knots (9.3 km/h), the ships sail a distance of 25 m during this time interval giving rise to a shot ('pop') interval of 25 m and a CMP interval of 12.5 m. The reflected wavefield from the sub-surface is recorded using all four streamers. By sequencing the firing, complete sub-surface coverage can be achieved across a swathe, typically 300 m wide.

Three-dimensional surveys are also achieved using a single streamer with one source array where the streamer is allowed to drift off the boat track line. The feathering of the streamer caused by cross-currents provides a swathe of coverage of the sub-surface (Figure 6.12). By careful monitoring of the positions of the streamer at regular time intervals, the positions of the active hydrophone groups can be determined relative to the source, and hence the locations of the respective midpoints can be calculated.

A further development in 3-D data acquisition has been Concentric Circle Shooting (CCS – trademark by Grant Tensor and patented by W. French) for improved imaging of salt domes. The first deployment of this method in Europe was in Block 29/2c in the North Sea in October 1991 by Mobil (Hird *et al.* 1993). Concentric Circle Shooting uses a single ship from which a dual streamer/dual source is deployed where each source array is fired alternately (the *flip-flop method*). The ship sails in a circle of a predetermined radius in order to acquire the sub-surface coverage needed. The radius of each circle is decreased until a minimum size is achieved which is then infilled using straight survey lines.

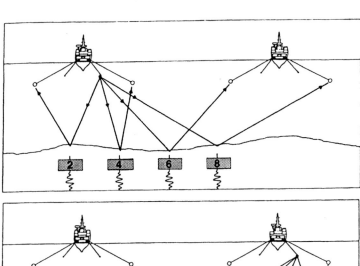

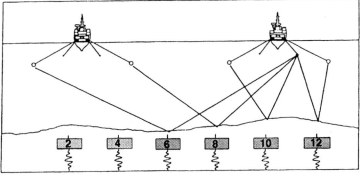

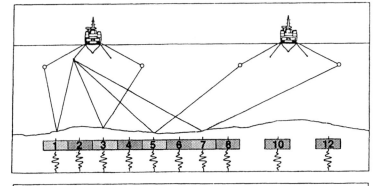

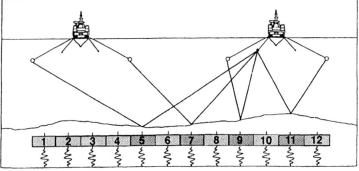

Figure 6.11 The sequence of firing four source arrays to obtain complete sub-surface coverage in the Quad/Quad system. This firing sequence is similar to that of a car's engine – in the sequence 1, 3, 4, 2. After four source arrays have been fired, 12 common midpoints (CMPs) have been recorded across a swathe of about 300 m. From Hansen *et al.* (1989), by permission

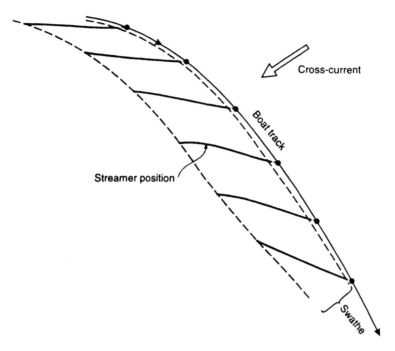

Figure 6.12 Feathered hydrophone streamer locations at discrete intervals of time showing the swathe covered by this geometry

In a salt diapir environment, the CCS method permits the acquisition of data in essentially the strike direction, which minimises non-hyperbolic moveout distortion of the shot records and maximises the energy returns along the entire length of the streamer (Reilly and Comeaux 1993). Another advantage of the circular shooting geometry is that it provides optimal imaging of any radial or non-linear fault pattern within the survey area.

Of critical importance to 3-D data acquisition is positional accuracy. While 3-D data acquisition has resulted in a 10-fold increase in seismic data collected, the amount of survey data has mushroomed 30-fold. The collection of navigational data, and the mathematical processing of information, for instance, from up to 48 compasses (12 per streamer), 6 radio positions and 92 acoustic ranges for each and every shot fired, is a non-trivial exercise. Consequently, there has been a substantial growth in the supply of specialist navigational control software.

6.2.5   Vertical seismic profiling (VSP)

Vertical seismic profiling (VSP) is effectively seismic surveying using boreholes. Seismic detectors are located at known levels within a borehole and shots fired at the surface, and vice versa. The VSP method utilises both the downgoing wavefield as well as the reflected and/or diffracted wavefield to provide additional information about the zone imaged by the method. In areas of structural complexity, seismic data acquired from only the surface may not provide the clarity of imaging necessary to resolve the detail required. Furthermore, additional information may be needed in areas now occupied by drilling rigs and oil platforms. As drilled holes already exist in these cases, it makes economic as well as technical sense to use them.

The basic layout of a VSP survey is shown schematically in Figure 6.13. There are several different types of VSP configurations such as the static VSP, with one string of detectors at a fixed position with a single shot at the surface (shown in Figure 6.13); single-level walkaway VSP, where shots are fired into one downhole detector from source points with increasing distance from the wellhead (i.e. walking away from the hole); and multilevel walkaway VSP, the same

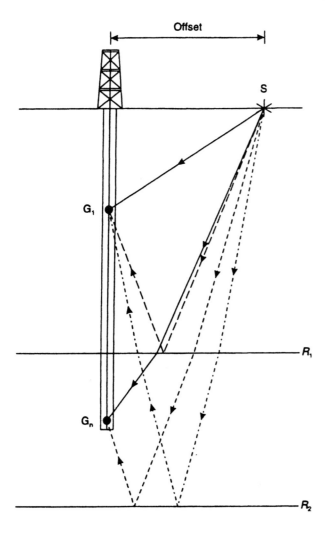

Figure 6.13 Schematic showing the principle of vertical seismic profiling (VSP). A string of geophones or hydrophones is suspended down a borehole and a shot is fired at S. Direct, reflected and diffracted wavefields are detected along the string of geophones between G_1 and G_n. R_1 and R_2 are successive reflectors

(A)

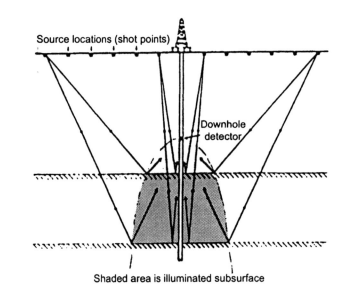

Figure 6.14 Single-level walkaway VSP with sample raypaths showing (A) a section through the inverted paraboloid zone of illumination, and (B) downgoing and upgoing wavefield signals recorded at a single detector. After Ahmed *et al.* (1986), by permission

(B)

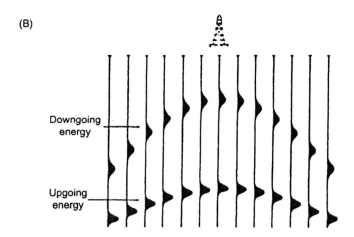

as for single-level walkaway VSP but with a string of downhole detectors over a wide range of depth levels.

In the case of a single-level walkaway VSP, a double wavefield situation is created in which it is difficult to differentiate between the downgoing and upgoing signals (Figure 6.14). If such a VSP is undertaken along radial transects around a wellhead, the zone of illumination has the form of an inverted paraboloid (shaded section in Figure 6.14A). In the case of a multilevel walkaway VSP survey, the corresponding schematic raypaths and associated common-shot point gathers are as shown in Figure 6.15. Use of the multilevel walkaway survey helps to resolve the problem of identifying the

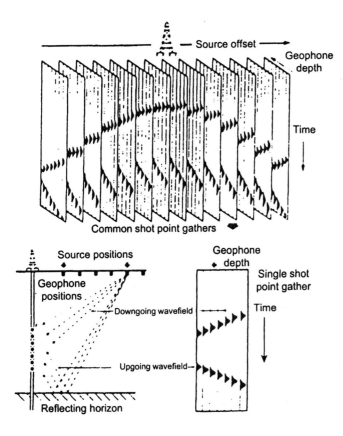

Figure 6.15 Multilevel walkaway VSP survey schematic and associated shot-point gathers. After Ahmed *et al.* (1986), by permission

downgoing and upgoing wavefields. An added advantage of the VSP method is that it can utilise shear waves as well as P-waves.

Incident P-waves can be mode-converted under suitable conditions to vertically polarised S-waves (SV) or S-waves can be generated specifically (e.g. Edelmann 1992). As the S-wave velocity is between 50 and 75% that of P-waves, the S-wave wavelength is significantly shorter, thereby increasing the resolution available. In addition, highly fractured zones tend to attenuate S-waves preferentially with respect to P-waves as the S-waves are more directly affected by changes in the rigidity modulus or density of the medium. Also, Poisson's ratio, estimates of porosity and attenuation (Q) can be computed giving more engineering attributes for the host rock (Ahmed *et al.* 1986). In addition, where a VSP survey is undertaken on land, and where S-waves can be generated specifically as well as P-waves, then the shear-wave splitting (polarisation into components) can be used to determine anisotropy of the host media at a resolution much less than the wavelength (e.g. Crampin 1985; Douma *et al.* 1990). Further details of VSP surveys have been given by Hardage (1985) and Maio *et al.* (1995), for example, and their use is discussed further in Section 6.4.2.

6.2.6   Cross-hole seismology: tomographic imaging

Of increasing importance in seismic investigations is the use of hole-to-hole (*cross-hole*) surveys, also known as *borehole tomography*. Two or more boreholes may be used simultaneously, with one hole being used to deploy a source at one or more levels, and the other borehole(s) to house geophone or hydrophone strings. The basic configuration is shown in Figure 6.16.

The essence of the method is that the raypath lengths between sources and receivers are known, and that by measuring the travel times along these paths the average velocity along each can be

(A)

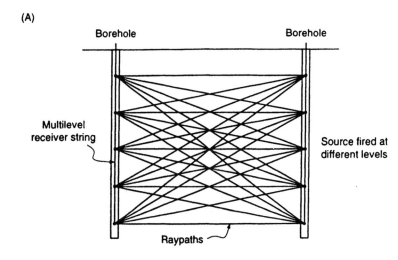

Figure 6.16 Cross-hole surveying. (A) Raypaths from a source fired at different levels, and (B) from a single source position, into a multilevel string of receivers

(B)

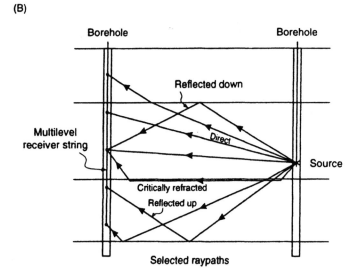

determined. The plane in between the source and receiver holes is divided into a grid of cells. Each constituent cell is assigned a velocity and the synthetic travel time through it along the portion of raypath intersecting the cell is calculated. By summing the incremental travel times of imaged cells along a given raypath and comparing the total against the measured travel time, the assigned velocities in the various cells can be changed iteratively so that the difference between the observed and modelled travel times is reduced to a minimum. As some cells are imaged by many raypaths, the iterative process can produce very good estimates of cell velocity. Consequently, any zones with anomalous velocity lying in the imaged plane between boreholes can be identified. The methods of producing tomographic images is based upon similar principles to those used for medical scanning in hospital radiological or orthopaedic departments.

Just as seismic velocities (P- and S-wave) can be determined, so too can Poisson's ratio and relative attenuation, thus providing important engineering information about the host materials. In applications for engineering site investigation, for example, the inter-borehole separation should be no more than 10–15 times the maximum size of target being imaged. In hydrocarbon surveys, the distance between boreholes may be of the order of 1 km.

A major advantage of the cross-hole survey is that high-frequency sources can be used and introduced to geological horizons at depths inaccessible from surface surveys owing to the attenuation of the high frequencies near the surface. Consequently, the resolution possible using this method is much better than with either surface or VSP surveys. A major disadvantage in the oilfield situation is the requirement to have simultaneous access to two boreholes, especially if they are part of a producing field. In engineering surveys, cross-hole boreholes are usually drilled specifically for the survey and the cost of drilling the holes has to be considered as part of the overall expense of the survey. Further details of cross-hole surveys are given, for example, by Stewart (1991), Tura *et al.* (1994) and Lines *et al.* (1993), and of the data processing methods by Hatton *et al.* (1986).

6.3 REFLECTION DATA PROCESSING

There are three principal stages in seismic data processing: deconvolution, stacking and migration. In addition, there are many auxiliary processes that are necessary to improve the primary processes. While maintaining a particular sequence of processing is essential for the principal stages, other 'commutative' processes can be undertaken at any stage; 'commutativity' means that the final results can be achieved irrespective of the order in which the processes are undertaken. The success of an individual process is dependent not only on the choice of most appropriate parameters but also on the effectiveness of the processing stages already completed (Yilmaz 1987). Conventional data processing is based on certain assumptions. Many of the secondary processes are designed to make data compatible with the assumptions of the three primary processes (Yilmaz 1987):

- Deconvolution assumes a stationary, vertically incident, minimum-phase source wavelet and reflectivity series containing all frequencies and which is noise-free.
- Hyperbolic moveout is assumed for stacking.
- Migration is based on a zero-offset (primaries only) wavefield assumption.

Although these assumptions are not exactly valid, drilling results have confirmed the accuracy of the processing, thus demonstrating the validity of these methods. This is also true because the various processes are in themselves robust and their performance is relatively insensitive to the underlying assumptions.

The following sections are designed to provide a brief overview of the main seismic data processing stages required to convert raw data into an interpretable seismic record. The main stages in data processing are shown in Figure 6.17 and are described briefly in turn. Seismic data processing has been discussed much more fully by Hatton *et al.* (1986) and by Yilmaz (1987), for example.

6.3.1 Preprocessing

6.3.1.1 *Demultiplexing*

Signals received from detectors are digitised at small increments in time, known as the *sampling interval*. In large-scale seismic surveys, this may be 2 ms or 4 ms over a total two-way travel time range of 6–7 seconds. For higher-resolution surveys the sample interval may be reduced to 0.25 ms, and in very-high resolution, such as in single-channel marine engineering surveys using pingers, the sample interval may be as short as 15 μs.

The format in which the data are stored is usually the industry-standard SEG (SEG stands for Society of Exploration Geophysicists). There are a number of recognised versions such as SEG-B and the more modern SEG-D. Details of SEG formats are given in the SEG publication *Digital Tape Standards*. However, some contractors use their own formats.

The recorder stores data in multiplexed time sequential form, where the record contains the first sample from each trace (trace 1, sample 1; trace 2, sample 1; trace 3, sample 1; ... last trace, sample 1; then trace 1, sample 2; trace 2, sample 2; etc.). While this is convenient

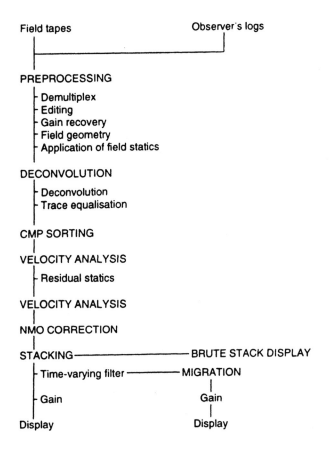

Figure 6.17 Basic data processing flowchart. After Yilmaz (1987), by permission

for recording purposes it is not so for processing. Consequently, the order of samples is changed by demultiplexing from time sequential order to trace sequential order (trace 1, all samples in order; trace 2, all samples in order, etc.). In this format all the samples for one trace are contained in one block and can be managed as a discrete dataset very easily. The recognised format of demultiplexed data is SEG-Y, which is supposed to be universally compatible. However, in engineering seismology, manufacturers of seismographs have developed instruments that follow the SEG-Y format but in their own modified forms, which can cause problems when transferring data to processing software. Individual data processing houses demultiplex data to the form of SEG-Y which is most appropriate for their processing system. Further details of SEG formats have been given by Hatton *et al.* (1986).

Once into trace-sequential format, the traces recorded from a single shot (Figure 6.18A) can be displayed as a *common shot gather*, for example. This is the standard display of a seismic record on an engineering seismograph. Traces purtaining to the same midpoint on

(A)

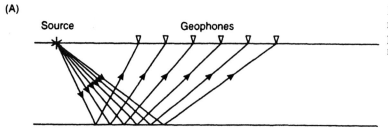

(B)

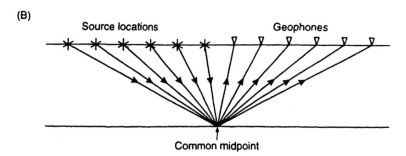

(C)

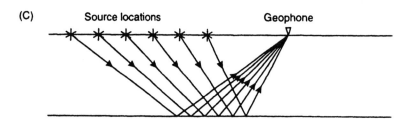

Figure 6.18 Layouts for (A) a common-shot gather; (B) a common midpoint gather; and (C) a common receiver gather

a reflector can be collated as a *common midpoint gather* (Figure 6.18B; see also Figure 6.3). Traces recorded at the same receiver but from different shot locations can be collated as *common receiver gathers* (Figure 6.18C).

6.3.1.2 *Editing*

Once the data have been demultiplexed, they can be edited to remove dead traces (i.e. those containing no information) or those with monocyclic waveforms caused by electronic interference (e.g. a sinusoidal waveform with constant amplitude and wavelength), to reverse the polarity of traces recorded the wrong way around, or those with excessive noise. Editing effectively is cleaning the dataset of bad traces that would otherwise pollute the quality of the remaining data if used in subsequent data processing.

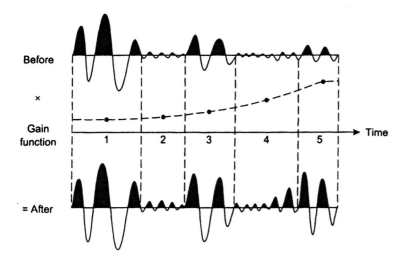

Figure 6.19 The application of a time-varying gain function to a waveform exhibiting spherical divergence, in order to recover signal amplitudes at later travel times. Gain functions are applied in discrete windows (labelled 1 to 5 shown here)

6.3.1.3 Gain recovery

In Section 4.4.1, it was explained how energy is lost as a seismic wave travels away from its source by geometric spreading or spherical divergence. As the rate of energy loss can be calculated, it can be compensated by applying a gain function to the data in order to restore the signal amplitudes with increasing two-way travel times. Just as geologically significant reflection signals are amplified by the application of a gain function, so too is noise (Figure 6.19); this is a disadvantage of the process.

6.3.2 Static corrections (field statics)

Static corrections are made to take into account the effect on travel times of irregular or changing topography and of near-surface velocity variations. In the preceding descriptions of survey layouts, it has been assumed that the source and receivers are all located at the ground (or water) surface. While this may be true of seismic surveys using an impact or vibrator source into a spread of surface-laid geophones, in many cases the source and/or the detectors may be located at different levels within the ground. To compensate for the different lengths of raypaths associated with a buried source to a spread of geophones over an irregular topography, corrections may have to be applied so that the processed data refer to a specified datum that is taken to be horizontal and to contain no lateral variations in seismic velocity. This process of referring to an arbitrary datum is also known as *datuming*. If these adjustments are not made, reflections on adjacent traces may be shifted in time, producing irregular and incoherent events. As these adjustments are applied to a fixed source–receiver geometry for a given survey, they are known as *static* corrections or *field statics*.

In addition to the source–receiver geometrical corrections, adjustments to travel times also have to be made to account for the near-surface weathered zone in land reflection surveys. Seismically this weathered zone refers to the portion of the sub-surface above the water table where pore spaces are filled by air rather than water. The seismic weathered layer does not necessarily conform to the depth of geological weathering. This layer may exhibit anomalously low velocities (from 150 to 800 m/s) and hence is often referred to as the *low-velocity layer* (LVL). Smaller-scale seismic refraction surveys are commonly undertaken in order to determine the thickness and velocity of the weathered zone. This information is used to determine the appropriate depths at which the shots should be placed below the base of the weathered layer, and to provide the travel-time corrections at each detector position in the main reflection survey.

It is important to consider the spatial consistency of any near-surface model derived throughout a survey area before computing statics corrections. Otherwise mis-ties between intersecting reflection lines may be introduced by inconsistent selection of appropriate statics. It is not currently feasible for the low-velocity layer to be defined precisely seismically. Consequently, errors or inaccuracies remaining in the data are known as *residuals*. These may lead to a reduction in the coherence of individual reflection events on seismic sections. A specific process of *residual static analysis* is undertaken during the pre-stack data processing to refine the statics corrections. Such processing can lead to substantial reduction in trace-by-trace jitter of individual reflection events.

The various static corrections, which are described in turn in the following three sections, have been reviewed succinctly by Marsden (1993a,b,c).

6.3.2.1 *Source–receiver geometry*

In a survey undertaken over an irregular surface, the differences in elevation of the source and the individual receivers have to be taken into consideration in order to compensate for the differences in travel times caused by the irregular topography. Given the situation shown in Figure 6.20A, a time correction can be calculated such that the final travel times then refer to a plane datum. The travel time taken from a reflector to a given geophone will be delayed by the difference in elevation divided by the seismic velocity of the near-surface layer. The greater the elevation of the geophone relative to the datum, the larger will be the time delay. Similarly, the time delay associated with a shot location above a given datum is simply the distance of the shot above the datum divided by the layer velocity. The total time delay due to elevation is given simply by the sum of the time delays of the shot and

Figure 6.20 (*opposite*) (A) Field static corrections to account for irregular topography. E_g and E_d are the elevations of geophones and the new datum, respectively. (B) (i) Uncorrected traces showing the amount of intertrace jitter, and (ii) traces corrected for differences in elevation shown in (A)

(A)

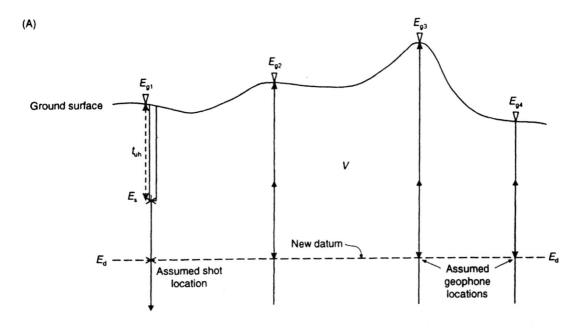

(B)

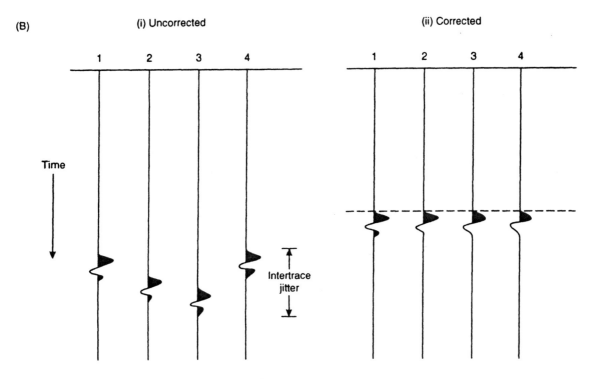

those associated with each geophone. The actual expressions for the time correction are shown in Box 6.2. The assumption is usually made that the downgoing and upgoing raypaths are virtually vertical.

Box 6.2   Static correction for elevation (see Figure 6.20)

Time delay due to the difference in elevation between source and datum (t_s) is given by:

$$t_s = (E_s - E_d)/V.$$

The time delay due to the difference in elevation between a given geophone and datum (t_g) is given by:

$$t_g = (E_g - E_d)/V.$$

Total static correction due to elevation (t_e) is given by:

$$t_e = t_s + t_g.$$

Velocity in the near-surface can be calculated from the measurement of the uphole travel time (t_{uh}) and the depth of the shot (h), such that:

$$V = d/t_{uh}.$$

E_s, E_g and E_d are the elevations of the shot, any geophone and the reference datum respectively.

An estimate of the average near-surface velocity can be obtained by measuring the uphole travel time from the shot to a geophone at the surface immediately above. As the shot depth (h) is known, and the uphole time (t_{uh}) is measured, the average near-surface velocity can be calculated very simply $(V = h/t_{uh})$.

Velocities derived from uphole shots are never likely to be exactly appropriate for the elevation component of field statics as they are still averaged values. Similarly, the average velocity defined over too large a vertical shot interval may not be locally representative. This is especially true in cases where the velocities change markedly over vertical distances significantly shorter than the vertical shot interval. Only one uphole survey may be undertaken in an entire survey area, in which case the velocities obtained for the near-surface materials may be totally unrepresentative away from the uphole survey hole. This applies particularly where Quaternary deposits are present in the near-surface.

Velocity can also be determined using refraction statics (see Section 6.3.2.3) and *data smoothing statics* methods. The latter method is automated and relies on statistical means of minimising the jitter in reflection events between adjacent traces.

Marine statics refer to the depth of the shot and the streamer below a given datum level, which is not the ambient sea level which fluctuates with tide. For example, a typical level in the UK is Ordnance Datum (Newlyn). Particularly in high-resolution engineering surveys, tidal corrections need to be made accurately as the water depth will change over a tidal cycle.

An additional correction is sometimes made to allow for the seabed topography. If a depression exists in the seabed, the additional depth of water over this provides a time delay that is manifest as a downwarping in deeper planar reflection events. For instance, a depression 7.5 m deep will add a time delay of 10 ms (assuming a P-wave velocity in seawater of 1500 m/s or 1.5 m/ms; two-way distance = 15 m, hence time delay = 15/1.5 = 10 ms).

In single-channel marine seismic surveys in shallow water, it is sometimes necessary to allow for the offset between the hydrophone streamer and source. This is because, with the shallow depth, the offset results in significant difference in the total raypath length compared with normal incidence. As all reflection events are considered assuming normal incidence, a time reduction is necessary to produce accurate two-way travel times to reflection events. This correction can also apply to constant-offset seismic reflection profiling on land, especially when the optimum offset is comparable to the depth to the first reflector. An alternative method which avoids the problems of source–receiver geometry in marine surveys is to determine the elevation of the seabed using echosounding. Distances to reflectors are determined from the two-way travel times using an assumed reduction velocity relative to the seabed.

6.3.2.2 *Feathering*

A further consideration in marine seismic reflection profiling is the position of the streamer relative to the source at any given time. When significant cross-currents occur, the streamer can drift off track to a considerable degree. This deviation from the boat track is known as *feathering* and the angle of the streamer to the boat track is the *feathering angle*. However, with streamers up to 6 km long, they do not keep straight. Consequently, the positions of the tail buoy and mid-streamer compasses need to be logged at regular time intervals. This information is needed to determine the appropriate common midpoints and sub-surface coverage (see also Section 4.6.2).

6.3.2.3 *Weathered layer*

Of particular importance to seismic reflection surveying is accurate determination of field statics. In addition to elevation corrections, consideration has to be given to the presence of a near-surface weathering zone that may have anomalously low velocities and can exhibit considerable variation in seismic velocities.

The weathering zone, when present, can range in thickness from only a few metres to several hundreds of metres. When this zone is only thin, source shots can be located relatively easily beneath the weathered layer. When the zone is very thick, this is no longer feasible owing to the costs of having to drill such deep shot holes and to the reduction in production rates that this would cause. Consequently, it is necessary to calculate the delay-time effect of the weathered zone.

Incorrect evaluation of the weathering layer can result in corruption of stacking (see Section 6.3.5) and may even introduce spurious features into deep reflectors. The importance of correct determination of the weathering layer has led to the development of a host of different methods (e.g. Hatherley *et al.* 1994) to account for the weathering layer with increasing sophistication, such as 2-D refraction tomography (e.g. Docherty 1992). Refraction statics are an important method of investigating the weathering layer and special refraction surveys are used.

While many of the different methods produce virtually identical statics corrections, the differences in the speed of computation can be very marked. For industrial applications, rapidity of calculation is an important factor, and methods that produce the correct best estimates of statics corrections in the least time are obviously preferred. As an example, Marsden (1993b) describes a test undertaken by Amoco in which a comparison was made between using the slope/intercept method and refraction tomography. The former method took three days to calculate, whereas the tomographic method took only three hours. One example of a rapid method of statics determination has been given by Zanzi and Carlini (1991). They demonstrated that if the calculations are undertaken after Fourier transformation of reflection data, then the calculation process can be substantially reduced without loss of accuracy.

6.3.3   Convolution and deconvolution

An initial waveform (W) propagating into the ground is filtered by layers with different acoustic impedances (which form a reflectivity time series denoted by R) through which the signal passes, resulting in a modified waveform (S) observed on a seismogram. This 'convolution' process is denoted by $S = W*R$, where $*$ represents the process of convolution. The seismic method generates a waveform whose initial shape should be known and the resulting seismogram S is measured. The principle of this process is shown in Figure 6.21. The only unknown is R. In order to unravel the seismogram to obtain this time series of ground properties, the seismogram needs to be 'deconvolved'. The main complication is that the time series R consists not only of primary reflection events, but also reverberation, ghosting,

Figure 6.21 (*opposite*) The principle of convolution:

Step 1: Convolve source wavelet and reflectivity series by multiplying the first sample of the source wavelet (1) by the first component of the reflectivity series (1) to give the first constituent of the output response

Step 2: Move the source wavelet array on one sample and convolve; hence $(1 \times \frac{1}{2}) + (1 \times (-\frac{1}{2})) = \frac{1}{2} - \frac{1}{2} = 0$

Step 3: Move the source wavelet array on one sample and convolve; hence $(\frac{1}{2} \times 1) + (\frac{1}{2} \times (-\frac{1}{2}) + (1 \times \frac{1}{2}) = \frac{1}{2} + (-\frac{1}{4}) + \frac{1}{2} = \frac{3}{4}$

Step 4: Move the source wavelet array on one sample and convolve; hence $(\frac{1}{2} \times (-\frac{1}{2}) + (\frac{1}{2} \times \frac{1}{2}) = -\frac{1}{4} + \frac{1}{4} = 0$

Step 5: Move the source wavelet array on the sample and convolve; hence $(\frac{1}{2} \times \frac{1}{2}) = \frac{1}{4}$

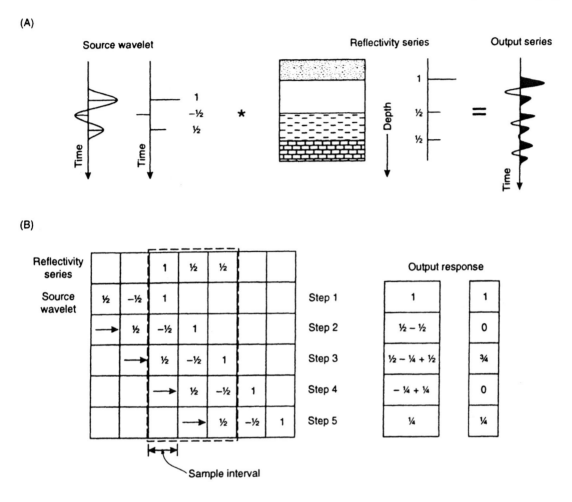

(A)

Source wavelet   Reflectivity series   Output series

(B)

Reflectivity series | Source wavelet | Step 1–5

diffractions, multiples and noise. Consequently the deconvolution process requires methods of removing or suppressing unwanted signals and of compressing the initial waveform to as close to a spike (Dirac) pulse as possible. By so doing, the geologically significant components of the time series R may then become clearer.

Deconvolution is an analytical procedure to remove the effects of previous filtering such as that arising from convolution. There are a number of different deconvolution processes:

- *Dereverberation* or *deringing* refers to the process by which ringing associated with multiple reflections within a water layer or other near-surface layer is strongly attenuated. Specific filters can be designed to remove certain simple types of reverberation, such as the Backus filter.
- *Deghosting* is designed to remove the effects of energy which leaves a source and travels upwards and is reflected back down to the

receiver. Ghosting is caused by signals bouncing off the undersurface of the water/air interface or the weathered layer where its lowermost boundary is well defined.

- *Whitening* (trace equalisation) adjusts the amplitudes of all frequency components to be the same within a bandpass. The effect of this is to make the amplitudes of adjacent traces comparable over some predetermined time interval. This may have the effect of worsening the signal-to-noise ratio in situations where the signal amplitude is already weak and where the window length across which the trace equalisation is applied is inappropriate.

The effect of each of the three processes described above is to shorten the pulse length on the processed seismograms, thereby improving the vertical resolution. The ultimate objective, which cannot as yet be achieved, is the compression of each waveform into a single spike (Dirac pulse, δ) such that each reflection is also a simple spike. By so doing, it should be feasible to determine the reflectivity series that defines the sub-surface geological stratigraphic sequences. The deconvolution operator (or inverse filter) (I) is designed by convolving it with a composite wavelet W to produce a spike pulse (i.e. $I*W = \delta$); see also Box 6.3. The designed inverse filter (I) can then be convolved with the seismogram trace S to produce the reflectivity series R (i.e. $I*S = R$). The type of filter that best achieves (as judged statistically) the reduction of the wavelet to a spike is known as a *Wiener filter*. This stage of deconvolution is referred to as *spiking* or *whitening deconvolution*. The latter name refers to the fact that in a spike, all the frequency components have the same amplitude.

Box 6.3   Deconvolution processes

A source wavelet (W) convolved (indicated by $*$) with a reflectivity series (R) produces the observed seismic trace (S), which is denoted:

$$R*W = S. \qquad (1)$$

A deconvolution operator (D) can be designed such that when it is convolved with the source wavelet (W) it produces a spike output (δ):

$$D*W = \delta. \qquad (2)$$

As seismic data processing aims to image the sub-surface geology, the objective of the deconvolution process is to unwrap the seismic trace to reveal the reflectivity series (R):

From Equation (1) above, $D*S = D*R*W = D*W*R$.
From Equation (2) above, $D*W*R = \delta*R = R$ (as $\delta = 1$).

When a deconvolution operator is changed as a function of travel time, instead of remaining the same for all travel times, the process is known as *time-variant deconvolution*. The need for such a time-variant function is that the frequency content of a wavelet changes along its raypath because of the preferential attenuation of higher-frequency components by the ground. Consequently, the wavelet increases in wavelength with travel time, i.e. the bulk of the energy occurs at decreasing frequencies with increasing travel time.

A method of removing multiples is to predict when they might be expected and then remove them. This is the process of *predictive deconvolution*. By knowing the arrival times of particular primary events, the arrival times of multiples can be predicted. For this method it is assumed that there is no systematic distribution of sub-surface reflectors, and that the composite waveform from an active seismic source has the bulk of its energy at the start of the pulse rather than somewhere down the tail of the waveform.

The use of predictive deconvolution leads on to a further process for identifying key primary or even secondary reflection events, namely *correlation techniques*. A multiple is a duplication of the primary reflection and hence should have the same or very similar characteristics as the primary event. For instance, the wavelet pulse shape associated with the seabed reflection is also evident in the first and successive multiples of the seabed reflection. If the wavelet shape of the seabed reflection is isolated, it can be moved along each trace and compared against the remaining seismogram. Where its shape is present in the seismogram at later travel times, the degree of correlation is high (a value of 1 indicates a perfect match). Successive multiples should be evident by having a high correlation coefficient. Correlating one signal against a time-shifted copy of itself is known as *autocorrelation*. Comparing one time series with another to obtain a quantitative measure of similarity is referred to as *cross-correlation*, of which autocorrelation is a special case.

6.3.4 Dynamic corrections, velocity analyses and stacking

The most critical parameter in seismic surveying, irrespective of the type or scale of application, is the determination of seismic velocity. It is this factor which is used to convert from the time domain (the seismogram) to the depth domain (the geological cross-section). Consequently, correct analysis of velocities is of vital importance.

In a horizontally layered sequence (Figure 6.22), where each layer has a diagnostic seismic 'interval' velocity V_i, the single-way travel time of a seismic wave through each layer is equal to the layer thickness (z_i) divided by the appropriate velocity (V_i) (see Box 6.4). Obviously, the two-way travel time is double this time. The overall travel time at normal incidence is thus the sum ($\sum$) of all the individual two-way travel times ($\sum(z_i/V_i)$). The average velocity V' is assumed to

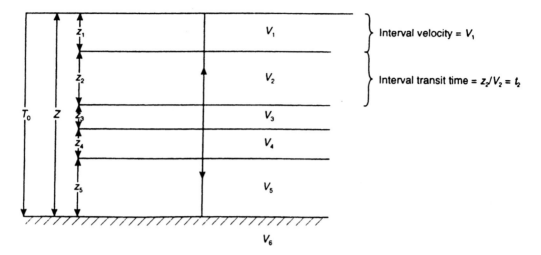

Figure 6.22 Definition of seismic velocity analysis terms (see Box 6.2). In the example illustrated, the total number of layers $(n) = 6$ and so $n - 1$ refers to the fifth layer

apply only to specific straight-raypath travel. In the case of normal incidence, the average velocity is simply the total raypath distance divided by total travel time. The weighted-average velocity is termed the root-mean-square (RMS) velocity (V_{RMS}) and applies to horizontal layers and normal incidence. Typically the RMS velocity is a few per cent higher than the average velocity. Neither of these two statistical parameters should be used uniquely as being diagnostic of the seismic velocity in a given material.

Box 6.4   Definition of velocity analysis terms (see Figure 6.22)

Interval velocity V_i = layer thickness z_i/interval transit time t_i:

$$V_i = z_i/t_i.$$

Average velocity V' is total raypath length (Z) divided by total travel time T_0:

$$V' = Z/T_0.$$

The root-mean-square velocity is:

$$V_{RMS} = [\sum V_i^2 t_i)/\sum t_i]^{1/2}$$

Two-way travel time of a ray reflected from the nth interface at a depth z is:

$$t_n = (x^2 + 4z^2)^{1/2}/V_{RMS}$$

where x is the offset distance.

————— *continued* —————

─── *continued* ───

Dix Formula: The interval velocity (V_{int}) over the *n*th interval is

$$V_{int} = \left[\frac{(V_{RMS.n})^2 \, t_n - (V_{RMS.n-1})^2 \, t_{n-1}}{(t_n - t_{n-1})} \right]^{1/2}$$

where $V_{RMS.n}$, t_n and $V_{RMS.n-1}$, t_{n-1} are the RMS velocity and reflected ray two-way travel times to the *n*th and (*n* − 1)th reflectors respectively.

The travel times associated with geophones at large offsets are greater than those at short offsets by virtue of the increased raypath distances (Figure 6.23). In the case of a horizontal reflector at a depth *z* below ground level, the difference in travel time at the largest offset from normal incidence is known as the *normal moveout* (NMO). From traces collated together in a common midpoint (CMP) gather, the same problem arises. The intention is to stack all the traces relating to the same CMP in order to improve the signal-to-noise ratio and thus emphasise the coherent signals at the expense of the incoherent

Figure 6.23 Selected raypaths and corresponding seismic traces illustrating the effect of normal moveout (NMO). After Dohr (1981), by permission

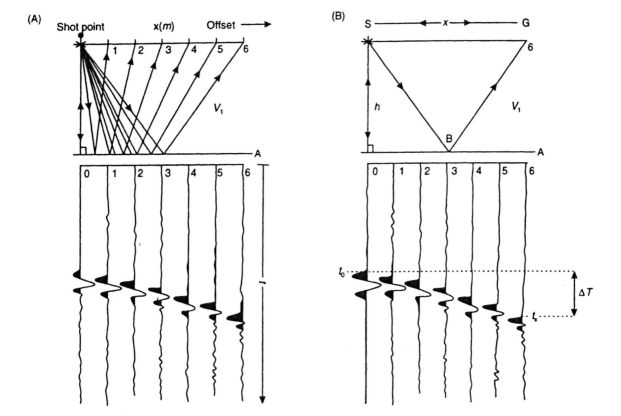

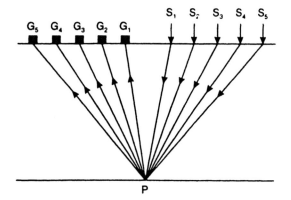

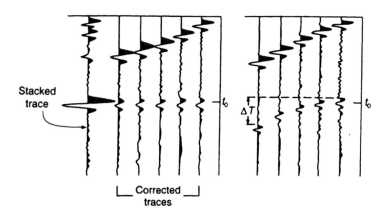

Figure 6.24   Given the source–receiver layout and corresponding raypaths for a common depth point spread, shown in (A), the resulting seismic traces are illustrated in (B), uncorrected (on the right), corrected (in the middle) – note how the reflection events now are aligned – and the final stacked trace. After Dohr (1981), by permission

signals minimised by destructive interference. In order to ensure that the stacking process is undertaken accurately, it is important to estimate the correct 'stacking' velocity. In the case of horizontal isotropic media, the stacking velocity is the same as the normal moveout velocity. However, when the reflector dips, it is important to carry out *dip moveout* to compensate for dip as well as the finite offset between source and detectors.

The calculation of NMO correction is given in Box 6.5. The larger the offset, the greater will be the NMO correction. The greater the two-way travel time, with a corresponding increase in velocity (as velocity tends to increase with depth as a general rule), the smaller will be the NMO correction at a given offset. The rate of increase in NMO correction with greater offset is hyperbolic (see Box 6.5), giving a curved reflection on a CMP gather. As velocity increases with depth, the size of the NMO correction (degree of curvature) decreases, i.e. effectively flattens at greater two-way travel times.

Just as raypaths increase in length with increasing offset from a single shot into a spread of detectors and give rise to normal moveout, the same situation occurs with the raypaths associated with a common midpoint gather (Figure 6.24). The whole point of acquiring multichannel seismic data, as opposed to single-channel data, is to be able to improve the signal-to-noise ratio and increase the resolution of geologically significant reflection events at the expense of incoherent noise. Consequently, the traces associated with a given CMP can be gathered and stacked. Obviously, in order to achieve the correct stacking, an appropriate NMO correction needs to be applied for each reflection event down the record. It is possible to determine the RMS velocity to a given reflector by plotting a $T^2 - X^2$ graph; the gradient of the straight-line segment appropriate for each selected reflection is $1/(V_{RMS})^2$.

The volume of data acquired in modern seismic surveys necessitates an automated method of determining the correct stacking velocities. One such method uses *constant-velocity gathers* (CVGs). Given a CMP gather, it is assumed that the RMS velocity is constant throughout the entire raypath. This RMS velocity is applied to the CMP gather to correct for NMO. A panel of CVGs is produced in which the RMS velocity applied to the CMP gather is increased by a given increment from one CVG to the next (Figure 6.25). Where the correct RMS velocity has been applied, any particular reflection events should appear horizontal. Where the RMS velocity is too low, the reflection, instead of being straight, curves upwards (smiles); where the RMS velocity is too fast, the reflection curves downwards

Figure 6.25 A constant-velocity gather (DVG) for seismic data at Shot Point 187. The same seismic data are shown in each panel, the only difference being the RMS velocity applied to the data (labelled at the top of each panel in ft/s). Three events have been encircled (A, B and C) – see text for details. The two-way travel time (in seconds) and RMS velocity (in ft/s) for the three events are, respectively: Event A at 0.8 s, 6400 ft/s; Event B at 1.7 s, 9600 ft/s; and Event C at 2.3 s, 11 000 ft/s. After McQuillin *et al.* (1984), with kind permission from Kluwer Academic Publishers

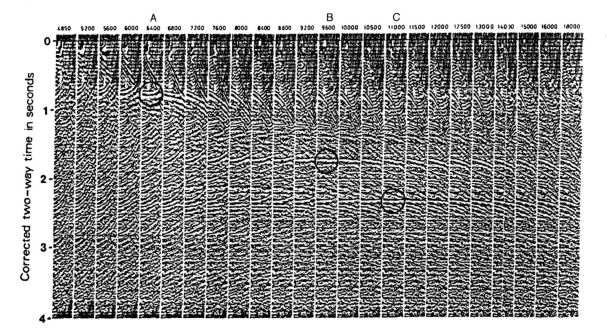

(frowns). By inspection of a panel of CVGs, the correct RMS velocities can be picked for particular reflection events.

Box 6.5 Normal moveout (NMO) correction (Figure 6.23)

For a medium with a seismic velocity V and a depth to a reflector z, the two-way travel time S–B–G is:

$$t_x = t_{SB} + t_{BG}.$$

The raypath $SB = BG$ can be determined using Pythagoras's theorem such that:

$$BG^2 = (x/2)^2 + z^2, \text{ so } BG = \{(x/2)^2 + z^2\}^{1/2}.$$

The total travel time is thus $t_x = (2/V).\{(x/2)^2 + z^2)\}^{1/2}$ as time $= x/V$. Substituting for z (from the two-way travel time for S–A–S), we obtain:

$$t_x = (2/V).\{(x/2)^2 + (Vt_0/2)^2\}^{1/2} \tag{1}$$

and

$$t_x^2 = x^2/V^2 + t_0^2 \quad \text{(where } t_0 = 2z/V\text{).} \tag{2}$$

Equation (2) is the equation of a straight line of the form '$y = m, x + c$', where m is the gradient and c is the intercept. By plotting a graph of t_x^2 against x^2, it can be seen that the gradient $m = 1/V^2$ and the intercept $c = t_0^2$, so V and t_0 can be obtained.

Equation (1) can be rewritten in hyperbolic form as:

$$(V^2 t^2)/4z^2 - x^2/4z^2 = 1. \tag{3}$$

At zero offset $(x = 0)$, equation (3) reduces to $t_0 = 2z/V$.

Equation (3) can be rewritten making time the subject: $t^2 = 4z^2/V^2 + x^2/V^2$. In turn this can be expressed in binomial form, such that:

$$t = (2z/V).\{1 + (x/2z)^2\}^{1/2} = t_0 \{1 + (x/Vt_0)^2\}^{1/2}. \tag{4}$$

By binomial expansion, and limiting the case to small offsets with $x/Vt_0 \ll 1$, and with t_n at $x = x_n$, this equation reduces to:

$$t_n \approx t_0 \{1 + \tfrac{1}{2}(x_n/Vt_0)^2\}. \tag{5}$$

Moveout is the difference in travel time at the two different offsets, such that:

$$t_2 - t_1 \approx (x_2^2 - x_1^2)/2V^2 t_0.$$

In the case of the *normal moveout correction* (ΔT), the difference in times is relative to zero offset $(x = 0)$, so the last equation becomes:

$$\Delta T = t_x - t_0 \approx x^2/2V^2 t_0.$$

Three picked events (labelled A, B and C) are encircled in Figure 6.25. As the amount of NMO decreases with increasing travel time, so too does the sensitivity of the velocity analysis. This can be seen by comparing the curvature of the events on panels either side of the ones picked for each of the three events. For event A at 0.8 seconds, the event is flattest for a RMS velocity of 6400 ft/s. The slight upturned tail of the event suggests that even this velocity is slightly too slow. Panels for 6000 ft/s and 6800 ft/s clearly show a smile and a frown respectively. For event B at 1.7 seconds, the amount of curvature on the adjacent panels is much smaller, making the choice of velocity harder. However, for event C at 2.3 seconds, velocities of 10 500 ft/s and 11 500 ft/s appear to be virtually indistinguishable from that chosen (11 000 ft/s). Reflections occurring at shorter two-way travel times will have lower RMS velocities, and those further down the records will have larger RMS velocities. It is thus possible to pick the correct RMS velocities for all principal reflection events on a given CMP gather to produce a vertical velocity profile.

It is usual to display the velocity information as a scaled *velocity semblance spectrum* (Figure 6.26). *Semblance* is a measure of the coherence of the stacking process; when it equals 1 it implies perfect selection of the normal moveout correction. *Coherence* is a measure of the degree of fit of a theoretically derived hyperbolic curve at a given travel time for a chosen RMS velocity. Scaled semblance profiles are usually plotted alongside any velocity spectrum (Figure 6.26).

It is normal practice to provide velocity analyses at regular but widely separated intervals along a seismic reflection line so as to ensure that the stacking is undertaken properly. For example, velocity analyses may be undertaken every 3 km along a survey line while shot intervals are as small as 50 m or less. By compiling a display of velocity profiles along a section (Figure 6.27), it is possible to identify (1) anomalous velocity picks due to incorrect velocity analysis, and (2) zones where velocity changes are associated with significant geological structures. When particular marker horizons have been identified, it is possible to carry out *horizontal velocity analysis* (HVA), by which velocity information is obtained at every CMP position along a survey line. HVA is usually undertaken only where there are considered to be significant and important velocity changes due to geological structure. The improved knowledge of the lateral velocity changes can be input into the stacking process to improve the overall visualisation of the structure being imaged (Figure 6.28).

Two further methods of velocity analysis have been developed. One is the use of *travel-time tomography* and the other is *pre-stack depth migration* (PDM) (Lines *et al.* 1995; Whitmore and Garing 1993). The objective of travel-time tomography is to match modelled travel times obtained by ray tracing to interpreted travel times produced prior to stacking. The main difficulty with the method is obtaining reliable travel-time measurements. Plane-wave pre-stack depth migration is

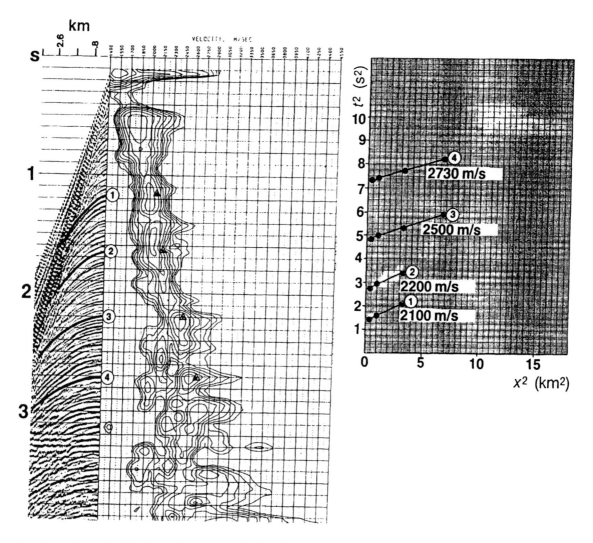

Figure 6.26 The $(t^2 - x^2)$ velocity analysis applied to a CMP gather. The triangles on the velocity spectrum (centre panel) represent velocity values derived from the slopes of the lines shown on the graph at the right. From Yilmaz (1987), by permission

based on the principle that plane-wave depth migration should be independent of the angle of plane-wave illumination if the velocities are correct. PDM uses an estimate of the velocity field to map plane-wave gathers into depth for each of the plane-wave orientations. When the data are sorted on the basis of 'common reflection points', the migrated depth images should not be dependent upon angle. A migrated event should appear flat as a function of plane-wave angle when the correct velocity has been applied (Figure 6.29). If velocities are too slow, the migrated event will have an upward curvature (smile); if the velocities are too fast, the curvature will be downwards (frown).

There are other forms of stacking which constitute migration processes, such as diffraction stacking, and some are discussed briefly in Section 6.3.6.

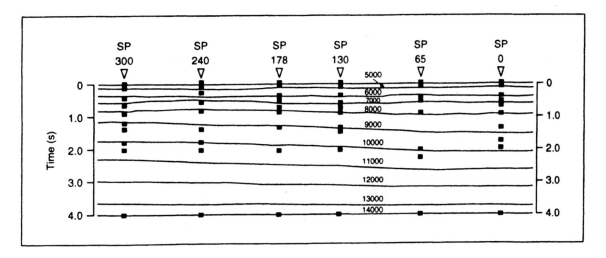

Figure 6.27 Constant-velocity profile along a seismic line. Velocity analyses have been carried out at the shot points indicated and the velocities are labelled in units of ft/s. From McQuillin *et al.* (1984), with kind permission from Kluwer Academic Publishers

6.3.5 Filtering

The term 'filtering' covers a very wide range of data processing techniques whose purpose is to modify a waveform or change its frequency composition in some controlled manner. If an input = $f(A) + f(B)$, then the output is equal to output (A) plus output (B). Algebraically, given an 'impulse response function' $[F]$ (i.e. a filter) and an input $[G]$ (a waveform), then the output $[H]$ is described by $H = G * F$, where $*$ represents the process of convolution.

A filter can be designed to remove reverberations within a dataset associated, for example, with a ringy seabed, such as may occur when overcompacted glacial clays form the seabed. The effect of such a filter designed to remove unwanted components of a signal is shown in Figure 6.30. A successfully designed filter, when applied to the data, should produce the required output. However, in the example provided, there is no noise present, which is very unrealistic. With noise, the filter is never perfect but may reduce unwanted signals substantially, if not remove them entirely.

Other types of filter are used to remove specific frequency elements, such as low-frequency groundroll or high-frequency jitter. To remove such band-limited signals, specific filters are designed to operate over the required frequency ranges. For example, to remove groundroll, a low-cut filter can be used (i.e. cuts out the low-frequency signals associated with the groundroll). To reduce high-frequency jitter, where the jitter occurs at frequencies significantly greater than those associated with the primary signals, a high-cut filter can be used. Sometimes, electricity mains cause interference at 50 to 60 Hz. To eradicate this, a 'notch filter' is used which attenuates those frequencies most strongly.

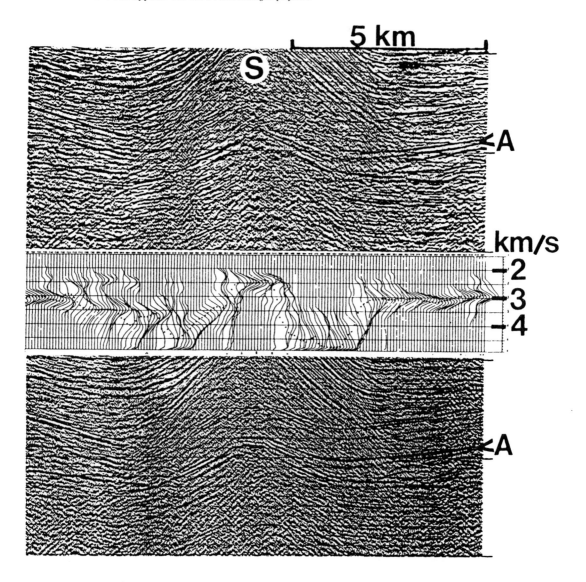

Different shaped filters are shown in Figure 6.31. A bandpass filter allows frequencies within the specified range (the bandwidth) to pass without attenuation by the filter; all other frequencies are affected by the filter. Similarly a low- or high-cut filter selectively removes the low or high frequencies, respectively. A low-cut filter is thus the same as a highpass filter, and a high-cut filter is the same as a lowpass filter.

As has been mentioned previously, high frequencies are progressively attenuated by the filter response of the ground. Consequently, it may be inappropriate to apply the same filter design to signals at later travel times as was applied to those at short travel times. This requires

Figure 6.28   A CMP stacked section obtained by sparsely spaced conventional velocity analysis (top) and by horizontal velocity analysis (HVA) (bottom). The HVA for horizon A below the salt dome (S) is shown in the centre. From Yilmaz (1987), by permission

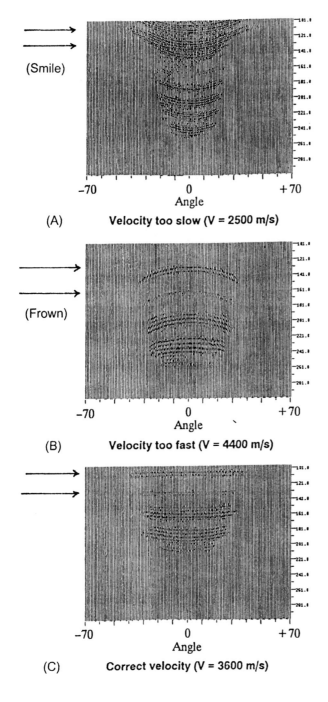

(Smile)

(A) **Velocity too slow (V = 2500 m/s)**

(Frown)

(B) **Velocity too fast (V = 4400 m/s)**

(C) **Correct velocity (V = 3600 m/s)**

Figure 6.29 (A) Migration panel for CRP showing a 'smile' caused by a low velocity estimate in layer 1 (arrowed). (B) Migration panel for CRP showing a 'frown' created by a high velocity estimate in layer 1. (C) Migration panel for CRP shows a flat event in the case of a correct velocity. The depth of the reflected event is not dependent upon offset – a requirement for the correct velocity. From Lines *et al.* (1993), by permission

a filter whose design changes with increasing travel time, i.e. a time-variant filter. A seismic record may be divided into a discrete number of time sections (windows), such as from 0 to 2 seconds, 2–3 seconds, 3–4 seconds and 5–6 seconds. A different filter design may be selected

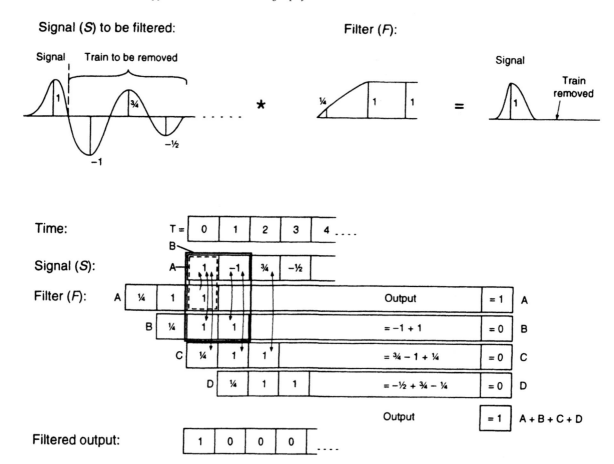

for each of these four time windows. It is important, however, that the change in design of filters between adjacent windows is not so extreme as to cause significant variations in the character of dipping events across the window frame.

Instead of using an impulse response function in the frequency domain, a filter can be expressed in terms of a series of sine and cosine waves of different frequencies and amplitudes in the time domain. The final response is the sum of all the constituent sine and cosine waves. The analysis of time series data to provide this information is called *Fourier analysis*. Consider a function $f(t)$ over the time interval $t = 0$ to $t = T$. It can be expressed as a constant plus the sum of all cosine waves plus the sum of all sine waves. Each of the constituent waves is expressed as an amplitude multiplied by the sine or cosine as a function of time (see Box 6.6). More detailed discussion of Fourier analysis is beyond the scope of this book. Further details can be found in texts by McQuillin *et al.* (1984), Yilmaz (1987), Hatton *et al.* (1986), among many others.

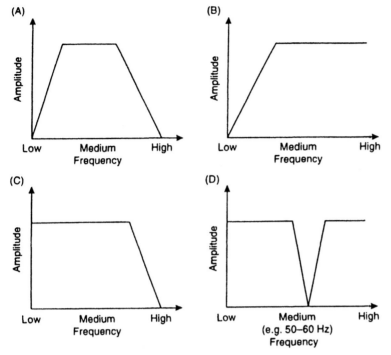

(A)

(B)

(C)

(D)

Figure 6.30 (*opposite*) Filtering out a reverberant signal:

At each relevant time sample (i.e. at $T = 1$ or $2, \ldots$) the signal amplitude S is multiplied by the corresponding segment of the filter F. Hence for $T = 1$, the calculation is simply of the form signal x filter $(S*F)$ such that output $= [1 \times (-1)] + (1 \times 1) = -1 + 1 = 0$. Taking this a stage at a time, we have:

Stage A: At $T = 0$, the first element of the filter (1) is multiplied by the corresponding sample of the signal (1), hence the output $= 1$

Stage B: At $T = 1$, the first element of the filter (1) is multiplied by the corresponding sample of the signal (-1) giving a value of -1. This is added to the product of the second element of the filter (1) and its corresponding sample of the signal (1) to give a value of 1 and on overall output of $-1 + 1 = 0$

Stage C: As for Stage B but shifted by one time sample ($t = 2$)

Stage D: As for Stage C but shifted by one time sample ($t = 3$)

Figure 6.31 (*above left*) Types of filter: (A) bandpass; (B)) low-cut (high-pass); (C) high-cut (lowpass); and (D) notch

Box 6.6   Fourier description of a waveform in the time domain

A function $f(t)$ over the time interval $t = 0$ to $t = T$ can be expressed by:

$$f(t) = a_0 + \sum_{j=1}^{\infty} a_j \cos(2\pi jt/T) + \sum_{j=1}^{\infty} b_j \sin(2\pi jt/T)$$

where a_0 is a constant. The amplitude A and phase ϕ at each frequency is given by:

$$A_j = (a_j^2 + b_j^2)^{1/2} \text{ and } \phi_j = \arctan b_j/a_j.$$

6.3.6   Migration

Migration is the process of restoring the geometrical relationships between seismic events on a time section to take account of structural complexities and steeply dipping interfaces. A steeply dipping reflector gives rise to a seismic event on a stacked seismic section that is not in its correct geometrical position (see Figure 6.32). The purpose of the migration process is to place a given seismic event in its correct geometrical position on the time section. A further benefit of migration

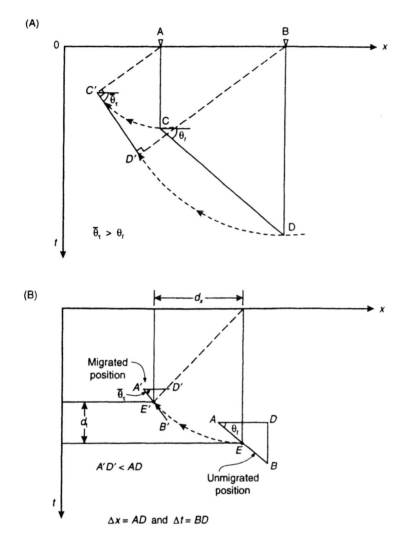

Figure 6.32 The principle of migration. (A) dipping reflector C–D on a stacked section is migrated to its correct geometry $C'-D'$ (B) The migration process moves an event (E) by a lateral distance d_x and vertically by d_t. The gradient of the event increases from θ_t to $\bar{\theta}$. After Yilmaz (1987), by permission

Figure 6.33 (*opposite*) The principle of the diffraction stack or Kirchhoff migration. (A) A diffractor lies at a depth h vertically below a receiver and gives rise to a diffraction event (B) on a seismic reflection record. (C) A diffraction stack hyperbola for a known velocity is matched against the observed diffraction. Given a correlation along the hyperbola, all the same-phase events are summed and the observed diffraction is shrunk back to a point (D). Courtesy of BP

is the collapse of diffraction hyperbola to their points of origin, and the consequent clarification of seismic events (for example, seismic rollover) associated with discontinuities within interfaces such as at fault planes. As a common midpoint stack smears dipping events (see Section 6.2.2), *dip moveout* (DMO) processing is commonly undertaken before stacking; DMO is sometimes considered to be a migration process.

There are a number of different migration procedures, such as Kirchhoff (Claerbout 1985), finite-difference, frequency–wavenumber ($f-\kappa$) domain (Stolt 1978), and turning-wave migration; with the exception of the last method they are described and discussed in considerable detail by Berkhout (1984) and Yilmaz (1987), among others. Turning-wave migration or imaging has been described by

Claerbout (1985), Hale *et al.* (1991), Meinardus *et al.* (1993), and Ratcliff *et al.* (1991), among others.

Time migration from a stacked section (i.e. zero offset) produces a migrated time section and is appropriate in the absence of strong lateral velocity variations. When such velocity gradients are present, time migration does not produce an accurate image of the sub-

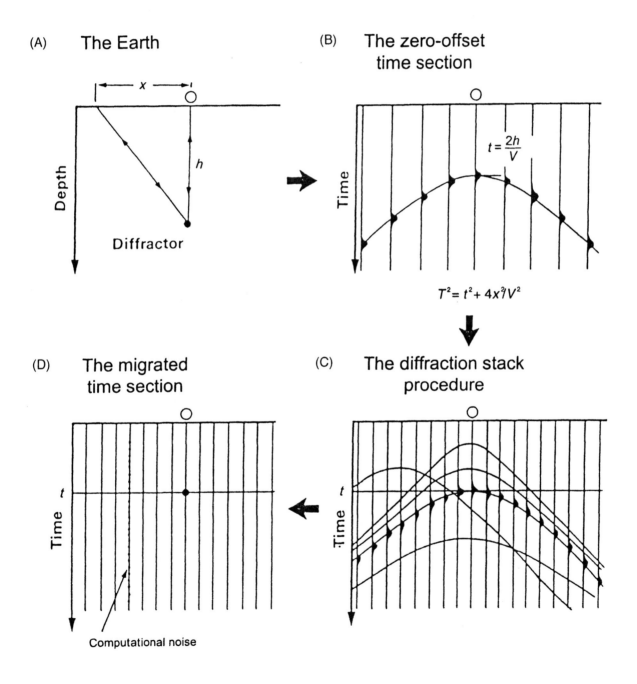

surface. Instead, depth migration should be used, the output of which is a migrated depth section. Ray-tracing is used particularly to generate depth sections.

The basis of the migration method is Huygens' secondary source principle. This is, that any point on a wavefront can act as a secondary source producing circular wavefronts when viewed in two dimensions, or spherical wavefronts in three dimensions. The process of migration is to collapse the apparent secondary wavefronts to their points of origin. For example, a point source generates a diffraction hyperbola (see Figure 6.33). By scanning an appropriate number of adjacent traces, the hyperbola can be collapsed back to the point source by the migration process. The number of adjacent traces scanned is known as the *migration aperture*. When the seismic velocity is slow, the curvature of the diffraction hyperbola is steep, and consequently the number of traces required is small. As the hyperbola curvature decreases (flattens) with increasing seismic velocity, the migration aperture also has to increase (Figure 6.34). Incomplete collapse of diffraction hyperbolae occurs if insufficient aperture widths are chosen. It is also important to ensure that the migration aperture is large enough to capture all the required data for a given geological target.

The horizontal and vertical displacements between the un-migrated and migrated sections can be calculated (see Box 6.7). At a two-way travel time of 4 s and with a seismic velocity of 4000 m/s, the horizontal (in km) and vertical displacements (seconds) are 6.4 km and 1.6 s respectively. Even at a two-way travel time of 1 s and with a seismic velocity of 2500 m/s the displacements are 625 m and 0.134 s. It is thus obvious that the degree of displacements is very significant, and that even small errors in the migration process can have large practical implications when choosing potential wellsites.

Box 6.7   Vertical and horizontal displacements through migration (see Figure 6.32)

Horizontal (distance) and vertical (time) displacements d_x and d_t and dip angle after migration (θ_τ) can be expressed in terms of material velocity V, two-way travel time t and apparent dip (θ_t) of the reflection as measured on an unmigrated section using the following expressions:

$$d_x = (V^2 t \tan \theta_t)/4$$
$$d_t = t\{1 - [1 - (V^2 \tan^2 \theta_t)/4]^{1/2}\}$$
$$\tan \theta_\tau = \tan \theta_t / [1 - (V^2 \tan^2 \theta_t)/4]^{1/2}$$
$$\tan \theta_t = \Delta t / \Delta x.$$

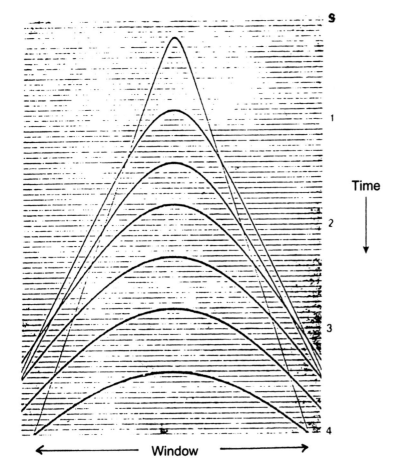

Figure 6.34 Diffraction hyperbolae calculated using increasing velocities. A slow velocity results in a tight hyperbola (uppermost curve) while that associated with the fastest velocity has both the broadest window and the flattest of the curves. After Yilmaz (1987), by permission

The migration process has three basic effects when compared with events on unmigrated sections. Migration (a) steepens reflections, (b) shortens reflections, and (c) moves reflections up-dip (see Figure 6.32).

Kirchhoff migration has a tendency to generate noise but handles all required dips and can cope with horizontal and vertical velocity variations. Finite-difference migration produces little noise and can also cope with velocity variations; but in the case of steep dips the computational time is lengthy and hence expensive. The f–κ migration handles all dips very quickly but has difficulty in coping with velocity variation in the cross-dip situation. It also only works for constant-velocity models so there is a requirement to undertake comprehensive velocity–depth conversions.

Kirchhoff migration uses a hyperbola in 2-D or hyperboloid in 3-D. For pseudo 3-D migration, the migration process is run in two stages that are basically a 2-D migration in the x-direction and then a 2-D migration in the y-direction. This process is known as '2D2' and may require some 600 traces for one migration run; in contrast a full

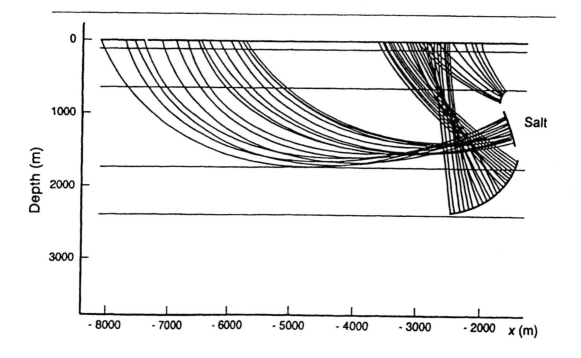

3-D migration for the same coverage might require 70 000 traces with a correspondingly longer calculation time and greater costs.

Turning-wave migration is of especial importance in imaging the overhanging sides of salt diapirs (Hale *et al.* 1991). Turning waves travel downwards initially, pass through a turning point and travel upwards before being reflected from the salt–rock interface (Figure 6.35). Figure 6.36A shows a stacked section from the Gulf of Mexico (Meinardus *et al.* 1993) after application of dip moveout and has three events highlighted; event 3 arises from turning waves at an apparent distance of over 5 km from the salt dome. The salt–rock interface giving rise to these reflections lies at a two-way travel time of 1.3 s on the turning-wave migrated section (Figure 6.36B) compared with 4.5 s on the TV DMO section. The improvement in clarity of imaging is quite spectacular. The scale of geometrical repositioning achieved by migration (of the order of kilometres) in large-scale hydrocarbon exploration surveys demonstrates the importance of the method in providing much more reliable and realistic images of the sub-surface in structurally complex areas. However, for small-scale engineering investigations, migration may not be either technically or financially justifiable (Black *et al.* 1994).

Along with major developments in migration methods (whether pre- or post-stack), advances in computing hardware have facilitated the processing of the vastly increased amount of data. Of major significance has been the application of 'massively parallel processors'

Figure 6.35 The overhang face of the salt dome on the right edge is recorded with turning waves as indicated by their raypaths. From Meinardus *et al.* (1993), by permission

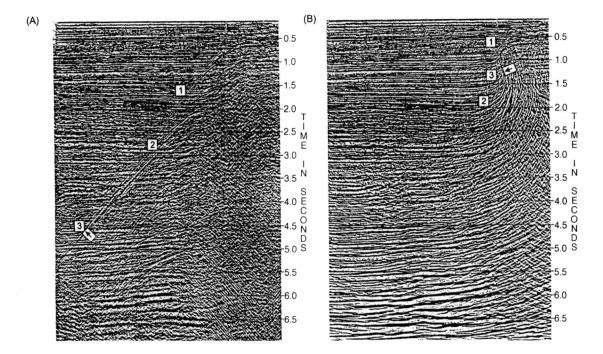

(A)

(B)

TIME IN SECONDS

TIME IN SECONDS

(e.g. the Cray T3D system). Undoubtedly, developments in computer technology will continue to have major effects on the capabilities of the seismic processing industry.

While all of the above discussion has been in relation to seismic data, migration processes can also be applied to ground penetrating radar data (see Chapter 12). The issues concerning velocity determination in order to translate from the time domain to the depth domain apply equally to radar surveys.

6.4 CORRELATING SEISMIC DATA WITH BOREHOLE LOGS AND CONES

It is of great importance to constrain geophysical models with independent information where possible so as to produce an interpretation which represents what is actually present within the ground. The proof of the model interpretation is in the drilling! However, geophysical interpretation is an interative process. Working with a forward model, an initial interpretation is obtained. Given independent data, the model can be refined or constrained. Given yet more information, the interpretation can be revised yet further. With each iteration, the model should be representing the actual ground structure more realistically each time. In the following section, several different methods of correlation are described briefly, ranging from

Figure 6.36 (A) Stack of a seismic reflection record in the Gulf of Mexico after the application of time-variant dip moveout (DMO). Three events are indicated by numbers. (B) The same section after turning-wave migration. The difference in position of event (3) (arrowed) between the two sections is very marked, with the event being restored to the salt overhang on the migrated section. From Meinardus *et al.* (1993), by permission

the use of boreholes, the comparison with synthetic seismograms, through to correlation in engineering surveys with Dutch cone logs.

6.4.1   Sonic and density logs, and synthetic seismograms

A well-established method of correlation is to use borehole geophysical logs to produce reflectivity series for comparison with the seismic sections. Significant primary reflections should occur where there is significant contrast in acoustic impedance. Borehole sonic and density logs provide the information from which acoustic impedances can be derived over vertical distances much shorter than the seismic wavelength.

A sonic logging tool consists of two transmitters, one at each end of the tool, and two pairs of detectors located in between the transmitters. A sound wave is generated from one source and is refracted along the borehole wall and the critically refracted waves are detected at each of one pair of receivers separated by a known distance. A second signal is then generated from the other source in the opposite direction and the critically refracted arrivals are detected by the second pair of receivers. The instrumentation measures the travel time taken for the signal to travel through the distance of the borehole wall equivalent to the separation between the pairs of receivers. The travel time from shooting in one direction is averaged with that obtained for the reverse direction to make allowance for the tilt of the tool with respect to the borehole wall. The average single-way transit time is then logged for that interval.

Typically the transit time is recorded in units of μs/ft. To convert these times to a seismic velocity, a simple conversion factor can be used: given the single-way transit time (T) in μs/ft, then the seismic velocity V in km/s is given by $V = 304.79/T$. For example, if $T = 90\,\mu$s/ft, then $V = 304.79/90 = 3387$ km/s. Consequently, the sonic data give a measure of the seismic velocity. The density log gives a corrected reading in Mg/m^3. If the numerical values of each of these two logs is cross-multiplied (i.e. velocity times density), then a log of the acoustic impedance Z is produced.

From an inspection of the acoustic impedance log, significant interfaces across which there is a marked contrast in impedance can be readily identified. Such interfaces can also be correlated with the other available borehole geophysical and geological logs to provide a lithological as well as a chronological stratigraphy. The acoustic impedance log can be used to derive a vertical reflectivity series. That is, across each interface the reflection coefficient can be determined. This can be used with an artificial wavelet to generate a synthetic seismogram that can be compared directly with the observed seismogram. The added advantage with direct correlation with borehole logs is that there is an actual measurement of depth against which the seismic section can be constrained.

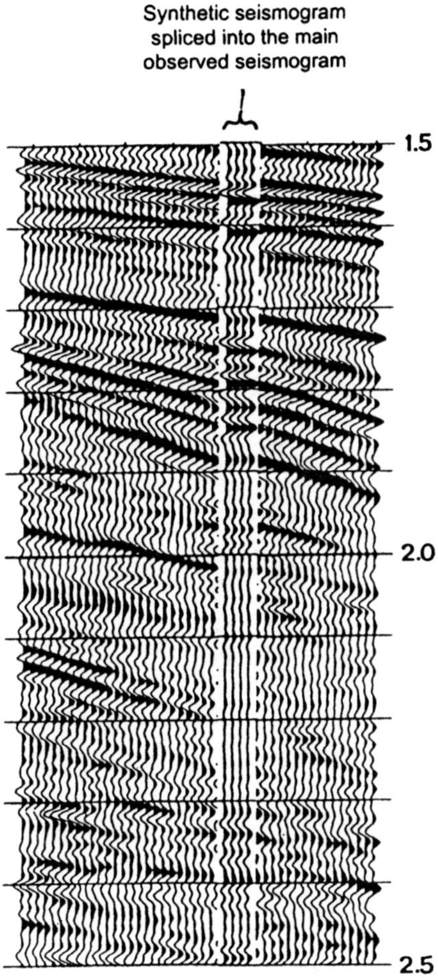

Synthetic seismogram
spliced into the main
observed seismogram

Figure 6.37   Part of a seismic record from the Campos Basin, offshore Brazil, into which a synthetic seismogram has been inserted in order to demonstrate the correlation between the observed and the synthetic data. From Verela *et al.* (1993), by permission

An example of the use of a synthetic seismogram to correlate with measured seismic data is shown in Figure 6.37 (Varela *et al.* 1993). The strong reflections seen in the segments of seismic section match those in the synthetic seismogram very well.

6.4.2   Correlation with Dutch cones

In shallow land and marine engineering applications, it is possible to correlate seismic reflection records with Dutch cone and standard penetration test (SPT) logs. These comprise two types of surveys as follows.

A Dutch *cone penetrometer test* (CPT) consists of vertical insertion into the ground of an instrumented 3.6 to 4.4 cm diameter cone on the end of an extension rod (Figure 6.38). Typically, the platform from

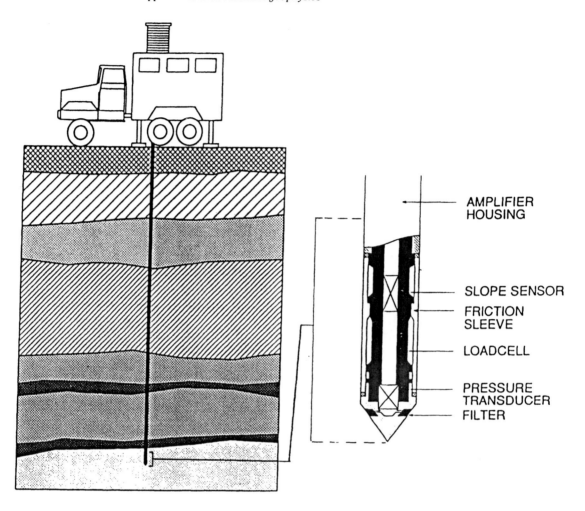

Figure 6.38 Cone penetration testing (CPT) system with a piezocone tip in detail. From Hunt (1992), by permission

which the cone is deployed is a truck with a cabin, which houses all the instrumentation and the hydraulic rams which drive the cone and rods into the ground (Figure 6.38). By pushing the rods into the ground at a constant rate (e.g. 2 cm/s), the resistances on the cone tip and sleeve can be measured and displayed as a function of depth (Figure 6.39). Vertical resolution is of the order of 2 cm. Different types of cones can be deployed such as the piezocone (tip and sleeve friction, dynamic water pressure), soil, groundwater and vapour sampling cones (Meigh 1987), and electrical conductivity (e.g. Tonks *et al.* 1993) and seismic cones.

Whereas all the usual cone tips measure physical or chemical parameters in the immediate vicinity of the instrumented cone, the seismic cone works on a different principle. Here, only the geophone is placed within the cone head. The seismic source remains at the ground surface (Figure 6.40A). For example, a large piece of heavy-

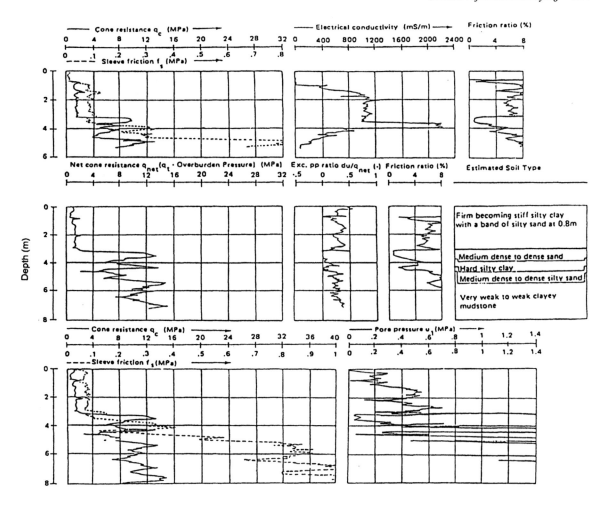

Figure 6.39 Examples of the output from cone penetration testing using a piezocone and a conductivity cone. From Hunt (1992), by permission

duty timber (such as a railway sleeper) is anchored to the ground to ensure good ground coupling. A sledge hammer is struck against one end of the timber to generate a shear wave that propagates down into the ground where it is detected by the stationary cone sensor. Another hammer blow is applied but to the opposite end of the timber to produce a shear wave of opposite polarity to the first. The use of two opposite-polarity shear waves helps the identification of the onset of the shear wave (Figure 6.40B). The two waveforms are logged against the position of the sensor depth. The cone and rods are pushed an incremental distance further into the ground (typically 0.25 m or 0.5 m) and the two surface seismic shots are fired again. The difference in total travel time measured between adjacent sensor positions can be inverted to give an S-wave velocity averaged over the incremental distance plotted as a function of depth. The process is repeated until the required depth has been reached or it is not possible to push the

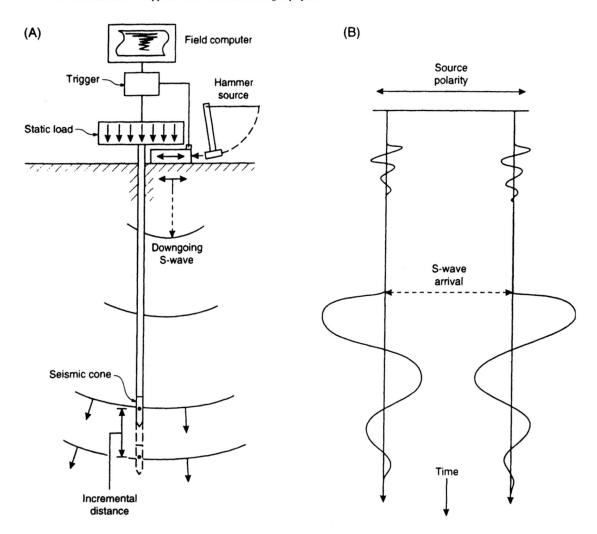

(A)

Field computer

Trigger

Static load

Hammer source

Downgoing S-wave

Seismic cone

Incremental distance

(B)

Source polarity

S-wave arrival

Time

rods any further into the ground. Alternatively, and more conventionally, the sledge hammer is impacted on to a rubber baseplate to generate P-waves.

Averaged P-wave velocities can be obtained in the same way as for S-waves. Zones of higher-velocity material produce short travel times while the converse is true for materials with slow elastic-wave velocities. Given the development of a uniform seismic source, other parameters may be determinable using this method (relative attenuation, Poissons' ratio, elastic moduli, etc.). The use of two seismic cones deployed side-by-side would permit the equivalent of cross-hole seismic tomography. One cone tip would have to be replaced by an appropriate source.

A further method which can be used with the seismic cone is the equivalent of the walkaway VSP. If the surface source is moved to

Figure 6.40 Seismic cone test. (A) Schematic of the survey components. (B) The first breaks of S-waves with opposite polarities can be clearly identified. The polarity of the downgoing P-waves is not reversed

larger offsets, with the sensor located at the same vertical depth, then the slant angle raypath geometry can be explored. Additionally, the source can be moved around the cone position at a common offset to give azimuthal information. This may be particularly beneficial in materials that possess a strongly developed anisotropic fabric. Three-dimensional seismic cone tomography has yet to be fully developed commercially. The basic method is only just starting to be accepted amongst the engineering community, but the potential for its usage is considerable. Similarly, the scope for development of other active cone tips (e.g. EM, electrical resistivity, etc.) and deployed in 'walkaway' profiling and cross-hole tomography is enormous. The combination of 'clever' cone sensors and surface geophysical methods is perhaps one of the most exciting areas of development in environmental geophysics and ground investigation.

In the drilling of boreholes, whether on land or afloat, *standard penetration testing* is a very basic method of testing the hardness of the material being drilled through. In essence it is a measure of the number of blows of a shell-and-auger rig tool to penetrate a given distance, usually 10 cm. The larger the number of blows, the harder is the material. Similarly, the use of *dynamic probing* works on a similar principle – a rod is driven continuously into the ground and the distance driven and the energy imparted are measured (e.g. Russell and Gee 1990). Consequently, if a borehole is constructed through materials in which there are distinct changes in hardness, the interfaces across which hardness changes should also be evident in seismic reflection records. Associated with changes in hardness should be differences in acoustic impedance. Thus, where hardness changes, so acoustic impedance should also. By correlating a good-quality high-resolution marine seismic reflection record with SPT logs, significant reflections should be readily identifiable at interfaces identified with the SPTs. For example, in a commercial marine resource evaluation project in Hong Kong, SPT logs proved to be very valuable in aiding the interpretation of seismic reflection records to derive a general seismic stratigraphy for the area. The degree of correlation between the seismic records and SPT logs was extremely good.

Once the reflections on the seismic sections have been correlated with borehole and cone data, the seismic time series record becomes interpretable in geological terms. Depths, interval seismic velocities, seismic characteristics associated with particular lithological units, etc., can be determined with confidence. With the advent of more detailed quantitative analysis of high-resolution single-channel seismic data, further information about the geotechnical parameters of the soils and sediments present can be obtained. Research is currently ongoing as to how this can be achieved (e.g. Haynes *et al.* 1993).

6.5   INTERPRETATION

6.5.1   Vertical and horizontal resolution

Determination of resolution within seismic surveying depends upon four factors: quality of raw data, frequency bandwidth, source response characteristics, and the nature of the vertical sequence of possible reflectors. Each of these is discussed briefly.

The acquisition of raw seismic data has to be undertaken in such a manner that the quality of the records obtained is suitable for the task in hand. This requires two stages: the first is to ensure that the correct equipment is specified to start with, and the second, that the equipment used is operated properly. There are still too many occasions when the wrong kit has been used or the right equipment has been used badly. No amount of processing will transform substandard data into high-quality records.

Another point is that it is essential that the appropriate frequency range system is used for the objective of the survey. There is no point using a lower-frequency sparker system if shallow penetration and high resolution are required. Conversely, the selection of a pinger, with its high frequency and restricted bandwidth, is inappropriate to obtain even moderate penetration through gravels. In essence, however, resolution (see Section 4.5.5 and Table 4.6) is dependent upon frequency. High-frequency sources provide higher resolution than low-frequency sources. The Rayleigh resolution limit is equal to one-quarter of the dominant wavelength of the incident waveform. The Widness limit is an eighth of the wavelength. This has been discussed in more detail by, for example, Phadke and Kanasewich (1990).

The third consideration is *source response*. In hydrocarbon exploration, seismic contractors are obliged to record the seismic source pulse. It is seldom done in shallow engineering surveys. With the increased usage of digital acquisition systems in engineering type surveys, and with the sophistication of analysis now available, it is becoming increasingly important that the source pulse should be recorded on each and every survey. By having a knowledge of the shape and duration of the source pulse, the subsequent interpretation of the seismic record is made more realistic, if not easier. Source pulse shape and duration are fundamental aspects of interpreting seismic (and ground penetrating radar) records (Reynolds 1990a). If a source pulse is ringy, so too will be the reflections; if the source comprises three bars, for example, so will the subsequent reflections, although the tails of later reflections may become more subdued and less obvious. If the source pulse is 'clean', then the reflections arising from this are also likely to be sharp. When many reflectors lie in close proximity to each other, the ability to resolve between them is determined in part by the cleanness of the source signal. It is more

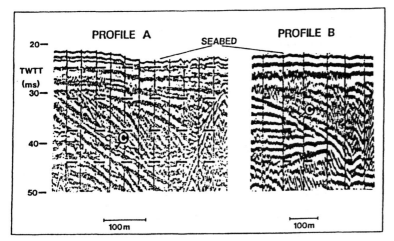

Figure 6.41 Two sparker records acquired over the same ground but using different frequency bandwidth filters. From Reynolds (1990a or b), by permission

difficult to resolve units with a ringy source than with a clean one as the tail of the earlier reflection interferes with the onset of the next and the two wavetrains may coalesce to cause a complex and unresolvable mess.

To some extent the cleanness of the source can be improved by filtering the data during acquisition (Figure 6.41) and with post-processing. It is important that, during the setting up of any acquisition system, the source be optimised to produce the sharpest signal possible. It should also be remembered that source quality drifts, and periodic checks of the source characteristics should be made during each day of a survey to ensure consistency.

Related to this third factor is the directionality of the source used in marine surveys. Spatial smearing occurs owing to the finite size of the source array: the source is not infinitessimal but has a finite footprint. Consequently, the in-line and cross-line responses may be different and these need to be considered, especially in 3-D surveys. Roberts and Goulty (1990) have provided an example where directional deconvolution of the signature from a marine source array has been achieved in conjunction with pre-stack migration or dip moveout (DMO) processing. A particular benefit of undertaking this processing sequence is that shallow, dipping reflections have better lateral continuity and frequency content. Consequently, the processing takes into account a source characteristic and improves the lateral resolution.

The fourth factor to be considered is the very nature of the sequence of reflectors present within the sub-surface. Flat-lying, widely spaced reflectors are much easier to image than are those that are closely spaced and steeply dipping. The resolvability of thinly bedded strata is of great concern to the hydrocarbon industry and much effort has been made to improve the techniques used for unravelling such sequences. The minimum thickness of a bed that gives rise to an identifiable reflection is known as the *detectable limit* or the *limit of visibility*. Numerically, this is taken as being equivalent to one thirtieth of the dominant wavelength of the incident waveform.

Horizontal resolution of unmigrated seismic sections is usually taken as being equivalent to the width of the first Fresnel zone (see Section 4.3.1). If horizontal stacking of traces has been carried out, then this may lead to smearing of features with a consequent loss of horizontal resolution. This effect is also present on some ground penetrating radar sections. It is always advisable for those involved in the interpretation of seismic and radar sections to be familiar with the methods of data acquisition and processing used in the production of their sections.

In the case of steeply dipping reflectors, careful consideration needs to be given to the geometry of features in real space with respect to unmigrated versus migrated sections. Horizontal resolution in these situations is largely a function of the efficiency and accuracy of the migration process.

6.5.2   Identification of primary and secondary events

A fundamental aspect of interpreting time sections (seismic or ground penetrating radar) is being able to distinguish between primary events (i.e. caused by the local sub-surface features) and secondary events that are artefacts of the data-acquisition geometry and instrumentation (i.e. multiples, reverberations, instrument noise, interference, etc.). In the vast majority of surveys, great effort is made to remove or reduce secondary events. In the case of analogue shallow seismic reflection surveys undertaken over water, multiples and off-section reflections (from harbour walls, moored boats, etc.) can mask primary information that is of greatest interest. Under such circumstances, the benefits of the survey are very limited. However, modern digital acquisition systems have been developed for use in exactly these circumstances and provide methods whereby particularly the water-bed multiple can be suppressed (Haynes *et al.* 1993). Primary events that would otherwise have been masked then become more visible and hence more interpretable. As a general rule, the gradient of a multiple of a dipping event is always steeper than its primary reflection, with second multiples steeper than the first multiple, and third multiples steeper than the second and so on (Figure 6.42). Where the major problem often lies is in differentiating multiples in the case of horizontally bedded units when the multiples are parallel to the primary events. Under these circumstances it can be difficult to distinguish between the two types of event.

Since the late 1980s, attention has been paid to the information content of multiples. It has been noted that, in some surveys, discrete areas exhibit anomalously high reflection strengths giving rise to

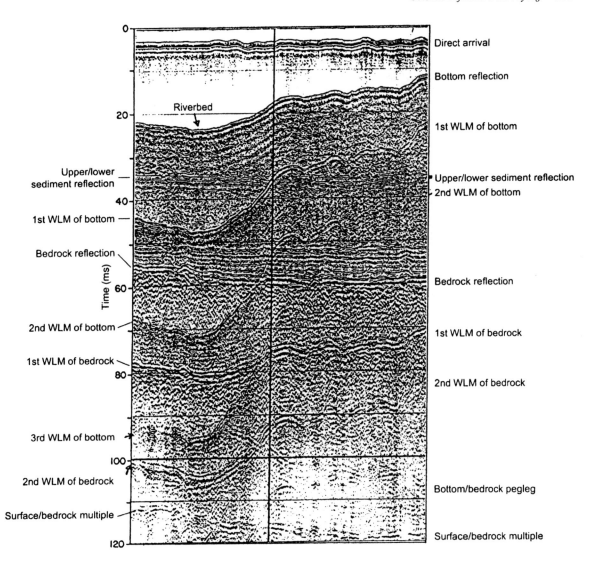

Labels on figure (left side, top to bottom): Riverbed; Upper/lower sediment reflection; 1st WLM of bottom; Bedrock reflection; 2nd WLM of bottom; 1st WLM of bedrock; 3rd WLM of bottom; 2nd WLM of bedrock; Surface/bedrock multiple

Labels on figure (right side, top to bottom): Direct arrival; Bottom reflection; 1st WLM of bottom; Upper/lower sediment reflection; 2nd WLM of bottom; Bedrock reflection; 1st WLM of bedrock; 2nd WLM of bedrock; Bottom/bedrock pegleg; Surface/bedrock multiple

Time (ms) axis: 0, 20, 40, 60, 80, 100, 120

strong multiples. The areas concerned tend to be found in rivers, bays and estuaries, and particularly those affected by industrial effluent. In other areas, the absence of multiples is indicative of highly attenuating material such as cellulose (from wood pulp mills) or decaying organic matter.

McGee (1990) has analysed multiples from a number of high-resolution surveys in order to determine what information can be gleaned about the near-surface materials. Figure 6.43 shows an analogue seismic record across the Saint Clair River at the Canadian–US border near Detroit where waste material had been dumped at the base of a rock outcrop. A discontinuous reflection from the

Figure 6.42 Analogue seismic profile across the Saint Clair River, on the Canada–US border, north of Detroit, showing all types of multiples for a two-layer situation. WLM = water-layer multiple. After McGee (1990), by permission

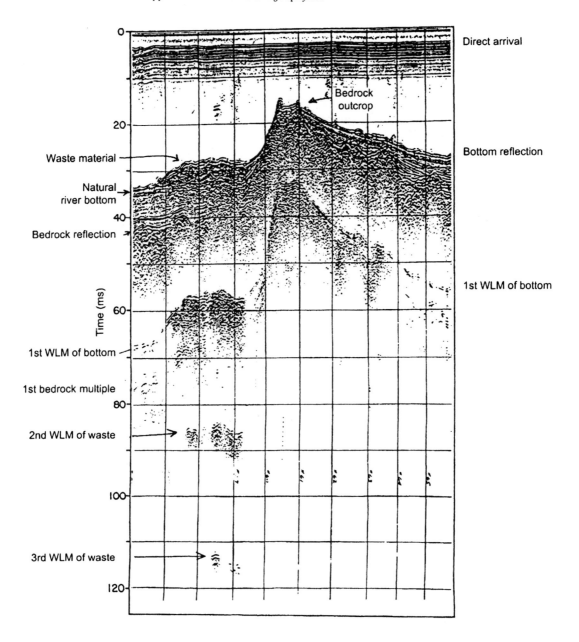

Direct arrival

Bedrock outcrop

Waste material

Natural river bottom

Bedrock reflection

Bottom reflection

1st WLM of bottom

1st WLM of bottom

1st bedrock multiple

2nd WLM of waste

3rd WLM of waste

Time (ms)

seafloor is faintly visible below the waste material. The water-layer multiple decays more rapidly over the waste material than elsewhere. This indicates that the reflectivity of the waste is greater than that of both the natural riverbed sediments and the bedrock outcrop.

Figure 6.44 shows another profile across the Saint Clair River from Sarnia, Ontario, Canada, to Port Huron, Michigan, USA. The concave downward event at 23 ms near the centre-line is a reflection from the top of a 6 m diameter railway tunnel that passes obliquely beneath

Figure 6.43 Analogue seismic profile across the Saint Clair River near Detroit, USA, showing reflections and multiples arising from a waste dump adjacent to a bedrock outcrop. After McGee (1990), by permission

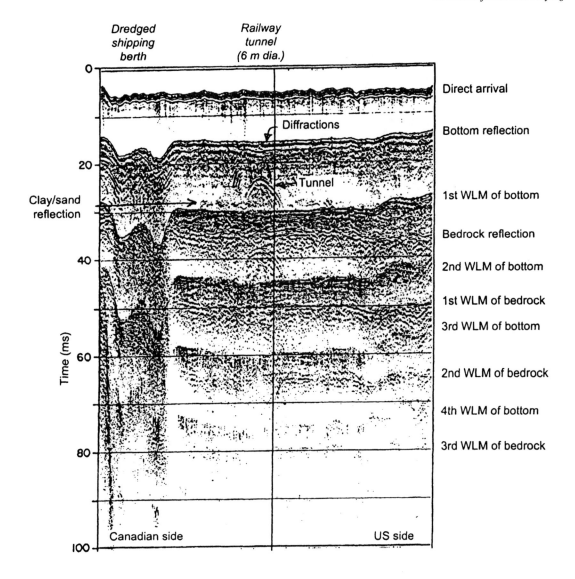

Figure 6.44 Analogue seismic profile across the Saint Clair River in the vicinity of a railway tunnel. After McGee (1990), by permission

the river. The dredged area on the Canadian side is associated with a commercial wharf. The soil there is soft clay over firm clay. On the US side the soil is sand over firm clay. Except for these locations the tunnel was excavated in firm clay containing large boulders and lenses of sand and gravel. It is thought that the diffractions seen just below the riverbed reflection have been caused by these inhomogene-ities. The lower part of the tunnel lies in a sand layer, the top of which gives rise to the weak reflection a couple of milliseconds above the first water-layer multiple (WLM) of the river bottom. The strong reflection beneath has been produced by the top of bedrock, the Detroit Sandstone. The presence of water-layer multiples of second

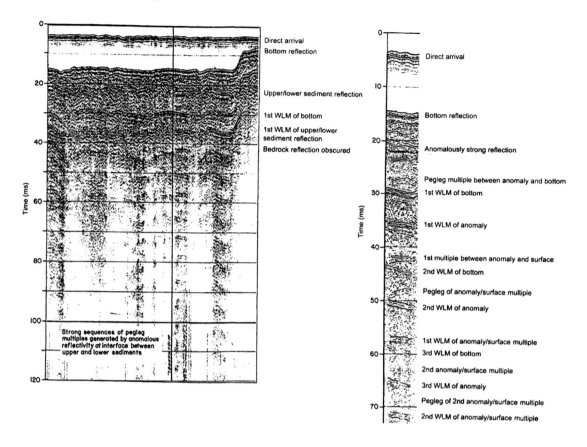

The left-hand labels read:

Direct arrival
Bottom reflection

Upper/lower sediment reflection

1st WLM of bottom
1st WLM of upper/lower
sediment reflection
Bedrock reflection obscured

Strong sequences of pegleg
multiples generated by anomalous
reflectivity at interface between
upper and lower sediments

Time (ms) axis: 0, 20, 40, 60, 80, 100, 120

The right-hand labels read:

Direct arrival

Bottom reflection

Anomalously strong reflection

Pegleg multiple between anomaly and bottom
1st WLM of bottom

1st WLM of anomaly

1st multiple between anomaly and surface
2nd WLM of bottom

Pegleg of anomaly/surface multiple

2nd WLM of anomaly

1st WLM of anomaly/surface multiple
3rd WLM of bottom

2nd anomaly/surface multiple

3rd WLM of anomaly

Pegleg of 2nd anomaly/surface multiple

2nd WLM of anomaly/surface multiple

Time (ms) axis: 0, 10, 20, 30, 40, 50, 60, 70

Figure 6.45 On the left is a digital seismic profile across the Saint Clair River showing numerous patches of anomalous reflectivity indicated by pegleg multiples. On the right is an enlargement of the portion located immediately to the right of the mid-channel line, showing the multiple sequences in detail. After McGee (1990), by permission

and higher order is obvious, but the tunnel produces only one multiple. Considering that it is lined with cast iron, this is somewhat surprising. However, the lack of higher-order multiples associated with the tunnel was thought by McGee to be due to its curvature, causing reflected wavefronts to diverge.

Figure 6.45 shows another sequence across the Saint Clair River in which several areas of pegleg multiples are evident. A portion of this record has been enlarged to show the multiple sequences in more detail. Several orders and combinations of water-layer and pegleg multiples can be identified on the basis of their arrival time. Slight differences in apparent dip between the water-bottom reflection and the anomalous reflection become greatly exaggerated by the higher-order events. Soft grey clay over firm blue clay was found in a borehole drilled nearby. The interface between the two clays was encountered within 0.3 m of the depth calculated from the reflection time by assuming that the seismic velocity in both the water and the upper soil was 1500 m/s. The postulated cause of the higher reflectivity in certain locations was an anomalously large density contrast between the two clays.

Higher reflectivity is often observed associated with channel lag deposits in rivers or palaeo-channel infills. As the sediment coarseness increases from fine sands on the channel rim to gravels and even cobbles at the lowermost point, so the reflectivity increases. Consequently, higher-order multiples often arise from the gravel part of the lag deposits, with the reflectivity decaying laterally and in sympathy with the decrease in sediment grain size. Commonly, gravel bands give rise to pulse broadening too, so producing a characteristic reflection pattern.

McGee (1990) has also argued that, when the data are recorded digitally and with sufficient dynamic range, the amplitudes of the multiples can yield valuable information. For example, in water-saturated and weakly consolidated sediments, the seismic velocity in most of the materials is approximately 1500 m/s. Any reflections that arise do so, therefore, from a contrast in density only. If the reflection coefficient can be computed from the digitally recorded data for the water–seabed interface, then the density of the soil can be estimated. Using this, and the measured decay in subsequent amplitudes, and still assuming a saturated soil velocity of 1500 m/s, the densities of deeper soil horizons can be estimated. In areas of anomalous high reflectivity, the increase in density of the material causing the effect can be calculated.

With advances in computer facilities, it is becoming possible to acquire digital data with high enough dynamic range (e.g. at least 16-bit) and fast enough sampling rates (> 25 kHz; 40 µs). The data can be processed using commercially available workstations (e.g. a Sun IPX workstation running Sierra Geophysics ISX/SierraSeis software) to utilise the full processing power normally reserved for hydrocarbon exploration surveys (Haynes *et al.* 1993). It is likely that sophisticated processing of digital high-resolution seismic profiling data will yield estimates of parameters of direct interest to geotechnical engineers and engineering geologists, as outlined in Figure 6.46.

6.5.3 Potential interpretational pitfalls

Pitfalls in seismic interpretation have been discussed in detail by Tucker and Yorston (1973), Tucker (1982) and Steeples and Miller (1994). The most basic difficulty is translating from the time-domain seismic 'image' to the geological cross-section and the biggest uncertainty is in determining the seismic velocities and their lateral and vertical variations. Just two aspects are highlighted here, both of which are related to velocity.

The first aspect is apparent dip on a seismic reflector produced by velocity variation within a layer (Figure 6.47A). As the velocity increases or decreases along the horizon, so the travel time through that material decreases or increases respectively, even though the reflector is horizontal.

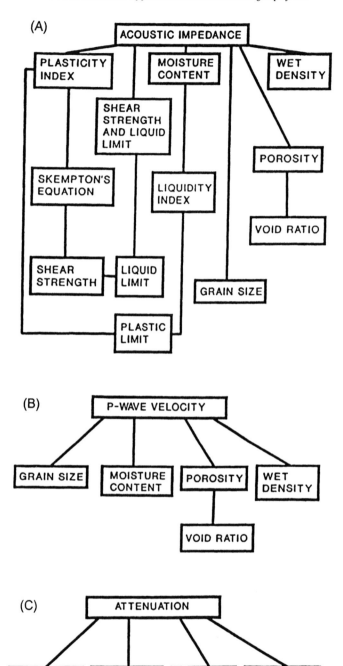

Figure 6.46 Schematics showing the empirical relationships between the geotechnical properties of a marine or estuarine sediment and its seismo-acoustic properties: (A) acoustic impedance; (B) P-wave velocity; and (C) attenuation. From Haynes *et al.* (1993), by permission

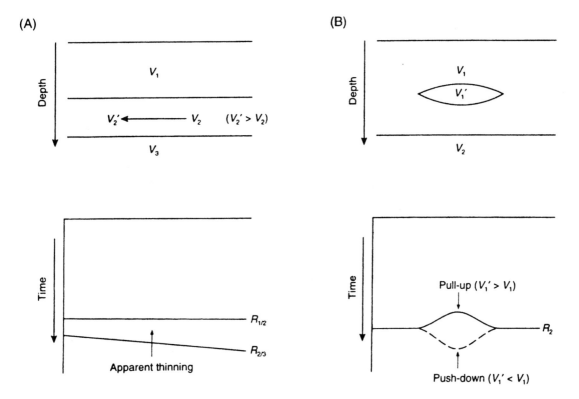

(A)

(B)

Figure 6.47 Depth and corresponding time sections to illustrate (A) apparent thinning of a horizon due to lateral velocity changes; and (B) pull-up and push-down of a lower reflection due to the presence of a lens with a higher or lower velocity, respectively, relative to that of surrounding material. These interpretational features are relevant to both seismic reflection profiling and to ground penetrating radar surveys

The second aspect is apparent bulges or cusps in an otherwise planar reflector. These are caused by velocity 'pull-ups' or 'push-downs' due to areas with an anomalous velocity above the reflector (Figure 6.47B). A 'pull-up' occurs when the overlying velocity is faster than that adjacent to it so the travel time through the anomalous zone is shorter than elsewhere. Conversely, velocity 'push-down' occurs when the overlying velocity is lower than that adjacent, giving rise to a longer travel time.

One other common pitfall in interpretation, particularly of marine seismic profiles, is the 'apparent thickness' of the seabed sediment. Often the seabed reflection comprises a train of pulses that masks inclined reflections that actually reach to the seabed surface. A common mistake is to calculate the period of the seabed reflection (in milliseconds) from the first break to the end of the wavetrain and then compute an apparent thickness. This is completely wrong. No apparent thickness can be determined at all unless the thickness of the sediment cover exceeds that derived from the seabed wavetrain. For example, given a period of 5 ms two-way travel time for the seabed reflection wavetrain, and a seafloor sediment P-wave velocity of 1500 m/s, the seabed sediment layer would have to be thicker than 1.88 m before the base of the seabed sedimentary layer could be imaged. Any thinner than this, and the base of the sediment layer would be masked by the tail of the seabed reflection wavetrain.

6.6   APPLICATIONS

6.6.1   High-resolution seismic profiling on land

In this section a number of examples are given to demonstrate the efficacy of the technique under suitable conditions. The surveys described range from the mapping of buried channels and of glacial stratigraphic sequences, and the evaluation of permitting (licensing) hazardous-waste injection. These case histories illustrate how well seismic reflection surveys can image the sub-surface as well as indicating different applications.

The first example, provided by Pullan and Hunter (1990), describes the imaging of a buried rock valley beneath flat farmland at Dryden, Ontario, Canada. An in-hole shotgun source was used with the optimum offset of 15 m between source and receiver. High-frequency (100 Hz) geophones were used and data were recorded through a 300 Hz highpass filter on the seismograph. The recorded data were amplified using automatic gain control (AGC) and filtered with a digital bandpass filter (240–800 Hz) before final display. The minimum depth of visible reflections is around 7.5 m, about half the typical optimum offset distance.

The processed seismic section is shown in Figure 6.48 in which the buried rock valley is obvious. The geology of the site comprises flat-lying units of sand and clay (Quaternary) above Precambrian quartz pegmatite bedrock. The natural groundwater level is at only 1 m depth. Consequently, the seismic source was fired into fully

Figure 6.48 Unmigrated optimum offset reflection section from Dryden, Ontario, showing a steep-sided bed-rock valley filled with clay and sand deposits; the section is around 500 m long. From Pullan and Hunter (1990), by permission

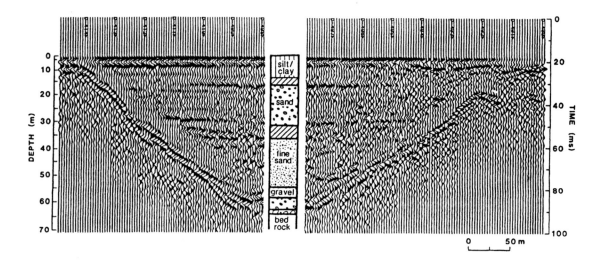

water-saturated fine-grained sediments–ideal conditions for high-resolution seismic reflection profiling. Despite the site being completely flat, the bedrock topography is marked–bedrock lies at a depth of only 15 m at each end of the 500 m long seismic line, but is about 65 m below ground level in the middle of the rock valley. The vertical geological sequence was determined from a borehole drilled as part of a Quaternary mapping project in the area, and a simplified log is spliced into the section for the sake of correlation. It can be seen that the major interfaces have been imaged successfully.

The depth scale shown in Figure 6.48 has been derived from a detailed velocity analysis but the image does not show the exact geometry for the bedrock reflector because of the steep dip of the rock valley sides. Pullan and Hunter (1990) estimated that the error in the depth scale with respect to the valley sides is only 6–8%, so the unmigrated section provides a reasonable image of the buried rock valley. This accuracy is certainly adequate for the siting of boreholes for engineering or groundwater purposes.

The second example, also provided by Pullan and Hunter (1990), is an 850 m long optimum offset section from Shawville, Quebec, Canada, which crosses the proto-Ottawa River valley. The present-day Ottawa River lies 5 km to the west of the site. The seismic section (Figure 6.49) shows the buried valley to be about 500 m wide and over 100 m deep. The channel is incised into Precambrian marble and gneiss, and is infilled with glacio-lacustrine sediments.

In comparison with the previous example, the source used in this case was a 16 lb (7 kg) sledge hammer, with four strikes at the baseplate to generate sufficient energy to produce a sensible bedrock reflection. The optimum offset used here was 60 m, which is much larger than in the previous example because of the significant airwave generated by the hammer striking the baseplate. There was virtually no airwave generated in the first case as the shot was fired within the

Figure 6.49 Optimum offset reflection section from Shawville, Quebec, showing a broad (ca. 500 m wide) buried river channel, part of the proto-Qtawa River. From Pullan and Hunter (1990), by permission

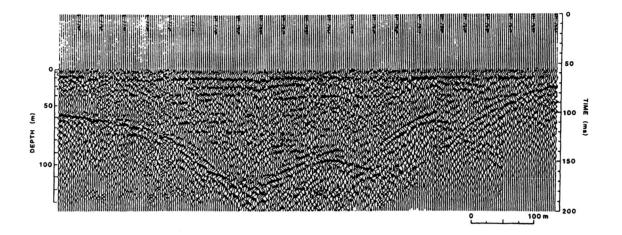

ground. The data were filtered using 100 Hz geophones and a 100 Hz highpass filter on the seismograph, followed by filtering with a digital bandpass filter (100–400 Hz) during subsequent processing.

The section in Figure 6.49 gives a general indication of the form of the buried channel. Several diffraction events (produced by knolls or depressions on the bedrock surface) can also be seen on the section, producing a more rounded impression of the bedrock topography. The section is presented here to demonstrate that penetration of up to 100 m (achieved using just a hammer source) coupled with the benefit of the optimum offset technique produced a good-quality image of the bedrock surface.

The third example, provided by Slaine *et al.* (1990), was acquired during a comprehensive investigation of the overburden and bedrock in the vicinity of the Ontario Waste Management Corporation's (OWMC) proposed site for a hazardous waste treatment facility. It was also claimed to have been the first commercial seismic reflection survey for the evaluation of a proposed hazardous waste site in Canada. The objective of the seismic survey was to determine the continuity of sub-surface overburden conditions between borehole locations on-site and beyond the site boundaries.

Stratigraphic control was provided by geologic logs from 19 boreholes. The geology of the site was found to consist of four Quaternary glacio-lacustrine/glacial units. The Upper and Lower Glacio-lacustrine units (silt–clay) and the Halton Unit, an interbedded till, have a total thickness of 35 m. These units overlie a lower dense sandy-silt till (Wentworth (Lower) Till), commonly 1–2 m thick, which in turn overlies the Guelph Dolostone (dolomitised limestone) bedrock.

The seismic survey was undertaken using a 12-gauge in-hole shotgun source, with 100 Hz geophones buried at 0.5 m with a geophone interval of 5 m, and a 300 Hz highpass filter on the engineering seismograph. The data were processed by filtering with a bandpass filter (500–1000 Hz), corrected for surface statics, and displayed after applying AGC. Velocity analyses were also undertaken in order to determine the lateral and vertical variability in velocity. However, it was found that, as the sediments were predominantly water-saturated, there was no systematic variation in velocity with depth and that a value of 1600 m/s was representative of the overburden. Further details of the velocity analyses used are given by Slaine *et al.* (1990). The depth scale for the optimum offset section was calculated using the formula:

$$\text{Depth } (m) = \tfrac{1}{2}|(VT)^2 - X^2|^{1/2}$$

where $V = 1600$ m/s, T is the two-way travel time in seconds, and X is the optimum offset (m). Depths to interfaces determined from the seismic sections were on average within 4% of the depths found by

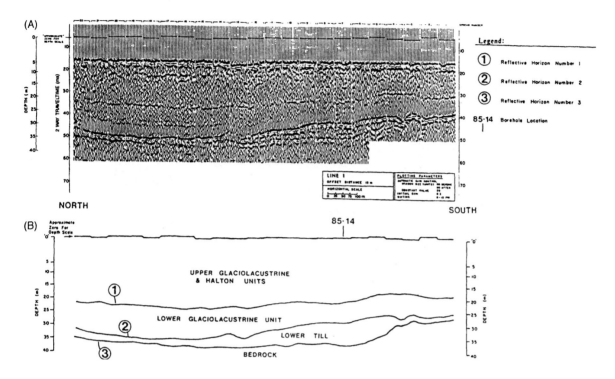

Figure 6.50 (A) Seismic and (B) geologic cross-sections along an 850 m long optimum offset reflection section acquired in Ontario. From Slaine *et al.* (1990), by permission

drilling. The position of the water table for each segment of the seismic section is represented in Figure 6.50 by a horizontal straight line at around 5 ms.

Three laterally-continuous reflections were identified on the sections, one of which is shown in Figure 6.50. From the correlation with the borehole logs, it was found that the three principal reflections were associated with interfaces between the units as shown in the figure. It was found, therefore, that the seismic survey met the original objectives and provided information about the sub-surface materials that was not readily obvious from the borehole logs alone.

Miller *et al.* (1995) have examined the vertical resolution achievable using high-quality seismic reflection profiling at a site in southeastern Kansas, USA. A 12-fold common depth point profile was designed to image geometric changes greater than 2 m at depths from 25 m to 100 m in a depositional shelf-to-basin environment dominated by limestones, sandstones and shales. At the time of the survey the water table was more than 10 m deep.

The seismic data were acquired with a 15-bit EG&G Geometrics 2401 seismograph using a half-millisecond sampling rate, and 100 Hz low-cut and 500 Hz high-cut analogue filters. The source used was a downhole 0.50-calibre shotgun and receivers consisted of three 40 Hz geophones at 1 m spacing to form a group centred on each

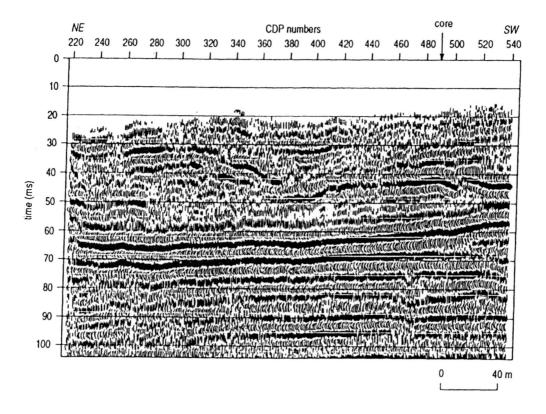

0   40 m

Figure 6.51 Uninterpreted seismic reflection profile at a site in south-eastern Kansas. From Miller *et al.* (1995), by permission

station with a station interval of 2.5 m. The optimum recording window had a source-to-nearest-receiver group offset of 12.5 m and a source-to-farthest-receiver group offset of 70 m, giving a total spread length of 58 m. Details of the data processing and modelling are discussed in detail by Miller *et al.* (1995).

The uninterpreted seismic record is shown in Figure 6.51. A borehole, Clarkson #2, is located at CDP 494, and the geological cross-section and corresponding geophysical logs (gamma ray and neutron density) are shown in Figure 6.52. A model derived for the seismic section was consistent with the borehole control from Clarkson #2, and the velocities and densities correlate with the lithologies encountered in the borehole core. The derived model is shown in Figure 6.53 with the interpreted version of Figure 6.51 shown in Figure 6.54. The buried channel has been imaged quite clearly, as have several of the principal rock units.

A comparison of Figure 6.51 with Figure 6.48 is quite striking. Both were acquired over the same travel time range (ca. 100 ms) with comparable systems although the survey shooting configurations were different (12-fold CDP profiling versus the use of the optimum offset technique). Two other prime differences should be noted:

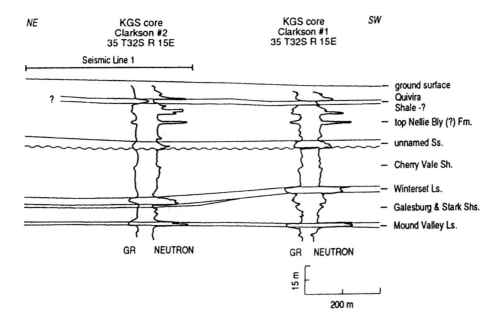

NE   KGS core   KGS core   SW
              Clarkson #2   Clarkson #1
              35 T32S R 15E   35 T32S R 15E

        Seismic Line 1

                                                    — ground surface
    ?   — Quivira
                                                      Shale -?
                                                    — top Nellie Bly (?) Fm.

                                                    — unnamed Ss.

                                                    — Cherry Vale Sh.

                                                    — Winterset Ls.

                                                    — Galesburg & Stark Shs.

                                                    — Mound Valley Ls.

        GR   NEUTRON   GR   NEUTRON

                                    15 m ⌐

                                    └─────────────┘
                                         200 m

Figure 6.52 Geological cross-section based on borehole lithological core and geophysical (gamma ray and neutron density) logs; the zone corresponding to the seismic line in Figure 6.51 is shown shaded. From Miller *et al.* (1995), by permission

- In the present example the water table was deep and shots were fired in the unsaturated zone, in contrast with the fully water-saturated situation in the earlier example.
- In the present case the geology is mainly a cyclic thinly bedded sequence of the Late Carboniferous, compared with unlithified Quaternary sediments.

The vertical resolution apparent in Figure 6.48 is about 2–3 m whereas that determined for the present case is about 7 m for a wavelength of approximately 21 m at a depth of 70 m. This latter value corresponds to one-third wavelength resolution rather than the more usual one-quarter wavelength. This comparison demonstrates that high-resolution seismic reflection profiling is better suited to sites with fine-grained, saturated near-surface conditions.

In the last example in this section, the scale of investigation is increased. In the USA, deep-well injection is used to dispose of hazardous liquid waste products. It is usual to use well-log data to assess the viability of such disposal, particularly with respect of the presence of faults that could act as fluid conduits allowing hazardous waste to migrate beyond its intended destination. Underground sources of drinking water could easily become contaminated if such unforeseen leakage were to occur. In the present case, discussed in detail by Zinni (1995), seismic surveys have been used in conjunction with well-log data to help delineate known sub-surface faults and to locate previously unknown faults in St John the Baptist Parish, Louisiana, USA. Municipal water wells near Ruddock supply

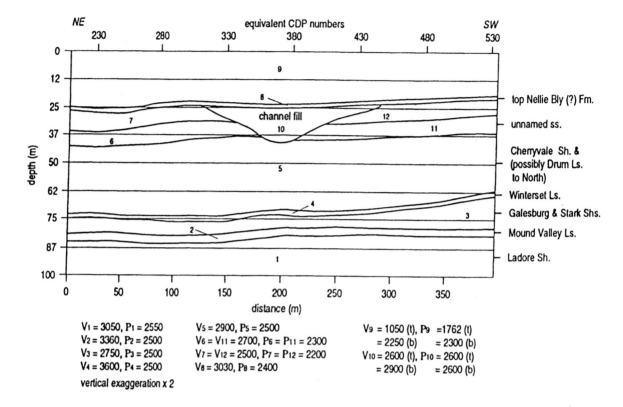

Figure 6.53 Two-dimensional geological model derived from the seismic survey and correlated with borehole data. For this model, seismic velocity (V in m/s) and density (P) in kg/m^3) are defined at each interface and at the top (t) and bottom (b) of the channel fill. From Miller *et al.* (1995), by permission

freshwater from the Covington aquifer, the principal local source of drinking water. A chemical manufacturer has been disposing of liquid hazardous waste into a zone encompassing the Covington aquifer (Figure 6.55) for 30 years via four injection wells. A recent deep geological investigation in the area had suggested that the Covington aquifer might be connected hydrologically with two of these hazardous-waste-injection reservoirs. Under the local authority's regulations pertaining to contamination of any Underground Sources of Drinking Water (USDW), the confining zone(s) of a hazardous-waste-injection reservoir should be free of bisecting transmissive faults and fractures. The seismic investigation was used to assess the structural, and hence hydrogeological, conditions in relation to licensing liquid hazardous-waste disposal. The geographical layout of the various wells and seismic line 1094 is shown in Figure 6.56.

The seismic survey was completed using between 24- and 30-fold coverage with asymmetrical split-spread geophone arrays with geophone group and shot intervals of 67 m and 134 m, respectively, and both hydrophone group and shot intervals of 67 m (for lake recording). The energy source was dynamite. Processing included dip moveout and migration.

The seismic record for line 1094 (shown in Figure 6.57) reveals four previously unknown faults, three of which (A, C and D in Figre 6.56)

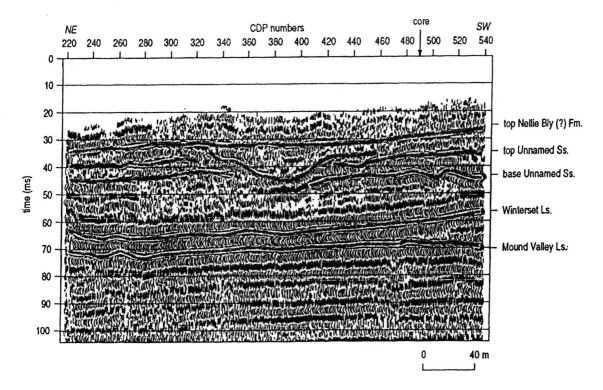

Interpreted version of the 12-fold CDP stacked section shown in Figure 6.51. From Miller *et al.* (1995), by permission

bisect the injection reservoirs and the Covington aquifer. The presence of these faults significantly increases the possibility of hydrologic communication by providing paths, such as fault planes, along which fluids can migrate. Alternatively or in addition, sand units could have been juxtaposed against other sand units permitting leakage to occur across the contacts. Furthermore, the upper injection reservoir was found to be part of a fluvial system that in places had scoured down into the Covington aquifer. Sand had evidently been deposited directly from the injection reservoir interval on the Covington aquifer sands. Consequently, the injection reservoir was thought by Zinni to be in hydrologic communication with the Covington aquifer.

If vertical migration of liquid hazardous waste occurs along the fault planes following injection into the underground reservoirs, the water quality of the Covington aquifer could be adversely affected. Similarly, other shallow freshwater aquifers could also be jeopardised. Without the use of the seismic survey, the risk of contaminating underground sources of drinking water would not have been assessed properly.

6.6.2 High-resolution seismic profiling over water

There are very few commercially available digital-data acquisition systems or streamers available with specifications adequate to

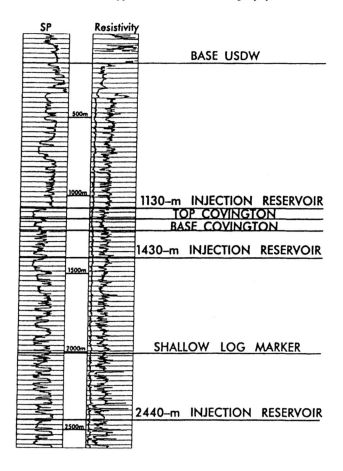

Figure 6.55 Borehole logs at Dupont # 8 Pontchartrain Works (for the location, see Figure 6.56). This well is the deepest of the four injection wells (total depth 2592 m), and has penetrated the 1130, 1430 and 2440 m injection reservoirs and the Covington aquifer. The base of the Underground Source of Drinking Water (USDW) is at 245 m depth. From Zinni (1995), by permission

provide the quality of field data necessary if detailed and quantitative analyses of the data are to be undertaken. Systems that do have the correct specification tend to exist in research institutes, although a couple of commercial systems are now available (e.g. GeoAcoustics Sonar Enhancement System and Elics – Delph 1).

An example of what can be achieved with equipment with a good specification is shown in Figure 6.58, from a survey at Lake Windermere in Cumbria, UK (Haynes *et al.* 1993). The system used was a Carrack SAQ-V that was capable of sampling up to 12 channels with a total throughput of 48 kHz (i.e. 3 channels at 16 kHz, 12 channels at 4 kHz, etc.). The system is PC-based, making it small enough to be deployable on small boats for use in very shallow water surveys. The section illustrated in Figure 6.58 was obtained using an ORF 'Geopulse' Model 5819B source, operated at 280 J with a trigger length of 0.5 s and a sweep length of 150 ms. The signals were received by an EG&G Model 265 hydrophone streamer and recorded using a Dowty Waverley Model 3700 thermal linescan recorder.

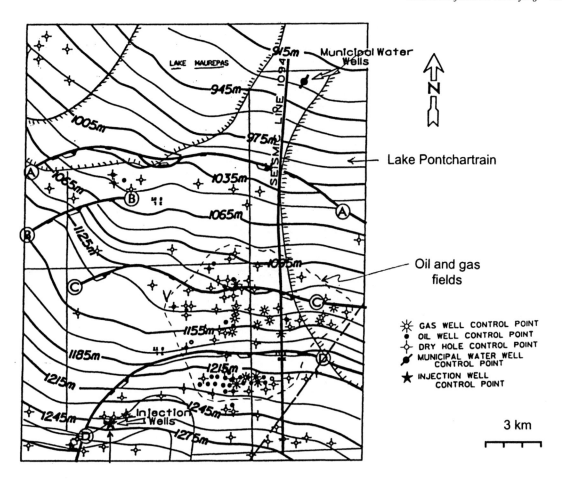

Figure 6.56 Location map of the survey and well-field area, Louisiana, USA. The position of each of the municipal water wells, injection wells, velocity survey well (V), and the approximate outline of the Frenier, Laplace and Bonnette Carre oil and gas fields, are marked. After Zinni (1995), by permission

Lake Windermere is an ideal location to undertake a very-high-resolution seismic survey as on a calm day the lake surface is mirror-like and acoustically it is very quiet. The data obtained are of very high quality and, on the section shown, reflectors can be resolved to better than 20 cm down to a sediment depth in excess of 15 m. Four different sediment types can be distinguished: lake-bed mud, slumped and varved (seasonally laminated) clays, and glacial till.

In estuarine and marine engineering investigations, a major difficulty encountered is the presence of shallow gas accumulations thought to be associated with the decay of organic matter within near-surface sediments or to have migrated from depth. In some cases, the gas escapes to the surface of the water and vents naturally. In other, more dramatic cases, the gas release may be triggered by drilling activity leading to blowouts which can be highly dangerous. Drilling rigs and boats have been lost through degassing of sediments. Past gas escapes from the seafloor are commonly characterised by

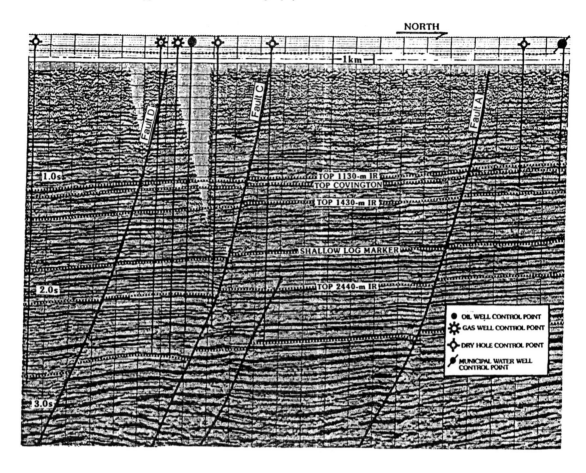

Figure 6.57 Seismic line 1094 (location marked in Figure 6.56). Faults A, C and D intersect the 1130, 1430 and 2440 m injection reservoirs (IR) and the Covington aquifer. From Zinni (1995), by permission

seabed pockmarks or scars. The issue of shallow gas has had to be addressed extremely seriously by both the hydrocarbon industry and increasingly the engineering community.

A detailed discussion of seabed pockmarks and associated seepages has been given by Hovland and Judd (1988). Seismically, shallow gas is manifest by poor definition, acoustic opacity, rapid signal attenuation and masking of underlying reflection events. Only a small quantity of gas, 0.1% of sample volume as bubbles, is sufficient to decrease the P-wave velocity in affected sediments by up to one-third that of a normally saturated and bubble-free equivalent sediment.

There are three recognised forms of shallow gas accumulations as seen on high-resolution seismic profiles (Taylor 1992): (1) gas curtains (the most common); (2) gas blankets; and (3) gas plumes. The lateral extent and acoustic characteristic of each of these is listed in Table 6.1.

An example of the seismic expression of each of these three forms of shallow gas accumulation is shown in Figure 6.59. In each case, the surveys were acquired using a Uniboom sub-bottom profiler.

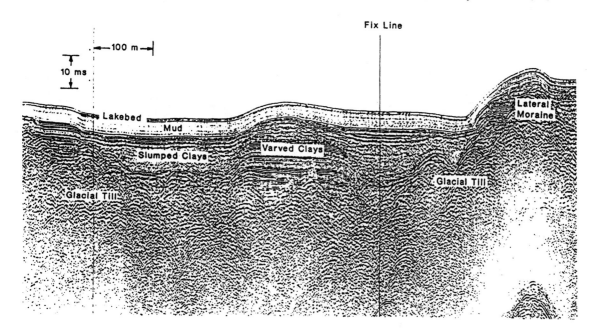

Table 6.1 Classification of types of shallow gas accumulations (Taylor 1992)

Name	Lateral extent	Acoustic characteristics
Plume	< 50 m	Edges distinct, apparent vertical connection to source
Curtain	100–500 m	Edges distinct, no apparent vertical connection to source
Blanket	> 500 m	Edges often difficult to resolve; no apparent vertical connection source

Figure 6.58 Example of high-resolution seismic data obtained at Lake Windermere, Cumbria, UK, using a Boomer-type source. From Haynes *et al.* (1993), by permission

Taylor (1992) reported that in one survey, in the Firth of Forth, Scotland, both a Uniboom and a 30 in^3 sleeve-gun failed to produce sufficient energy to propagate through a shallow gas accumulation. Instead, a marine refraction experiment was undertaken with a 10 in^3 airgun as a source and bottom-mounted hydrophones on the far side of the gassified sediments in order to undershoot the gas. Refracted headwaves from bedrock were recorded despite the presence of the shallow gas.

While the acquisition of high-quality digital data using high-specification systems is extremely desirable, it is not always either practicable or financially justified as perceived by clients. It is possible, however, even with basic analogue systems, to produce reliable

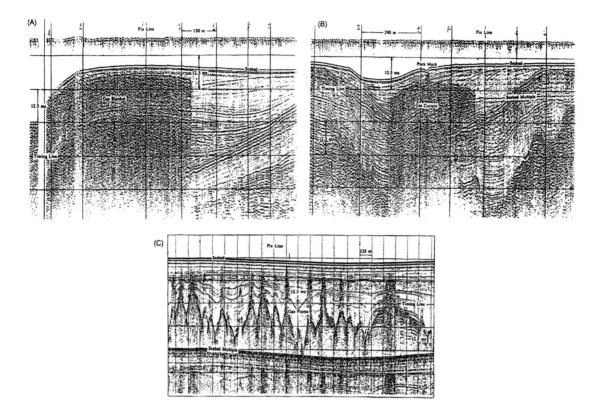

information about the vertical and lateral distribution of geological interfaces. One such example is a series of analogue surveys in Plymouth Sound, Devon, UK, using a Boomer™ surface-tow source and a sparker to map an infilled rock valley (Reynolds 1990a). Two boreholes were drilled within Plymouth Sound to provide vertical geological correlation. An example of an analogue sparker section with details from one of the boreholes is shown in Figure 6.60. The sedimentary sequence consisted of contemporary estuarine muds overlying fine–coarse sands with gravel lenses. The lower interface of the sand unit was marked by a prominent, partially cemented shelly gravel overlying organic muds finely laminated in parts with silts. The base of the sediments was marked by coarse gravel overlying weathered Devonian slate at a depth of 26.6 m below the seabed. The upper sand unit was characterised by dipping reflections (Figures 6.60 and 6.41) thought to be associated with point bar foresets. The top of the organic muds, which was dated provisionally at 8300 BP on the basis of microfossils, could be traced laterally over a substantial part of the central portion of Plymouth Sound, and thus served as an isochron. The sediments infilling the rock valley formed during a period of rising sea level following the end of the last ice age. The

Figure 6.59 Uniboom sub-bottom profiles showing (A) a gas blanket, and (B) a gas curtain in the Firth of Forth, Scotland; and (C) gas plumes in the north-east Irish Sea. From Taylor (1992), by permission

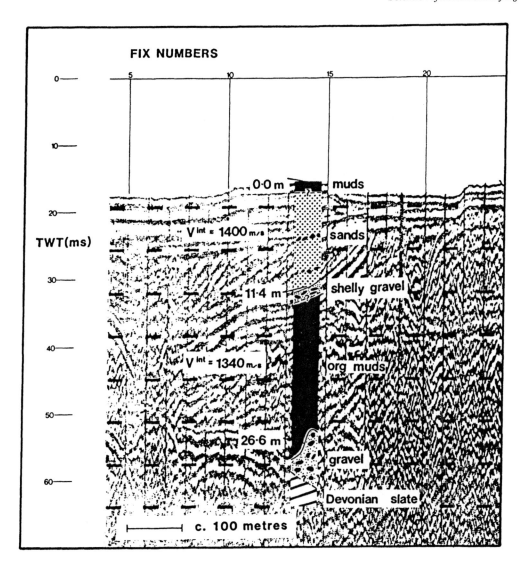

FIX NUMBERS

interfaces between the principal units mirror the temporal changes in the proto-River Tamar as it flowed through Plymouth Sound.

Using a grid of survey lines, posted two-way travel time maps of the principal laterally extensive sub-surface reflections were produced (Figure 6.61) from the analogue records. The travel-time data were contoured using a simple graphical package (SURFER from Golden Software Inc., USA) on an IBM PC-XT microcomputer. From these, isometric projections were also produced for each reflection and are shown stacked in Figure 6.62. The travel times were normalised to a time datum 60 ms below Ordnance Datum (Newlyn) for the graphical display. In Figure 6.62, the knoll evident on the eastern side of the valley is bedrock, which is why there are few changes in this region

Figure 6.60 Analogue sparker record over a buried rock valley within Plymouth Sound, Devon, with the main geological sequences as found by drilling. Two preliminary average interval velocities were determined for the main sequences corresponding to the depth ranges 0–11.4 m and 11.4–26.6 m using the observed two-way travel (TWT) times and the measured depths below seabed to the corresponding interfaces from the borehole lithological log. From Reynolds (1990a), by permission

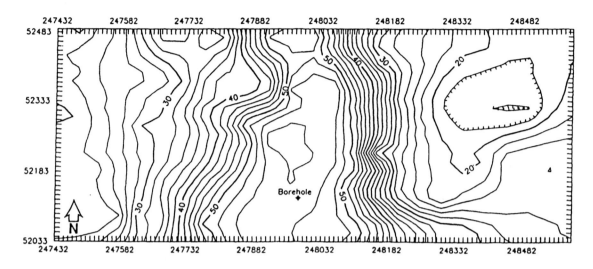

Figure 6.61 Posted two-way travel time map to bedrock from a part of Plymouth Sound. Values along the *x* and *y* axes are Universal Transverse Mercator (UTM) coordinates. The location of the borehole (detailed in Figure 6.60) is also shown. From Reynolds (1990a), by permission

between the lower three projections. However, the main sediment deposition was on the western side and the gradual shallowing of the channel can be seen clearly.

The last example in this section is of a three-dimensional seismic reflection survey undertaken in the River Scheldt in Antwerp, Belgium, to map a clay diapir as part of a geotechnical investigation (Henriet *et al.* 1992). The clay diapir in Rupelian Clay had been discovered during a survey in 1982 close to a site proposed for a metro tunnel under the river. The 1982 survey was shot using an EG&G Uniboom source, and a sample seismic section is shown in Figure 6.63. The diapir has an apparent diameter of 60 m. The reflections bulge upwards with a vertical amplitude increasing from a few decimetres at 50 m depth to a few metres at about 25 m depth. Higher horizons are pierced by the diapir.

Many reflections from within the Rupelian Clay are prominent as clusters of diffraction hyperbolae (marked with solid triangles in Figure 6.63). The diffractions are associated with large concretions (septarian nodules) with diameters in the range between 0.5 m and 1 m and thicknesses of 0.2–0.3 m. The clay diapirs themselves were not thought to constitute a hazard as their geotechnical properties had been found to be comparable to those of undisturbed clay. However, the formation of the diapiric structures was known to drag some septarian nodules vertically. The distribution of the nodules was important as it influenced the choice of drilling equipment for the tunnelling.

Following the successful 1982 survey, a more detailed investigation was undertaken in 1990 using a true 3-D data-acquisition system (Figure 6.64). The receiver configuration consisted of an array of 12 purpose-built microstreamers, each with two hydrophone groups 1 m apart, that were towed from a modified Hobiecat catamaran from

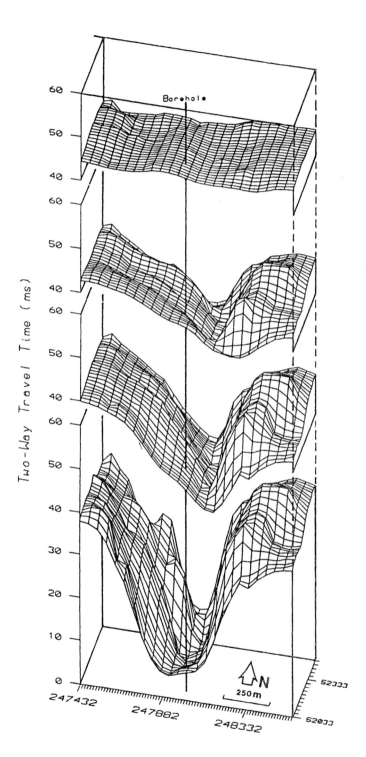

Figure 6.62 Stacked isometric projections of normalised two-way travel time to, in order of increasing depth: seabed, internal reflector in the upper sand unit, the shelly gravel layer between the upper sand and underlying organic clays, and bedrock. From Reynolds (1990a), by permission

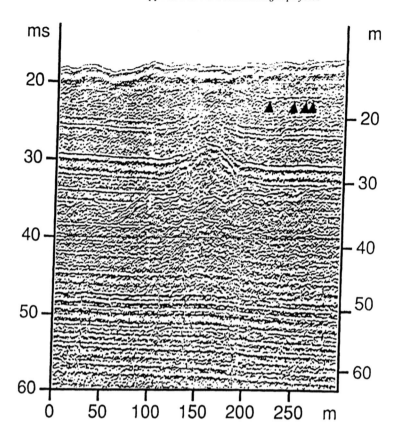

Figure 6.63 Uniboom profile from the 1982 survey across a clay diapir in Rupelian Clay under the River Scheldt, Antwerp, Belgium. Diffraction hyperbolae (indicated by solid triangles) are associated with calcareous septarian nodules within the clay. The riverbed reflection occurs at about 18 ms. The effect of the diapir is evident at 150 m, especially around 30 ms. From Henriet *et al.* (1992), by permission

which the source was also deployed. The distance between each streamer was 1 m. Each shot generated 24 seismic traces in a swathe 5.5 m wide centred along the boat track. The source–receiver catamaran was towed about 10 m behind the *RV Belgica*. The shot interval was 1 m. Positioning accuracy was to within decimetres and was necessarily accurate to permit $1 \times 1\,\text{m}^2$ binning. Two seismic sources were used: an EG&G Uniboom and a modified Sodera 15 in^3 watergun lashed horizontally to a fender at 0.2 m depth.

In order to achieve the required shot interval and firing rate, a very slow ground speed was required (between 0.5 and 2 m/s). This could only be achieved realistically by sailing against the current, which in the river Scheldt is very strong. It was thus impractical to turn the 45 m long survey vessel for fear of disrupting the towed acquisition system. Consequently, the vessel sailed against the current acquiring data *en route*. At the end of the survey line, the vessel moved across to the next track line and decreased her sailing speed to less than that of the river current, effectively sailing backwards while still maintaining headway into the current and keeping the towed array well stretched behind the vessel.

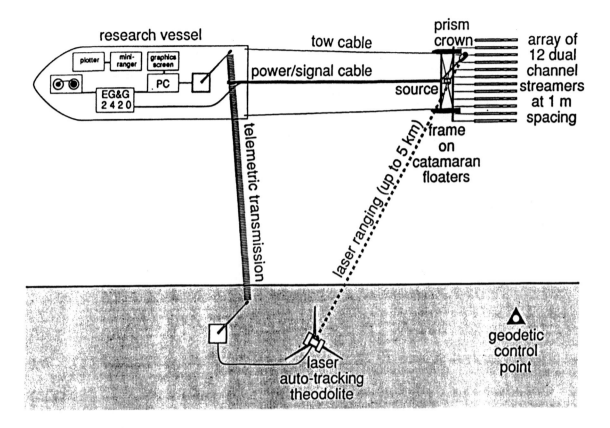

Figure 6.64 Layout of the system used in the 1990 three-dimensional survey at Antwerp, Belgium. From Henriet *et al.* (1992), by permission

The actual survey took three days to complete and about 20 000 shots were digitally recorded, translating into 250 000 common midpoints. The survey area was $50 \times 180 \, m^2$ in area with at least 5-fold and in places in excess of 10-fold bin coverage. Details of the data processing have been discussed by Henriet *et al.* (1992).

It was demonstrated during this survey that the characteristics of the two sources were very similar and the data were stacked together – not something that is normally recommended. The record length for the watergun was 60 ms and about 80 ms for the Uniboom. The combined record is shown in Figure 6.65. The obvious change in signal character around 60 ms is due to the differences in record length between the two sources. Slight updoming in the reflection at 58 ms is evident. The main clay diapir is in the centre of the record. Variation in signal character between the two sources was apparently due to differences in bin coverage. It is important, therefore, when interpreting seismic records by comparison, that such factors are taken into consideration if misinterpretation is not to happen. This case history has demonstrated that 3-D seismic surveys can be translated down to the scale of geotechnical surveys and can provide valuable information with a high degree of spatial resolution.

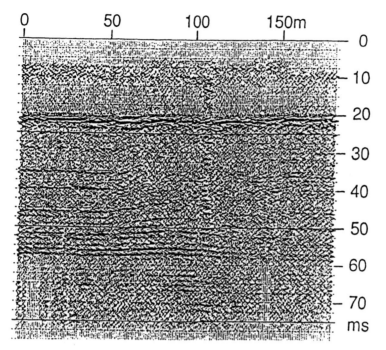

B.P.filter: 500 - 1400 HZ

Figure 6.65 Composite stacked section combining both watergun data from 0 to 60 ms and Uniboom data from 0 to the end of the record across a clay diapir at Antwerp, Belgium. Slight updoming of the reflection at 58 ms is evident. The seismic responses from the main diapir are in the centre of the record. From Henriet *et al.* (1992), by permission

6.6.3   Geophysical diffraction tomography in seismic palaeontology

In 1979, two hikers discovered partially exposed bones from the tail of a hitherto unknown species of dinosaur in the high desert of New Mexico (Witten *et al.* 1992). Excavation began in 1985 in the Morrison Formation, a 145 million years old (Late Jurassic) sandstone well known for its dinosaur remains further north in the USA, but until then, unrecognised in New Mexico. By 1986, it was realised that the bones were from a relative of *Diplodocus* with a projected length of at least 33.5 m (110 ft). David Gillette, who first described the dinosaur, coined the informal name 'seismosaurus' for the new dinosaur, alluding to its considerable size making the earth shake when it moved. The geological setting of the bones indicated that the body had suffered from rigor mortis before burial and that the animal's carcass had come to rest on a bend in a local river on a point bar and had been buried subsequently by fine sand. The sandstone surrounding the skeleton is massive but predominantly homogeneous, thus providing an excellent medium for remote sensing.

The 1985–86 field excavations indicated that the skeleton extended into a cliff face but the orientation and extent of the bones could only be conjectured. The projected orientation of the skeleton was at

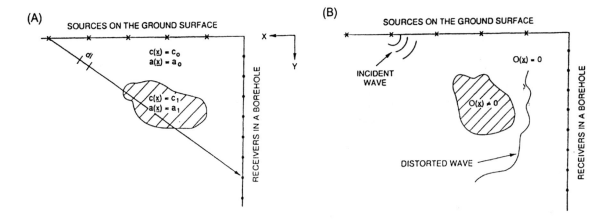

a level of at least 2.5 m (8 ft) beneath the surface. Traditional excavation using shovel, hammer and chisel would have been extremely slow and would have involved the removal of hundreds of tonnes of material. In April 1988, and again in July 1989, the site was investigated using geophysical diffraction tomography (GDT).

Geophysical diffraction tomography can take two forms: back-projection or back-propagation, the concepts of which are illustrated schematically in Figure 6.66. For both methods, it is assumed that a series of receivers is located down a borehole and a series of sources is placed along the ground surface. In back-projection, it is assumed that the energy emanating from a source travels through the subsurface in straight lines or as rays from source to a given receiver whose distance from the source is known exactly. The measured travel time of the first arriving signal and its amplitude are related to local variations in wave speed or wave attenuation. Reconstructing an image from variations in measured parameters is achieved mathematically by using data from many raypaths, which intersect the same unit cells, to iterate towards a solution, which is displayed as a pixelated image (i.e. each cell is assigned a value from the iteration which is displayed as one pixel in the final image). Back-projection has been used for imaging inhomogeneities such as karst features, imaging the Earth's mantle, and in producing electrical images from arrays of electrodes. Back-projection assumes straight raypaths and ignores any refraction effects.

A preferred alternative is back-propagation, which is reputed to provide sharper images and take into account refraction effects. The field system is exactly the same as in back-projection but the effect being sought here is the distortion of the wavefront caused by inhomogeneities within the ground being imaged. The distortion of the waveform in both time and space are measured at the receivers. A benefit of the back-propagation method is that it utilises information contained in the received signals regarding the constructive and

Figure 6.66 Schematic of the concepts of (A) back-projection imaging, and (B) back-propagation imaging. From Witten and King (1990a), by permission

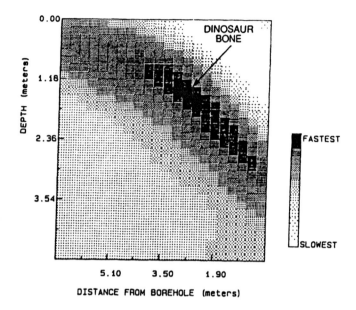

Figure 6.67 Tomographic image displaying a velocity anomaly interpreted as being due to a dinosaur bone in a near-homogeneous sandstone formation, New Mexico. The interpretation was confirmed by subsequent excavation. From Witten and King (1990a), by permission

destructive interference associated with multiple raypaths between source and receiver. The methods are described in more detail by Witten and King (1990a).

The GDT survey undertaken in April 1988 produced one image displaying a feature that was believed to be due to a dinosaur bone (Figure 6.67). In early July 1989, a dorsal vertebra and partial rib were exposed at the location indicated by the geophysical survey.

The field acquisition layout is shown in Figure 6.68. Approximately 1 m of overburden was removed from a rectangular plot about 9 m × 12 m, of which a flat L-shaped area was used in the GDT survey. A series of four boreholes (A–D in Figure 6.68) were constructed. They were 6 m deep, lined with PVC casing capped at the bottom, and the annulus around the outside of the casing within the borehole was backfilled with sand. Each borehole was filled with water and a hydrophone receiver array with active elements at 0.15 m intervals deployed down the borehole. Source positions were defined along radial lines emanating from the boreholes, with the first source position 1.2 m from the borehole and successive positions at intervals of 0.6 m. The source consisted of a commercially available seismic Betsy gun. Further details of the data acquisition and processing have been described by Witten *et al.* (1992).

The results of the GDT survey are summarised in Figure 6.69. Dinosaur bones identified by image reconstruction are shaded black, those found by direct excavation are shaded grey and the postulated positions of undiscovered bones, based upon experience of the disposition of other dinosaur skeletons that have

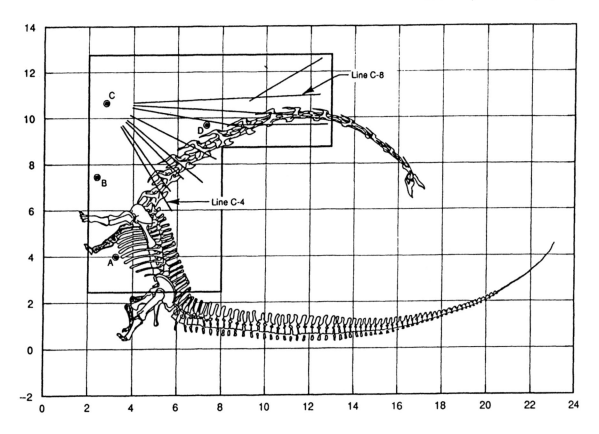

Figure 6.42 Illustration of the L-shaped study area and the horizontal extent and positions of the 10 offset VSP survey lines around the remains of the 'seismosaurus' skeleton. Grid cells are 2 m × 2 m. The borehole positions are indicated by the letters A–D. From Witten *et al.* (1992), by permission

been found with a rigor mortis posture, are unshaded. One of the reasons why the method worked at this site is that the material surrounding the bones was found to have a P-wave velocity of only 450 m/s, in stark contrast to that of the bones which had a velocity of 4500 m/s.

The use of GDT as part of this palaeontological investigation has been beneficial in the location of significant parts of the skeleton of this giant dinosaur. It has also demonstrated the usefulness of the technique in being able to image objects successfully.

The GDT method has also been applied to imaging artificial buried objects such as metal drums, pipes and tunnels (e.g. King *et al.* 1989; Witten and King 1990a,b). One such example is illustrated in Figure 6.70. A field test was conducted under controlled conditions by burying waste-simulating targets at known locations and then imaged using back-propagation imaging. The targets were 55-gallon metal drums with a diameter of 0.61 m and 1 m tall, either empty or water-filled, and plastic bags containing styrofoam packing pellets. The positions of the targets relative to a borehole are shown in Figure 6.70A. An array of 29 uniformly spaced hydrophones spanning a depth interval of 0.61 m to 4.9 m was located in a cased monitoring

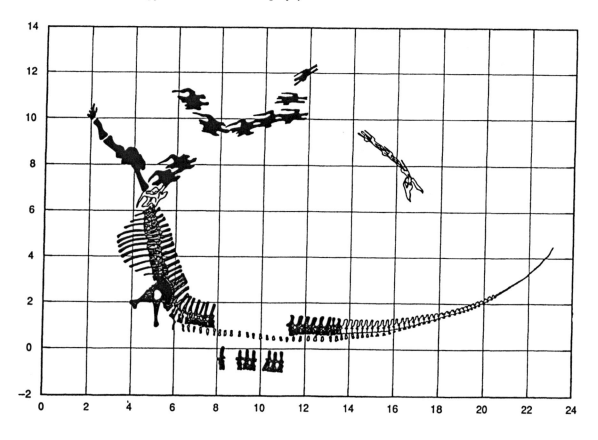

well. Source positions were established along two lines, one of which is shown in Figure 6.70. The sources were located over a distance of 1.8–14.6 m from the well. The resulting image is shown in Figure 6.70B in which the water-filled drum is evident as a faster-velocity anomaly, and the empty drum and the styrofoam are associated, weakly, with low-velocity zones. Note that in the image, there is a velocity anomaly which smears obliquely across the image from the location of the drum towards the base of the well. This is an artefact of the GDT method and such elongation of features occurs in the predominant direction of incident wave propagation. In this case, the sources were at the surface and the waves travelled down towards the well.

Figure 6.69 Interpretation of skeletal position based on known bone locations and the results of the GDT surveys. Bones identified by image interpretation are shaded black; those found by excavation are shaded grey; bones conjectured as being present are unshaded. The skeletal drawing is schematic and is not intended to demonstrate anatomical details. From Witten *et al.* (1992), by permission

6.6.4 Forensic seismology

The impact of the main segments of the Boeing 747 aircraft which crashed at Lockerbie in Scotland on 21 December 1988 as a result of a bomb blast on board was equivalent to a small earthquake measuring 1.3 ML on the Richter scale (Redmayne and Turbitt 1990). The event was detected by a series of seismometers at Eskdalemuir, part of

(A)

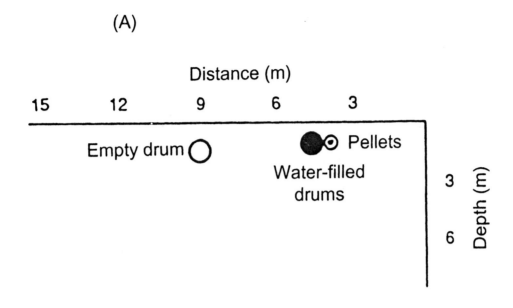

(B)

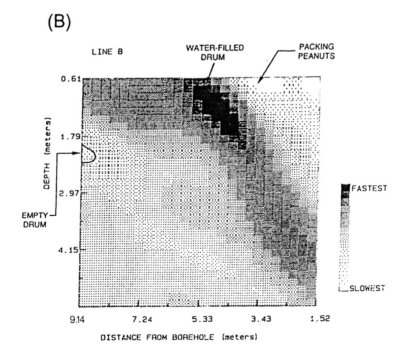

a wider network established and monitored routinely by the Global Seismology Unit of the British Geological Survey. As the location of the impact was known, it was possible to calibrate and check their algorithms for locating epicentres. Their best fit for the Lockerbie impact was only some 300 m in error.

Figure 6.70 A GDT survey undertaken at a test site over buried known targets: (A) the layout of the targets, and (B) a tomographic image obtained along the line in (A). From Witten and King (1990b), by permission

Records from these seismological stations are analysed to provide the approximate location of an aircrash event to within an area of about 1 km². In remote and largely unpopulated areas covering many hundreds of square kilometres, such information can help to target the search area and reduce the time taken to locate a crashed aircraft. The impact at Lockerbie was also used to calibrate the relationship between the energy of a falling mass and the resulting seismic magnitude of the impact. This kind of information is useful in the search for meteorite impacts.

Information arising from events recorded by the seismological network is passed to the British Ministry of Defence at Aldermaston. Analysis of the records from the event at Lockerbie indicated that the crash could have been felt up to 1 km away from the point of impact and that structural damage from seismic waves alone could have been caused to buildings up to 100 m away.

Two other implications arise from the analysis of seismic waves generated from aircrashes. The first is that the magnitude of the peak particle velocity can be estimated, and hence sensitive buildings and structures can be designed to withstand such ground disturbance. The second is that the knowledge of the force of impact, determined using the seismic evidence in conjunction with other information, combined with a study of how a particular plane behaves on impact, can help to improve aircraft design.

Section 3
ELECTRICAL METHODS

Chapter 7

Electrical resistivity methods

7.1   INTRODUCTION

Electrical resistivity methods were developed in the early 1900s but have become very much more widely used since the 1970s, due primarily to the availability of computers to process and analyse the data. These techniques are used extensively in the search for suitable groundwater sources and also to monitor types of groundwater pollution; in engineering surveys to locate sub-surface cavities, faults and fissures, permafrost, mineshafts, etc.; and in archaeology for mapping out the areal extent of remnants of buried foundations of ancient buildings, amongst many other applications. Electrical resistivity methods are also used extensively in downhole logging. For the purposes of this chapter, applications will be confined to the use of direct current (or very-low-frequency alternating current) methods.

Electrical resistivity is a fundamental and diagnostic physical property that can be determined by a wide variety of techniques, including electromagnetic induction. These methods will be discussed in their respective chapters. That there are alternative techniques for the determination of the same property is extremely useful as some methods are more directly applicable or more practicable in some circumstances than others. Furthermore, the approaches used to determine electrical resistivity may be quite distinct – for example, ground contact methods compared with airborne induction techniques. Mutually consistent but independent interpretations give the interpreter greater confidence that the derived model is a good approximation of the sub-surface. If conflicting interpretations result, then it is necessary to go back and check each and every stage of the data acquisition, processing and interpretation in order to locate the problem. After all, the same ground with the same physical properties should give rise to the same model irrespective of which method is used to obtain it.

7.2   BASIC PRINCIPLES

7.2.1   True resistivity

Consider an electrically uniform cube of side lenght L through which a current (I) is passing (Figure 7.1). The material within the cube resists the conduction of electricity through it, resulting in a potential drop (V) between opposite faces. The resistance (R) is proportional to the length (L) of the resistive material and inversely proportional to the cross-sectional area (A) (Box 7.1); the constant of proportionality is the 'true' *resistivity* (symbol: ρ). According to Ohm's Law (Box 7.1) the ratio of the potential drop to the applied current (V/I) also defines the resistance (R) of the cube and these two expressions can be combined (Box 7.2) to form the product of a resistance (Ω) and

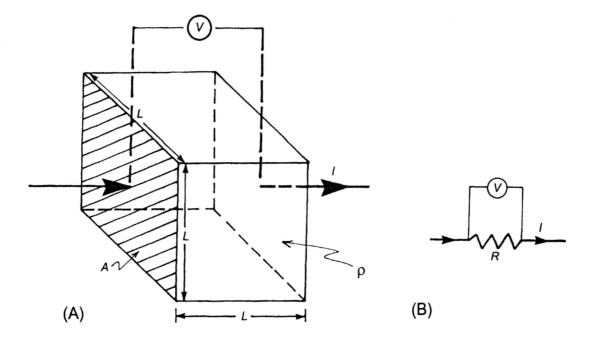

(A)   (B)

Figure 7.1 (A) Basic definition of resistivity across a homogeneous block of side length L with an applied current I and potential drop between opposite faces of V. (B) The electrical circuit equivalent, where R is a resistor

a distance (area/length; metres); hence the units of resistivity are ohm-metres (Ω m). The inverse of resistivity ($1/\rho$) is conductivity (σ) which has units of siemens/metre (S/m) which are equivalent to mhos/metre (Ω^{-1} m^{-1}). It should be noted that Ohm's Law applies in the vast majority of geophysical cases unless high current densities (J) occur, in which case the linearity of the law may break down.

If two media are present within the resistive cube, each with its own resistivity (ρ_1 and ρ_2), then both proportion of each medium and their geometric form within the cube (Figure 7.2) become important considerations. The formerly isotropic cube will now exhibit variations in electrical properties with the direction of measurement (known as *anisotropy*); a platey structure results in a marked anisotropy, for example. A lower resistivity is usually obtained when measured parallel to limitations in phyllitic shales and slates compared with that at right-angles to the laminations. The presence and orientation of elongate brine pockets (with high conductivity) strongly influence the resistivity of sea ice (Timco 1979). The amount of anisotropy is described by the *anisotropy coefficient*, which is the ratio of maximum to minimum resistivity and which generally lies in the range 1–2. Thus it is important to have some idea of the form of electrical conductors with a rock unit. Detailed discussions of anisotropy have been given, for example, by Maillet (1947), Grant and West (1965) and Telford *et al.* (1990) (see also Section 7.3.3).

Box 7.1   True resistivity (see Figure 7.1)

Resistance (R) is proportional to length (L) divided by area (A):

$$R \propto L/A.$$

This can be written as $R = \rho L/A$, where ρ is the true resistivity.

Ohm's Law

For an electrical circuit, Ohm's Law gives $R = V/I$, where V and I are the potential difference across a resistor and the current passing through it, respectively.

This can be written alternatively in terms of the electric field strength (E; volts/m) and current density (J; amps/m^2) as:

$$\rho = E/J \, (\Omega \, m)$$

Box 7.2   Resistivity

$$\rho = \frac{VA}{IL} \, (\Omega/m)$$

There are three ways in which electric current can be conducted through a rock: electrolytic, electronic (ohmic) and dielectric conduction. *Electrolytic conduction* occurs by the relatively slow movement of ions within an electrolyte and depends upon the type of ion, ionic concentration and mobility, etc. *Electronic conduction* is the process by which metals, for example, allow electrons to move rapidly, so carrying the charge. *Dielectric conduction* occurs in very weakly conducting materials (or insulators) when an external alternating current is applied, so causing atomic electrons to be shifted slightly with respect to their nuclie. In most rocks, conduction is by way of

Figure 7.2 Three extreme structures involving two materials with true resistivities ρ_1 and ρ_2. After Grant and West (1965), by permission

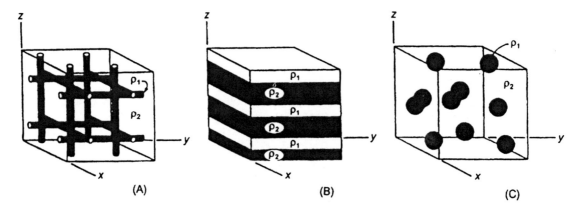

(A)   (B)   (C)

pore fluids acting as electrolytes with the actual mineral grains contributing very little to the overall conductivity of the rock (except where those grains are themselves good electronic conductors). At the frequencies used in electrical resistivity surveying dielectric conduction can be disregarded. However, it does become important in 'spectral induced polarisation' and in 'complex resistivity' measurements (see Chapter 9).

The resistivity of geological materials exhibits one of the largest ranges of all physical properties, from $1.6 \times 10^{-8}\,\Omega\,\text{m}$ for native silver to $10^{16}\,\Omega\,\text{m}$ for pure sulphur. Igneous rocks tend to have the highest resistivities; sedimentary rocks tend to be most conductive, largely due to their high pore fluid content; and metamorphic rocks have intermediate but overlapping resistivities. The age of a rock also is an important consideration: a Quaternary volcanic rock may have a resistivity in the range $10\text{--}200\,\Omega\,\text{m}$ while that of an equivalent rock but Precambrian in age may be an order of magnitude greater. This is a consequence of the older rock having far longer to be exposed to secondary infilling of interstices by mineralisation, compaction decreasing the porosity and permeability, etc.

In sedimentary rocks, the resistivity of the interstitial fluid is probably more important than that of the host rock. Indeed, Archie (1942) developed an empirical formula (Box 7.3) for the effective resistivity of a rock formation which takes into account the porosity (ϕ), the fraction (s) of the pores containing water, and the resistivity of the water (ρ_w). Archie's Law is used predominantly in borehole logging. Korvin (1982) has proposed a theoretical basis to account for Archie's Law. Saline groundwater may have a resistivity as low as $0.05\,\Omega\,\text{m}$ and some groundwater and glacial meltwater can have resistivities in excess of $1000\,\Omega\,\text{m}$.

Resistivities of some common minerals and rocks are listed in Table 7.1, while more extensive lists have been given by Telford *et al.* (1990).

Box 7.3 Archie's Law

$$\rho = a\phi^{-m}s^{-n}\rho_w$$

where ρ and ρ_w are the effective rock resistivity, and the resistivity of the pore water, respectively; ϕ is the porosity; s is the volume fraction of pores with water; a, m and n are constants where $0.5 \leqslant a \leqslant 2.5, 1.3 \leqslant m \leqslant 2.5$, and $n \approx 2$.
The ratio ρ/ρ_w is known as the Formation Factor (F).

Some minerals such as pyrite, galena and magnetite are commonly poor conductors in massive form yet their individual crystals have high conductivities. Hematite and sphalerite, when pure, are virtual insulators, but when combined with impurities they can become very

Table 7.1   Resistivities of common geologic materials

Material	Nominal resistivity (Ω m)
Sulphides:	
Chalcopyrite	$1.2 \times 10^{-5} - 3 \times 10^{-1}$
Pyrite	$2.9 \times 10^{-5} - 1.5$
Pyrrhotite	$7.5 \times 10^{-6} - 5 \times 10^{-2}$
Galena	$3 \times 10^{-5} - 3 \times 10^{2}$
Sphalerite	1.5×10^{7}
Oxides:	
Hematite	$3.5 \times 10^{-3} - 10^{7}$
Limonite	$10^{3} - 10^{7}$
Magnetite	$5 \times 10^{-5} - 5.7 \times 10^{3}$
Ilmenite	$10^{-3} - 5 \times 10$
Quartz	$3 \times 10^{2} - 10^{6}$
Rock salt	$3 \times 10 - 10^{13}$
Anthracite	$10^{-3} - 2 \times 10^{5}$
Lignite	$9 - 2 \times 10^{2}$
Granite	$3 \times 10^{2} - \times 10^{6}$
Granite (weathered)	$3 \times 10 - 5 \times 10^{2}$
Syenite	$10^{2} - 10^{6}$
Diorite	$10^{4} - 10^{5}$
Gabbro	$10^{3} - 10^{6}$
Basalt	$10 - 1.3 \times 10^{7}$
Schists (calcareous and mica)	$20 - 10^{4}$
Schist (graphite)	$10 - 10^{2}$
Slates	$6 \times 10^{2} - 4 \times 10^{7}$
Marble	$10^{2} - 2.5 \times 10^{8}$
Consolidated shales	$20 - 2 \times 10^{3}$
Conglomerates	$2 \times 10^{3} - 10^{4}$
Sandstones	$1 - 7.4 \times 10^{8}$
Limestones	$5 \times 10 - 10^{7}$
Dolomite	$3.5 \times 10^{2} - 5 \times 10^{3}$
Marls	$3 - 7 \times 10$
Clays	$1 - 10^{2}$
Alluvium and sand	$10 - 8 \times 10^{2}$
Moraine	$10 - 5 \times 10^{3}$
Sherwood sandstone	100–400
Soil (40% clay)	8
Soil (20% clay)	33
Top soil	250–1700
London clay	4–20
Lias clay	10–15
Boulder clay	15–35
Clay (very dry)	50–150
Mercia mudstone	20–60
Coal measures clay	50
Middle coal measures	>100
Chalk	50–150
Coke	0.2–8
Gravel (dry)	1400
Gravel (saturated)	100
Quaternary/Recent sands	50–100

Table 7.1   (*continued*)

Material	Nominal resistivity (Ω m)
Ash	4
Colliery spoil	10–20
Pulverised fuel ash	50–100
Laterite	800–1500
Lateritic soil	120–750
Dry sandy soil	80–1050
Sand clay/clayey sand	30–215
Sand and gravel	30–225
Unsaturated landfill	30–100
Saturated landfill	15–30
Acid peat waters	100
Acid mine waters	20
Rainfall runoff	20–100
Landfill runoff	< 10–50
Glacier ice (temperate)	$2 \times 10^6 – 1.2 \times 10^8$
Glacier ice (polar)	$5 \times 10^4 – 3 \times 10^5$ *
Permafrost	$10^3 – > 10^4$

* $-10°C$ to $-60°C$, respectively; strongly temperature-dependent. Based on Telford *et al.* (1990) with additional data from McGinnis and Jensen (1971), Reynolds (1987a). Reynolds and Paren (1980, 1984) and many commercial projects.

good conductors (with resistivities as low as $0.1\,\Omega$ m). Graphite dispersed throughout a rock mass may reduce the overall resistivity of otherwise poorly conducting minerals. For rocks that have variable composition, such as sedimentary rocks with gradational facies, the resistivity will reflect the varying proportions of the constituent materials. For example, in northern Nigeria it is possible, on the basis of the interpreted resistivities, to gauge whether a near-surface material is a clayey sand or a sandy clay. Resistivities for sandy material are about $100\,\Omega$ m and decrease with increasing clay content to about $40\,\Omega$ m, around which point clay becomes the dominant constituent and the values decrease further to those more typical of clay: well-formed and almost sand-free clay has a value in the range $1–10\,\Omega$ m (Reynolds 1987a).

The objective of most mordern electrical resistivity surveys is to obtain true resistivity models for the sub-surface because it is these that have geological meaning. The methods by which field data are obtained, processed and interpreted will be discussed later.

The *apparent resistivity* is the value obtained as the product of a measured resistance (R) and a *geometric factor* (K) for a given electrode array (see Section 7.3.2), according to the expression in Box 7.2. The geometric factor takes into account the geometric spread of electrodes and contributes a term that has the unit of length (metres). Apparent resistivity (ρ_a) thus has units of ohm-metres.

7.2.2   Current flow in a homogeneous earth

For a single current electrode implanted at the surface of a homogeneous medium of resistivity ρ, current flows away radially (Figure 7.3).
The voltage drop between any two points on the surface can be
described by the potential gradient $(-\delta V/\delta x)$, which is negative
because the potential decreases in the direction of current flow. Lines
of equal voltage ('equipotentials') intersect the lines of equal current at
right-angles. The current density (J) is the current (I) divided by the
area over which the current is distributed (a hemisphere; $2\pi r^2$), and so
the current density decreases with increasing distance from the current
source. It is possible to calculate the voltage at a distance (r) from
a single current point source (Box 7.4). If, however, a current sink is
added, a new potential distribution occurs (Figure 7.4) and a modified
expression is obtained to describe the voltage at any point (Box 7.5).

Figure 7.3 (A) Three-dimensional
representation of a hemispherical
equipotential shell around a point
electrode on a semi-infinite, homogeneous medium. (B) Potential decay
away from the point electrode

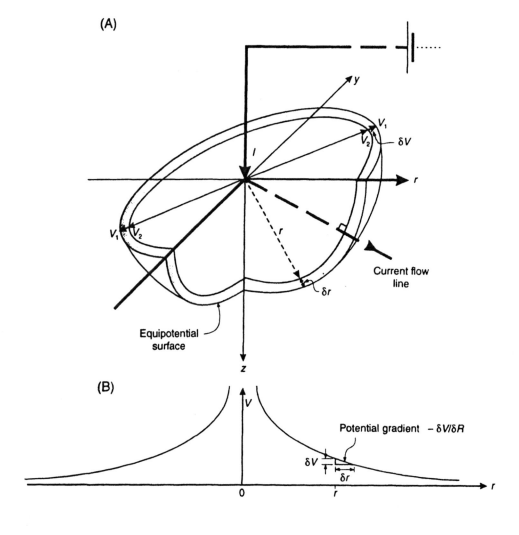

Box 7.4 (See Figure 7.3)

The potential difference (δV) across a hemispherical shell of incremental thickness δr is given by:

$$\frac{\delta V}{\delta r} = -\rho . J = -\rho \frac{I}{2\pi r^2}.$$

Thus the voltage V_r at a point r from the current point source is:

$$V_r = \int \delta V = -\int \rho \frac{I}{2\pi r^2} \delta r = \frac{\rho I}{2\pi} . \frac{1}{r}.$$

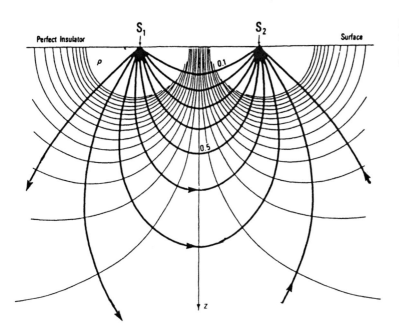

Figure 7.4 Current and equipotential lines produced by a current source and sink. From van Nostrand and Cook (1966), by permission

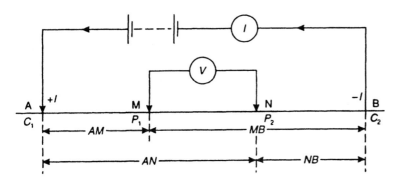

Figure 7.5 Generalised form of electrode configuration in resistivity surveys

Box 7.5 (See Figure 7.5)

For a current source and sink, the potential V_P at any point P in the ground is equal to the sum of the voltages from the two electrodes, such that: $V_P = V_A + V_B$ where V_A and V_B are the potential contributions from the two electrodes, $A(+I)$ and $B(-I)$.

The potentials at electrode M and N are:

$$V_M = \frac{\rho I}{2\pi}\left[\frac{1}{AM} - \frac{1}{MB}\right], \quad V_N = \frac{\rho I}{2\pi}\left[\frac{1}{AN} - \frac{1}{NB}\right].$$

However, it is far easier to measure the potential difference, δV_{MN}, which can be rewritten as:

$$\delta V_{MN} = V_M - V_N = \frac{\rho I}{2\pi}\left\{\left[\frac{1}{AM} - \frac{1}{MB}\right] - \left[\frac{1}{AN} - \frac{1}{NB}\right]\right\}$$

Rearranging this so that resistivity ρ is the subject:

$$\rho = \frac{2\pi\delta V_{MN}}{I}\left\{\left[\frac{1}{AM} - \frac{1}{MB}\right] - \left[\frac{1}{AN} - \frac{1}{NB}\right]\right\}^{-1}$$

7.3 ELECTRODE CONFIGURATIONS AND GEOMETRIC FACTORS

7.3.1 General case

The final expression in Box 7.5 has two parts, namely a resistance term (R; units Ω) and a term that describes the geometry of the electrode configuration being used (Box 7.6) and which is known as the *geometric factor* (K; units m). In reality, the sub-surface ground does not conform to a homogeneous medium and thus the resistivity obtained is no longer the 'true' resistivity but the *apparent resistivity* (ρ_a) which can even be negative. It is very important to remember that the apparent resistivity is not a physical property of the sub-surface

Box 7.6. **The geometric factor** (see Figure 7.5)

The geometric factor (K) is defined by the expression:

$$K = 2\pi\left[\frac{1}{AM} - \frac{1}{MB} - \frac{1}{AN} + \frac{1}{NB}\right]^{-1}.$$

Where the ground is not uniform, the resistivity so calculated is called the *apparent resistivity* (ρ_a):

$$\rho_a = RK, \text{ where } R = \delta V/I.$$

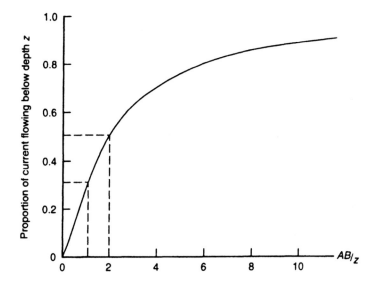

Figure 7.6 Proportion of current flowing below a depth z (m); AB is the current electrode half-separation

media, unlike the true resistivity. Consequently, all field resistivity data are apparent resistivity while those obtained by interpretation techniques are 'true' resistivities.

Figure 7.6 shows that, in order for at least 50% of the current to flow through an interface at a depth of z metres into a second medium, the current electrode separation needs to be at least twice – and preferably more than three times – the depth. This has obvious practical implications, particularly when dealing with situations where the depths are of the order of several hundreds of metres, so requiring very long cable lengths that can produce undesirable inductive coupling effects. For very deep soundings where the electrode separation is more than several kilometres, telemetering the data becomes the only practical solution (e.g. Shabtaie *et al.* 1980, 1982). However, it should be emphasised that it is misleading to equate the depth of penetration with the current electrode separation as a general rule of thumb in the region of a resistivity survey. This aspect is discussed in Section 7.3.3.

7.3.2 Electrode configurations

The value of the apparent resistivity depends on the geometry of the electrode array used, as defined by the geometric factor K. There are three main types of electrode configuration, two of which are named after their originators – Frank Wenner (1912a,b) and Conrad Schlumberger – and a range of sub-types (Table 7.2 and Figure 7.7). The geometric factors for these arrays are given in Box 7.7 and a worked example for the Wenner array is given in Box 7.8. Arrays highlighted in bold in Table 7.2 are those most commonly used.

Table 7.2　Electrode configurations (see also Figure 7.7)

Wenner arrays	**Standard Wenner**
	Offset Wenner
	Lee-partitioning array
	Tripotential (α, β and γ arrays)
Schlumberger array	**Standard Schlumberger**
	Brant array
	Gradient array
Dipole–dipole arrays	**Normal** (axial or polar)
	Azimuthal
	Radial
	Parallel
	Perpendicular
	Pole–Dipole
	Equatorial
	Square (special form of equatorial)

Figure 7.7　Electrode configurations used in electrical surveys

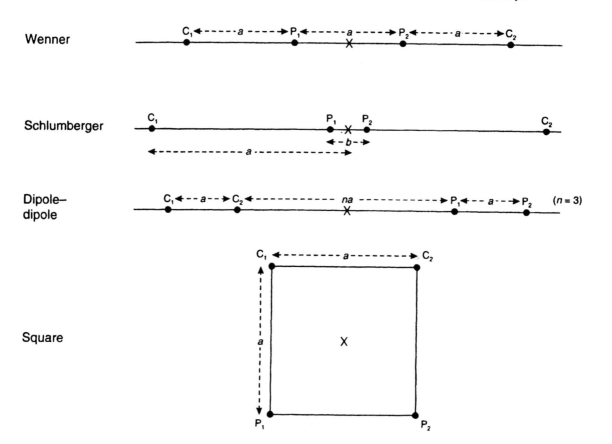

Dipole–dipole arrays have been used extensively by Russian geophysicists since 1950, and especially in Canada, particularly for 'induced polarisation' surveys (see Chapter 9) in mineral exploration, and in the USA in groundwater surveys (Zohdy 1974). The term 'dipole' is misapplied in a strict sense because the inter-electrode separation for each of the current or potential electrode pairs should be insignificant with respect to the length of the array, which it is not. However, the term is well established in its usage.

Box 7.7 Apparent resistivities for given geometric factors for electrode configurations in Figure 7.7

Wenner array: $\rho_a = 2\pi a R$ (alpha/beta arrays)
                         $\rho_a = 3\pi a R$ (gamma rays)

Two-electrode: $\rho_a = 2\pi s R$

Lee array: $\rho_a = 4\pi a R$

Schlumberger array: $\rho_a = \dfrac{\pi a^2}{b}\left[1 - \dfrac{b^2}{4a^2}\right] R; \quad a \geqslant 5b$

Gradient array: $\rho_a = 2\pi \dfrac{L^2}{a} \dfrac{1}{G} R$

where $G = \dfrac{1 - X}{(Y^2 + (1 - X)^2)^{3/2}} + \dfrac{1 + X}{(Y^2 + (1 + X)^2)^{3/2}}$

and $X = x/L, \; Y = y/L$

Dipole–dipole array: $\rho_a = \pi n(n + 1)(n + 2) a R$

Pole–dipole array: $\rho_a = 2\pi n(n + 1) a R$

Square array: $\rho_a = \pi a (2 + \sqrt{2}) R$

These different types and styles of electrode configuration have particular advantages, disadvantages and sensitivities. Factors affecting the choice of array type include the amount of space available to lay out an array and the labour-intensity of each method. Other important considerations are the sensitivity to lateral inhomogeneities (Habberjam and Watkins 1967a; Barker 1981) and to dipping interfaces (Broadbent and Habberjam 1971).

A graphic example of the different responses by the three main electrode configurations is given by so-called 'signal contribution sections' (Barker 1979) shown in Figure 7.8. These sections are contoured plots of the contribution made by each unit volume of the sub-surface to the voltage measured at the surface.

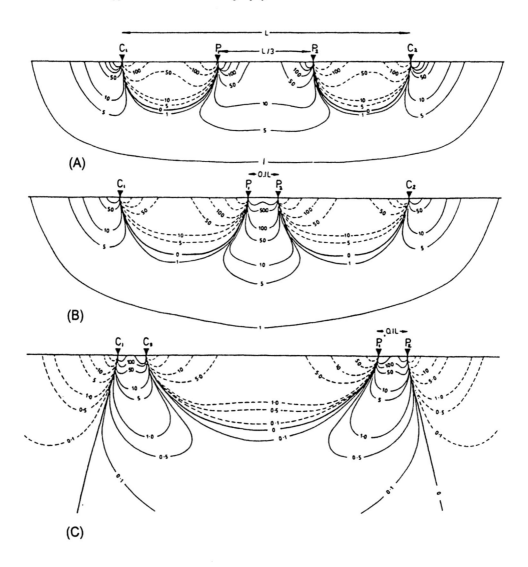

Box 7.8　Worked example of how to calculate a geometric factor

Using the expression previously defined in Box 7.6 (see also Figure 7.5), and substituting in the correct values for the *Wenner array*:

$$K = 2\pi \left[\frac{1}{a} - \frac{1}{2a} - \frac{1}{2a} + \frac{1}{a} \right]^{-1} = 2\pi \left[\frac{2}{a} - \frac{2}{2a} \right]^{-1}$$

$$= 2\pi a.$$

Hence, as $\rho_a = KR$, $\rho_a = 2\pi a R$.

Figure 7.8　Signal contribution sections for: (A) Wenner, (B) Schlumberger and (C) dipole–dipole configurations. Contours indicate the relative contributions made by discrete volume elements of the sub-surface to the total potential difference measured between the two potential electrodes P_1 and P_2. From Barker (1979), by permission

Figure 7.8A shows the signal contribution for a Wenner array. In the near-surface region, the positive and negative areas cancel each other out and the main response, which originates from depth, is largely flat (see the 1 unit contour). This indicates that for horizontally, layered media, the Wenner array has a high vertical resolution. The Schlumberger array has almost as high a vertical resolution, but note that the form of the signal contribution at depth is now concave upwards (Figure 7.8B). For the dipole–dipole array (Figure 7.8C), the lobate form of the signal contribution indicates that there is a poor vertical resolution and that the array is particularly sensitive to deep lateral resistivity variations, making it an unsuitable array for depth sounding (Bhattacharya and Patra 1968). Nevertheless, this sensitivity can be utilised in resistivity profiling (see Section 7.4.3).

A modified electrode array (Lee partitioning array) was devised by Lee (Lee and Schwartz 1930) in an attempt to reduce the undesirable effects of near-surface lateral inhomogeneities. An alternative tri-potential method was proposed by Carpenter (1955) and by Carpenter and Habberjam (1956) which combined the apparent resistivities obtained for the alpha, beta and gamma rays (Figure 7.9). The

Figure 7.9 Wenner tripotential electrode configurations for $N = 2$. x is the fixed interelectrode separation, and the active electrode separation is $2x$. From Ackworth and Griffiths (1985), by permission

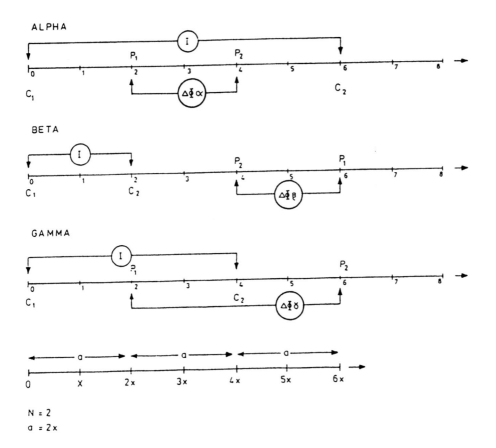

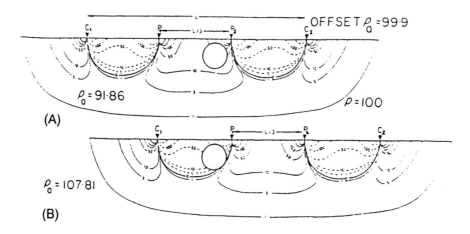

method has been discussed further by Ackworth and Giffiths (1985). A smoothing technique using the tripotential method was produced by Habberjam and Watkins (1967a).

An alternative technique, called the *Offset Wenner* method (Barker 1981), has been readily adopted for its ease of use. The method is extremely simple in concept. Figure 7.10 shows a single contribution section for a standard Wenner array. A conducting sphere buried in a semi-infinite homogeneous medium with true resistivity of $100\,\Omega\,m$ is located in a positive region of the signal contribution section (Figure 7.10A). The corresponding apparent resistivity, calculated using an exact analytical method (Singh 1976), is $91.86\,\Omega\,m$. Offsetting the Wenner array one spacing to the right (Figure 7.10B), the previously positive areas are now negative and vice versa, and the buried sphere is located in a negative region resulting in an apparent resistivity of $107.81\,\Omega\,m$. The average of these two apparent resistivities is $99.88\,\Omega\,m$, thereby reducing the error due to a lateral inhomogeneity from around $\pm 8\%$ to only 0.1%.

One array that is seldom used, but which has two major advantages, is the square array. This is a special form of the equatorial dipole–dipole array for $n = 1$. The square array is particularly good for determining lateral azimuthal variations in resistivity. By swapping P_1 and C_2, the square is effectively rotated through $90°$ and thus the apparent resistivity can be determined for two orthogonal directions. For ground that is largely uniform, the two resistivities should be the same, but where there is a difference in resistivity due to a form of anisotropy (*transverse anisotropy* as it is measured only in the $x - y$ plane), the two resistivities will differ. The ratio of the two resistivities is an indication of the transverse anisotropy. Profiles and maps of transverse anisotropy can be interpreted qualitatively to indicate anomalous ground. The second advantage of the square array is that it lends itself to rapid grid mapping. By moving two electrodes at

Figure 7.10 (A) Signal contribution section for a Wenner array with a conducting sphere (negative K) in a positive region in a medium with resistivity $100\,\Omega\,m$. (B) Offset Wenner electrodes in which the sphere is now in a negative region. Distortion of contours due to the presence of the sphere is not shown. From Barker (1981), by permission

Table 7.3 Comparison of dipole–dipole, Schlumberger, square and Wenner electrode arrays

Criteria	Wenner	Schlumberger	Dipole–dipole	Square
Vertical resolution	√√√	√√	√	√√
Depth penetration	√	√√	√√√	√√
Suitability to VES	√√	√√√	√	×
Suitability to CST	√√√	×	√√√	√√√
Sensitivity to orientation	Yes	Yes	Moderate	No
Sensitivity to lateral inhomogeneities	High	Moderate	Moderate	Low
Labour intensive	Yes (no*)	Moderate (no*)	Moderate (no*)	Yes
Availability of interpretational aids	√√√	√√√	√√	√

$\sqrt{}$ = poor; $\sqrt{}\sqrt{}$ = moderate; $\sqrt{}\sqrt{}\sqrt{}$ = good; × = unsuitable
* When using a multicore cable and automated electrode array

a time, the square can be moved along the transect. By increasing the dimensions of the square, and thus generally increasing the depth penetration and repeating the same survey area, three-dimensional models of the resistivity distribution can be obtained. Of all the electrode configurations, the square array is the least sensitive to steeply dipping interfaces (Broadbent and Habberjam 1971) and thus it can cope in situations where the sub-surface media are not horizontally-layered. Being a particularly labour-intensive field method, it is best restricted to small-scale surveys where the electrode separation is only of the order of a few metres. This technique has particular value in 3-D mapping of buried massive ice and in shallow archaeological investigations, for example.

A general guide to the suitability of the dipole–dipole, Schlumberger, square and Wenner electrode configurations is given in Table 7.3. An important consideration for the suitability of a given array is the scale at which it is to be deployed. For example, a square array is not really appropriate for depth sounding ('vertical electrical sounding'; VES) or for 'constant separation traversing' (CST) with a large square side; whereas it is perhaps better than either the Wenner or Schlumberger arrays for applications concerned with very shallow depths (< 2 m), such as in archaeological investigations. While the main electrode configurations are now well established in their major applications, small-scale mini-resistivity surveys have yet to realise their full potential.

7.3.3 Media with contrasting resistivities

A geological section may show a series of lithologically defined interfaces which do not necessarily coincide with boundaries identified electrically. For example, in an unconfined sandstone aquifer, there is a capillary zone above the water table making the boundary from 'dry' to 'saturated' a rather diffuse one. Furthermore, different lithologies can have the same resistivity and thus would form only one electric unit.

A geoelectric unit is characterised by two basic parameters: the *layer resistivity* (ρ_i) and the *layer thickness* (t_i) for the *i*th layer ($i = 1$ for the surface layer). Four further electrical parameters can be derived for each layer from the respective resistivity and thickness; these are called the *longitudinal conductance* (S_L; units mS); *transverse resistance* (T; units $\Omega\,m^2$); *longitudinal resistivity* (ρ_L; units $\Omega\,m$); and *transverse resistivity* (ρ_T; units $\Omega\,m$). They are defined in Box 7.9 for the model shown in Figure 7.11. The sums of all the longitudinal conductances and of the transverse resistances for a layered ground are called the Dar Zarrouk 'function' and 'variable', respectively. (The curiously named Dar Zarrouk parameters were so called by Maillet (1947) after a place near Tunis where he was a prisoner of war.)

The importance of the longitudinal conductance for a particular layer is that it demonstrates that it is not possible to know both the true layer conductivity (or resistivity) *and* the layer thickness, so giving rise to layer *equivalence*. For example, a layer with a longitudinal conductance of 0.05 mS can have a resistivity of $100\,\Omega\,m$ and thickness 5 m. Layers with the combination of resistivity $80\,\Omega\,m$ and thickness 4 m, and $120\,\Omega\,m$ and 6 m, are all equivalent electrically. Equivalence needs to be considered during interpretation of sounding curves (see Section 7.5.4) and generally in the interpretation of electrical data, whether obtained by contact electrical or electromagnetic induction methods.

Box 7.9   Dar Zarrouk parameters

For a given layer:

Longitudinal conductance	$S_L = h/\rho = h.\sigma$
Transverse resistance	$T = h.\rho$
Longitudinal resistivity	$\rho_L = h/S$
Transverse resistivity	$\rho_T = T/h$
Anisotropy	$A = \rho_T/\rho_L.$

For *n* layers:

$$S_L = \sum_{i=1}^{n} (h_i/\rho_i) = \frac{h_1}{\rho_1} + \frac{h_2}{\rho_2} + \frac{h_3}{\rho_3} + \cdots \frac{h_n}{\rho_n}$$

$$T = \sum_{i=1}^{n} (h_i\rho_i) = h_1\rho_1 + h_2\rho_2 + h_3\rho_3 + \cdots h_n\rho_n$$

Where a point current source is located close to a plane boundary between two homogeneous media, the lines of current flow (and hence

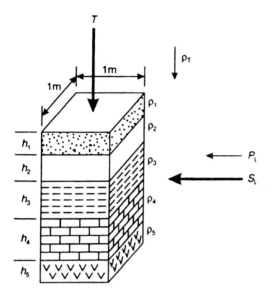

Figure 7.11 Thickness (h) and true resistivity (ρ) of component layers with an indication of the total longitudinal conductance (S_L) and total transverse resistance (T); subscripts L and T refer to longitudinal and transverse, respectively

of equipotential) are refracted at the boundary in proportion to the contrast in resistivity between the two media (Figure 7.12 and Box 7.10). In a homogeneous material with no boundaries in the vicinity, current flow lines are symmetrically radial. If a boundary is nearby, the current flow lines will become distorted (Figure 7.12B): current flowing towards a medium with a higher resistivity will diverge from the radial pattern and current densities adjacent to the boundary will decrease, whereas the current converges on approaching a medium with a lower resistivity with a consequent increase in current densities. The potential at a point adjacent to a plane boundary can be calculated using optical image theory (Figure 7.12C). If a current source of strength S is placed in one medium of resistivity ρ_1, the source's image point lies in the second medium of resistivity ρ_2 (where $\rho_2 > \rho_1$) but has a reduced strength kS, where k is dependent upon the resistivity contrast between the two media and lies in the range ± 1. This k factor is akin to the reflection coefficient in optics and in reflection seismology, and has the form given in Box 7.11. If the current passes from a lower resistivity medium to one with a higher resistivity, k is positive; if it passes into a medium with a lower resistivity, k is negative.

Box 7.10 (See Figure 7.12A)

Refraction of current flow at a plane boundary:
$$\tan\theta_2/\tan\theta_1 = \rho_1/\rho_2.$$

(A)

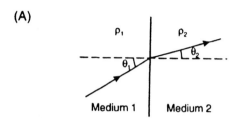

Figure 7.12 (A) Refraction of current flow lines, and (B) Distortion of equipotential and current flow lines from a point electrode across a plane boundary between media with contrasting resistivities (Telford *et al.* 1990). (C) Method of optical images for the calculation of a potential at a point (see text for details)

(B)

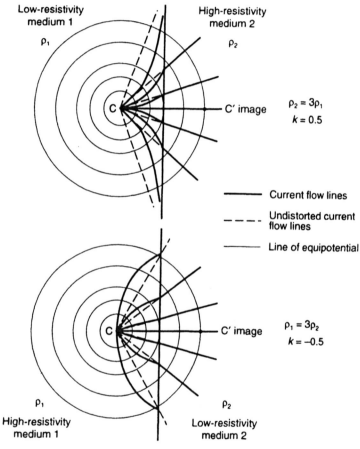

(C)

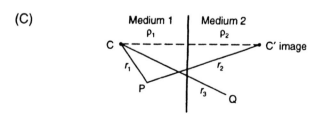

Box 7.11   Electrical reflection coefficient, *k* (see Figure 7.12C)

Potential at P:

$$V = \frac{I\rho_1}{4\pi}\left[\frac{1}{r_1} + \frac{k}{r_2}\right].$$

Potential at Q:

$$V = \frac{I\rho_2}{4\pi}\left[\frac{1+k}{r_3}\right].$$

At the interface $V = V'$ and $r_1 = r_2 = r_3$. Hence:

$$\frac{\rho_1}{\rho_2} = \frac{1-k}{1+k}$$

or

$$k = \frac{\rho_2 - \rho_1}{\rho_2 + \rho_1}$$

In the case when the boundary is vertical, different types of anomaly will be produced dependent upon the electrode configuration used and whether it is developed at right-angles or parallel (broadside) to the boundary. Examples of the types of anomalies produced are illustrated in Figure 7.13. The cusps and discontinuity in (A), (B) and (C) are due to the positioning of the electrodes relative to the vertical boundary with each cusp occurring as one electrode crosses the boundary. In the case of the Wenner array, it can be explained in detail (Figures 7.13D and E) as follows.

As the array is moved from the high-resistivity medium towards the low-resistivity medium (case (i) in Figure 7.13), the current flow lines converge towards the boundary, increasing the current density at the boundary but decreasing the potential gradient at the potential electrodes. The apparent resistivity gradually falls from its true value until a minimum is reached when the current electrode C_2 is at the boundary (ii). Once C_2 has crossed into the low-resistivity unit (iii), the current density increases adjacent to the boundary but within the low-resistivity medium, causing the potential gradient between potential electrodes to rise. When the entire potential electrode dipole has crossed the boundary (iv), the current density is highest in the high-resistivity medium, causing the potential gradient across $P_1 - P_2$ to fall dramatically. With the current electrode C_1 now into the low-resistivity unit (v), the current adjacent to the boundary is divergent. This results in an elevated potential gradient between P_1 and P_2 which falls to a normal value when the entire collinear array is sufficiently far away from the boundary. At this point the current flow is radial once more. The magnitude of the cusps and discontinuities is

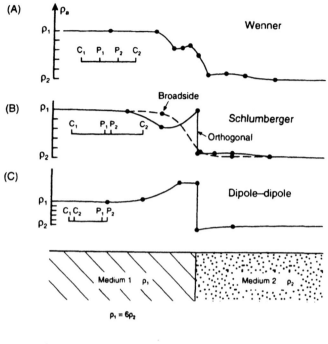

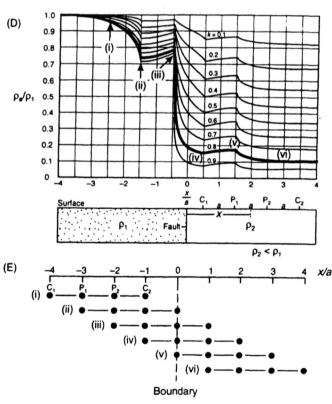

Figure 7.13 Apparent resistivity profiles measured over a vertical boundary using different electrode arrays: (A) Wenner (with its characteristic W-shaped anomaly), (B) Schlumberger, and (C) dipole–dipole. After Telford *et al.* (1990), by permission of Cambridge University Press. (D) Profile shapes as a function of resistivity contrast. From van Nostrand and Cook (1966), by permission. (E) Plan view of successive moves of a Wenner array with electrode positions indicated for six marked points in (D)

Figure 7.14 (*opposite*) Apparent resistivity profiles across a thin vertical dyke using (A) a dipole–dipole array and (B) a Wenner array. (C) Computed normalised resistivity profiles across a thin vertical dyke with different resistivity contrasts. After van Nostrand and Cook (1976), by permission

(A)

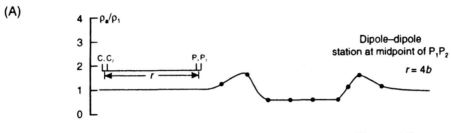

(B)

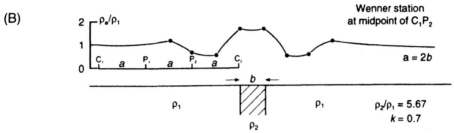

(C)

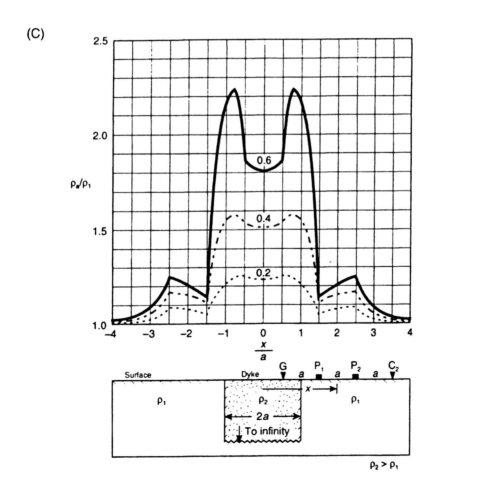

dependent upon the resistivity contrast between the two media. However, if the array is orientated parallel to the boundary such that all electrodes cross it simultaneously, the cusping is reduced. (Figure 7.13B, dashed line).

The anomaly shape is similarly varied over a vertical dyke (Figure 7.14) in which the width of the anomaly varies not only with electrode configuration but also with the ratio of the dyke width to electrode separation. An example of an apparent resistivity profile across a buried hemicylindrical valley is shown in Figure 7.15. The field and modelled profiles are very similar. The mathematical treatment of apparent resistivity profiles across vertical boundaries is discussed in detail by Telford *et al.* (1990).

The important consideration arising from this discussion is that different array types and orientations across an identical boundary between two media with contrasting resistivities will produce markedly different anomaly shapes. Thus interpretation and comparison of such profiles based simply on apparent resistivity maxima or minima can be misleading; for example, if maxima are used to delimit targets, a false target could be identified in the case of maxima doublets such as in Figure 7.14. Similar complex anomalies occur in electromagnetic data (Section 9.4.3). Furthermore, the type of anomaly should be anticipated when considering the geological target so that the appropriate electrode configuration and orientation can be chosen prior to the commencement of the survey.

Figure 7.15 Apparent resistivity profiles across a buried valley: (A) theoretical profile, and (B) measured profile. After Cook and van Nostrand (1954) (see Parasnis 1986)

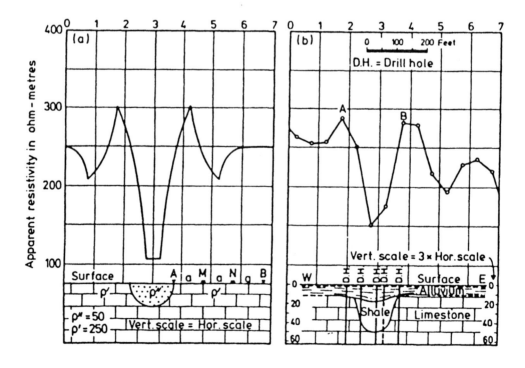

7.4   MODES OF DEPLOYMENT

There are two main modes of deployment of electrode arrays. One is for depth sounding (to determine the vertical variation of resistivity) – this is known as *vertical electrical sounding* (VES). The other is for horizontal traversing (horizontal variation of resistivity) and is called *constant separation traversing* (CST) (also called 'electrical resistivity traversing', ERT). In the case of multi-electrode arrays, two forms are available. Microprocessor-controlled resistivity traversing (MRT) is used particularly for hydrogeological investigations requiring significant depths of penetration. Sub-surface imaging (SSI) or two-dimensional electrical tomography is used for very high resolution in the near-surface in archaeological, engineering and environmental investigations.

7.4.1   Vertical electrical sounding (VES)

As the distance between the current electrodes is increased, so the depth to which the current penetrates is increased. In the case of the dipole–dipole array, increased depth penetration is obtained by increasing the inter-dipole separation, not by lengthening the current electrode dipole. The position of measurement is taken as the mid-point of the electrode array. For a depth sounding, measurements of the resistance ($\delta V/I$) are made at the shortest electrode separation and then at progressively larger spacings. At each electrode separation a value of apparent resistivity (ρ_a) is calculated using the measured resistance in conjugation with the appropriate geometric factor for the electrode configuration and separation being used (see Section 7.3). The values of apparent resistivity are plotted on a graph ('field curve') the x- and y-axes of which represent the logarithmic values of the current electrode half-separation ($AB/2$) and the apparent resistivity (ρ_a), respectively (Figure 7.16). The methods by which these field curves are interpreted are discussed in detail in Section 7.5.

In the normal Wenner array, all four electrodes have to be moved to new positions as the inter-electrode spacings are increased (Figure 7.17A). The offset Wenner system has been devised to work with special multicore cables (Barker 1981). Special connectors at logarithmically spaced intervals permit a Wenner VES to be completed by using a switching box which removes the necessity to change the electrode connections physically. Note that the offset Wenner array requires one extra electrode separation to cover the same amount of the sub-surface compared with the normal Wenner array. When space is a factor, this needs to be considered in the survey design stage.

In the case of the Schlumberger array (Figure 7.17C), the potential electrodes ($P_1 P_2$) are placed at a fixed spacing (b) which is no more

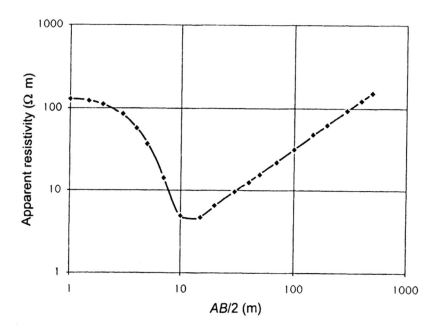

Figure 7.16   A vertical electrical sounding (VES) showing apparent resistivity as a function of current electrode half-separation (*AB*/2)

than one-fifth of the current-electrode half-spacing (*a*). The current electrodes are placed at progressively larger distances. When the measured voltage between P_1 and P_2 falls to very low values (owing to the progressively decreasing potential gradient with increasing current electrode separation), the potential electrodes are spaced more widely apart (spacing b_2). The measurements are continued and the potential electrode separation increased again as necessary until the VES is completed. The tangible effects of so moving the potential electrodes is discussed at the end of Section 7.4.4. A VES using the Schlumberger array takes up less space than either of the two Wenner methods and requires less physical movement of electrodes than the normal Wenner array, unless multicore cables are used.

The dipole–dipole array is seldom used for vertical sounding as large and powerful electrical generators are normally required. Once the dipole length has been chosen – i.e. the distance between the two current electrodes and between the two potential electrodes – the distance between the two dipoles is then increased progressively (Figure 7.17C) to produce the sounding. The square array is rarely used for large-scale soundings as its setting out is very cumbersome (Figure 7.17E). The main advantage of the electrode configuration is the simplicity of the method when setting out small grids. In small-scale surveys investigating the three-dimensional extent of sub-surface targets, such as in archaeology, the square sides are of the order of only a few metres.

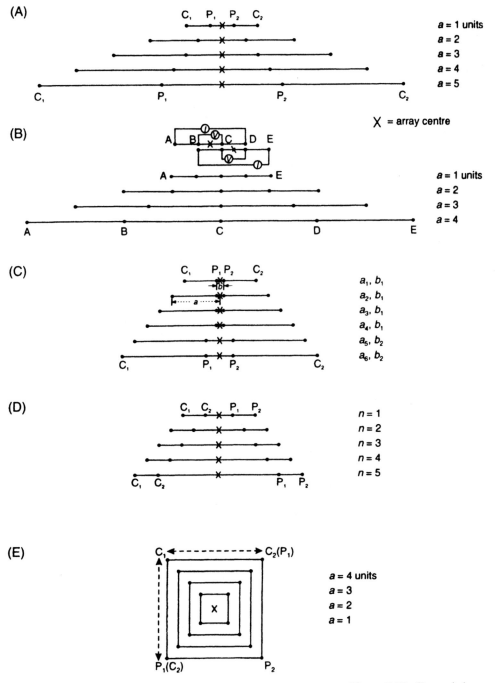

Figure 7.17 Expanded arrays (with successive positions displaced for clarity) for: (A) Wenner, (B) offset Wenner, (C) Schlumberger, (D) dipole–dipole and (E) square arrays

7.4.2 Automated array scanning

In 1981 Barker published details of the offset Wenner array using multicore cables and multiple electrodes for VES investigations. In 1985, Griffiths and Turnbull produced details of a multiple electrode array for use with CST. This theme was developed by van Overmeeren and Ritsema (1988) for hydrogeological applications and by Noel and Walker (1990) for archaeological surveys. For deeper sounding, where multicore cabling would become prohibitively heavy, the cable is wound into 50 m sections on its own drum with an addressable electronic switching unit and power supply mounted in the hub of each cable reel. The switching units are controlled by a laptop computer which can switch any electrode to either of two current or two potential cables which connect the entire array of drum reels. This system is known as the *microprocessor-controlled-resistivity traversing system* (Griffths *et al.* 1990).

In van Overmeeren and Ritsema's *continuous vertical electrical sounding* (CVES) system, an array of multiples of 40 electrodes is connected to a microprocessor by a multicore cable. Usiing software control, discrete sets of four electrodes can be selected in a variety of electrode configurations and separations and a measurement of the resistance made for each. Instead of using one cable layout for just one VES, the extended electrode array can be used for a number of VES, each one offset by one electrode spacing. If the first VES is conducted with its centre between electrodes 15 and 16, for example, the next VES will be centred between electrodes 16 and 17, then 17 and 18, 18 and 19, and so on. A field curve is produced for each sounding along the array and interpreted by computer methods (see Section 7.5.3) to produce a geo-electric model of true layer resistivities and thickness for each VES curve. When each model is displayed adjacent to its neighbour, a panel of models is produced (Figure 7.18) in which the various resistivity horizons can be delimited. It is clear from Figure 7.18D that the CVES interpretation is closest to the known physical model compared with those for either the tripotential alpha or beta/gamma ratio sections (Shown in Figure 7.18B and C respectively). This particular method requires special equipment and associated computer software, but it highlights a novel application of both field method and data analysis to improve the resolution of shallow resistivity surveys.

In sub-surface imaging (SSI), typically 50 electrodes are laid out in two strings of 25, with electrodes connected by a multicore cable to a switching box and resistance meter. The whole data acquisition procedure is software-controlled from a laptop computer. Similar products have been produced, such as the LUND Automatic Imaging System (ABEM), and MacOhm 21 (DAP-21) Imaging System (OYO), and the Sting/Swift (Advanced Geosciences Inc.), among others.

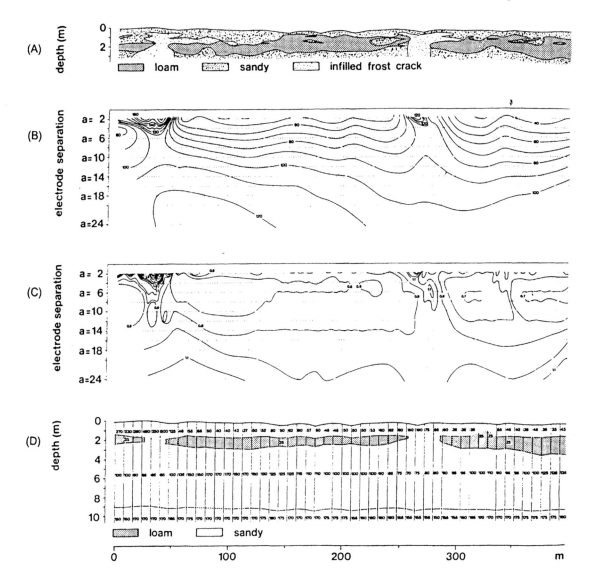

Figure 7.18 High-resolution soil survey using a scanned array. (A) Soil section determined by shallow hand-drilling. Pseudo-sections obtained using (B) Wenner tripotential alpha and (C) beta/gamma arrays. (D) Continuous vertical electrical sounding results with true resistivities indicated. From van Overmeeren and Ritsema (1988), by permission

As with van Overmeeren and Ritsema's CVES method, a discrete set of four electrodes with the shortest electrode spacing ($n = 1$; see Figure 7.19) is addressed and a value of apparent resistivity obtained. Successive sets of four electrodes are addressed, shifting each time by one electrode separation laterally. Once the entire array has been scanned, the electrode separation is doubled ($n = 2$), and the process repeated until the appropriate number of levels has been scanned. The values of apparent resistivity obtained from each measurement are plotted on a pseudo-section (Figure 7.19) and contoured. The methods of interpretation are described in more detail in Section 7.5.6. There are considerable advantages in using SSI or equivalent

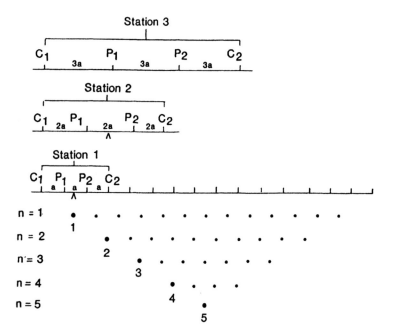

Figure 7.19 Example of the measurement sequence for building up a resistivity pseudo-section. Courtesy of Campus Geophysical Instruments Ltd.

methods. With multicore cable and many electrodes, the entire array can be established by one person. The acquisition of apparent resistivity data is controlled entirely by the software whose parameters are selected at the outset. By changing the inter-electrode spacing between electrodes, the vertical and horizontal resolutions can be specified to meet the objectives of the survey. For example, the horizontal resolution is defined by the inter-electrode spacing, and the vertical resolution by half the spacing. For example, using a 2 m inter-electrode spacing, the horizontal and vertical resolutions are 2 m and 1 m, respectively, for the pseudo-section display. Whether sub-surface features can be resolved at a comparable scale is determined also by the lateral and vertical variations in true resistivity.

7.4.3   Constant-separation traversing (CST)

Constant-separation traversing uses a manul electrode array, usually the Wenner configuration for ease of operation, in which the electrode separation is kept fixed. The entire array is moved along a profile and values of apparent resistivity determined at discrete intervals along the profile. For example, a Wenner spacing of say 10 m is used with perhaps 12 electrodes deployed at any one time at 5 m intervals. Alternate electrodes are used for any one measurement (Figure 7.20) and instead of uprooting the entire sets of electrodes, the connections are moved quickly and efficiently to the next electrode along the line, i.e. 5 m down along the traverse. This provides a CST profile with

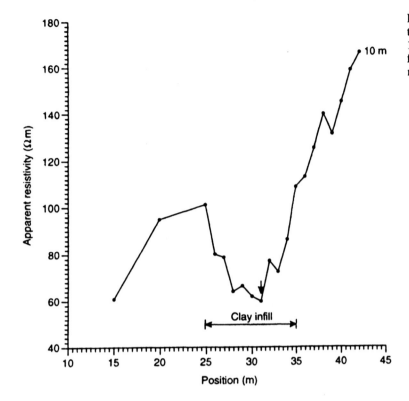

Figure 7.20 A constant-separation traverse using a Wenner array with 10 m electrode spacing over a clay-filled solution feature (position arrowed) in limestone

electrode separation of 10 m and station interval of 5 m. The values of apparent resistivity are plotted on a linear graph as a function of distances along the profile (Figure 7.20). Variations in the magnitude of apparent resistivity highlight anomalous areas along the traverse.

Sörensen (1994) has described a 'pulled array continuous electrical profiling' technique (PA–CEP). An array of heavy steel electrodes, each weighing 10–20 kg, is towed behind a vehicle containing all the measuring equipment. Measurements are made continuously. It is reported that 10–15 line kilometres of profiling can be achieved in a day. The quality of results is reported to be comparable to that of fixed arrays with the same electrode geometry.

7.4.4   Field problems

In order for the electrical resistivity method to work using a collinear array, the internal resistance of the potential measuring circuit must be far higher than the ground resistance between the potential electrodes. If it is not, the potential circuit provides a low-resistance alternative route for current flow and the resistance measured is completely meaningless. Most commercial resistivity equipment has an input resistance of at least 1 MΩ, which is adequate in most cases.

In the case of temperature glacier ice, which itself has a resistivity of up to 120 MΩ m, a substantially higher input resistance is required (preferably of the order of $10^{14}\,\Omega$).

Electrical resistivity soundings on glaciers are complicated by the fact that ice does not conduct electricity electronically but by the movement of protons within the ice lattice and this causes substantial polarisation problems at the electrode–ice contact. Consequently, special techniques are required in order to obtain the relevant resistivity data (Reynolds 1982).

Perhaps the largest source of field problems is the electrode contact resistance. Resistivity methods rely on being able to apply current into the ground. If the resistance of the current electrodes becomes anomalously high, the applied current may fall to zero and the measurement will fail. High contact resistances are particularly common when the surface material into which the electrodes are implanted consists of dry sand, boulders, gravel, frozen ground, ice or laterite. If the high resistance can be overcome (and it is not always possible), there are two methods that are commonly used. One is to wet the current electrodes with water or saline solution, sometimes mixed with bentonite. The second method is to use multiple electrodes. Two or three extra electrodes can be connected to one end of the current-carrying cable so that the electrodes act as resistances in parallel. The total resistance of the multiple electrode is thus less than the resistance of any one electrode (see Figure 7.21 and Box 7.12). However, if this method is used, the extra electrodes must be implanted at right-angles to the line of the array rather than along the direction of the profile. If the extra electrodes are in the line of the array, the geometric factor may be altered as the inter-electrode separation (C_1–P_1–P_2–C_2) is effectively changed. By planting the electrodes at right-angles to the line of the array, the inter-electrode separation is barely affected. This problem is only acute when the current electrode separation is small. Once the current electrodes are sufficiently far apart, minor anomalies in positioning are insignificant. This also applies when laying out the survey line to start with.

Box 7.12   Resistances in parallel

Total resistance of multiple electrodes is R_T:

$$1/R_T = 1/R_1 + 1/R_2 + 1/R_3 + \cdots 1/R_n = \sum_{i=1}^{n}(1/R_i).$$

For example, if $r_1 = r_2 = 0.2R$ and $r_3 = r_4 = 0.5R$, then:

$$1/R_T = 1/0.2R + 1/0.2R + 1/0.5R + 1/0.5R + 1/R = 15/R.$$

Thus $R_T = R/15$, and R_T is much less than the lowest individual resistance ($= R/5$).

(A)

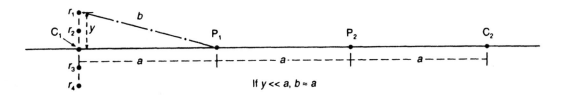

(B)

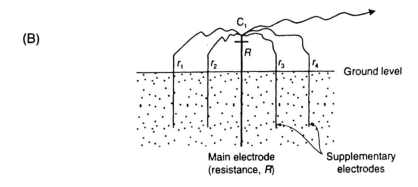

(C)

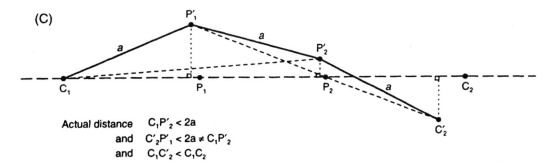

Actual distance $C_1P'_2 < 2a$
and $C'_2P'_1 < 2a \neq C_1P'_2$
and $C_1C'_2 < C_1C_2$

Ideally, a VES array should be expanded along a straight line. If it curves significantly and/or erratically (Figure 7.21C), and no correction is made, cusps may occur in the data owing to inaccurate geometric factors being used to calculate apparent resistivity values. Cusps in VES field curves are particularly difficult to resolve if their cause is unknown. Even if the apparent resistivity values have been calculated correctly with appropriately modified geometric factors, ambiguities may arise in the field curve which it may not be possible to model or interpret. In the case of CST data, if the correct geometric factors are used to derive the apparent resistivities, the CST profile may be interpreted normally. It always pays to keep adequate field notes in addition to recording the geophysical data so that appropriate corrections can be made with recourse to the correct information

Figure 7.21 (A) Supplementary electrodes planted in a straight line at right-angles to the main electrode have minimal effect on the geometric factor as long as the offset $y \uparrow a$. (B) Any number of additional electrodes act as parallel resistances and reduce the electrode contact resistance. (C) An out-of-line electrode array will give rise to erroneous ρ_a values unless the appropriate geometric factor is used. Shortened C_1C_2 produces elevated ΔV between P_1 and P_2 and needs to be compensated for by a reduced value of the geometric factor.

rather than to a rather hazy recollection of what may have been done in the field.

The presence of pipes, sand lenses or other localised features, which are insignificant in relation to the main geological target, can degrade the quality of the field data and thus reduce the effectiveness of any interpretation. If a conductive clay lens is present, for example, then when a current is applied from some distance away from it, the lines of equipotential are distorted around the lens and the current flow lines

(A)

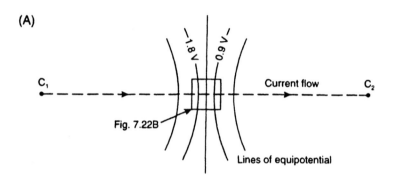

Figure 7.22 Distortion of current flow lines and equipotentials around an anomalous feature. The boxed area in (A) is enlarged to show detail in (B). The magnitude of equipotentials is for illustrative purposes only

(B)

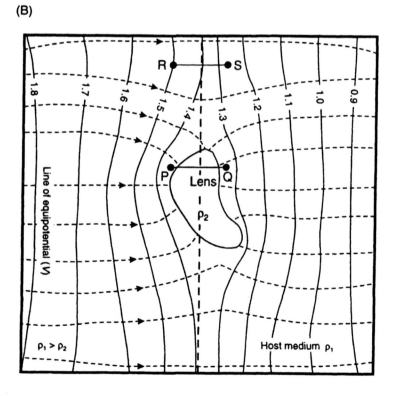

are focused towards the lens (Figure 7.22). The potential between P and Q (< 0.1 V) is obviously smaller than that measured between R and S (≈ 0.25 V) which are outside the field of effect of the lens. The apparent resistivity derived using this value of potential is lower than that obtained had the lens not been there, hence the occurrence of a cusp minimum (Figure 7.23A). If the lens has a higher resistivity than the host medium, the current flow lines diverge and the potential between P and Q becomes anomalously high and results in a positive cusp (Figure 7.23B).

Another feature which may occur on VES profiles is current leakage, particularly at large current electrode separations, when the array is aligned parallel to a conductive feature such as a metal pipe or a surface stream. The values of apparent resistivity become increasingly erratic owing to the voltage between the potential electrodes falling to within noise levels and tend to decrease in value (Figure 7.23C). If the position and orientation of a pipe is known, there should be no ambiguity in interpretation. There is no point in extending the VES once it is obvious current leakage is occurring.

A method of reducing the effects of these lateral inhomogeneities using the offset Wenner array has been described in Section 7.3.2. There is, however, no alternative method for the Schlumberger electrode configuration and cusps can be removed by smoothing the curve (dashed lines in Figure 7.23).

Figure 7.23 Distortion of Schlumberger VES curves due to (A) a conductive lens or pipeline, and (B) a resistive lens. After Zohdy (1974), by permission

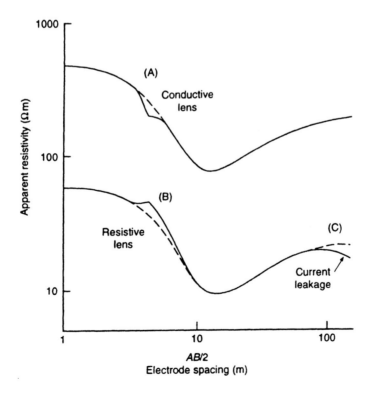

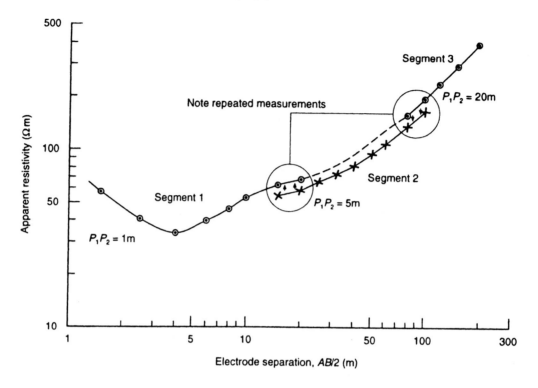

An additional but easily resolvable problem can occur with Schlumberger depth soundings. When the separation of the potential electrode pair is increased (b_1 to b_2 in Figure 7.17C), the contact resistance may change, causing a discrete step up or down of the next segment of the curve (Figure 7.24). Although the value of the apparent resistivity may change from the use of one electrode pair to another, the gradient of the change of apparent resistivity as a function of current electrode half-separation should remain the same. Consequently, the displaced segments can be restored to their correct values and the curve smoothed ready for interpretation. Segments at larger potential electrode separations should be moved to fit the previous segment obtained with a shorter electrode separation. So in Figure 7.24, segment 3 is moved down to fit segment 2 which is moved up to join on the end of segment 1. Measurements of resistance should be repeated at both potential electrode separation when crossing from one segment to the next. As all the electrodes are moved when a manual Wenner array is expanded, there is no discernible displacement of segments of the curve. Instead, the field curve may appear to have lots of cusps and blips through which a smooth curve is then drawn, usually by eye (Figure 7.25). An alternative, preferable, approach is to use the offset Wenner array (see Section 7.3.2) which improves the quality of the acquired field data (Figure 7.25).

Figure 7.24 Displaced segments on a Schlumberger vertical electrical sounding curve due to different electrode resistances at P_1 and P_2 on expanding potential electrode separations; segment 3 is displaced to fit segment 2 which is in turn displaced to fit segment 1 to produce a smoothed curve

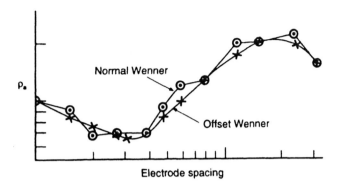

Normal Wenner

ρ_a

Offset Wenner

Electrode spacing

Figure 7.25 The difference in data quality that can be obtained by using an offset Wenner array in place of a normal Wenner array; the normal curve is more noisy

7.5 INTERPRETATION METHODS

Vertical sounding field curves can be interpreted qualitatively using simple curve shapes, semi-quantitatively with graphical model curves, or quantitatively with computer modelling. The last method is the most rigorous but there is a danger with computer methods to over-interpret the data. VES field curves may have subtle inflections and cusps which require the interpreter to make decisions as to how real or how significant such features are. Often a noisy field curve is smoothed to produce a graph which can then be modelled more easily. In such a case, there is little point in spending large amounts of time trying to obtain a perfect fit between computer-generated and field curves. As a general rule, depending on how sophisticated the field acquisition method is, layer thicknesses and resistivities are accurate to between 1% and 10%, with poorer accuracies arising from the cruder field techniques. Furthermore, near-surface layers tend to be modelled more accurately than those at depth, primarily because field data from shorter electrode separations tend to be more reliable than those for very large separation, owing to higher signal-to-noise ratios.

7.5.1 Qualitative approach

The first stage if any interpretation of apparent resistivity sounding curves is to note the curve shape. This can be classified simply for three electrical layers into one of four basic curve shapes (Figures 7.26A–D). These can also be combined to describe more complex field curves that may have several more layers. Note that the curve shape is dependent upon the relative thicknesses of the in-between layers (layer 2 in a 3-layer model; Figures 7.26C, D). The maximum angle of slope that the rising portion of a resistivity graph may have on a log–log graph is 45°, given the same scales on both axes (Figure 7.26A). If the field curve rises more steeply, then this suggests error in the data or that geometric effects due to steeply inclined horizons are distorting the data.

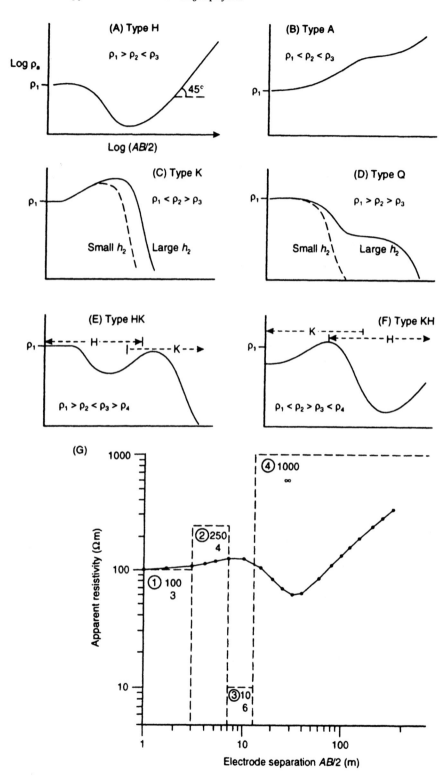

The *relative* magnitudes of the true resistivities obtained from the levels of the flat portions or shoulders of the graph are a useful starting point before more involved interpretation. For example, in Figures 7.26A and B, the only difference between the two models is the resistivity of layer 2. In Figure 7.26A, layer 2 resistivity is less than those for both layers 1 and 3. In Figure 7.26B, the second layer resistivity is between those for layers 1 and 3. In the case of Figure 7.26D, if the second layer is very thin (dashed line for small h_2) it may not be evident on the curve that this layer exists, i.e. its effects are 'suppressed'. 'Suppression' is discussed in more detail in Section 7.5.4.

From Figure 7.26G, it can be seen that the number of layers identified is equal to the number of turning points (TP) in the curve, plus one. The presence of turning points indicates sub-surface interfaces, so the number of actual layers must be one more than the number of boundaries between them. However, the coordinates of the turning points in no way indicate the depth to a boundary or provide specific information about the true resistivities (Figure 7.26G). From the curve shape alone, the minimum number of horizontal layers and the relative magnitudes of the respective layer resistivities can be estimated.

Figure 7.26 (*opposite*) Apparent resistivity curve shapes for different resistivity structures: (A) to (D) are three-layer models; (E) and (F) are four-layer models; (G) shows a block model for the layer resistivities and thicknesses and the resulting apparent resistivity curve. Neither the minimum nor the maximum apparent resistivities occur at electrode separations equivalent to the layer depths. To penetrate to bedrock, electrode separation should be about three times the bedrock depth for a Schlumberger array

7.5.2 Master curves

Interpretation of field curves by matching against a set of theoretically calculated master curves is based on the assumptions that the model relates to a horizontally stratified earth and that successively deeper layers are thicker than those overlying. Although this second assumption is rarely valid, the use of *master curves* does seem to provide a crude estimate of the physical model.

Synthetic curves for two-layer models can be represented on a single diagram (Figure 7.27), but for three-layer models the range of graphs is very large and books of master curves have been published (Mooney and Wetzel 1956; European Association of Exploration Geophysicists 1991). It is only practicable to use the master curves method for up to four layers. If more layers are present, the graphical approach is far too cumbersome and inaccurate. Three- and four-layer models can also be interpreted using master curves for two layers with the additional use of auxiliary curves (Figure 7.28) as outlined below.

The field data, smoothed and corrected as necessary, are plotted on a log–log graph on a transparent overlay at the same scale as the master curves. The overlay is placed on the master curves and, keeping the *x*- and *y*-axes of the two graphs parallel, the overlay is moved until the segment of the field curve at shortest electrode spacings fits one of the master curves and its *k* value is noted (Figure 7.29; in this case, $k = -0.3$). The position of the origin of the master curve is marked (A) on the overlay, which is then placed over the auxiliary curve sheet and the line for the same *k* value is traced on

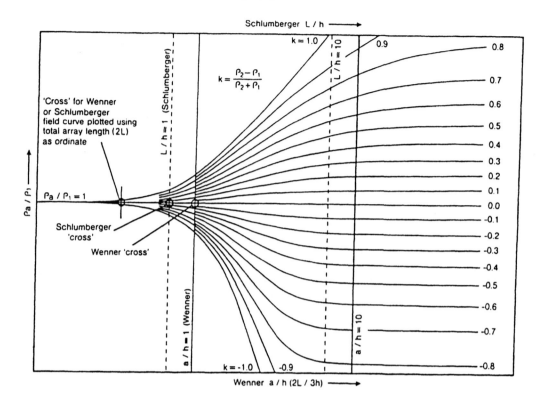

Figure 7.27 Two-layer master curves for Schlumberger and Wenner arrays. From Milsom (1989), by permission

to the overlay. The coordinates of A (which are read off the graph) are first estimates of the true resistivity and thickness of the top layer. Next, the overlay is replaced over the master curve sheet and moved so that the traced auxiliary curve for $k = -0.3$, always lies over the origin of the master curve until the next portion of the field curve is coincident with one of the master curves beneath and the new k value noted; in the example in Figure 7.29 the second k value is $+1.0$. When a fit has been obtained, the new position of the origin of the master curves (B) is marked on to the traced auxiliary curve. The overlay is returned to cover the auxiliary curves and the point B is placed over the origin of the auxiliary curves and the line corresponding to the new k value is again traced on to the overlay. The coordinates of B (also measured off the graph) are first estimates of the resistivity of the second layer and of the total depth to the second interface. The above process is then repeated, going from master curve to auxiliary curves and back again, noting the new k values in each case until the entire field curve has been matched against master curves. The final result should be, for a three-layer case as the example, two points of origin A and B giving ρ_1, t_1 and ρ_r and $(t_1 + t_2)$ and hence t_2. From the first k value, the resistivity of the second layer can easily be calculated (see Box 7.11), and from the second k value, ρ_3 can be estimated, thus completing the determination of the model parameters.

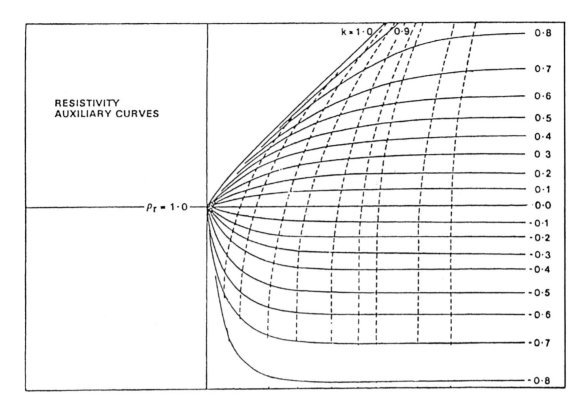

RESISTIVITY
AUXILIARY CURVES

$\rho_r = 1 \cdot 0$

Curve-matching is constrained to fit the field curve with one calculated from a very limited range of master curves. If the resistivity contrasts do not give rise to a k value that corresponds to one of the master curves, then the accuracy of the fitting (and of the subsequent interpretation) will be reduced. Furthermore, the use of master curves does not allow the problems of equivalence and suppression to be resolved (see Section 7.5.4). Interpretations obtained using this graphical approach should be regarded as crude estimates of the sub-surface layer parameters that can then be put into a computer model to obtain much more accurate and reliable results.

Figure 7.28 Auxiliary curves for a two-layer structure

7.5.3 Curve matching by computer

In 1971, Ghosh described a convolution method by which computers can be used to calculate master curves for vertical electrical soundings obtained using either a Wenner or Schlumberger array. The method uses what is called a 'linear digital filter', the details of which are given by Koefoed (1979).

The program synthesises an apparent resistivity profile for an n-layered model in which the variables are layer thickness and resistivity. Model profiles can then be compared with the field curves and

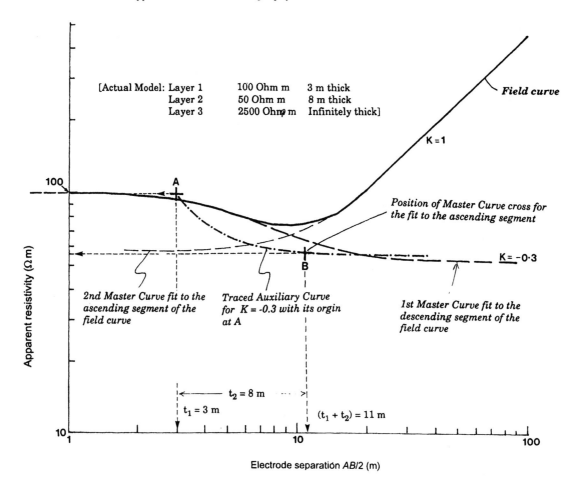

Figure 7.29 Fitting master and auxiliary curves to a Schlumberger VES curve; see text for details

adjustments to the layering and resistivity values can be made by trial and error to obtain as near correspondence as possible to the field curve.

However, in cases where a very good conductor underlies a comparatively resistive layer (where $\rho_1 > 20\rho_2$ and $-1 < k < -0.9$), Ghosh's method was found to produce inaccurate profiles owing to the structure of the filter, which had too few coefficients to track a rapidly falling resistivity curve.

A computer program can easily be checked to see whether it produces erroneous profiles by obtaining an apparent resistivity curve for a simple two-layer model where the resistivity contrast is at least 20:1 with the lower layer being the more conductive. If the program is unable to cope with such a contrast, the portion of the graph at larger electrode separations will appear to oscillate (Figure 7.30) rather than pass smoothly to the true resistivity of the second layer.

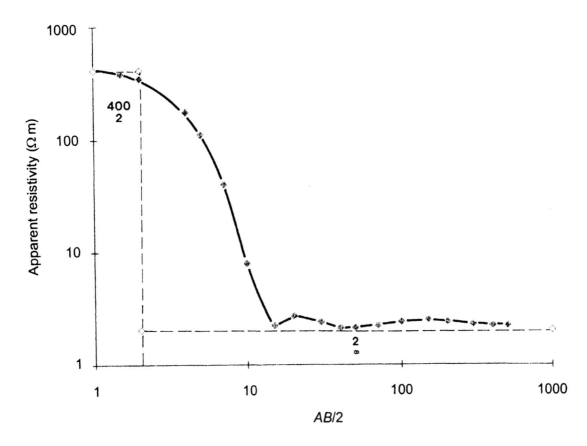

Much work has gone into the design of linear digital filters to overcome this computational difficulty (e.g. O'Neill and Merrick, 1984), and now modern software packages can cope with even the most extreme resistivity contrasts. Although these packages can deal with as many as 25 layers, 2–6 layers are usually adequate to describe the sub-surface. By increasing the number of layers beyond this, the length of time to produce an acceptable fit is increased dramatically (as there are so many more combinations of layer parameters to try) and, more often than not, the field data do not justify such a level of discrimination and may lead to over-interpretation.

As with master curves, it is always best to fit segments of the field curve at shortest electrode separations first and then to work progressively to greater separations. Once the top layers have been resolved, it is then easier to obtain good estimates of the parameters for the lower layers. The geoelectric basement (the lowest layer) is taken to be semi-infinite in depth and only the layer resistivity is required. Some computer packages display both the field and model curves simultaneously and may produce statistical parameters to describe the closeness of the fit. Optimisation of the interpretation can

Figure 7.30 Oscillating tail produced by a computer program that has insufficient coefficients in its linear digital filter to cope with a resistivity contrast of more than 20:1

be achieved automatically by successive iterations to reduce the degree of misfit until it falls within a specified and acceptable statistical limit.

A major advantage of the computer approach is that it provides an opportunity to investigate the problems of equivalence and suppression quickly and efficiently. See Section 7.5.4 for a more detailed discussion.

With the more sophisticated computer packages, some care has to be taken as to the method by which the convolution method is undertaken as different results may be produced (Figure 7.31). Having said that, as long as the user of the computer program is aware of its advantages and disadvantages, then good and reliable interpretations will be obtained. The danger, as with all uses of computers, is that, for come inexplicable reason, computer-generated results may appear to have greater credibility than those produced by more traditional means, which is not necessarily justified. There is evidently an increasing and undesirable tendency for people to plug data into a computer package, produce a result without thinking about the methodology or of experimental errors or about the geological appropriateness of the model produced. As with all the tools, computers must be used properly. With the availability of laptop computers, at least preliminary interpretation of field curves

Figure 7.31 Effects of different computational methods on the scale of resistivity contrasts with which the various types of computer program can cope without producing erroneous data. O'Neill and Merrick (1984), by permission

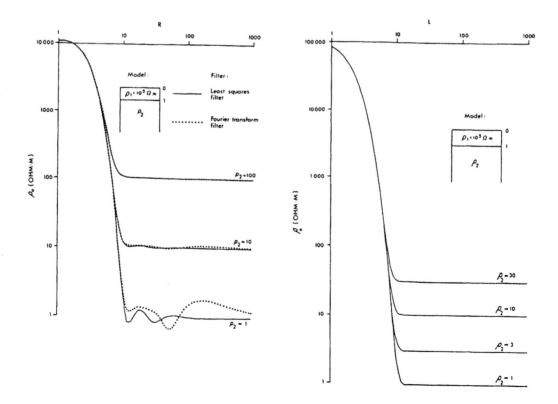

Table 7.4 Material types and their respective resistivity ranges for Kano State, northern Nigeria (Reynolds 1987a)

Material	Resistivity range (Ω m)
Sandy soil with clay	60–100
Clayey sand soil	30–60
Clay	10–50
Weathered biotite granite	50–100
Weathered granite (low biotite)	50–140
Fresh granite	750–8000

Table 7.5 Typical VES interpretation from Kano State, northern Nigeria (Reynolds 1987a)

Layer	Resistivity (Ωm)	Material type
1	95	Very sandy soil
2	32	Clay
3	75	Weathered granite*
4	3500	Fresh granite*

*Granite type also determined from local geological mapping and from local hand-dug wells.

can be done while still in the field, with master and auxiliary curves being used as a backup in case of problems with the computer.

Once an acceptable layer model has been produced for each vertical electrical sounding, the models can be displayed side-by-side much like borehole logs. Between the various VES models, correlations are made between layers with comparable resistivities to build up a two-dimensional picture of both the vertical and lateral variations in resistivity. This can be extended into third dimension so that maps of the individual layer thicknesses, akin to isopachyte maps, can be produced.

The final stage of a resistivity interpretation should be to relate each accepted VES model to the unknown local geology. Tables of resistivities, such as that in Table 7.1, or more geographically specific rock-type/resistivity ranges, can be used instead of referring to layer numbers. The interpretation can then be described in terms of rock units such as those listed in Table 7.4 which are from Kano State in northern Nigeria. Thus a typical resistivity interpretation would consist of perhaps four layers with resistivities as given in Table 7.5.

7.5.4 Equivalence and suppression

In the case of a three-layered model, if the middle layer is conductive relative to those around it, then current flow is focused through and virtually parallel to that layer. The longitudinal conductance S_L for this layer is $h_2/\rho_2 = $ constant, and as long as the thickness and resistivity are changed (within limits) so as to maintain that ratio, there will be no appreciable change in the resulting apparent resistivity curve. All the pairs of h_2/ρ_2 are electrically equivalent and no single pair of values is preferable to any other.

However, if computer interpretation of VES curves is undertaken, the range of h and ρ values can be determined so that estimates of the ranges of both thickness and true resistivity can be made. This in itself can be extremely valuable. For example, in northern Nigeria, the subsurface consists of at least three layers – soil, weathered granite and fresh granite. It has been found that weathered granite with a resistivity in the range 50–140 Ω m provides a reasonable source of water (given a saturated thickness > 10 m) to supply a hand-pump on a tubewell in a rural village. If the resistivity is less than 50 Ω m, this indicates that there is more clay present and the supply is likely to be inadequate. If the interpretation indicates a second-layer resistivity of 60 Ω m and thickness 12 m, which is characteristic of an acceptable water supply, this layer may be electrically equivalent to only 8 m of material (i.e. too thin) with a resistivity of 40 Ω m ($< 50\,\Omega$ m, thus probably clay rich). This combination could prove to be problematical for a reliable water supply. If, however, the computer models demonstrate that the lowest limit of thickness is 10.5 m and of resistivity is 55 Ω m, then the site could still be used. On the other hand, if the equivalent layer parameters are well into those for the thin clay-rich range, then it is better to try to select another site.

Similarly, if the middle layer is resistive in contrast to those layers surrounding it, then current tends to flow across the layer and thus the product of the layer resistivity and thickness (which is the transverse resistance, T; see Box 7.9) is constant. If, for example, a gravel layer with resistivity 260 Ω m and thickness 2 m is sandwiched between soil ($\rho_1 < 200\ \Omega$ m) and bedrock ($\rho_3 > 4500\,\Omega$ m), then $T_2 = 520\ \Omega\,\text{m}^2$. If the model thickness is reduced to only 1 m, the gravel layer resistivity must be doubled. Similarly, if the resistivity is reduced the thickness must be increased to compensate. Computer analysis can be used again to resolve the range of layer parameters which produce no discernible change in the apparent resistivity curve.

In the example illustrated in Figure 7.32A, a layer of resistivity 35 Ω m and thickness 2 m is sandwiched between an upper layer of resistivity 350 Ω m and thickness 6 m and the bedrock of resistivity 4500 Ω m (upper bar below curve). The longitudinal conductance S_L for layer 2 is 0.057 siemens (2 m/35 Ω m). The two other models depicted by horizontal bars are extreme values for the middle layer, but which have the same longitudinal conductance and which gave rise to model curves that are coincident with the one shown to better than 0.5%. The depths to bedrock thus range from 7.1 m to 10 m (in the models calculated) and there is no geophysical way of telling which model is 'right'. It is by considering whether a 10 cm thick highly

(A)

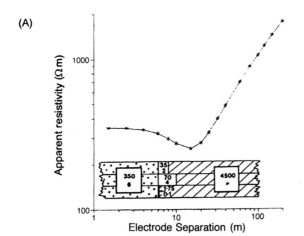

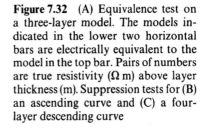

Figure 7.32 (A) Equivalence test on a three-layer model. The models indicated in the lower two horizontal bars are electrically equivalent to the model in the top bar. Pairs of numbers are true resistivity (Ω m) above layer thickness (m). Suppression tests for (B) an ascending curve and (C) a four-layer descending curve

(B)

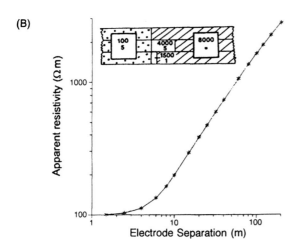

(C)

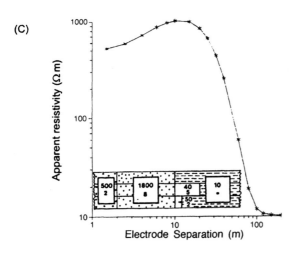

conductive horizon is more geologically reasonable than a far thicker, more resistive band, that the most appropriate model can be chosen.

An additional form of equivalence needs to be borne in mind at the interpretation stage and that is the equivalence between an isotropic and an anisotropic layer. This Doll–Maillet equivalence (see Maillet 1947) results in an overestimate of layer thickness in proportion to the amount of anisotropy which, if present, is of the order of 10–30%.

Apart from equivalence, the other interpretational problem is that of *suppression*. This is particularly a problem when three (or more) layers are present and their resistivities are ascending with depth (*A*-type curve; see Figure 7.26B) or descending with depth (*Q*-type curve; see Figure 7.26D). The middle intermediate layer may not be evident on the field curve and so its expression on the apparent resistivity graph is suppressed. The computer method is invaluable here to estimate (1) if there is a hidden layer present, and (2) if there is, its range of layer parameters. In Figure 7.32B, curves for three-layer models (middle and lower bars) with second layers of resistivity 1500 to 4000 Ω m and thicknesses 1 to 5 m are graphically indistinguishable from that of a two-layer model (top bar). For a layer to be suppressed, its resistivity should approach that of the one below so that the resistivity contrast between the top and the suppressed layer is comparable to that between the top and lowermost layers. The effects of missing such a suppressed layer can have major effects on the estimation of the depth to bedrock. In Figure 7.32C, a similar example is given but for descending resistivities. The curves for models with a suppressed layer (middle and lower bars) fit the three-layer case (top bar) to better than 1% and are indistinguishable graphically.

If the intermediate layers are thin with respect to those overlying, and if either equivalence or suppression is suspected, master curves will provide no solution. However, equivalent or suppressed layers can be modelled very effectively using computer methods in conjunction with a knowledge of what is geologically reasonable for the field area in question.

7.5.5 Inversion, deconvolution and numerical modelling

Zohdy (1989) produced a technique for the automatic inversion of resistivity sounding curves. Least-squares optimisation is used in which a starting model is adjusted successively until the difference between the observed and model pseudo-sections is reduced to a minimum (Barker 1992). It is assumed that there are as many sub-surface layers as there are data points on the field sounding curve (Figure 7.33) and that the true resistivity of each of these assumed layers is that of the corresponding apparent resistivity value. The mean depth of each layer is taken as the electrode spacing at which the apparent resistivity was measured multiplied by some constant. The value of this constant is one which reduces the difference between

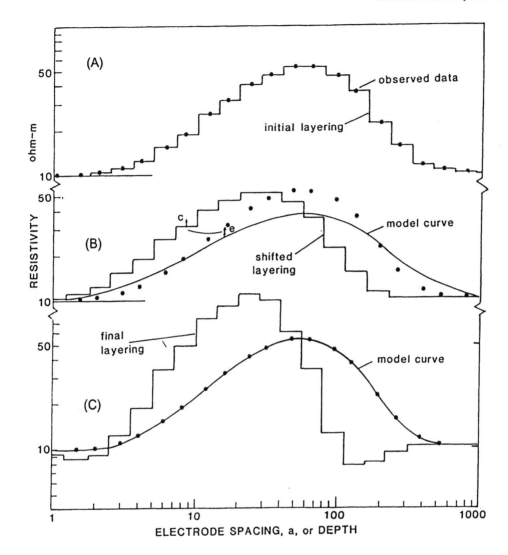

Figure 7.33 Automatic sounding inversion technique. (A) Observed data and initial layering. (B) Shifted layering and resulting model sounding curve. The difference (*e*) between the model and observed curves is used to apply a correction (*c*) to the layering. (C) The final layering and resulting model curve that is closely similar to the observed data. From Barker (1992), by permission

the observed and model resistivity curves to a minimum and is determined by trial and error.

The starting model is used to generate a theoretical synthetic sounding curve which is compared with the field data. An iterative process is then carried out to adjust the resistivities of the model while keeping the boundaries fixed. After each iteration the theoretical curve is recalculated and compared with the field data. This process is repeated until the RMS difference between the two curves reaches a minimum (Figure 7.33).

Zohdy's method has been developed by Barker (1992) for the inversion of SSI apparent resistivity pseudo-sections and more recently by using a deconvolution method (Barker, personal communication). Consequently, it is possible to produce fully automated inversions of SSI pseudo-sections. The final results are displayed also as pseudo-sections in terms of the variations in true resistivity with depth as a function of distance along the array. Examples of inverted pseudo-sections are given in Section 7.7.

In addition to the above inversion routines, others have been produced, often in association with particular equipment, and also as specific developments of true tomographic imaging (e.g. Shima 1990; Daily and Owen 1991; Noel and Xu 1991; Xu and Noel 1993). Commercially available imaging inversion packages are available from a number of sources and are related to a style of data acquisition and equipment. Packages vary from those which can operate easily on a laptop computer, more sophisticated processing may require the computational power of a workstation with full colour plotting facilities.

Finite-element forward modelling can be undertaken using commercially available software. The resistivity response for a two-dimensional model is calculated and displayed as a pseudo-section for comparison with the original field data. This approach is used to help generate realistic sub-surface geometries in definable model structures (e.g. Figure 7.34).

Figure 7.34 Final interpretation of faulted Triassic sequence in Staffordshire, UK. (A) Two-dimensional finite difference model. (B) Computed apparent resistivity pseudo-section. (C) Field data. (D) Geological interpretation based on (A) and additional information. From Griffiths *et al.* (1990), by permission

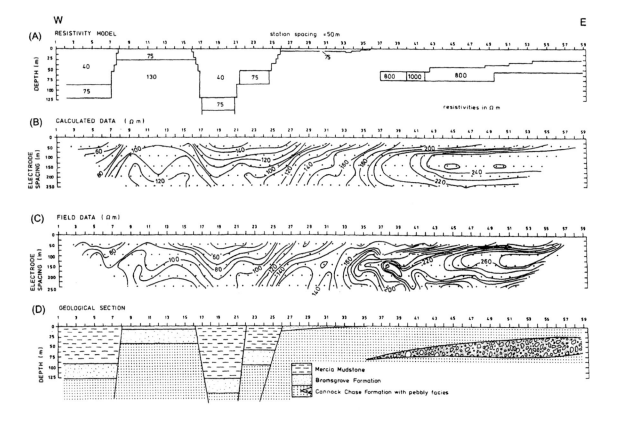

7.6   MISE-À-LA-MASSE METHOD

The mise-à-la-masse or 'charged-body potential' method is a development of the CST technique but involves placing one current electrode within a conducting body and the other current electrode at a semi-infinite distance away on the surface (Figure 7.35). The voltage between a pair of potential electrodes is measured with appropriate corrections for any self-potentials.

For an isolated conductor in a homogeneous medium, the lines of equipotential should be concentric around the conductor (Figure 7.36A). In reality, lines of equipotential are distorted around an irregularly shaped conductive orebody (Figure 7.36B) and can be used to delimit the spatial extent of such a feature more effectively than using the standard CST method. The mise-à-la-masse method is particularly useful in checking whether a particular conductive mineral-show forms an isolated mass or is part of a larger electrically connected orebody. In areas where there is a rough topography, terrain corrections may need to be applied (Oppliger 1984). There are no general rules that can be applied to mise-à-la-masse data. Each survey is taken on its own merits and a plausible model constructed for each situation, although Eloranta (1984), for example, has attempted to produce a theoretical model to account for the observed potential distributions.

There are two approaches in interpretation. One uses the potential only and uses the maximum values as being indicative of the conductive body. The other converts the potential data to apparent resistivities and thus a high surface voltage manifests itself in a high

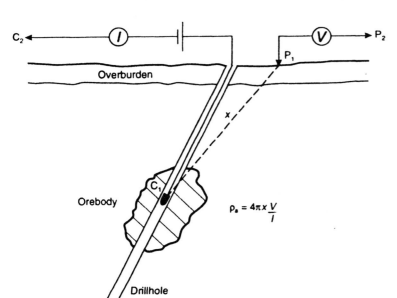

$$\rho_a = 4\pi x \frac{V}{I}$$

Figure 7.35 Positions of electrodes used in a mise-à-la-masse survey. One form of geometric factor is given where x is the distance between the C_1 electrode down the hole and the P_1 mobile electrode on the ground surface

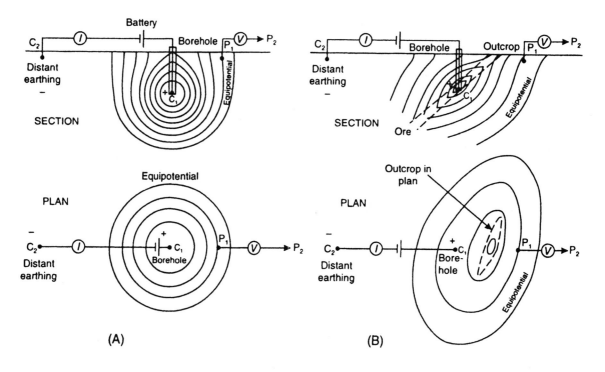

(A)   **(B)**

apparent resistivity ($\rho_a = 4\pi x\, V/I$, where x is the distance between C_1 and P_1).

An example of a mise-à-la-masse survey in ore prospecting in Sweden is shown in Figure 7.37. One current electrode was placed 65 m below ground, 89 m down an inclined borehole (borehole 33) and the surface potentials mapped (Figure 7.37A). As there were so many boreholes available in this survey, it was possible to determine the vertical distribution of potentials as well as the surface equipotential distribution (Figure 7.37B). Combining all the available data, it was possible to obtain a three-dimensional image of the potential distribution associated with the target orebody and thus delimit its size, strike and structure (Parasnis 1967).

A second example is shown in Figure 7.38, in which the effects of terrain on the surface potentials can be clearly seen. The positive electrode was placed at 220 m depth down a 1 km deep borehole (Figure 7.38A) and the surface potentials mapped (Figures 7.38B and C). Oppliger (1984) found that when terrain slopes exceeds 10°, surface electric potentials can be adversely affected. The terrain-corrected surface potentials are shown in Figure 7.38C. The main differences are that the low of 85 Ω m and the high of 168 Ω m both change in value (to 100 and 150 Ω m respectively) and, in particular, the orientation of the elongate apparent resistivity high is rotated through 30°. The ridge form of this resistive anomaly suggests that

Figure 7.36 (A) Concentric and symmetrical distribution of equipotential lines around a current electrode emplaced within a homogeneous medium. (B) Distortion of equipotentials due to the presence of an orebody. After Parasnis (1966), by permission

Figure 7.37 (*opposite*) (A) Map of surface potentials obtained in a mise-à-la-masse survey in Sweden with C_1 in borehole 33. The location of Line 2680S is also shown. (B) Potentials in a vertical section through profile 2680S. From Parasnis (1967), by permission

(A)

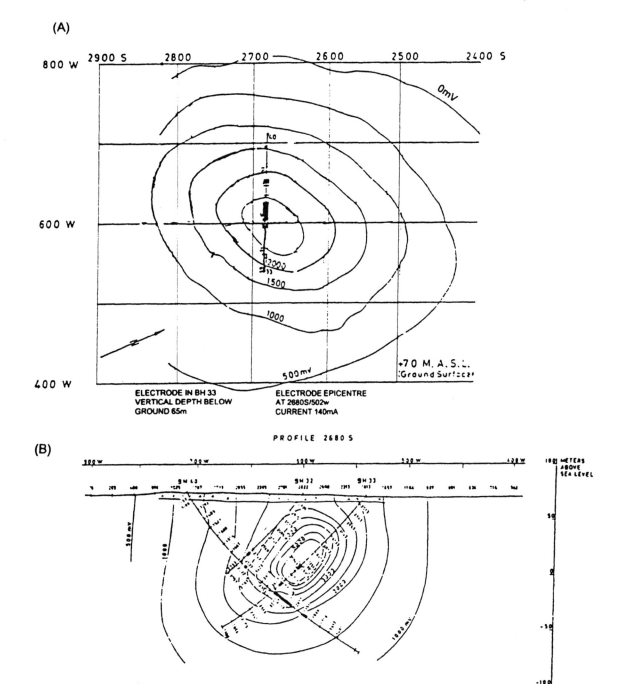

ELECTRODE IN BH 33
VERTICAL DEPTH BELOW
GROUND 65m

ELECTRODE EPICENTRE
AT 2680S/502w
CURRENT 140mA

PROFILE 2680 S

(B)

+   Actual position of current
electrode

−   Mineralisation with
Pb>0.5% Cu>0.1%

CURRENT 340 mA

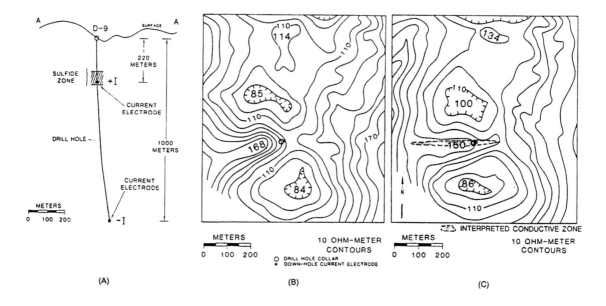

(A)   (B)   (C)

a conductive body is bounded to the north and south (as indicated by the marked lows) and extends a limited way in an east–west direction. This interpretation has been confirmed by other investigations.

7.7   APPLICATIONS AND CASE HISTORIES

7.7.1   Engineering site investigations

7.7.1.1   Sub-surface collapse features

In a small village in east Devon, a 5 m diameter hole appeared overnight in the middle of the road. The local water main had been ruptured and had discharged for over 12 hours and all the water had disappeared down a fissure into underlying limestone. Several of the local buildings started to crack badly, and on investigation it was found that the rafted foundations of several houses had broken and the houses were literally cracking open at the seams, resulting in the emergency evacuation of the residents.

A resistivity survey was initiated in order to determine the sub-surface extent of the problem prior to drilling. Fortunately, the front gardens of the houses affected were all open-plan so there was no difficulty in access, but space was at a premium. A series of constant-separation traverses was instigated using the Wenner array with electrode separations of 10, 15 and 20 m. The resulting apparent resistivity values were plotted as a contour map (Figure 7.39). It was

Figure 7.38 (A) Current electrode configurations. (B) Actual mise-à-la-masse apparent resistivities measured on the surface around inclined borehole D-9. Contours are every $10\,\Omega\,\mathrm{m}$. (C) Terrain-corrected apparent resistivities for the same survey with an interpreted conductive zone indicated. From Oppliger (1984), by permission

Figure 7.39 (*opposite*) (A) Apparent resistivity isometric projection obtained using constant-separation traverses with an electrode separation of 10 m. (B) Modelled microgravity profile that would be expected for the geological model shown in (C): interpreted depth to limestone constrained by drilling. A north–south profile is shown in Figure 7.20. The position of a clay-filled solution feature is arrowed

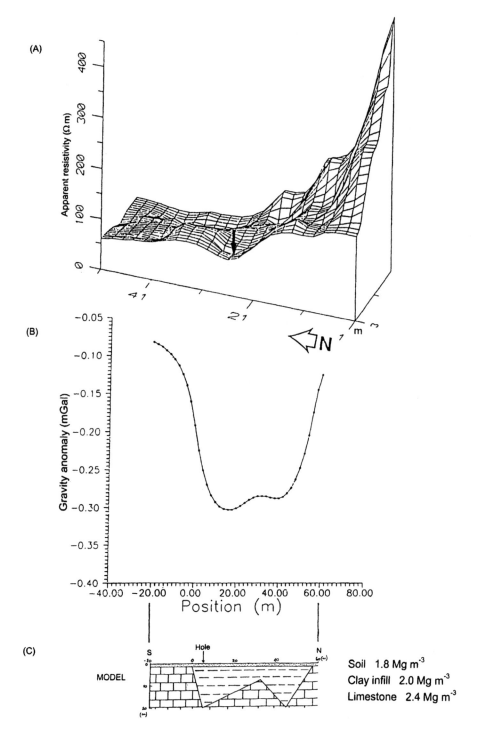

clear that where the hole had appeared, there was a deep infill of clay. It was this that had slipped through a neck of a fissure into a cave beneath, resulting in subsidence beneath the foundations of the houses and the rupture of the water main. The discharging water disappeared into this newly discovered cavern. The clay depth decreased uphill and suddenly increased again, indicating further clay-filled fissures. On drilling these resistivity anomalies, the depth to limestone was confirmed. One drillhole penetrated the cave but failed to locate the bottom; the cave was at least 20 m deep.

7.7.1.2   *Burial of trunk sewer*

A route for a proposed new trunk sewer in South Wales was investigated using electrical resistivity methods because access for drilling equipment was not possible. Both vertical electrical soundings and constant-separation traverses were used along the route and compared with available borehole data from the National Coal Board (Prentice and McDowell 1976). The material through which the sewer trench was to have been dug consisted of superficial deposits overlying Coal Measure sandstone and mudstone. The Coal Measure material was anticipated to be massive and strong and thus hard to excavate, while saturated superficial deposits and Coal Measure shales were thought to provide very unstable trench walls. The CST results using a Wenner array with 10 m electrode separation and 10 m station interval revealed locations where sandstone bedrock was interpreted to be close to the surface which would have required

Figure 7.40 Constant-separation traverse data obtained along the proposed route of a new trunk sewer in South Wales, with the interpreted geological section. After Prentice and McDowell (1976), by permission

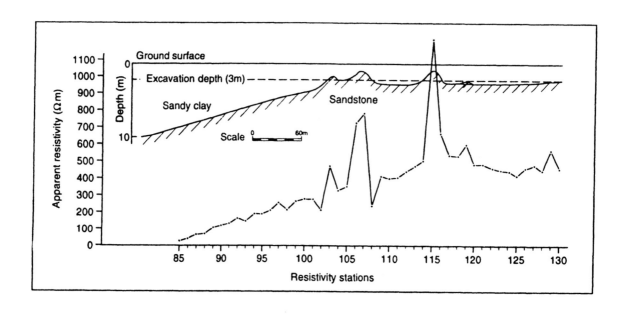

blasting for the excavation for the new sewer (Figure 7.40). Seismic refraction was also used to obtain acoustic velocities, which in turn were used to determine whether blasting or ripping techniques should be used in the excavation.

7.7.1.3 *Location of permafrost*

The presence of massive ground ice and frozen ground provides considerable problems to engineers involved in construction projects. First there are the difficulties in excavation, and secondly, substantial problems can emerge with the thawing of such affected ground. It is therefore vital that ice wedges and lenses, and the extent and degree of permafrost, can be determined well in advance.

Ice has a very high DC electrical resistivity in the range from $1\,M\Omega\,m$ to $120\,M\Omega\,m$ (Reynolds and Paren 1984) and therefore forms a particularly resistive target. A variety of geophysical profiles over a proposed road cutting near Fairbanks in Alaska are illustrated in Figure 7.41. Data obtained in the spring show more variability and resolution than when an active layer of thawed ground is present, as in the autumn measurements (Osterkamp and Jurick 1980). Other geophysical methods which are used successfully in this application are electromagnetic profiling, microgravity and ground radar surveying.

7.7.1.4 *Location of buried foundations*

As part of a trial survey in January 1993, electrical resistivity sub-surface imaging was used at a disused railway yard in order to locate old foundations concealed beneath railway ballast. Details of the geophysical survey have been described in more detail by Reynolds and Taylor (1994, 1995) and Reynolds (1995).

The SSI survey was carried out adjacent to a metal chain-link fence and an old diesel tank, and about 3 m from an existing building. It was thought that the remains of two former buildings might still be present beneath the railway ballast and the existing building. The site was totally unsuitable for electromagnetic profiling, because of the above ground structures. It was also unsuitable for ground penetrating radar owing to the coarse ballast and potentially conductive ash also found on site. Despite extremely high electrode contact resistances, a 25 m long array was surveyed with an inter-electrode separation of 1 m. This provided a vertical resolution of 0.5 m or better. The apparent resistivity data were filtered to remove noise spikes and displayed as a pseudo-section (Figure 7.42 A) which was inverted using a deconvolution technique (Barker, personal communication). The final pseudo-section of true resistivities against depth shows a general increase in resistivity with depth (Figure 7.42B). In particular, it revealed two areas of extremely high resistivity ($> 125\,000\,\Omega\,m$)

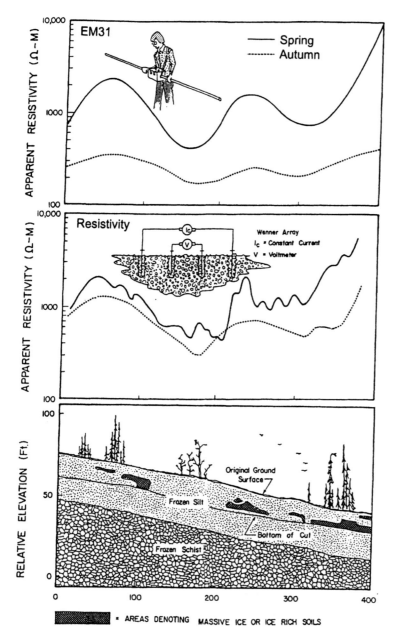

Figure 7.41 Massive ice and frozen ground in a sub-surface profile of a proposed road cut near Fairbanks, Alaska. Also shown are the spring and autumn survey data obtained using electrical resistivity constant-separation traversing and electromagnetic induction (EM31). Massive ground ice produces significant apparent resistivity highs. From Osterkamp and Jurick (1980), by permission

at a depth of about 1 m which had very flat tops to the anomalies. These were interpreted to be due to buried foundations. The main anomaly (between 6 and 11 m along the array) was found to correlate with the outline of one former building on an old plan. The second feature (starting at around 18 m) is thought to be due to the other old building. However, the location was found to be several metres further away from the first building than indicated on the plans. The

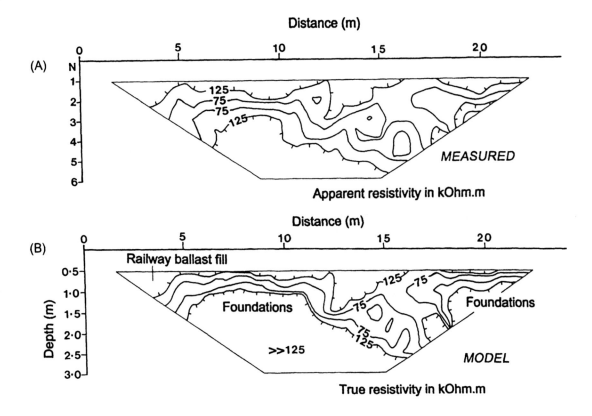

Figure 7.42 Electrical resistivity subsurface imaging pseudo-sections: (A) apparent resistivity profile, and (B) true resistivity–depth profile, over buried concrete slabs at 1 m depth. From Reynolds and Taylor (1995), by permission

depth to the foundation was thought to be reasonable as adjacent brick slabs excavated a few metres away were found at a comparable depth.

7.7.2   Groundwater and landfill surveys

7.7.2.1   *Detection of saline groundwater*

In the mid-1950s, a comprehensive electrical resistivity survey programme was initiated in order to map out saline groundwater in areas of the Netherlands below or at mean sea level. Figure 7.43 shows schematically the nature of the hydrogeology in the western part of the Netherlands. The vertical electrical soundings provided a means of obtaining information about the vertical distribution of fresh, brackish and saline water bodies and their areal extent (Figure 7.44).

Pockets of saline water were found which were thought to be remnants from before the fifteenth century after which time the present sea-dyke formed, cutting off the sea. To the west of Alkmaar, some 30 m of saline water was found above tens of metres of fresh water separated by an impermeable clay layer. Major demands for construction sand for the building of new roads and urbanisation

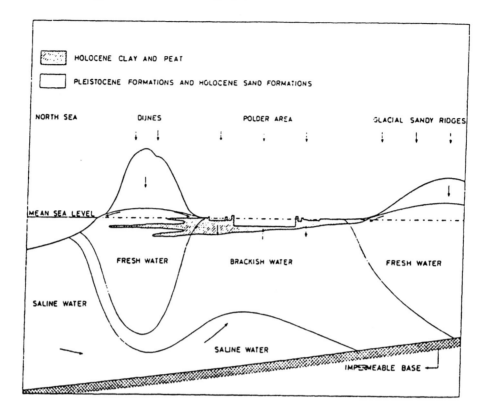

could have led to the extraction of sands with the demise of the clay barrier. This could have resulted in the mixing of the water bodies and the contamination of an otherwise potable water supply. Furthermore, correlation of resistivities from modelling of VES data with borehole information about groundwater chemistry has led to a relationship between chloride content and resistivity. Consequently, it is possible to determine chloride content of the groundwater from the resistivity data.

Figure 7.43 Schematic hydrogeological cross-section for the western part of the Netherlands. From van Dam and Meulenkamp (1967), by permission

7.7.2.2 Groundwater potential

In Kano state, northern Nigeria, an internationally funded aid programme was established in the 1980s to provide tubewells with handpumps for 1000 villages in rural areas. Village populations ranged from several hundred to no larger than 2000 people, but all were in very remote locations. Failure to obtain a reliable supply of water would have resulted in many of the villages being abandoned and the populations moving to the larger towns, thereby compounding the local problems of sanitation and health, education and employment, and the demise of rural culture and skills.

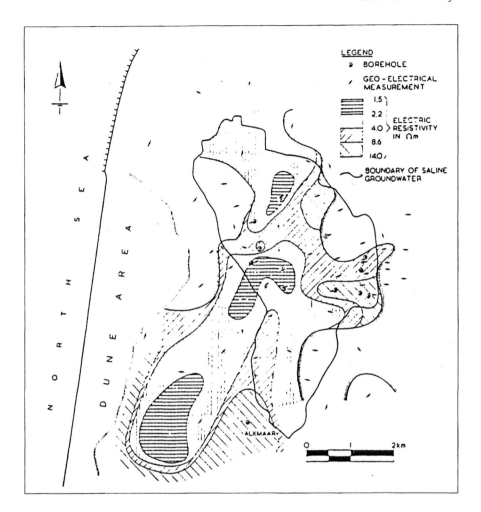

Figure 7.44 Distribution of resistivities of a sand layer with a saline groundwater boundary in Noord Holland as determined by many vertical electrical soundings. From van Dam and Meulenkamp (1967), by permission

It was first recommended that geophysics was unnecessary to locate groundwater; boreholes drilled anywhere would succeed. In practice, borehole failures were in excess of 82% of holes drilled, particularly in the southern areas. Geophysical methods were then called upon to improve the failure rate. Predominantly, vertical electrical soundings were used on sites selected following initial hydrogeological and photogeological inspection. Careful analysis of the VES data with the subsequent borehole information, led to the compilation of a database of typical formation resistivities and their likely hydrogeological potential. It became apparent that certain geographical regions had better and more easily resolved groundwater resources, and six drilling rigs were kept working on these sites. This provided a window of several months in which the more problematic sites could be investigated further until they, too, were ready for drilling.

Use of geophysics to help identify the groundwater potential in areas and to assist in the planning of drilling programmes led to the borehole failure rate falling to 17% of holes drilled and a saving to the project of £5 million – at least 10 times the cost of the geophysical surveys (Reynolds 1987).

7.7.2.3 Landfills

There is an increasing amount of interest in the use of high-resolution resistivity surveys in the investigation of closed landfills, particularly with respect to potential leachate migration. Both resistivity sounding and sub-surface imaging have been used very successfully.

There is no such thing as a typical landfill – some are extremely conductive, others are resistive relative to the surrounding media. There are many variables geophysically (Reynolds and McCann 1992) and care must be taken not to presume a particular geophysical response for any given site. For example, van Nostrand and Cook

Figure 7.45 Observed apparent resistivity profile across a resistive landfill using the Wenner array. From van Nostrand and Cook (1966), by permission

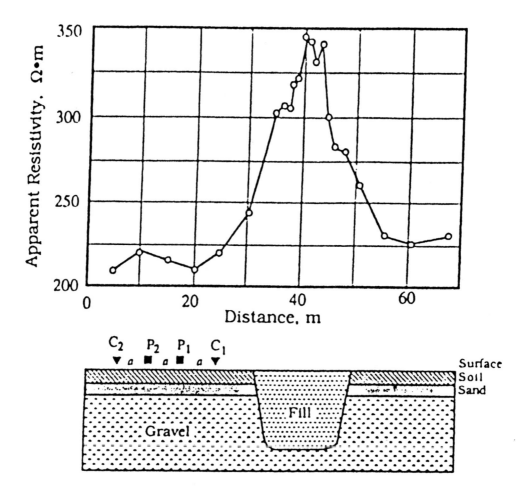

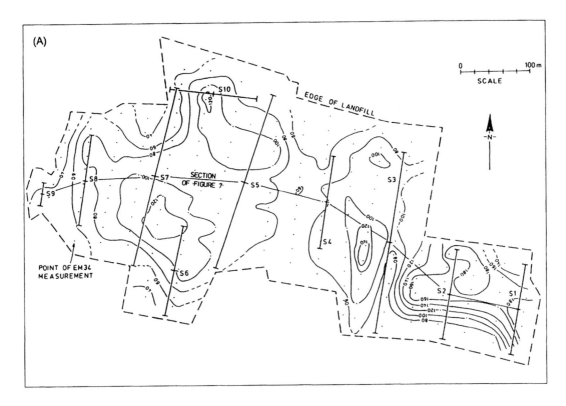

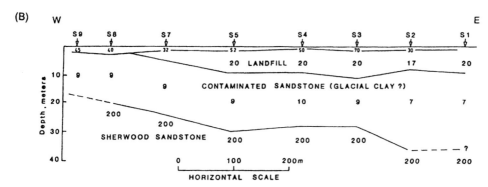

(1966) presented a very clear CST profile of apparent resistivity across a resistive landfill (Figure 7.45). Barker (1990) has shown resistivity sounding results across a landfill in Yorkshire in which the landfill is conductive (around $20\,\Omega\,m$ or less) over a contaminated substrate of sandstone (Figure 7.46). The offset Wenner method was used in this example.

Schlumberger soundings have been used by Monier-Williams *et al.* (1990) as part of a broader geophysical survey around the Novo Horizonte landfill in Brazil. Quantitative analysis of the soundings

Figure 7.46 (A) Contoured ground conductivity values over a landfill. Solid straight lines represent the positions and orientations of resistivity soundings. Contour interval is 20 ms/m. (B) Geoelectrical section across the landfill based on the soundings in (A). From Barker (1990), by permission

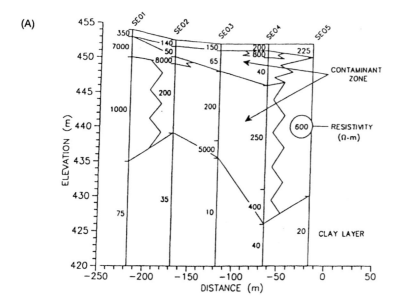

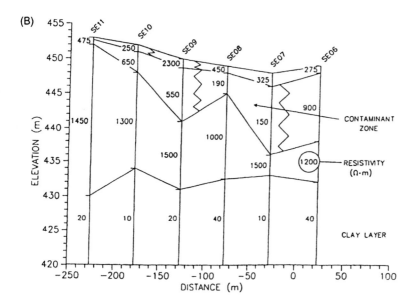

Figure 7.47 Two parallel resistivity sections based on the interpretation of Schlumberger soundings at the Novo Horizonte landfill site, Brazil. The profile in (A) is closer to the landfill than that shown in (B). The background resistivities above the basal clay are high; the lower values in the centre of the sections are assumed to be due to contamination. Note that the conductive zone in (B) is apparently more shallow than in (A). From Monier-Williams *et al.* (1990), by permission

and displays as resistivity panels have revealed significant zones with anomalously low resistivities (Figure 7.47). These have been interpreted as being contaminant plumes arising from the landfill. The displays shown in the figure are orientated parallel to the flank of the landfill, but at 10 m and 70 m distance away from it.

Sub-surface imaging pseudo-sections across a landfill are shown in Figure 7.48. The three panels illustrate the observed apparent

(A)

n = 6

(B)

30m

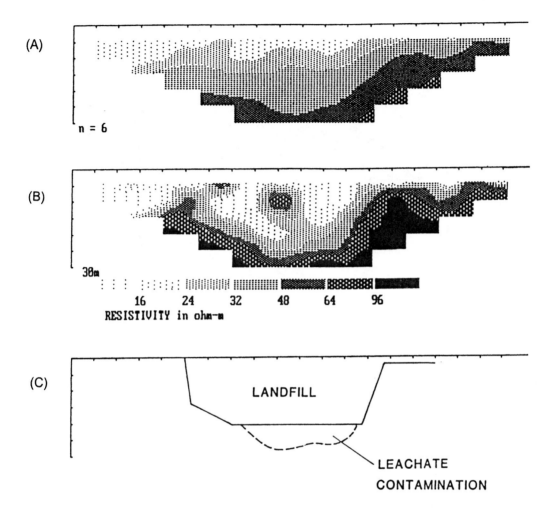

16   24   32   48   64   96

RESISTIVITY in ohm-m

(C)

LANDFILL

LEACHATE
CONTAMINATION

resistivity data, the inverted true resistivity–depth model and a schematic interpretation. In this case, the depth and geometry of the landfill were known at the outset. The zone of low resistivity associated with the saturated landfill extends more deeply than had been expected. This is interpreted as indicating the leakage of leachate through the base of the landfill (Barker 1992).

A further example of a sub-surface imaging pseudo-section inverted model is shown in Figure 7.49 (Reynolds 1995). The data were acquired over a closed shallow landfill constructed as a 'dilute and disperse' site over river gravels. The electrical image shows the thin capping material, the waste material and the basal gravels quite clearly. The image is entirely consistent with depths known from boreholes on site.

Figure 7.48 (A) Wenner apparent resistivity pseudo-section measured across a landfill. Electrode spacing = 10 m. (B) Resistivity depth section obtained after eight iterations. (C) Approximate section across the landfill based on existing information. From Barker (1992), by permission

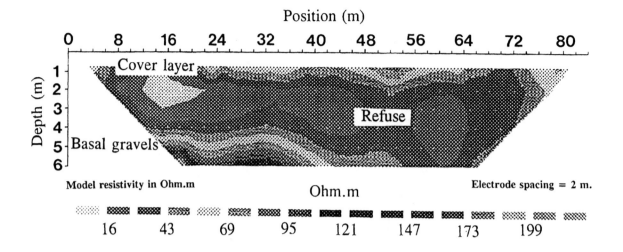

7.7.3   Glaciological applications

Electrical resistivity methods have been used since 1959 to determine glacier ice thickness. Measurements were first obtained on European glaciers on temperate ice (i.e. ice at its pressure melting point). In 1962, resistivity measurements were made on polar ice (i.e. ice well below its pressure melting point) and were found to be anomalously low by up to three orders of magnitude compared with temperate ice values. Whereas the electrical resistivity behaviour of polar ice is now reasonably understood (Figure 7.50; Reynolds and Paren 1984), the electrical behaviour of temperate ice is still poorly understood. In the 1970s a considerable amount of work was undertaken to develop field data acquisition in Antarctica. Interpretation methods were developed to yield information on vertical thermal profiles through the ice mass and whether or not ice shelves afloat on sea water were melting or freezing at their base. All of these data contribute to an understanding of ice dynamics (rate of ice movement, etc.) and the structure of the ice masses under study.

A series of vertical electrical soundings has been made on George VI Ice Shelf along a flow line of Goodenough Glacier which flows westwards from the Palmer Land Plateau in the Antarctic Peninsula (Figure 7.51). The field curves were modelled to take into account thermal effects and the resulting interpretations are shown in Figure 7.52. The estimated ice thicknesses and rates of bottom melting were in good agreement with those determined independently (Reynolds 1982).

Figure 7.49 Electrical resistivity pseudo-section acquired over a closed landfill in north Wales. From Reynolds (1995), by permission

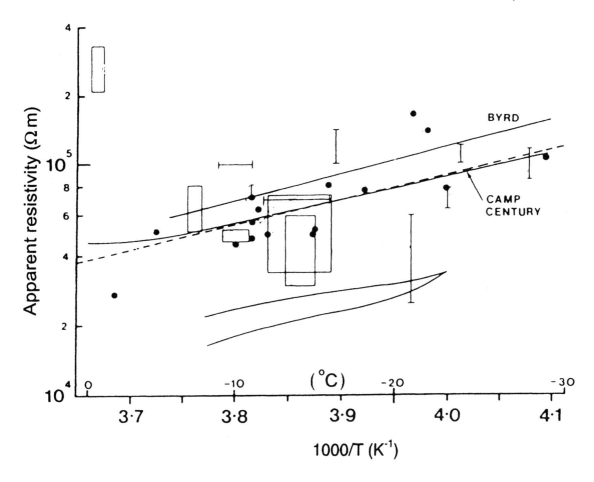

Figure 7.50 Resistivity of ice as a function of temperature. Mean values from georesistivity sounding of ice at 100m or deeper are plotted with estimated uncertainties against the estimated layer temperature from a wide variety of sources. Laboratory measurements on ice cores examined over a range of temperatures are shown by continuous lines. A regression line for the data is given by a dashed line. Given a particular temperature, the ice resistivity can be predicted within a factor of 2 or better. After Reynolds and Paren (1984), by permission

Other uses of resistivity measurements have been made by Haeberli and Fisch (1984) who drilled holes with a hot-water jet drill through Grubengletscher, a local glacier in Switzerland. By using a grounded electrode beyond the snout of the glacier and the drill tip as a mobile electrode, they were able to detect the point at which the drill tip broke through the highly resistive ice into the more conductive substrate (Figure 7.53A). Consequently, they were able to determine the ice thickness much more accurately than by using either the drilling or surface radio-echosounding. With debris-charged ice at the glacier base it is difficult to tell when the glacier sole has been reached judging by thermal drilling rates alone. The radio-echo-sounding depth measurements were found on average to be accurate to within 5%, but generally underestimated the depth. Electrodes were planted at the ice-bed interface at the ends of each of 14 boreholes and standard resistivity depth soundings were undertaken as if the glacier were not there (Figure 7.53B).

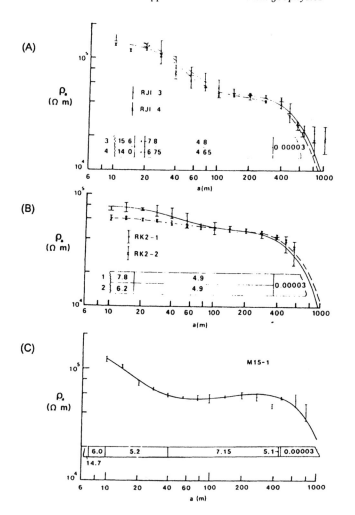

Figure 7.51 Apparent resistivity sounding curves obtained using a Schlumberger array at three sites along a glacial flow line on George VI Ice Shelf, Antarctica. In (A) and (B) two orthogonal soundings are shown at each site. Below each curve is the interpreted model in terms of true resistivities against depth within the ice sheet. The extremely low values of resistivity below the ice shelf indicate that it is afloat on sea water. Model resistivities are given in units of $10\,\text{k}\Omega\,\text{m}$. From Reynolds and Paren (1984), by permission

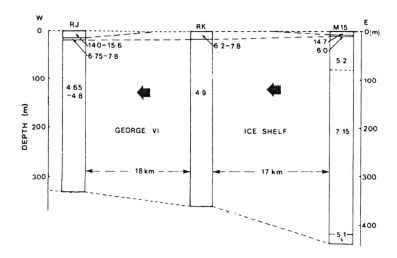

Figure 7.52 Resistivity structure through George VI Ice Shelf along a flow line. Resistivities are in units of $10\,\text{k}\Omega\,\text{m}$, and are plotted against depth within the ice sheet. From Reynolds and Paren (1984), by permission

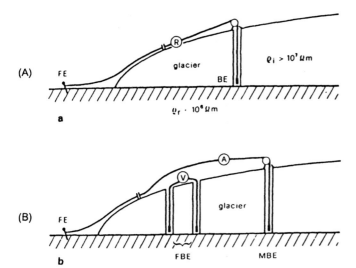

Figure 7.53 Electrode arrays used for the determination of the exact position of the glacier bed (A) and for resistivity soundings of glacier beds (B). FE = fixed electrode outside the glacier margins, BE = borehole electrode for determination of glacier-bed position, FBE = 'fixed' borehole electrodes for resistivity soundings of glacier beds (corresponding to potential electrodes MN in traditional surface soundings), MBE = 'moving' borehole electrode for resistivity soundings of glacier beds. R = resistivity meter, V = voltmeter, A = ammeter. From Haeberli and Fisch (1984), by permission

Haeberli and Fisch discovered that the Grubengletscher was underlain by over 100 m of unconsolidated sediments. It had been thought previously that the glacier was in direct contact with bedrock. This work has demonstrated that by using sub-glacial electrodes, significant new information can be obtained about the nature of the materials underlying the glacier. This can have considerable benefits when trying to understand the sub-glacial hydrogeological regime, for example. This is of particular importance at Grubengletscher because of two proglacial lake outbursts in 1968 and 1970 which caused considerable damage to the nearby village of Saas Balen.

7.8   ELECTROKINETIC (EK) SURVEYING IN GROUNDWATER SURVEYS

When a seismic wave travels through partially or fully saturated porous media, the seismic impulse effectively squeezes the rock, causing the pore fluid to move. This generates a small electrical or electrokinetic signal which can be detected using electrodes implanted at the ground surface, an idea first mooted by Thompson in 1936. Around the same time, Russian workers were experimenting with similar systems (e.g. Ivanov 1939). In 1959, Martner and Sparks reported a systematic study of seismoelectric coupling using explosives at various depths. They were the first to demonstrate that the conversion of seismic to electromagnetic energy at the water table could be detected using surface antennae.

A partially saturated vadose zone generates characteristically electrocapillary signals that are caused by the movement of air/water interfaces in pore throats, and electrophoretic signals that are caused by the displacement of bubbles within the pore fluid. However, fully saturated aquifers generate electrokinetic signals by the displacement of a single fluid (water or brine) when stimulated by a passing seismic wave. The difference between the two types of signal can be identified, leading to the determination of the depth to the water table in unconfined aquifers. The dry zone or basement or any non-aquifer rocks are characterised by the lack of any signal. The method is also referred to as seismoelectric or electroseismic surveying (Thompson and Gist 1993).

The basic data acquisition system consists of a seismic source, usually a sledge hammer or shallow explosives (Figure 7.54). The seismic impulse propagates into the ground and any electrokinetic signals generated by the passage of the P-wave are detected using an electrode-pair dipole with an inter-electrode separation of between 0.5 m and 10 m. Proprietary electronics and software enable the EK data to be converted to indicate permeability and cumulative flow in litres per second (GroundFlow Ltd).

Although the method is described here under 'electrical methods', it is not an active electrical technique. Neither is it purely a seismic method. The technique has been developed recently in the USA under the name Electro-Seismic Prospecting (Thompson and Gist, 1993)

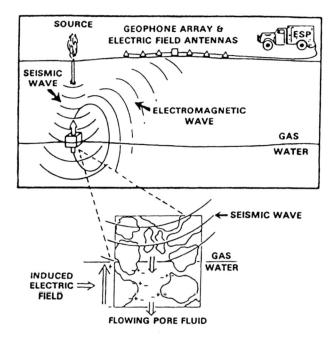

Figure 7.54 Schematic of electro-seismic prospecting as developed by Thompson and Gist (1993). Reproduced by permission

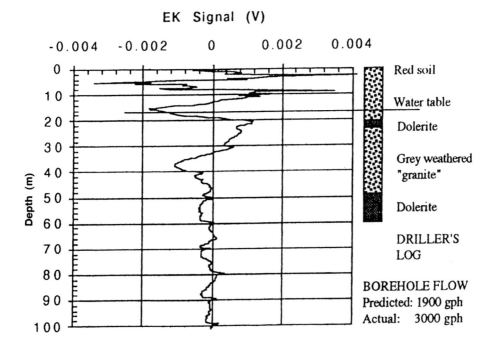

Figure 7.55 Typical electrokinetic signal at a borehole site in Zimbabwe. Courtesy of GroundFlow Ltd

for shallow (<1 km) hydrocarbon exploration, and in the UK by Clarke and Millar (GroundFlow Ltd; personal communication) for hydrogeological investigations, and by Butler *et al.* (1994) for mapping stratigraphic boundaries. Reference should be made to the various papers to see the fine differences between the various methods proposed by various workers.

An example of the use of EK surveying has been provided by GroundFlow Ltd, UK, and is shown in Figure 7.55. The figure shows the EK response at a borehole site in Zimbabwe. The determination of depth has been undertaken using an assumed P-wave velocity of 1250 m/s. Note the strong EK responses at and around the position of the water table. Strong electrocapillary signals from the vadose zone are evident above the water table, with electrokinetic signals derived from the saturated zone from 17 m to 50 m. At greater depths, there is very little signal response. The corresponding derived permeability–depth profile and cumulative flow responses are shown in Figures 7.56 A and B, respectively.

Comparison of predicted and actual borehole water flow rates has indicated that the EK-derived values provide a reasonable indication of the actual likely flow rates. However, this method in its modern form is only in its infancy. If the success of early trials is sustained, then this method promises to be a very useful additional tool in hydrogeological investigations.

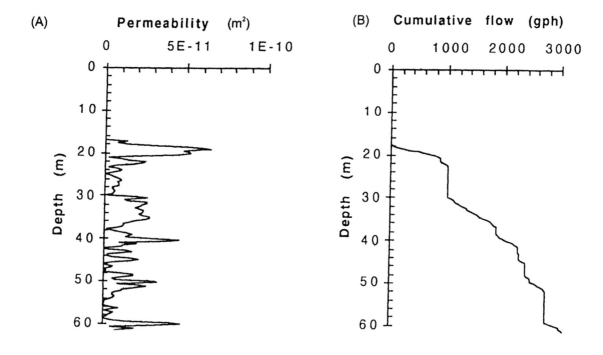

Figure 7.56 (A) Permeability–depth profile, and (B) predicted flow derived from electrokinetic data at a borehole site in Zimbabwe. Courtesy of GroundFlow Ltd

7.9   LEAK DETECTION THROUGH ARTIFICIAL MEMBRANES

The detection of leaks through man-made linings to lagoons and landfills has become increasingly important in the last few years. The intention of lining waste disposal sites is to ensure containment of the waste. However, small tears or cuts in the lining can end up with major leaks of contaminants. Various methods have been developed to try to locate holes in the linings so that they can be repaired or mitigation measures can be taken to remedy the contamination.

The general principle behind all the methods is that the artificial lining (high-density polyethylene (HDPE) geomembrane) is effectively a resistive barrier as long as there are no holes through it. A typical resistivity of an HDPE geomembrane is $> 10^7 \, \Omega \, \text{m}$.

Electrical current is passed between two electrodes, one of which is outside the membrane but in contact with the local groundwater, and one is within the waste or in a wet sand layer immediately overlying the geomembrane (Figure 7.57A). The same system can be used in water-filled lagoons (Figure 7.57B). Either a pair of potential electrodes or a roving single potential electrode (with the second one located with the external current electrode) is used to detect anomalous electrical potentials. These occur where electrical current is able to penetrate through holes in the geomembrane. Tears as small as

(A)

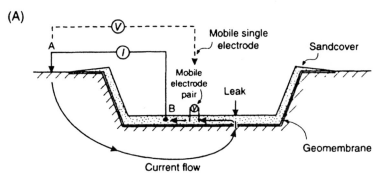

(B)

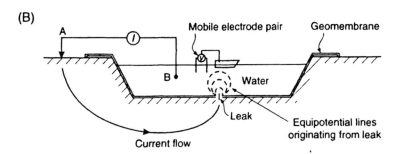

(C)

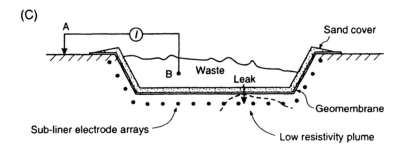

(D)

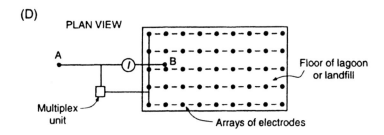

Figure 7.57 Leak detection systems in waste repositories or lagoons with man-made geomembrane liners. The survey layout in the case where (A) the liner is covered with a thin (0.5m) layer of sand and bentonite, and (B) the liner is filled with water. (C) A sub-liner leak detection system that uses permanently installed arrays of electrodes, with a typical plan layout shown in (D)

1 mm at a depth of 0.5 m beneath a sand layer have been located successfully.

In addition to post-construction leak detection, modern containment landfills are being built with sub-liner detection systems installed permanently. Arrays of electrodes connected by sealed multicore cables are buried at a depth of around 1 m below the geomembrane. A current circuit is provided by two electrodes as previously described. At the time of construction, measurements are made to check for holes in the liner so that they can be repaired before the disposal of any waste (Figures 7.57C and D). Instead of measuring potentials above the liner, as in the previous case, this technique permits the measurement of anomalous potentials *below* the liner. In addition, pseudo-sections of resistivity can be obtained by using the sub-liner arrays as sub-surface imaging arrays (see Section 7.4.2), without using the surface current electrode pair. At the construction of the disposal site, baseline measurements can be made when it can be assured that there are no leaks. Once waste has begun to be put into the facility, the sub-liner array can be monitored routinely and repeat pseudo-sections acquired.

The data are normalised so that they display changes in values relative to the baseline dataset. If a leak develops at a later stage in the facility's life, it can be identified from the routine monitoring. By using the array of sub-surface electrodes, the spatial distribution of any sub-liner pollutant plume can be determined so that remedial action can be taken. This may include constructing abstraction wells downstream of a leaking landfill so that the contaminated water can be pumped out and treated. Examples of such leak detection systems have been given by Mazác *et al.* (1990) and by Frangos (1994), among others. Many geophysical contractors are now offering remote leak detection surveys for containment waste repositories. The first landfill in the UK with a sub-liner leak detection system installed was constructed in 1995.

Chapter 8
Spontaneous (self) potential methods

8.1 INTRODUCTION

The *self-potential* or *spontaneous polarisation* (SP) method was devised in 1830 by Robert Fox who used copper-plate electrodes connected to a galvanometer to detect underground copper sulphide deposits in Cornwall, England. The method has been used since 1920 as a secondary tool in base metal exploration, characteristically to detect the presence of *massive* ore bodies, in contrast to the induced polarization method (see Chapter 9) which is used predominantly to investigate *disseminated* ore bodies. In recent years, the SP method has been extended to groundwater and geothermal investigations, and can also be used as an aid to geological mapping, for example, to delineate shear zones and near-surface faults.

The SP method ranks as the cheapest of the surface geophysical methods in terms of equipment necessary and amongst the simplest to operate in the field. Although the phenomenon of self-potentials is utilised more extensively in borehole well logging than in surface applications, the down-hole techniques will not be discussed further here.

8.2   OCCURRENCE OF SELF-POTENTIALS

The SP method is passive, i.e. differences in natural ground potentials are measured between any two points on the ground surface. The potentials measured can range from less than a millivolt (mV) to over one volt, and the sign (positive or negative) of the potential is an important diagnostic factor in the interpretation of SP anomalies, as described later.

Self-potentials are generated by a number of natural sources (next section), although the exact physical processes by which some are caused are still unclear. A summary of the common types of SP anomaly are listed in Table 8.1 with their respective geological sources. In addition to compositional variations, the geometry of geological structures can also create SP anomalies, and so the sources listed should only be used as a guide.

Table 8.1   Types of SP anomalies and their geological sources

Source	Type of anomaly
Mineral potentials	
Sulphide ore bodies (pyrite, chalcopyrite, pyrrhotite, sphalerite, galena)	
Graphite ore bodies	Negative ≈ hundreds of mV
Magnetite + other electronically conducting minerals	
Coal	
Manganese	
Quartz veins	Positive ≈ tens of mV
Pegmatites	
Background potentials	
Fluid streaming, geochemical reactions, etc.	Positive +/− negative ⩽ 100 mV
Bioelectric (plants, trees)	Negative, ⩽ 300 mV or so
Groundwater movement	Positive or negative, up to hundreds of mV
Topography	Negative, up to 2 V

Natural ground potentials consist of two components, one of which is constant and unidirectional and the other fluctuates with time. The constant component is due primarily to electrochemical processes, and the variable component is caused by a variety of different processes ranging from alternating currents induced by thunderstorms and by variations in the Earth's magnetic field, to the effects of heavy rainfall. In mineral exploration, components of the SP are called the *mineral potential* and the *background potential*, respectively. This terminology belies the usefulness of the so-called 'background' potentials as these can be used in geothermal and hydrogeological investigations as the main measured anomaly. The processes by which some of these potentials are generated are discussed briefly in the next section.

8.3   ORIGIN OF SELF-POTENTIALS

The common factor among the various processes thought to be responsible for self-potentials is groundwater. The potentials are generated by the flow of water, by water acting as an electrolyte and as a solvent of different minerals, and so on. The types of potentials are listed in Table 8.2 (with their alternative names where appropriate) and their mathematical definitions are given in Box 8.1. There are three ways of conducting electricity through rocks: by dielectric, electrolytic and electronic (ohmic) conduction. The electrical conductivity (σ, the inverse of resistivity) of porous rocks therefore depends on their porosity (and the arrangement of the pores) and on the mobility of water (or other fluids) to pass through the pore spaces (hence dependent upon ionic mobilities and solution concentrations, viscosity (η), temperature and pressure).

Table 8.2   Types of electrical potentials

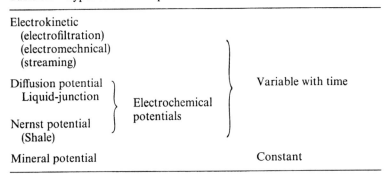

8.3.1 Electrokinetic potentials

An electrokinetic potential (E_k) forms as a result of an electrolyte flowing through a capillary or a porous medium, the potential being measured across the ends of the capillary (see Ahmad 1961). The potentials arising from this process are alternatively referred to as *electrofiltration, electromechanical* or *streaming* potentials.

Box 8.1   Electrical potentials

Electrokinetic:

$$E_k = \frac{\varepsilon\mu C_E \delta P}{4\pi\eta}$$

where: ε, μ and η are the dielectric constant, resistivity and dynamic viscosity of the electrolyte respectively; δP is the pressure difference; and C_E is the electrofiltration coupling coefficient.

Diffusion potential:

$$E_d = -\frac{RT(I_a - I_c)}{nF(I_a + I_c)}\ln(C_1/C_2)$$

where: I_a and I_c are the mobilities of the anions (+ve) and cations (−ve) respectively; R is the Universal Gas Constant ($8.314\,\text{J K}^{-1}\,\text{mol}^{-1}$); T is absolute temperatue (K); n is ionic valence; F is Faraday's Consant ($96\,487\ \text{C mol}^{-1}$); C_1 and C_2 are the solution concentrations.

Nernst potential:

$$E_N = -\frac{RT}{nF}\ln(C_1/C_2)$$

when $I_a = I_c$ in the diffusion potential equation.

Figure 8.1 *(opposite)* (A) Idealised electrofilteration SP profiles and maps for the following models (from Schiavone and Quarto 1984): (i) a vertical boundary with upwelling on the right; (ii) pumping from a well (injection into the well would produce the opposite sense of anomaly); (iii) horizontal boundary flow. Interfaces are marked by a contrast in streaming potential coefficients (C_1 and C_2). (B) Example of the SP anomaly produced by pumping from a well (after Semonov 1980)

According to Helmholtz's law, the flow of electric current is related to the hydraulic gradient and a quantity known as the electrofiltration coupling (or streaming potential) coefficient (C_E), which takes into account the physical and electrical properties of the electrolyte and of the network through the medium through which the electrolyte has passed. It is also important that the water flows parallel to either a geological boundary (Fitterman 1978, 1979a) or to its free surface (i.e. the water table).

Graphs of electrokinetic potentials obtained for different geological situations with characteristic values of C_E for each geological unit are given in Figure 8.1A (Schiavone and Quarto 1984). The potentials

(A)

(i)

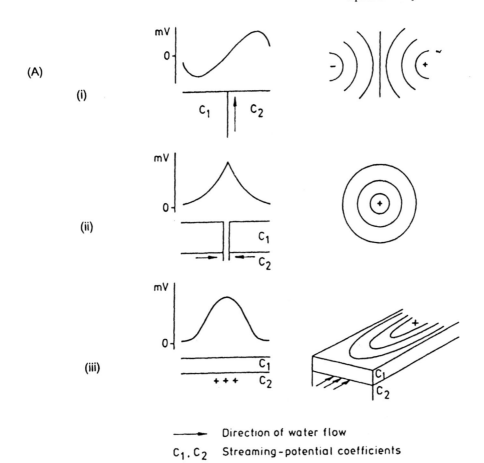

(ii)

(iii)

———➤ Direction of water flow

C_1, C_2 Streaming-potential coefficients

(B)

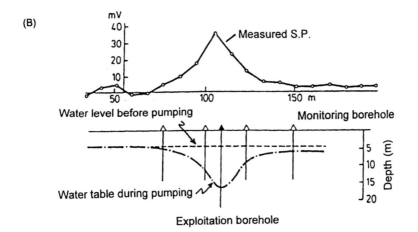

tend to increase in positiveness with the direction of water flow as the electric charge flows in the opposite direction. Consequently, negative charge flows uphill and can result in spectacular SP anomalies on topographic highs. Gay (1967) reported a potential of -1842 mV measured on a mountain top near Hualgayoc, Peru (mineralisation: *alunite); Nayak (1981) measured a value of -1940 mV on a hill of unmineralised quartzites in Shillong, India; Corwin and Hoover (1979) reported a value of -2693 mV on the peak of Adagdak Volcano, Adak Island, Alaska. This topographic effect requires a correction to be applied (see Section 8.5) particularly where slope angles exceed 20°.

Superimposed on the topographic flow generation of self-potentials can be potentials of the order of 5 mV which are caused by the brief but rapid percolation of water such as from heavy rainfall (Fournier 1989). Variation in soil moisture content may also produce locally variable SP signals (Corwin and Hoover 1979) with the electrode in the wetter soil often becoming increasingly positive (Poldini 1939). The small but measurable SP anomaly lasts only as long as the water flow. Potentials of the order of tens of millivolts can be induced artificially through pumping groundwater (Semenov 1980). The hydraulic gradient is increased by water abstraction, thereby increasing the rate of water flow towards the borehole; hence a positive anomaly is observed (Figure 8.1B).

A further factor that needs to be taken into account is *thermoelectric coupling*, which is the production of a potential difference across a rock sample throughout which a temperature gradient is maintained. The effect may be caused by differential thermal diffusion of ions in the pore fluid and of electrons and donor ions in the rock matrix (Corwin and Hoover 1979), a process called the *Soret Effect* (Heikes and Ure 1961).

8.3.2   Electrochemical potentials

Transient background *diffusion (or liquid-junction) potentials* (E_d) up to tens of millivolts may be due to the differences in the mobilities of electrolytes having different concentrations within groundwater. For this mechanism to explain the continued occurrence of background potentials, a source capable of maintaining imbalances in the electrolytic concentrations is needed, otherwise the concentrations differences will disappear with time by diffusion.

The *Nernst (shale) potential* (E_N) occurs when there is a potential difference between two electrodes immersed in a homogeneous solution and at which the concentrations of the solutions are locally different. It can be seen from Box 8.1 that the form of the equation for the Nernst potential is a special case of that for the diffusion potential and can easily be combined to form the *electrochemical potential*. For a solution of sodium chloride (NaCl) at 25°C with a ratio of concen-

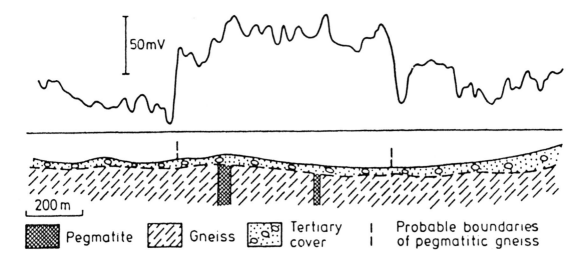

Legend:
▓ Pegmatite   ▨ Gneiss   ▒ Tertiary cover | Probable boundaries of pegmatitic gneiss

200 m

Figure 8.2 An SP profile across pegmatite dykes in gneiss. From Semenov (1980), by permission

trations of 5:1, the electrochemical potential is about $\pm 50\,mV$. The Nernst potential is of particular importance in well logging, in which case it is referred to as the *shale potential*. It can also be seen from Box 8.1 that the electrochemical potential is directly dependent upon the concentration differences (C_1/C_2) and temperature. The higher the temperature and the greater the concentration differences, the larger the electrochemical potential will be. For this reason, the measurement of self-potentials is important in the exploration for geothermal resources (e.g. Corwin and Hoover 1979) where the temperatures are obviously elevated and the concentrations of salts within the groundwater are also likely to be high (see also Section 8.6.3).

Further electrochemical potentials are attributable to adsorption of anions on to the surface of veins of quartz and pegmatite and are known as *adsorption (or zeta) potentials*. For example, an anomaly of up to $+100\,mV$ has been measured over vertical pegmatitic dykes within gneiss (Figure 8.2) (Semanov 1980; cited by Parasnis 1986). In addition, adsorption potentials may account for the observed anomalies over clays where the solid–liquid double layer may generate a potential.

8.3.3 Mineral potentials

Of most importance in the use of SP in mineral exploration is the *mineral potential* such as that associated with massive sulphide ore bodies. Large negative anomalies can be observed particularly over pyrite and chalcopyrite and other good electronic conductors. (Mineral potentials have also been observed over sphalerite, somewhat surprisingly, as it is a poor conductor.)

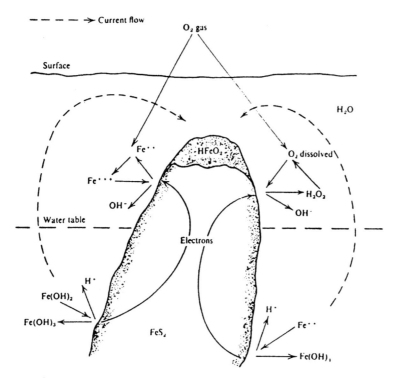

Figure 8.3 Physicochemical model proposed by Sato and Mooney (1960) to account for the self-potential process in a massive sulphide orebody. Reproduced by permission

Sato and Mooney (1960) have provided the most complete explanation of the electrochemical processes (Figure 8.3) which cause the observed self-potential anomalies, although no hypothesis is yet able to account for all the observed mineral potentials. Where a sulphide orebody straddles the water table, a cathodic electrochemical half-cell forms by the chemical reduction of the ions in the surrounding electrolyte, i.e. they gain electrons. In contrast, below the water table, an anodic electrochemical cell operates in which oxidation is dominant and ions lose electrons. The role of the massive orebody is to permit the flow of electrons from the lower half of the orebody to the upper half. The net result of this process is that the upper surface becomes negatively charged (hence the negative SP anomalies) and the lower half becomes positively charged. However, the fact that this hypothesis does not explain all the occurrences of self-potentials indicates that the actual physical processes are more complicated and not yet fully understood (Kilty 1984). This is even truer when more than one orebody is involved resulting in complex SP anomaly shapes (e.g. Becker and Telford 1965).

Where clay overlies a massive sulphide orebody, the mineral potential may be suppressed to the extent that no anomaly is observed (Telford *et al.* 1990). This may be a result of the adsorption potential (which tends to be positive) having the same amplitude as the mineral potential (negative polarity) and thus cancelling each other out.

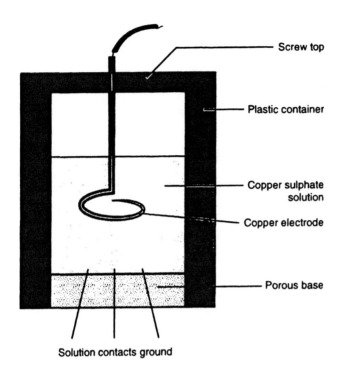

Figure 8.4 Cross-section through a porous pot electrode

Screw top

Plastic container

Copper sulphate
solution

Copper electrode

Porous base

Solution contacts ground

8.4 MEASUREMENT OF SELF-POTENTIALS

The measurement of self-potentials is very simple. Two non-polaris-able porous-pot electrodes are connected to a precision multimeter with an input impedance greater than 10^8 ohms and capable of measuring to at least 1 mV. Each electrode is made up of a copper electrode dipped in a saturated solution of copper sulphate which can percolate through the porous base to the pot (Figure 8.4) in order to make electrical contact with the ground. Alternatively, a zinc electrode in saturated zinc sulphate solution or silver in silver chloride can be used.

There are two field techniques, both of which are carried out at right-angles to the suspected strike of the geological target. The *potential gradient method* uses two electrodes, at a fixed separation, typically 5 m or 10 m, between which the potential difference measured is divided by the electrode separation to give a potential gradient (mV/m). The point to which this observation applies is the midpoint between the two electrodes. The two porous pots are leap-frogged along the traverse but care has to be taken to ensure that the correct polarity of the potential is recorded. The procedure in the *potential amplitude method* is to keep one electrode fixed at a base station on unmineralised ground and to measure the potential difference (in mV) between it and the second one which is moved along the traverse. This removes the problems of confusing polarity and accumulating errors. Care should be taken to ensure that the temperature of the electrolyte in the mobile pot does not differ significantly from that in the reference electrode, or else a potential difference will be produced. The temperature coefficient for copper–copper sulphate is about 0.5 mV/°C (about 0.25 mV/°C for silver–silver chloride electrodes).

As mentioned above, the self-potential consists of a static and a variable alternating component. The latter, which can have frequencies typically in the range of 5–10 Hz, is caused by atmospheric effects and its long period component may have amplitudes that are of the same order as the static mineral potential. Where this signal is present, the mineral potential can be resolved by taking measurements along the same profile at different times of the day and averaging the results. Electrical noise can also result if measurements are made too soon after heavy rain or too close to running surface water, as streaming potentials may then swamp any mineral potentials.

The maximum depth of sensitivity of the SP method is around 60–100 m, depending on the depth to the orebody and the nature of the overburden.

Self-potential measurements can also be made over water to measure streaming potentials. The porous pot electrodes are enclosed in special containers so that they can be towed through water without causing serious loss of electrolyte from the pots. This method will only work where there is little current flow (lateral of vertical) within the water column (Ogilvy *et al.* 1969); the amplitude of any SP anomaly obtained within a saline water body (resistivity $0.3-1\,\Omega\,m$) tends to be very small.

8.5   CORRECTIONS TO SP DATA

Self-potentials measured over a large area (of the order of many square kilometres) may have a regional trend due to 'telluric' currents (see Section 11.5) of $\geqslant 100\,mV/km$. The mineral potential may be superimposed upon this regional gradient. Thus, to interpret the anomaly due to mineralisation, its anomaly has to be isolated in much the same way as gravity residuals are obtained (see Section 2.6.1). Regional corrections must be applied before any adjustment for topography is made. However, for a local survey whose length is small in comparison with the regional wavelength, removing regional trends is usually unnecessary.

The association of negative anomalies with topographic highs has already been mentioned (Section 8.3.1). Telluric currents are also affected by changes in elevation. The combined effects are extremely hard to quantify explicitly but can be corrected for in a general manner as prescribed by Yüngül (1950) and discussed further by

Bhattacharya and Roy (1981) and Bhattacharya (1986). If the surface slope of a survey area is large (> 20°), the SP minimum may be well displaced from its cause and subsequent drilling may miss the orebody completely. In any attempts to correct data for either regional or topographic effects, the SP anomaly for an individual polarised body should be isolated. If the observed anomaly is due to the superposition of a number of anomalies from different geological sources (and hence with different shapes and polarities), corrections cannot be carried out and the locations of the tops of the orebodies must be taken as very approximate and alternative geophysical methods used to try to delimit the geological structure more explicitly.

It may also be necessary to make allowance for the effects of bioelectric potentials caused by vegetation. Passing from bare ground into an area of vegetation can cause a negative potential of several hundred millivolts, comparable to a mineral potential due to a sulphide orebody. Basic field observations should clarify the situation.

8.6 INTERPRETATION OF SELF-POTENTIAL ANOMALIES

SP anomalies are often interpreted qualitatively by profile shape, amplitude, polarity (positive or negative) and contour pattern (see Figure 8.1). The top of an orebody is then assumed to lie directly beneath the position of the minimum potential. If the axis of polarisation (i.e. the axis between the cathode and anode on the orebody) is inclined from the vertical, the shape of the profile will become asymmetrical with the steepest slope and positive tail both lying on the downdip side (Figure 8.5).

Complications arise when two or more geological features give rise to superimposed SP anomalies. One such example (Nayak 1981) is shown in Figure 8.6. An anomaly over graphitic phyllites is characteristically large (− 740 mV) owing to mineral electrochemical potentials. A second anomaly (− 650 mV) has been produced by electrokinetic potentials associated with water flow through permeable disintegrated conglomerates. However, if similar-sized bodies are present but with different dips, the resultant anomaly can be used to resolve between them. Consider two graphite bodies in gneiss (Figure 8.7) in two different models (Meiser 1962). The first is where the graphites dip towards each other in a synclinal structure, in which case the negative centres associated with each polarisable body are well separated, resulting in an anomaly with two negative minima. The second is where the graphite bodies dip away from each other in an anticlinal structure, in which case the two negative centres are very close together and may even combine to form one large negative minimum. The separation between the two minima is equal to the separation of the tops of the graphite bodies.

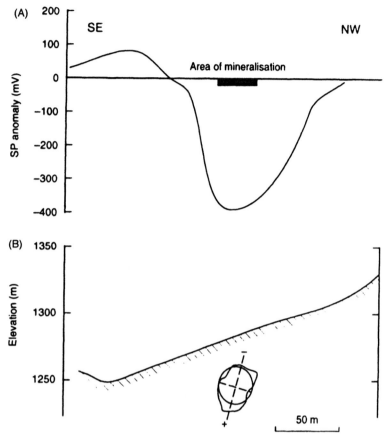

Figure 8.5 (A) Weiss SP anomaly in Ergani, Turkey, with the causative orebody shown schematically in (B). Note that the axis of polarization is inclined uphill. After Yüngül (1950), by permission

The next level of interpretation is to approximate the shape of the orebody to one of known geometry, usually a sphere or an inclined rod, with an assumed direction of polarisation. The direct or forward approach is to calculate the corresponding electrical potential for the model and to compare the synthetic anomaly with that observed. The theoretical basis of quantitative interpretations of SP anomalies over a polarised sphere is attributed to Petrovski (1928) and developed by de Witte (1948), over a bar by Stern (1945) and over a dipping plate by Meiser (1962). Other forms of model and revised methods of calculation are currently being developed (e.g. Hongisto 1993). The model is then adjusted until the two anomaly profiles agree within prescribed statistical limits. While this method may work for very well constrained data, if the actual geological feature causing the SP anomaly does not conform to a given geometric shape, the problem becomes very much more complicated mathematically and numerical methods of computation are required (Fitterman 1979b). However, the time and expense of such an approach may not be justified; it may be more prudent to use another geophysical method, such as induced

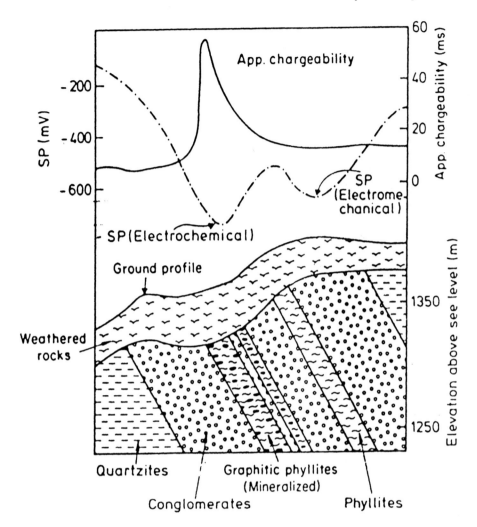

Figure 8.6 Two SP minima with different causes: one produced by electrochemical processes associated with mineralised graphite phyllites, and one caused by electrokinetic processes due to the flow of water in per-meable disintegrated conglomerates. From Nayak (1981), by permission

polarisation or microgravity, for instance, to delimit the causative body.

The inverse approach is to manipulate the observed anomaly to produce a model. This method may be used to estimate the size of the geological feature prior to other, more detailed, geological and geophysical investigations (Sill 1983). The approach is to assume that the geological feature conforms to a given geometric form (Figure 8.8) for which depths to the centre of the body may be estimated using a half-width technique. Unfortunately this method is notoriously inaccurate – the most serious limitation of this approach being that the width of the anomaly may be more indicative of the physical breadth rather than the depth of the body and so depth estimates may be in error by as much as $\pm 100\%$. Examples of actual graphite bodies

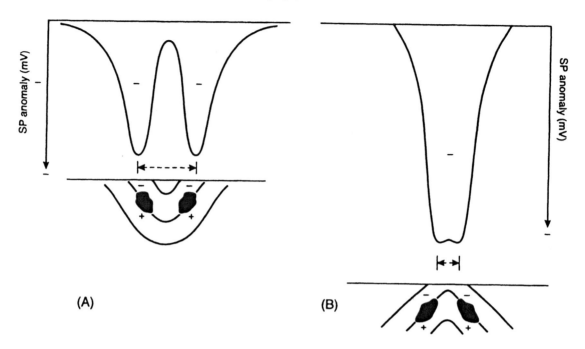

and their respective SP anomalies are shown in Figure 8.9. The observed anomaly in Figure 8.9B is the envelope of the anomalies due to the individual geological components A–D.

Figure 8.7 SP anomalies due to (A) two graphite bodies with axes of polarisation inclined away from each other (in syncline), and (B) inclined towards each other (in anticline). After Meiser (1962), by permission

8.7   APPLICATIONS AND CASE HISTORIES

8.7.1   Geothermal

The hydrogeological regimes associated with geothermal fields are often complex. Water bodies can have highly differing temperatures and salinities, and be highly mobile (e.g. Cioni *et al.* 1992). Consequently, streaming potentials may be well developed and hence may be measured using the SP method (see also Anderson and Johnson (1976) and Zohdy *et al.* (1973)).

8.7.1.1   *Roosevelt Hot Springs, Utah, USA*

A self-potential profile carried out across the Dome Fault Zone, Roosevelt Hot Springs, Utah, is shown in Figure 8.10 (Corwin and Hoover 1979). Alunite and pyrite occur in the zone, both of which normally produce negative polarity anomalies that may be evident on the profile within 1 km west of the reference electrode position. The area within 1 km to the east of the reference electrode has a positive anomaly of about +80 mV which is thought to be due to the

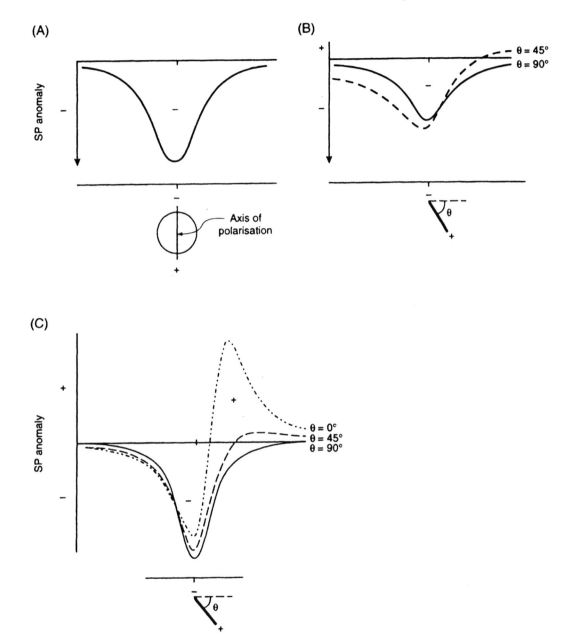

(A)

(B)

(C)

Figure 8.8 Self-potential anomalies associated with (A) a sphere, (B) a dipping plate (Parasnis 1986), and (C) a dipping rod (Telford *et al.* 1990)

geothermal activity. Comparison of the thermal gradient profile with the SP transect indicates that the axis of the thermal gradient anomaly is coincident with the position of the reference electrode. The negative potentials associated with the mineralised areas within the zone may be degrading the anomaly due to the geothermal activity. The geothermal SP anomaly results from streaming

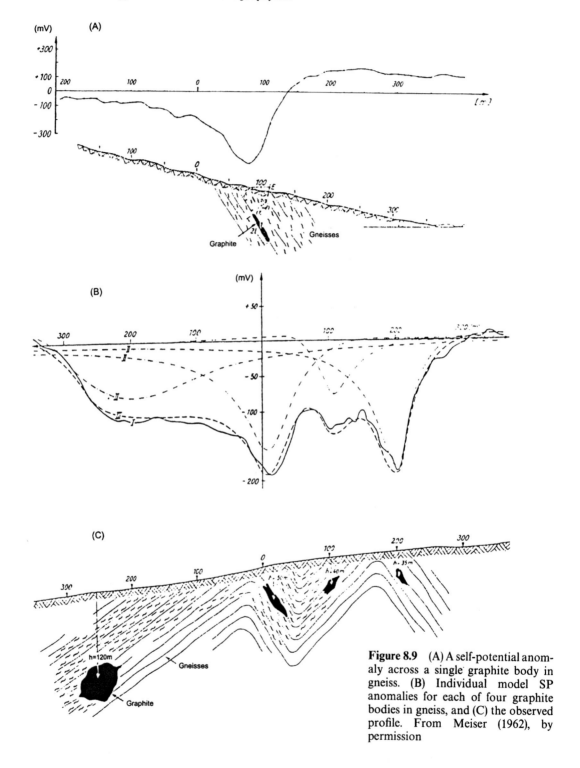

Figure 8.9 (A) A self-potential anomaly across a single graphite body in gneiss. (B) Individual model SP anomalies for each of four graphite bodies in gneiss, and (C) the observed profile. From Meiser (1962), by permission

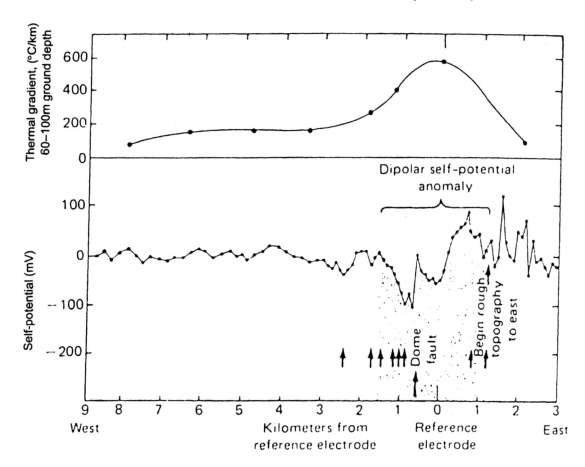

Figure 8.10 Thermal gradient and self-potential profiles over the Dome Fault Zone, Roosevelt Hot Springs, Utah. Arrows denote points at which mapped faults cross the SP survey line. From Corwin and Hoover (1979), by permission

potentials being driven by the convective cells within such a zone and also from elevated diffusion potentials due to the higher temperatures.

8.7.1.2 *Cerro Prieto geothermal field, Mexico*

Corwin and Hoover (1979) have reported on a significant SP anomaly of some 150 mV peak-to-peak amplitude associated with the Cerro Prieto geothermal field in Mexico (Figure 8.11). The anomaly is centred over the Hidalgo Fault which is thought to provide a major conduit for geothermal fluids. The actual geology of the field is still not fully understood, but it is clear that the production zone, which generates about 75 MW of electrical energy, is located between the maximum and minimum points of the SP anomaly which are 8 km apart (Fitterman and Corwin 1982). The width of the SP anomaly is possibly owing to several faults acting as geothermal conduits rather than just the Hidalgo Fault.

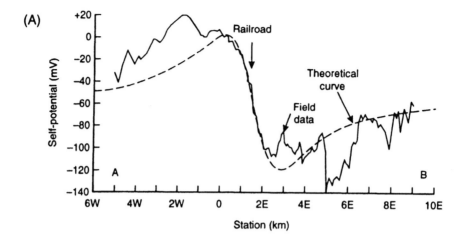

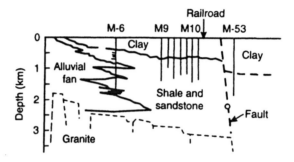

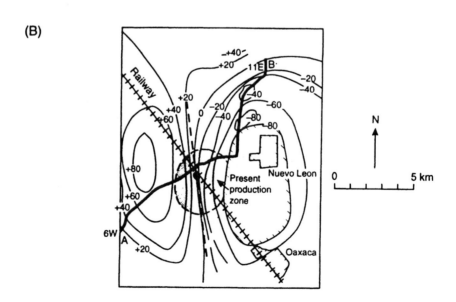

8.7.2 Location of massive sulphide ore bodies

8.7.2.1 *Kimheden orebody, Skellefte, northern Sweden*

The Kimheden orebody, which is mainly pyrite, occurs in steeply dipping sheets along a ridge in moraine-covered sericite–quartzite. The tops of the ore fragments are at no more than about 10 m depth. The SP contours (Figure 8.12) show several clear linear features which correlate extremely well with the known position of the

Figure 8.11 (*opposite*) (A) Self-potential anomaly along the profile A–B over the Cerro Prieto Geothermal Field, Mexico, with a simplified geological cross-section. After Corwin and Hoover (1979), by permission. (B) Self-potential map over the same field (profile line A–B marked) showing a distinct positive–negative couplet with the geothermal production area being midway between the two parts of the anomaly. After Fitterman and Corwin (1982), by permission

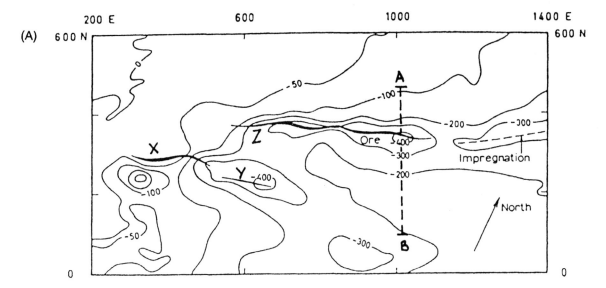

(A)

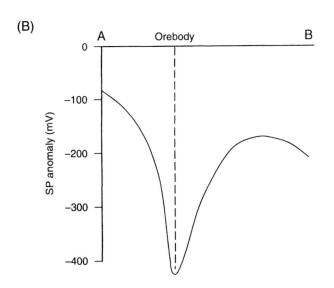

(B)

Figure 8.12 (A) Self-potential map of the Kimheden pyrite orebody in northern Sweden. From Parasnis (1966), by permission. (B) The SP anomaly across profile A–B. Map contours are in mV

orebodies (Parasnis 1966) with one exception. The most western orebody (X) was found to have a particularly high resistivity which may explain the absence of any anomaly (see Section 8.3.3).

8.7.2.2   Copper ore at Chalkidiki, northern Greece

The Chalkidiki area of northern Greece consists of gneisses and schists that are strongly sheared and intruded by mineralised granitic–granodioritic and ultrabasic structures (Figure 8.13). Pyrite, galena and sphalerite are associated with the acidic intrusions, and magnetite and chromite with the basic intrusions. Copper minerals and pyrite have resulted from Tertiary volcanic activity and are particularly associated with lava flows of trachytes, andesites and porphyritic granodiorite, and have been concentrated within a dense network of fractures and faults. There are three recognisable zones of copper mineralisation of which the shallowest consists of an oxidised leached zone (about 1% Cu). Below this is a zone of secondary enrichment 2–3 m thick in which copper concentrations are as high as 20% (malachite, azurite, cuprite, etc.). The lowest zone, which represents the primary mineralisation of chalcopyrite, pyrite, bornite and *syngenetic magnetite, begins at around 20–30 m depth and extends down to at least 300 m below surface. The magnetite within the volcanic rocks gives rise to distinctive magnetic anomalies, the maxima of which occur just within the margins of the trachyte dome. The minima of the self-potential anomalies occur specifically over the copper orebodies. Where the oxidisation of the copper sulphide is marked, which produces a low magnetic susceptibility, there is a corresponding cusp in the magnetic profile. This example also highlights the complementary use of two different geophysical methods (based on Zachos (1963) and Parasnis (1966)).

8.7.2.3   Sulphide orebody at Sariyer, Turkey

A classic, and oft cited, example of a self-potential anomaly over a sulphide orebody is the one given by Yüngül (1954) for a complex orebody at Sariyer in Turkey (Figure 8.14A). Chalcopyrite and pyrite occur in varying concentrations within a massive deposit within andesite and below Devonian schist. The area is characterised by a steep surface gradient which, if not corrected for, displaces the SP minimum downhill. The orebody comprises four regions, of which the one furthest downhill is pyritised and the three remaining zones have decreasing concentrations of copper from 14% on the downhill side to 1–2% on the upslope side. Each of these zones may be represented by a sphere whose SP anomaly contributes to the total anomaly observed (Figure 8.14B). The inflection points present on the observed profile can be modelled by changing the separation between the various model spheres.

Figure 8.13 *(opposite)* (A) Map of the solid geology and self-potential values (in mV) in Chalkidiki, northern Greece. (B) SP and magnetic total field anomaly (in nT) profiles r–r' and Δ–Δ. From Zachos (1963), by permission

(A)

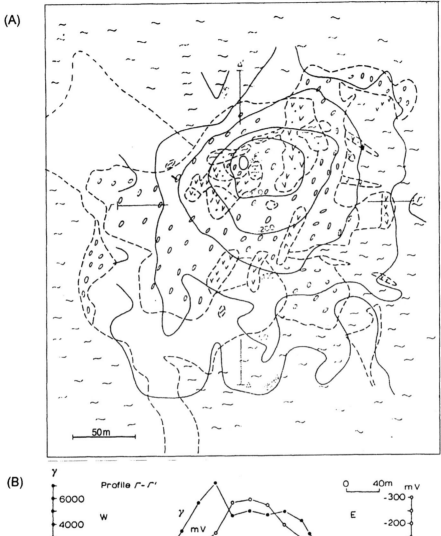

50m

(B)

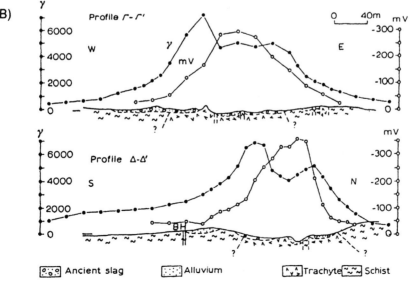

(A)

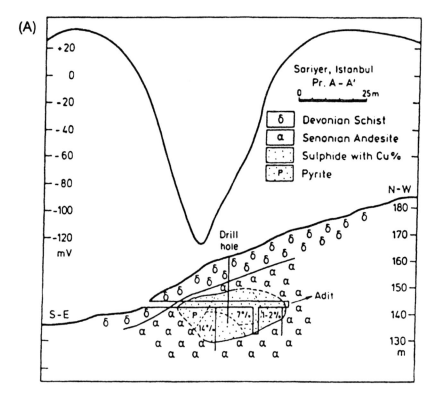

(B)

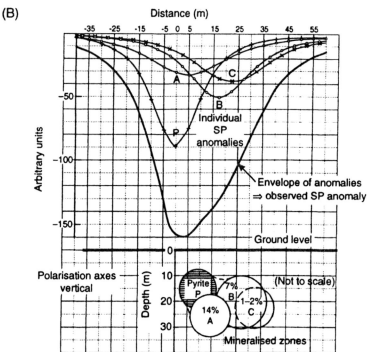

Figure 8.14 (A) Self-potential anomaly across a pyrite orebody at Sariyer, Turkey. The borehole is located at the location of the topographically corrected SP minimum. (B) An equivalent model assuming each segment of the orebosy conforms to a sphere with its axis of polarisation vertical with the corresponding individual SP anomalies and their envelope. After Yüngül, (1950), by permission

This example demonstrates that subtle inflection points within an anomaly may be extremely important and should not be dismissed as being insignificant. In addition, the field data must be of very high quality in order to isolate such subdued features. For another example of the use of SP data in pyrite orebody investigations, see Logn and Bölviken (1974).

8.7.3 Hydrogeology

Geothermal applications involving groundwater have already been discussed (Section 8.7.1) and the use of SP measurements in groundwater borehole testing were mentioned in Section 8.3.1. Other examples in hydrogeology are the use of SP measurements to detect the sites of leakages associated with man-made and natural dams (Ogilvy *et al.* 1969; Bogoslovsky and Ogilvy 1970a,b; Butler and Llopis 1990; Jansen *et al.* 1994), percolation of fresh groundwater through quartz gravels (Nayak 1981), and in the study of groundwater movement (Schiavone and Quarto 1984), for example.

A recent example of a more regional SP survey has been given by Fournier (1989) who has used the method in conjunction with electrical resistivity surveys to investigate the hydrogeology of volcanic aquifers in the Chaîne des Puys area of central France. He was able to delimit the catchment area of each spring, and found that an axial graben was the main aquifer and identified possible zones vulnerable to pollution. He also found that the water table provided the main source of SP anomalies and that their shape indicated the form of the water table (Figure 8.15). A geological cross-section of part of northern Chaîne des Puys, with its corresponding SP profile, is shown in Figure 8.16. Note that the negative maximum coincides with the topographic high. For the area, Fournier was able to categorise his SP anomalies in terms of hydrogeological significance (Table 8.3). It remains to be seen how applicable these associations are for other geographic areas.

Associated with the hydrogeological applications of SP measurements, is their possible use in earthquake prediction where the active fault planes are at very shallow depths and where instrumentation is both feasible and viable. It has been noted previously (Rikitake 1976) that natural electric fields within the ground may change prior to an earthquake, and therefore could act as sensible precursors to some types of seismic events. Renata (1977) has attempted to argue that these electrical changes may be due to electrokinetic potentials associated with stress buildup/relaxation processes.

Further evidence of what is now called the 'seismoelectric response' has been presented by Butler *et al.* (1994). In an experiment using a sledge hammer source and electric field receivers, an electromagnetic signal was observed when the seismic wave impinged upon a boundary between organic-rich fill and impermeable glacial till. The

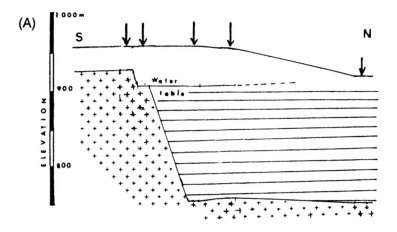

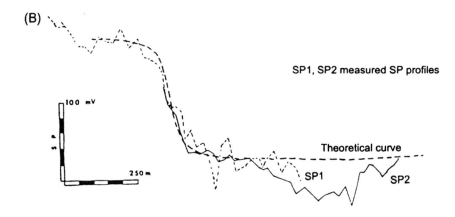

depth to the interface (confirmed by drilling) was from 1 m to 3 m. It is believed that the electrical response was a transient streaming potential produced by a seismically induced flow of pore water at the interface. The critical combination in this experiment was the juxtaposition of a permeable material and an aquitard at shallow depth. It is postulated that this technique could be used to help map aquitards or the boundaries of permeable formations where they terminate against an impermeable material.

Figure 8.15 (A) Geological cross-section of part of the axial graben in the Chaîne des Puys, central France, as derived from vertical electrical soundings (arrows). (B) The corresponding self-potential anomaly. From Fournier (1989), by permission

8.7.4 Landfills

Steep-sided landfills containing significant volumes of highly conductive leachate which may leak through the margins are known to generate significant SP anomalies (Coleman 1991). The reasons for this are primarily twofold. First, there is an ionic imbalance in concentrations each side of the landfill boundary, with high values within the landfill (due to the leachate) and low values outside (within

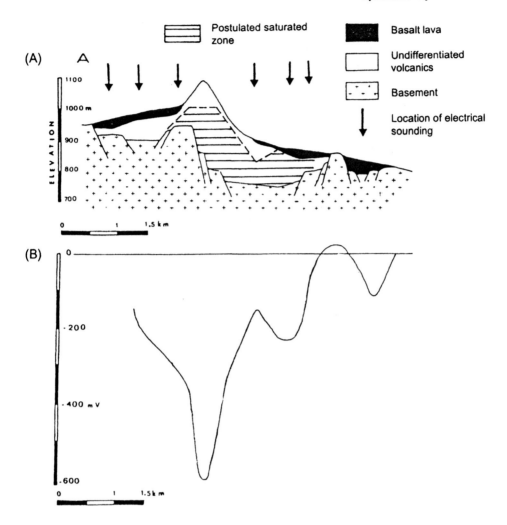

Table 8.3   Hydrogeological interpretation of SP anomalies (Fournier 1989)

Aspect	Location	Hydrogeological significance
High horizontal SP gradient	On flat topography	Lateral limit of an unconfined acquifer
	On a volcano flank	Ascent of water table in volcanic cone
SP minimum	On flat topography	Palaeovalley axis
	On a volcano flank or summit	Crest of the water table and underground watershed line
SP maximum	Above an unconfined acquifer	Depression of the water table due to better drainage
	Between two unconfined acquifers	Watershed line due to crest of the impervious basement

Figure 8.16 (A) Geological cross-section across a volcanic dome in the northern part of the Chaîne des Puys, central France, as derived from vertical electrical soundings (arrowed. (B) The corresponding self-potential anomaly. From Fournier (1989), by permission

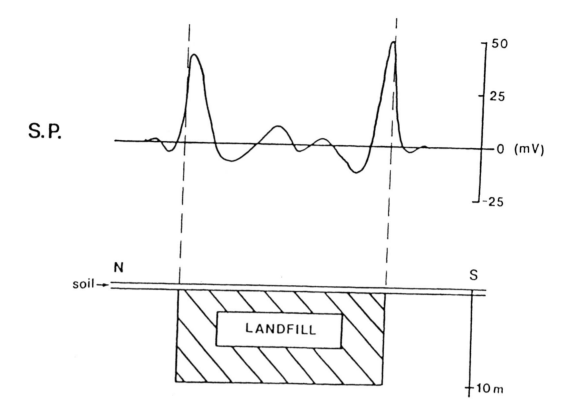

Figure 8.17 SP anomaly over a closed landfill, showing the typically larger anomalies associated with the landfill boundaries compared with those observed in the interior. From Coleman (1991), by permission

the natural groundwater). In order to equilibrate the imbalance, diffusion occurs. As ions (which are charge carriers) move, their movement constitutes an electromotive force – an electrochemical potential. Secondly, if the leachate physically flows from the landfill outwards, a streaming potential may be generated. Consequently, if both electrochemical and streaming potentials are generated, it is highly probable that a measurable SP anomaly would be present at the landfill boundaries.

Coleman (1991) has produced an example of SP anomalies associated with the margins of a landfill (Figure 8.17). While there are small SP anomalies within the landfill, the largest SP events are associated with the boundaries, where the ionic imbalances are greatest and the rate of flow of leachate is most pronounced.

8.7.5 Leak detection within embankments

The SP method has been used to detect leaks in earth dam embankments for decades (e.g. Bogoslovsky and Ogilvy, 1970a,b). Although less well publicised, marine or boat-borne SP surveys have been made for at least 60 years. Where groundwater flows through such a structure by finding the path of least resistance (e.g. piping), electrokinetic

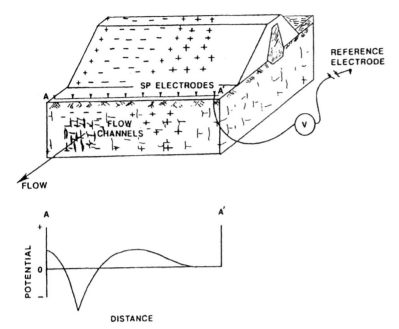

Figure 8.18 Schematic of the concept of SP anomalies generated by features associated with seepages through earth dams. From Butler and Llopis (1990), by permission

streaming potentials may be generated with sufficient magnitude to be detectable. The concept of the generation of SP anomalies by leaks is shown schematically in Figure 8.18. Negative charges are associated with locations where leaks enter a dam, or above seepage paths where the flow is generally horizontal or descending. In contrast, positive anomalies may occur where the flow is generally ascending and where surface seepage takes place. Consequently, the presence of either or both a negative and/or positive anomaly may be physically significant.

Butler and Llopis (1990) have described an example of an SP survey to examine possible leaks at the Mill Creek Dam and Reservoir, Washington, USA (Figure 8.19). Since its first test filling in 1941, considerable loss of stored water has been noticed. Attempts to change or stop the seepage have been unsuccessful. A concrete cutoff wall with flanking grout curtains was installed. The concrete wall was built on top of a massive basalt which underlies laterally variable conglomerate. Since the construction of these structures, the reservoir had not been filled but seepage was still evident.

In 1984, a geophysical investigation was instigated to detect anomalous seepage before, during and after test filling of the reservoir. Two SP electrode arrays comprising 85 metallic rods (copper-clad steel grounding rods) at 15 m spacings were installed two months before the first set of SP readings and four months before the first test filling. The reference electrode was located upstream. During the test, the reservoir level was raised by 10 m. The SP profile acquired along

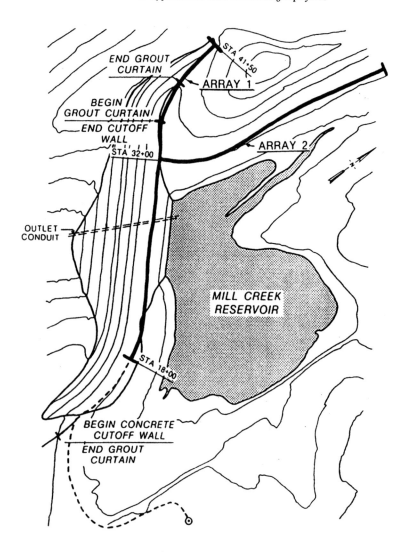

Figure 8.19   Plan of Mill Creek Dam and Reservoir, Washington, USA, showing the locations of SP survey lines 'Array 1' and 'Array 2', grout curtains and the concrete cutoff wall. From Butler and Llopis (1990), by permission

Array 1 (see Figure 8.19) before the filling of the reservoir is shown in Figure 8.20. Several notable features are evident:

- There is an anomaly of about $-380\,\text{mV}$ associated with a $1.07\,\text{m}$ diameter outlet conduit located about $20\,\text{m}$ below the electrode array.
- Three separate zones along Array 1 are distinguished by different base levels (indicated by dashed lines).
- The boundaries between zone 1/zone 2 and zone 2/zone 3 coincide approximately with the dam/right abutment contact and the end of the grout curtain, respectively. The anomaly associated with the boundary between zone 1 and zone 2 is thought to be due to a lateral change in material type. The zone 2/zone 3 boundary is

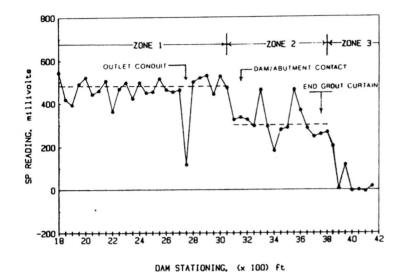

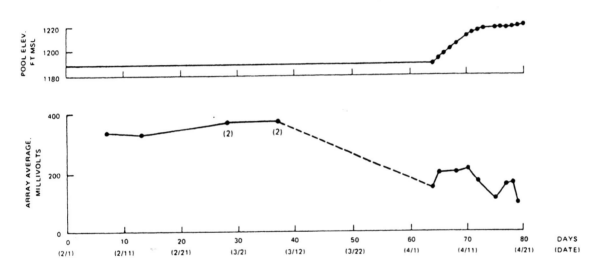

Figure 8.20 Array-1 SP data acquired prior to the start of the filling of the Mill Creek Reservoir in Figure 8.19. From Butler and Llopis (1990), by permission

Figure 8.21 Comparision of the reservior water level with SP array average as a function of time (see Figure 8.19). Raising of the water level started on the 5th day of April (4/5). From Butler and Llopis (1990), by permission

caused by a lateral change in the groundwater flow regime as a result of the presence of the cutoff wall and the grout curtain.

As a test of the SP response to the raising of the level of the reservoir, values of the array averages responded immediately to the increase in water level, as indicated by the data in Figure 8.21. No SP values were measured between 9 February and 5 March; the test fill began on 5 March. Note that the trend in SP anomaly with time is negative.

Further analysis of the SP data yielded an indication of several locations, most notably at location 18 (Figure 8.20) and at 30 + 50.

(A)

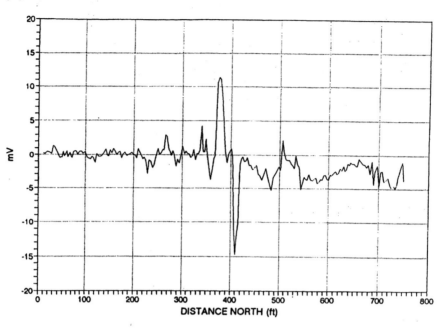

(B)

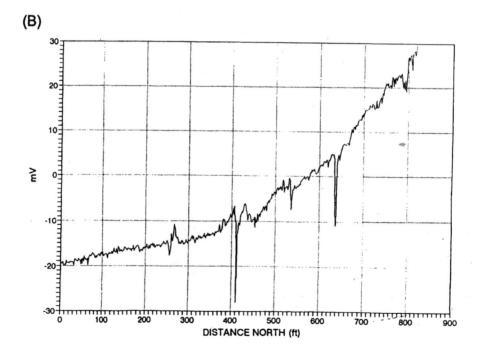

The cause at station 18 was thought to be seepage occurring under the cutoff wall in a very localised zone. The postulated seepage at 30 + 50 coincides with the end of the cutoff wall/grout curtain. The SP data, although not conclusive in themselves, provided valuable spatial and temporal data which aided a better understanding of the hydrological conditions of the site.

Marine surveys have been undertaken typically with a pair of electrodes (gradient array) behind a boat towed at the surface or at depth above the water bottom. Alternatively, using an electrode fixed on the water bottom, a mobile electrode is dragged along the bottom. Occasionally, a borehole SP logging tool is used as the mobile tool. Other configurations include a pair of electrodes separated by only 1.25 m, suspended beneath a boat or lowered to the water bottom and are used with a third, fixed electrode planted by a diver remote from the profile line. An alternative to the surface-towed gradient arrays is the benthic gradient array where two electrodes are towed behind a weighted towfish and are dragged along the water bottom in direct contact with the sediment. A development of this is the use of a fixed electrode, implanted by deploying the weighted electrode from the boat and letting it sink under gravity, and towing a weighted eel with a second electrode in its hind part, also in contact with the sediment.

In a recent study (Jansen *et al.* 1994) it was found that the benthic single roving electrode configuration produced much higher quality data than using a benthic gradient array. Examples of data acquired using the two different methods over the same profile are shown in Figure 8.22. In the benthic gradient array dataset (Figure 8.22A), one main negative anomaly can be seen at location 410 ft north. The same feature is evident on the benthic single roving electrode data (Figure 8.22B). However, an additional feature is clearly seen at location 640 ft north. Other features are evident on both datasets but appear to be less significant but do demonstrate the degree of repeatability, despite using a different electrode configuration. The reason the potential increases with the roving electrode is related to the changing distance from the remote electrode. In the gradient array, it is the potential measured across the pair of electrodes at a separation of 40 ft (12.2 m). The units of measurement should be given as mV/unit length (e.g. per dipole length). In the roving electrode case, it is not a complicated task to remove the effects of distance on the data to residualise the data in order to highlight anomalous zones.

Figure 8.22 (*opposite*) Examples of boat-borne SP profiling along a test line in a fresh-water filled reservoir, using (A) a benthic gradient array with a 40 ft (12.2 m) electrode spacing, and (B) a benthic single roving electrode with remote reference electrode. From Jansen *et al.* (1994), by permission

Chapter 9
Induced polarisation

9.1   INTRODUCTION

The phenomenon of *induced polarisation* (IP) is reported to have been noted by Conrad Schlumberger as early as 1912. The induced polarisation method has been used since the late 1940s, having been developed during the Second World War by William Keck and David Bliel as part of a US Navy project to detect mines at sea (Grow 1982). One aspect of IP, known as the *overvoltage effect*, has been known about since the nineteenth century.

In the 1980s there were considerable advances in instrumentation, and sophisticated techniques such as *complex resistivity* and *spectral IP* (Pelton *et al.* 1978) were developed, although there is still much research to be done to relate geological causes to the observed

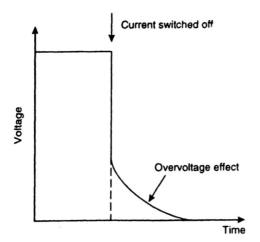

Figure 9.1 The overvoltage effect produced by induced polarisation after an applied current is switched off

geophysical data. Thus, there are difficulties in attempting quantitative interpretation.

The main current application of IP prospecting is in the search for disseminated metallic ores and, to a lesser extent, groundwater and geothermal exploration. Since the early 1990s, there has been an increased interest in the possible use of IP methods in environmental applications, although the work is still very much at a research stage.

Measurements of induced polarisation are made using conventional electrical resistivity electrode configurations (see Chapter 7) involving two current and two non-polarisable potential electrodes. When the applied current is switched off, the voltage between the potential electrodes takes a finite and measurable time (seconds to several minutes) to decay to zero (Figure 9.1) because the ground temporarily stores charge (i.e. becomes polarised) and acts somewhat like a capacitor. When the current is switched back on, the voltage does not peak instantaneously but builds up over the same time period (the *rise-time*) to its maximum applied value. The voltage decay and rise-time are dependent upon both instrumental and geological factors, and are thus diagnostic of the nature of the ground, as will be discussed later.

There are four systems of induced polarisation measurement. *Time domain* (or *pulse transient*) techniques measure the overvoltage as a function of time; and in *frequency domain* methods the apparent resistivity is measured at two or more different frequencies (usually lower than 10 Hz–Patella and Schiavone 1977). In the *phase domain* technique, the phase-lag between the applied current and the measured voltage is diagnostic of the nature of the sub-surface mineralisation. In *spectral IP*, phase and magnitude are measured over a range of frequencies from 10^{-3} to 4×10^3 Hz.

The induced polarisation method is an active one because voltages, which can be as high as several thousand volts in time-domain surveys, are applied to the ground in order to generate measurable overvoltages. The equipment used is similar to, but much more elaborate than, that employed in electrical resistivity work. The induced polarisation method excites (induces) a response in the ground which is dependent upon the distribution and nature of mineral grains present, and is most effective when the mineral grains are disseminated rather than combined in a massive form, as explained in the next section.

9.2 ORIGIN OF INDUCED POLARISATION EFFECTS

The exact causes of induced polarisation phenomena are still unclear, but the two main mechanisms that are reasonably understood are *grain (electrode) polarisation (overvoltage)* and *membrane (electrolytic) polarisation*, both of which occur through electrochemical processes.

9.2.1 Grain (electrode) polarisation

Grain or electrode polarisation occurs by the same process that results in self-potentials (see Chapter 8). If a metal electrode is placed in an ionic solution without a voltage being applied, charges with different polarities separate, resulting in the establishment of a potential difference between the electrode and the solution (Figure 9.2). The total magnitude of the potential is the Nernst potential and the adsorbed layer gives rise to the zeta potential (see Section 8.3.2). When a voltage is applied, the ionic balance is disturbed; this causes a current to flow, which in turn changes the potential difference between the electrode and the solution. When the applied voltage is removed, the ionic balance is restored by the diffusion of ions.

In the geological situation, current is conducted through the rock mass by the movement of ions, within groundwater, passing through interconnected pores or through the fracture and micro-crack structure within the rock. When an electronically conducting grain (e.g. a metal sulphide) blocks a flow channel, charge builds up (Figure 9.3) as in the electrochemical cell; this opposes the current flow and the grain becomes polarised, so creating a potential difference across the grain. On switching off the applied voltage, the ions diffuse back through the electrolytic medium and the potential difference acrosss the grain reduces to zero in a finite time, giving the characteristic overvoltage decay measured in time-domain systems.

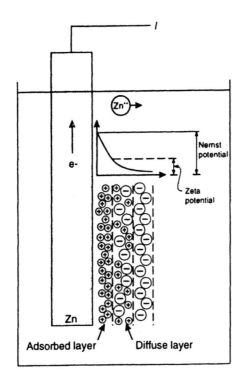

Figure 9.2 The phenomenon of electrode polarisation with the physical processes by which the Nernst and zeta potentials are obtained. From Beck (1981), by permission

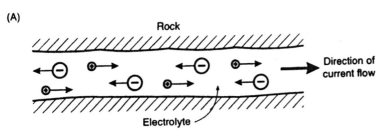

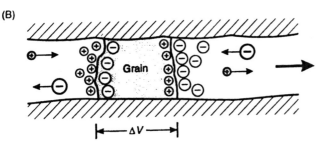

Figure 9.3 Grain (electrode) polarisation. (A) Unrestricted electrolytic flow in an open channel. (B) Polarisation of an electronically conductive grain, blocking a channel

Grain polarisation is essentially a surface phenomenon and this is why *disseminated* ores (with correspondingly large total surface areas) produce a significant IP response. (Sometimes an IP response is obtained over a halo of disseminated ore around a massive orebody.) Although it is the individual electronically conducting mineral grains

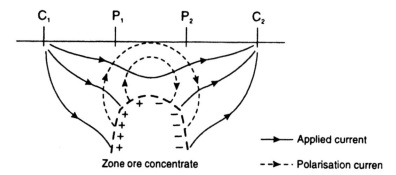

C_1 P_1 P_2 C_2

Zone ore concentrate

→ Applied current

--▶--· Polarisation curren

Figure 9.4   Macroscopic effect of grain polarisation over a disseminated ore body

that become polarised, complete zones with significant concentrations of ore will also take on a net polarisation; this results in a macroscopic polarisation current flow in the ground which is measured as the IP response. For the Wenner electrode configuration (see Section 7.3.2) in Figure 9.4, current polarises the zone of ore concentration which, when the applied current is turned off, generates the transient polarisation current that is measured at the surface. The current field is more complicated for the more commonly used dipole–dipole electrode array (see Section 7.3.2). The factors affecting the rate at which the ionic balance is restored are extremely complex and may depend upon the pore shape and size, rock structure, permeability, electrolytic conductivity and ionic concentration, and on the electronic conductivity of the mineral grain. Bornite, cassiterite, chalcopyrite, galena, graphite, limenite, magnetite, pyrite, pyrolusite and pyrrhotite, all exhibit strong IP responses as they have high electronic conductivities. The sulphides sphalerite, cinnabar and stibnite have low electronic conductivities and do not produce significant IP responses. For the same reason, they tend to produce only minimal, if any, self-potentials.

9.2.2   Membrane (electrolytic) polarisation

An IP response measured over rocks that do not contain sulphide mineral grains can be indistinguishable from the IP overvoltage effect over rocks containing low concentrations of disseminated ores, especially if traditional methods of time- and frequency-domain measurement are used. However, the more modern spectral and phase-domain IP systems may provide sufficiently diagnostic results.

There are two causes of membrane or electrolytic polarisation. One is by the constriction within a pore channel and the other is associated with the presence of clay within pore channels, such as in an impure sandstone. There is a net negative charge at the interface between most rock minerals and pore fluids. Positive charges within the pore

(A)

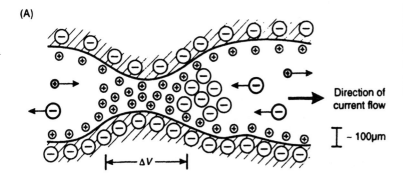

Direction of
current flow

$\underline{\text{I}}$ ~ 100μm

(B)

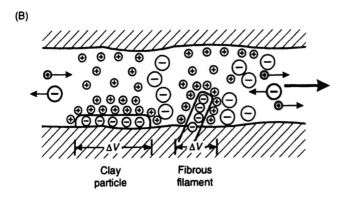

Clay
particle

Fibrous
filament

Figure 9.5 Development of membrane polarisation associated with (A) a constriction within a channel between mineral grains, and (B) negatively charged clay particles (Fraser *et al.* 1964) and fibrous elements along the sides of a channel

fluid are attracted to the rock surface and build up a positively charged layer up to about 100 μm thick, while negative charges are repelled. Should the pore channel diameter reduce to less than this distance, the constriction will block the flow of ions when a voltage is applied. Negative ions will leave the constricted zone and positive ions will increase their concentration, so producing a potential difference across the blockage (Figure 9.5A). When the applied voltage is switched off, the imbalance in ionic concentration is returned to normal by diffusion, which produces the measured IP response.

The second cause of membrane polarisation is the presence of clay particles or filaments of fibrous minerals, both of which tend to have a net negative charge. Positive ions are attracted to them, producing a positively charged cloud within the pore space. When a voltage is applied, positive charges can move between these similarly charged clouds but the negatively charged ions are blocked, which produces a difference in ionic concentration (Figure 9.5B). When the applied voltage is switched off, the imbalances in ionic concentration decay to normal levels by diffusion, so causing a measurable IP response.

9.2.3 Macroscopic processes

Various attempts have been made to explain induced polarisation in a quantitative way (e.g. Bertin and Loeb 1976) and to describe the phenomena qualitatively (e.g. Shuey and Johnson 1973; Sumner 1976). It is vital for interpretational purposes that any hypothesis should be able to explain the shape of the overvoltage decay curve with time and the variation of resistivity and phase with frequency. Indeed, the time-decay curve behaves in a complex manner proportionately to t^{-n}, where t is the time since the current was turned off – rather than being an exponential decay. Arguments involve the current density within and the dielectric constant (see Chapter 10) of the geologic material, but the induced polarisation phenomenon cannot be explained simply in terms of the dielectric constant of the rock. Models, which tend to be two-dimensional, fail to account adequately for three-dimensional reality, although Hohmann (1975) has attempted 3-D modelling. Some models have succeeded in identifying electric-circuit analogues in terms of resistor–capacitor systems (Bertin and Loeb 1976), but these are not explanations of the *physical* processes within the ground which give rise to induced polarisation.

9.2.4 Ionic processes

Just as macroscopic theories have failed to provide an adequate solution, so too have those that consider processes on the microscopic level. Forces acting on the ions and their resultant motions have been considered under diffusion theory which dates back to the end of the last century. The amplitude of the overvoltage in time-domain measurements should change as the square-root of the frequency but, according to experiments, it does not. If electrical forces are considered in addition to general diffusion, which takes into account the electric field and the ionic mobility, then it is possible to obtain dielectric parameters which are much closer to those derived from actual measurements. The effect of thermal agitation has also been considered and consequent mathematical modelling does reproduce the time-dependence of the overvoltage decay. Satisfactory physical explanations have yet to be found (Schufle 1958; Wong 1979).

9.3 MEASUREMENT OF INDUCED POLARISATION

Current is applied to the ground by means of two current electrodes, and the induced-polarisation effect is measured between two potential electrodes, most commonly in a dipole–dipole array. Sometimes Wenner, Schlumberger (gradient) and pole–dipole arrays are used. (For electrode array types, see Section 7.3.2.) Subsequent discussion of this method will be based on the use of the dipole–dipole array.

Electrode spacings are commonly tens to several hundred metres, but in broad reconnaissance surveys – for which the Schlumberger array tends to be used – the spacings can be even larger. The type of equipment used is bulkier and more elaborate than that used for resistivity surveys, and also depends on the type of IP survey being conducted. A transmitter is used to generate the applied current input into the ground, and the polarisation effects are detected by a receiver comprising non-polarisable porous-pots connected to the potential electrodes (see Chapter 8). Profiles are undertaken with fixed electrode spacings in much the same way as in the constant-separation traversing of electrical resistivity surveys (Chapter 7). More details of the field arrangements are given by Telford *et al.* (1990) and Milsom (1989).

9.3.1 Time-domain measurements

When a current is applied to the ground and switched off a few moments later, an overvoltage decay results (Figure 9.6). The total magnitude of the observed voltage (V_O) is equal to the actual voltage (V) due to the applied current plus a polarisation voltage (V_P) caused by the polarisation processes (Section 9.2). When the applied current is switched off, the voltage drops instantaneously by the amount V, leaving a residual voltage (the overvoltage) (V_P) which decays with time. One measure of the IP effect is the ratio V_P/V_O which is known as the *Chargeability* and is usually expressed in terms of millivolts per volt or per cent.

Instrumentally, it is extremely difficult to measure V_P at the moment the current is switched off, so it is measured at a fixed time (typically 0.5 s) after cutoff. Measurements are then made of the decay of V_P over a very short time period (0.1 s) after discrete intervals of time (also around 0.5 s). The integration of these values with respect to time gives the area under the curve (Figure 9.6), which is an alternative way of defining the decay curve. When the integral is divided by V_O, the resultant value is called the *apparent chargeability* (M_a) and has units of time (milliseconds) (see Box 9.1).

The *true* chargeability is virtually impossible to measure in a field situation as each layer within the ground will have its own absolute value of chargeability and of true resistivity. What is measured is a complex function of all the true resistivities and absolute chargeabilities for all the media being sampled within the range of the equipment. A short charging period will produce a lower IP response than a long charging period (Figure 9.7).

(A)

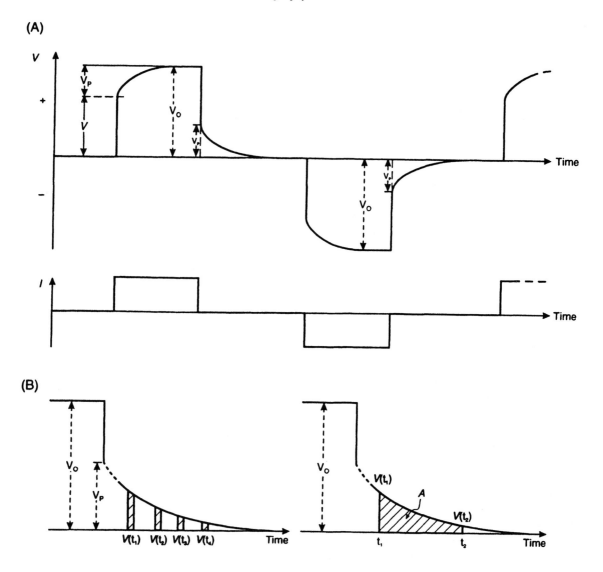

For given charging and integration periods (which differ between makes of IP equipment), the measured apparent chargeability is a diagnostic parameter that can be interpreted qualitatively in terms of the sub-surface geology. For a charging time of 3 s, an integration period of 1 s and a 1% volume concentration, chalcocite has a value of M_a of 13.2 ms, which is more than twice that of bornite (6.3 ms) and slightly higher than that of graphite (11.2 ms). In contrast, magnetite has a value of only 2.2 ms, and hematite has zero apparent chargeability.

Figure 9.6 (A) Application of a pulsed current with alternate polarity, and the consequent measured voltage showing the effect of the overvoltage (V_P) and the rise-time on the leading edge of the voltage pulse. (B) Two forms of measurement of the overvoltage at discrete time intervals $V(t_1)$, etc., and by the area beneath the overvoltage curve (A)

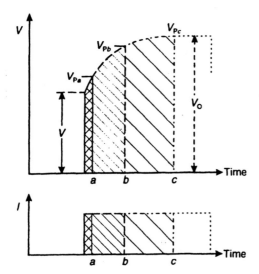

Figure 9.7 Increasing the charging time (*a* to *c*), which decreases the frequency of measurement, has the effect of increasing the overvoltage (V_{Pa} to V_{Pc}). Consequently, the apparent resistivity at lower frequency (e.g. at *c*) is greater than that at a higher frequency (e.g. at *a*)

Box 9.1   Chargeability (see Figure 9.6)

Chargeability:

$$M = V_P/V_O \text{ (mV/V or \%)}$$

where V_P is the overvoltage, and V_O the observed voltage with an applied current.

Apparent chargeability:

$$M_a = \frac{1}{V_O} \int_{t_1}^{t_2} V_P(t)\, dt = \frac{A}{V_O}$$

where $V_P(t)$ is the overvoltage at time t, and the other terms are as defined in Figure 9.6B.

The major advantage of integration and normalising by dividing by V is that noise from cross-coupling of cables and from background potentials is reduced. Care has to be exercised in selecting appropriate time intervals to maximise signal-to-noise ratios without reducing the method's diagnostic sensitivity.

9.3.2   Frequency-domain measurements

In frequency-domain (variable frequency) induced-polarisation studies, the apparent resistivity is measured at two frequencies less than 10 Hz (e.g. 0.1 and 5 Hz, or 0.3 and 2.5 Hz) using the same electrode array as in time-domain and direct-current resistivity measurements. The apparent resistivity at low frequency (ρ_{a0}) is greater than that at a higher frequency (ρ_{a1}) for reasons which can be appreciated by reference to Figure 9.7. For a short charging time the measured overvoltage is appreciably lower (V_{Pa}) than that obtained for longer charging times $(V_{Pb}$ and $V_{Pc})$.

The length of the decay is too short to be determined, so the total amplitude of voltage is measured with respect to the applied current, giving a value of the resistance (R) which, when multiplied by the appropriate geometric factor (see Section 7.3.2), is the apparent resistivity. If the current is switched in polarity, and on and off, with a time delay comparable to the length of the charging time, then this is the same as applying an alternating current signal at a given frequency $(f$ hertz). The shorter the charging and delay times, the higher the frequency, and so the apparent resistivity at low frequency is greater than that at a higher frequency. The two apparent resistivities are used to determine the *frequency effect* (FE) (unitless), which can be expressed alternatively as the *percentage frequency effect* (PFE) (units: %) (Box 9.2).

Box 9.2   Frequency effect (*FE*)

Frequency effect:

$$FE = (\rho_{a0} - \rho_{a1})/\rho_{a1} \text{ (unitless)}$$

where ρ_{a0} and ρ_{a1} are the apparent resistivities at low and higher frequencies respectively, and $\rho_{a0} > \rho_{a1}$.

Percentage frequency effect:

$$PFE = 100 \, (\rho_{a0} - \rho_{a1})/\rho_{a1} = 100 \, FE.$$

The frequency effect in the frequency-domain is equivalent to the chargeability in the time-domain for a weakly polarisable medium where FE is very much less than 1.

Marshall and Madden (1959) modified the expression involving the frequency effect to produce the *metal factor* (MF) (or the *metal conduction factor*) (Box 9.3). It is thought by some geophysicists that metal factor data delineate disseminated sulphide zones more effectively than frequency effect data.

Box 9.3   Metal factor (*MF*)

Metal factor:

$$MF = A(\rho_{a0} - \rho_{a1})/(\rho_{a0}\rho_{a1}) \text{ (units: siemens/m)}$$
$$= A(\sigma_{a1} - \sigma_{a0})$$

where ρ_{a0} and ρ_{a1} are the apparent resistivities, and σ_{a0} and σ_{a1} are the apparent conductivities $(= 1/\rho_a)$ at low and higher frequencies respectively; $\rho_{a0} > \rho_{a1}$ and $\sigma_{a0} < \sigma_{a1}$; and $A = 2\pi \times 10^5$.

_____ *continued* _____

——— *continued* ———

Alternatively, the metal factor is given by:

$$MF = A \times FE/\rho_{a0} = A \times FE/\rho_{a0} = FE/\rho_{a0} = A \times FE \times \sigma_{a0}$$

where FE is the frequency effect.

Although disseminated orebodies can be located using IP data, chargeability, frequency effect and metal factor do not give a good indication of the relative amount of the metallic mineralisation within the source of the IP response. It is necessary to go to more elaborate methods such as spectral IP, and even then the estimates obtained are not unambiguous.

(One further frequency-domain method that has been used but which has been superseded by the spectral IP method is 'phase IP'. Only one frequency is necessary. Induced polarisation is identified by the presence of a phase lag (ϕ) between the applied current and the polarisation voltage measured.)

9.3.3 Spectral IP and complex resistivity

Spectral IP is another name for complex resistivity, which in turn is related to the measurement of the dielectric properties of materials. An overview of the theoretical basis and interpretation of spectral IP (complex resistivity) data is given by Pelton *et al.* (1983).

The same field arrangement is used as in time- and traditional variable-frequency surveys but the equipment is considerably more sophisticated. It is important that the measurements be as precise as possible and that any noise be eradicated (see next section).

The magnitude of the complex resistivity ($|Z(\omega)|$) and phase (ϕ) of the polarisation voltage are measured over a wide range of frequencies (0.3 to 4 kHz) of applied current, which results in a diagnostic IP response spectrum (Figure 9.8). The frequency dependence is usually plotted as a binary function in the form of logarithms to base 2 rather than to base 10. The behaviour between the lower and upper frequency limits is known as the 'relaxation' of the electrical system (Shuey and Johnson 1973) and the entire dispersion can be defined if four electrical parameters are known (Box 9.4), namely, the DC resistivity (ρ_0), the IP chargeability (M), the time constant of the IP response (τ; also known as the relaxation time in dielectric studies), and the exponent of the angular frequency (ω).

Box 9.4 Spectral IP response

$$|Z(\omega)| = \rho_0\left[1 - M\left(1 - \frac{1}{1 + (i\omega\tau)^c}\right)\right]$$

where ρ_0 is the DC resistivity, M is the IP chargeability, ω is the angular frequency, τ is the time constant, and $i = \sqrt{-1}$.

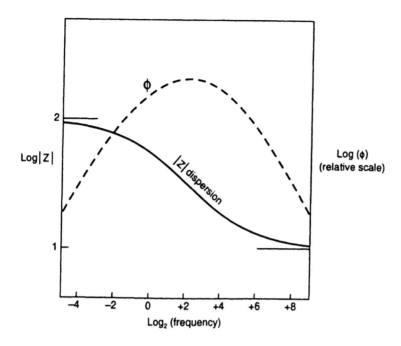

Figure 9.8 A typical IP spectral response (Pelton *et al.* 1983)

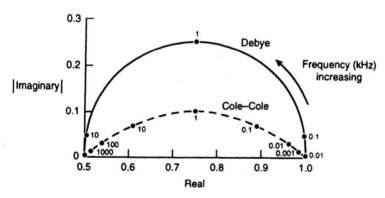

Figure 9.9 Cole–Cole relaxation spectra for Debye and Cole–Cole dispersions for $\tau = 1/2\pi$ and $c = 0.5$ (Pelton *et al.* 1983)

One form of relaxation commonly used is the *Cole–Cole relaxation spectrum* (Figure 9.9), named after its originators Cole and Cole (1941). This can be characterised by the critical frequency (F_c) which is the specific frequency at which the maximum phase shift is measured (Box 9.5). Note that this frequency is completely independent of resistivity. Phase angle and the critical frequency increase with increasing chargeability (Figure 9.10).

Box 9.5   Critical frequency (F_c)

$$F_c = [2\pi\tau(1 - M)^{1/2c}]^{-1}$$

where τ is the time constant, and M the IP chargeability.

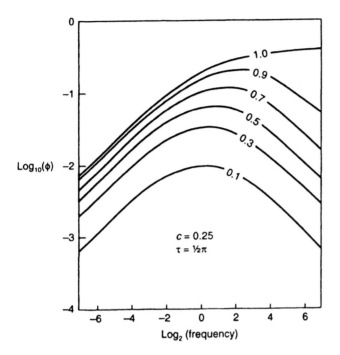

Figure 9.10 Phase angle curves for a typical Cole–Cole model with different chargeability ($M = 0.1$–1.0) (Pelton *et al.* 1983)

The fitting of two or more Cole–Cole dispersions to field spectra, a process known as *SIP inversion* (Song and Vozoff 1985), is the means by which the various intrinsic dielectric parameters are obtained. The 'texture' of mineralisation, which is characterised by the grain size and grain-size distribution of the polarisable particles within each group of ore grains but which is less dependent on the type of metallic minerals present, dictates the behaviour of the time constant (τ) and the frequency exponent (c). Where the polarisable mineralisation is coarse-grained, the relaxation time (τ) is large and the critical frequency (F_c) is small (conversely for fine-grained mineralisation). Similarly, massive sulphide orebodies have a distinctly different chargeability–relaxation time behaviour compared with graphite (Hallof and Klein 1982), and magnetite has a distinct chargeability–relaxation time behaviour compared with that of pyrrhotite (Pelton *et al.* 1978) (Figure 9.11). The maximum value of c is 0.5, and for most massive sulphides c is in the range 0.25–0.35 (Hallof 1983). Examples of the phase behaviour for several cases with various grain-size populations are illustrated in Figure 9.12. Where two grain-size distributions overlap or where a wide range of grain-sizes is distributed continuously, values of c fall to between 0.1 and 0.2 and the spectral curve is very flat. Computer analysis of such flat spectral curves is insensitive to variations in both c and τ and the estimations of grain-size distributions are likely to be ambiguous. Phase-angle spectra associated with different types of orebody are shown in

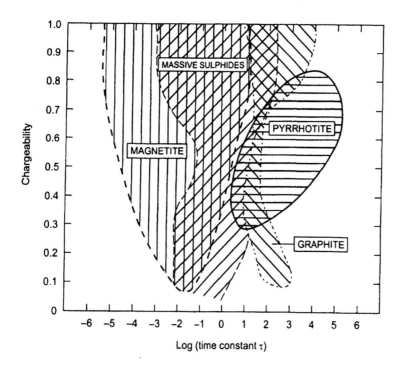

Figure 9.11 Chargeability as a function of relaxation time constant τ. After Hallof and Klein (1992), and Pelton *et al.* (1978), by permission

Figure 9.13. If chargeability–τ data plot in an overlapping field (such as magnetite and massive sulphides in Figure 9.11), then the phase-angle spectrum can be used to differentiate between the two ore types (Pelton *et al.* 1978).

Spectral IP parameters can be used to determine the mineralisation texture of an orebody and so separate out zones of primary mineralisation which consist of veinlets of ore from those of disseminated ore. Where long zones of sulphide or oxide iron formations occur, spectral IP can indicate where base metal concentrations increase by variations in the spectral 'texture'. In addition, changes in 'texture' along a pyrite zone may indicate the presence of gold, copper or zinc compared with barren areas (Hallof 1983).

The interpretation of spectral IP remains limited by the continued lack of understanding of the dielectric properties of rocks. The behaviour of the various dielectric parameters, such as the relaxation time, may vary with different physical conditions, such as with temperature (Saint-Amant and Strangway 1970; Ogilvy and Kuzmina 1972; Reynolds 1985). Assuming that dielectric parameters behave isothermally may result in misleading interpretations. One example of where this temperature dependence may be of practical advantage has been given by Reynolds (1985, 1987b). The relaxation time for glacier ice is strongly temperature-dependent. Temperate

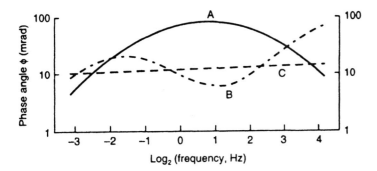

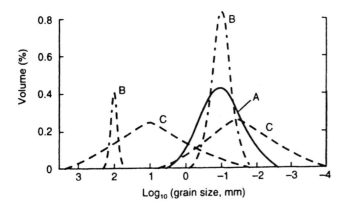

Figure 9.12 Theoretical spectral plots of phase with three models of grain size distributions. Data from Hallof (1983), by permission

glacier ice (i.e. ice at its pressure melting point throughout) has a diagnostically different temperature-dependence of relaxation time compared with that for polar glacier ice (i.e. ice well below its pressure melting point) (Figure 9.14). Some glaciers have both temperate and polar ice zones, with important implications for glacier ice flow and hazard evaluation, which can be differentiated by the dielectric measurement of relaxation time of ice samples. Spectral IP measurements have yet to be made over glaciers but the information obtained should prove to be immensely useful glaciologically. Spectral IP could also have applications in mapping out and distinguishing between massive ice bodies and interstitial ice in permafrost.

9.3.4 Noise reduction and electromagnetic coupling

To attain the high level of accuracy required for spectral IP, noise and electrical distortion of the IP response must be minimised before interpretation can proceed. There are four types of extraneous signal, of which three (current electrode variations, self-potentials and telluric currents) are regarded as 'noise' in IP work, and the remaining one

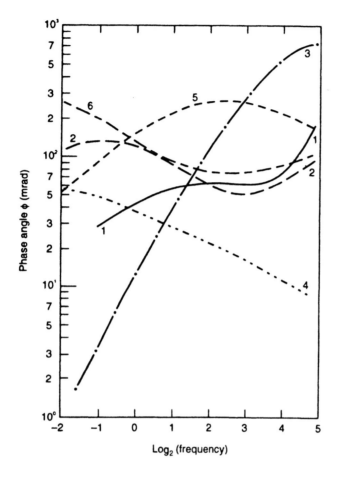

1. Porphyry copper
   (low concentration)

2. Porphyry copper
   (high concentration)

3. Magnetite

4. Pyrrhotite

5. Massive sulphide

6. Graphite

(electromagnetic inductive coupling) is a distortion that occurs at higher frequencies and has to be estimated and removed from the data (Hallof 1974; Hallof and Pelton 1980).

The noise signals can be constrained within the instrument by the use of filters. Current electrode variations due to changing current flow and the frequency of the applied current cause distortions within the waveform of the applied current. To compensate for this, a multi-channel approach is taken to measure voltage. Up to six pairs of potential electrodes are used to measure the IP voltage simultaneously, and the waveform of each pair is compared directly with that of the applied current so that the exact magnitude and phase of the signal waveform can be obtained for each of the six channels.

The most problematical effect is that due to electromagnetic coupling which occurs particularly at large electrode separations and at higher frequencies. The highest frequencies used in IP work overlap with the lower frequencies used by electromagnetic prospecting methods where *induction* becomes important. If a wire carrying

Figure 9.13 Characteristic phase-angle plots for different types of mineralisation. Data from Pelton *et al.* (1978), by permission

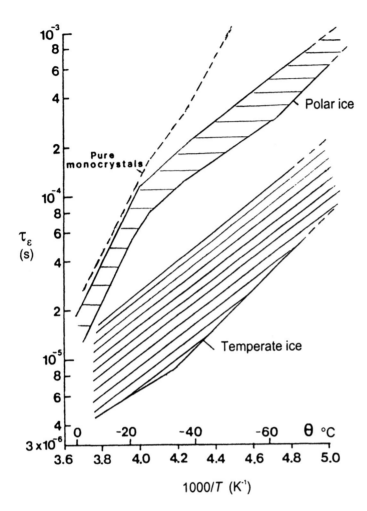

Figure 9.14 Temperature dependence of relaxation time constant τ for temperate and polar glacier ice, illustrating that τ can be used to differentiate between the thermal regimes of the two ice types. After Reynolds (1985), by permission

a current lies close to and parallel to another wire, a current will be induced into the second wire, such as happens close to power grid cables. If this induction occurs in wires connected to potential electrodes, spurious voltages will be measured. It is best to keep current-carrying cables well away from potential cables or to ensure that the two types of cable cross each other at right-angles. Furthermore, the current-carrying cables can induce a current into the ground which will distort the IP response. Fortunately, the frequency dependence of the electromagnetic coupling is recognisably different from that of the IP response and, if high enough frequencies are used, the coupling effects can be measured and then determined for the whole frequency range and removed from the data (Hohmann 1973; Wynn and Zonge 1975; Sumner 1976; Rathor 1977; Pelton *et al.* 1978; Hallof 1983). Pipelines have also been recognised as contributing to noise in IP surveys (Parra 1984).

9.3.5   Forms of display of IP data

Induced polarisation data for a given electrode separation, particularly when measured in the time domain, are commonly plotted as simple profiles and maps of chargeability (see Figure 8.6), metal factor or percentage frequency effect. In frequency-domain methods it is more usual to plot the data in the form of a pseudo-section (Figure 9.15).

Note that the spacing of the data points is dependent upon the integer value of n in dipole–dipole arrays and not on the actual dipole length. However, the increasing depth of penetration is implicit in the larger values of n and, as a guide, the depth of penetration is approximately equal to $a/2$ for $n = 2$; $\approx a$ for $n = 3$; and $\approx 2a$ for $n = 4$, where a is the dipole length. Each data value is plotted at the intersection of two lines drawn at 45° to the midpoint of each dipole used in the measurement. Two positions are indicated, for example, in Figure 9.15. For $n = 1$, and dipoles 1–2 and 3–4, the measured data

Figure 9.15   How to plot a pseudo-section. For a dipole–dipole array with current and potential electrodes at 1–2 and 3–4 respectively ($n = 1$), the measuring point is plotted at A; for dipoles at 4–5 and 8–9 ($n = 3$), the data value is plotted at B

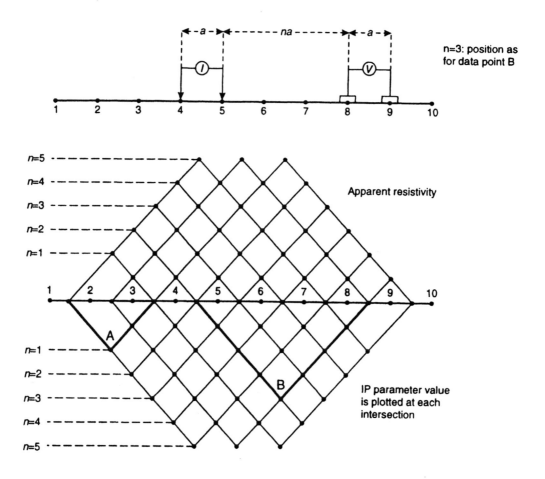

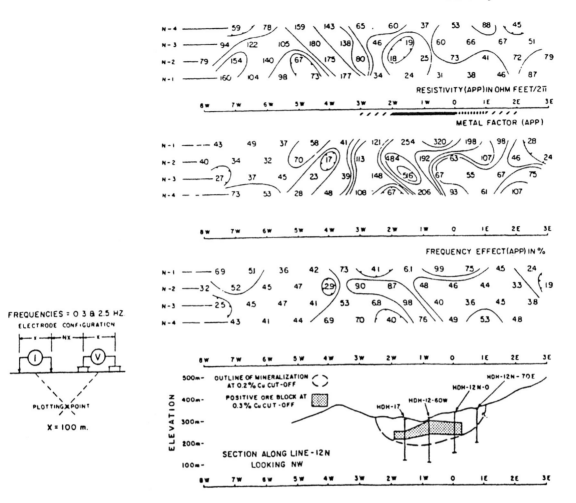

value is plotted at A; for $n = 3$ and dipoles 4–5 and 8–9, the position is at B. It is usual for the apparent resistivity pseudo-section to be drawn inverted above those for other IP parameters. Apparent resistivity, which is often plotted in units of $\Omega\,m/2\pi$, can also be plotted the correct way up as convenient. The plotted values can then be contoured. The surface projection of zones with IP anomalies are indicated respectively by solid, dashed, or oblique dashed bars along the top of the pseudo-section to indicate the location of definite, probable and possible mineralisation (Figure 9.16).

The interpretation of such displays tends to be by qualitative comparison with those obtained for theoretical or experimental scale models. Examples of the apparent resistivity, apparent metal factor and apparent phase for two models are given in Figure 9.17. In both cases the patterns in all three IP parameters are symmetrical about the central position and form patterns in the form of an inverted V.

Figure 9.16 Apparent resistivity, metal factor and frequency effect profiles over the Hinobaan deposit whose geological cross-section is shown schematically in the lower diagram. A dipole length of 100 m and frequencies of 0.3 and 2.5 Hz were used. From Pelton and Smith (1976), by permission

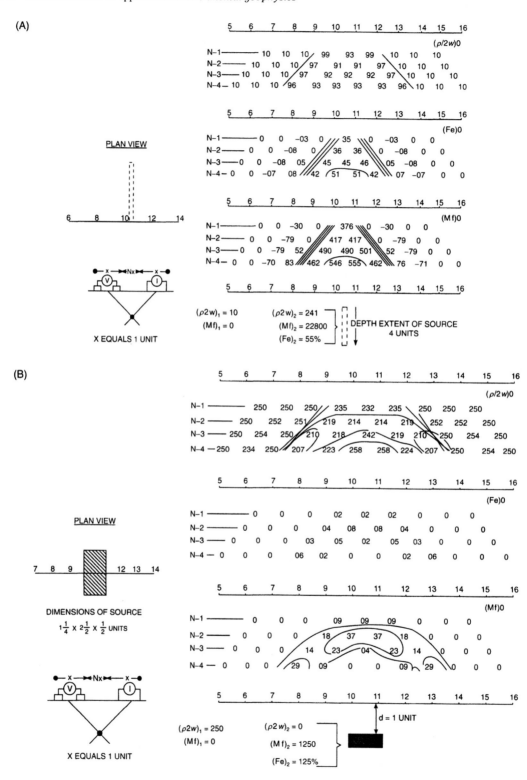

The apparent resistivity data over the horizontal rectangular block exhibit two 'lows' which correspond to end-effects of the block. Note also the lozenge shape anomaly 'high' in the apparent metal factor over the block model. From these two sets of data, the apparent

Figure 9.17 (*opposite*) Two model experiments to illustrate the apparent resistivity, frequency effect and metal factor over (A) a thin vertical prism, and (B) a horizontal rectangular block. From Hallof (1967), by permission

Figure 9.18 (*left*) Three forms of pseudo-section display of spectral IP data: (A) apparent resistivity at 1 Hz; (B) phase angle at 0.11 Hz; and (C) ratio of spectral parameters at three frequencies. From Hallof (1982), by permission. (D) Details of each plotting point, showing that the data are normalised to the mid-frequency value which always plots at 1,1 on the graphs (where the two axes cross)

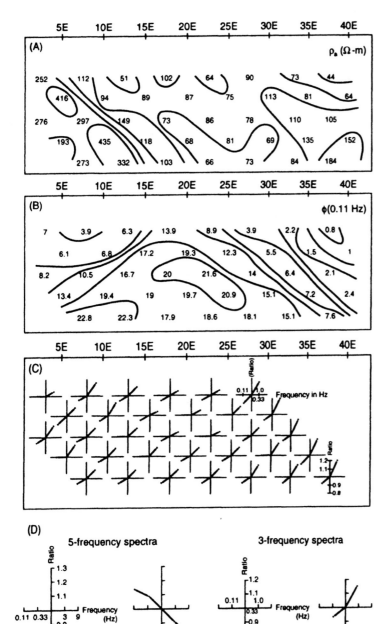

544   *An introduction to applied and environmental geophysics*

frequency effect is less sensitive to the model and less diagnostic than the apparent metal factor.

Another form of display is the use of three-frequency spectra which are plotted on a modified form of a pseudo-section (Figure 9.18). Spectral IP parameters measured at up to five frequencies, for example 0.11, 0.33, 1.0, 3.0 and 9.0 Hz, are abstracted from the total frequency spectrum and normalised with respect to the value at mid-frequency (i.e. ratio = 1.0); i.e. for five frequencies the parameters are normalised to the value at 1.0 Hz; for three frequencies (0.11–1.0 Hz), they are normalised to the value at 0.33 Hz. The normalised values are then plotted against the frequency (Hallof and Pelton 1981; Hallof 1982)(Figure 9.18D). The shape of the graph at each plotting position can be categorised into 'concave-up' and 'concave-down'(Fraser *et al.* 1964) or into specific and recognisable groupings based on the slope of the graph (Zonge and Wynn 1975; Hallof and Pelton 1981).

The most substantive development of both the display and interpretation of spectal IP is in the computer modelling of observed values of magnitude and phase according to Cole–Cole relaxation models. These permit the recognition and subsequent removal of inductive coupling from the spectra; and in the course of this, the four key IP parameters (ρ_0, M, τ and c) which define the IP response can be determined very accurately. An example of a spectral IP amplitude–phase diagram is given in Figure 9.19. It can be seen that

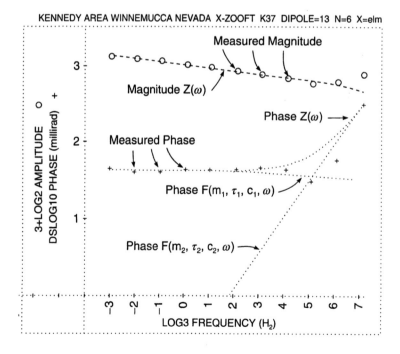

Figure 9.19 Example of a spectral IP amplitude–phase diagram. The steep gradient at high frequencies is due to inductive coupling (dispersion 1) while the data at lower frequencies with almost flat graphs (dispersion 2) can be inverted to give the electrical parameters of the ground. From Hallof 1982), by permission

at low frequencies the inductive coupling has little effect, whereas at high frequencies it becomes dominant.

9.4   APPLICATIONS AND CASE HISTORIES

9.4.1   Base metal exploration

A good example is the IP response over a disseminated copper-ore zone known as the Copper Mountain Orebody in Gaspe Area, Quebec, Canada (Figure 9.20). The orebody has been very well defined by many boreholes and the mineralisation has been evaluated as an average 4% metallic mineralisation comprising both pyrite and chalcopyrite. The increasing thickness of weathering towards the south is picked out by lower apparent resistivities. Low apparent resistivities, high apparent frequency effect and high apparent metal factor all coincide on $n = 3$ and 4 at locations between 46 and 49N, which also correlates with known sulphide mineralisation.

The IP method can produce valuable information about more massive types of mineralisation in cases where electromagnetic

Figure 9.20 Example of induced polarisation data for the Copper Mountain area, Gaspe, Quebec in Canada. From Hallof (1967), by permission

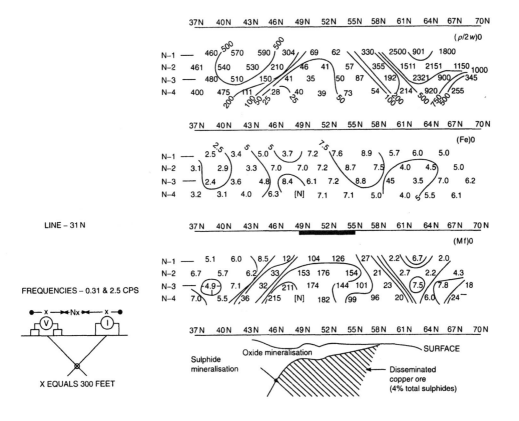

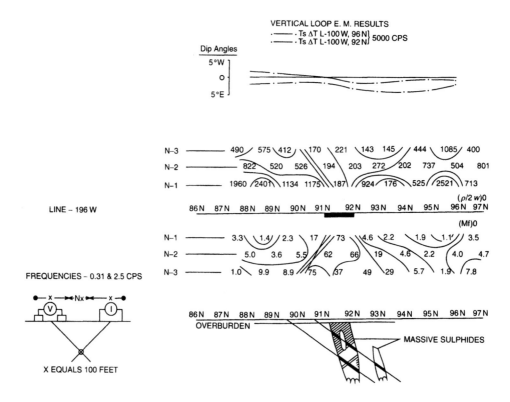

surveying fails. This is due to the mineralisation occurring in veinlets, which do not permit the formation of strong eddy currents in response to an induced electromagnetic field, rather than forming a totally massive orebody. One example of this is for Heath Steele Property, New Brunswick. IP data (apparent resistivity and apparent metal factor) and vertical loop EM data are presented in Figure 9.21, in which it is clear that the EM response is very weak in contrast to IP. The characteristic inverted V shape can be seen in the apparent metal factor data, indicating that the mineralisation is in near-vertical structures, as indicated by subsequent drilling.

Recent case histories of the application of spectral IP in the investigation of six Proterozoic sulphide, oxide and graphitic gneiss deposits in Finland have been described by Vanhala and Peltoniemi (1992). They found that in deposits with large differences in texture, such as graphitic gneiss and coarse-grained disseminated sulphide, could be separated on the basis of their diagnostic phase-spectra constants. They also found a good correlation between the observed grain size (from thin sections) and the grain size calculated from the apparent, field-survey phase spectra in the case of homogeneous disseminated textures.

Figure 9.21 Apparent resistivity and apparent metal factor data over a massive sulphide orebody at the Heath Steele Property, Newcastle, New Brunswick, compared with vertical loop EM data which show very poor resolution in contrast with the IP data. From Hallof (1967), by permission

9.4.2 Geothermal surveys

Geothermal sources often have high fluid mobility within circulating groundwater. This gives rise to low resistivities which can be detected using IP methods. The results on an IP survey over a possible geothermal field in Kenya are shown in Figure 9.22 (Hallof and Pelton 1981), which illustrates low apparent resistivity values for $n = 5$ and 6 between locations 14 and 17. Computer analysis of these data produced a geologically plausible model of a poorly defined but major conductor at a depth of about 350–400 m and a width of about 250 m. Although the survey did not delineate the possible geothermal centre absolutely, it did provide a constrained target for further, more detailed, investigations.

Zohdy *et al.* (1973) have described a combined resistivity, self-potential and induced polarisation survey over a vapour-dominated geothermal field in the Yellowstone National Park, Wyoming, USA (Figure 9.23). The broad SP anomaly is caused by water upwelling (as

Figure 9.22 Observed and calculated apparent resistivity data obtained during IP surveys of low-resistivity (0.3 Ω m) geothermal source in Kenya. From Hallof and Pelton (1981), by permission

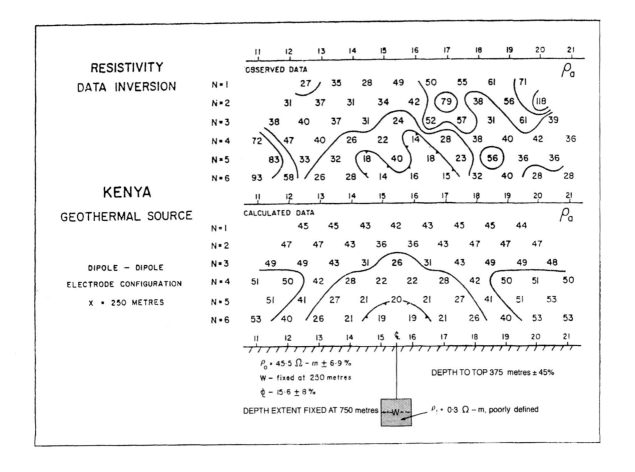

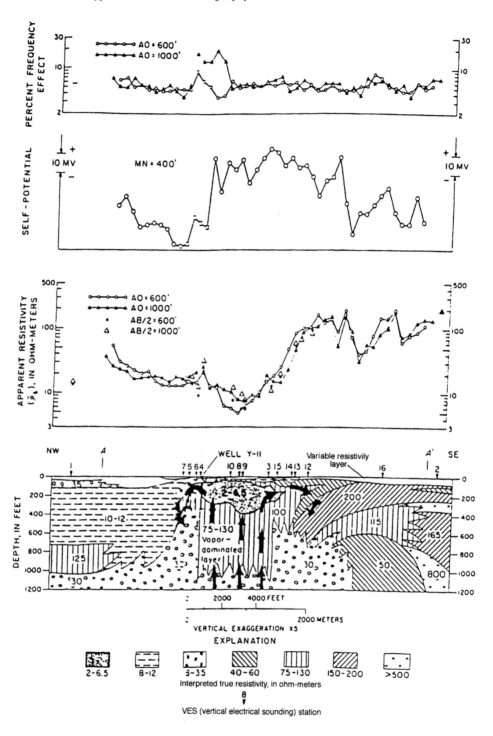

indicated by the arrows in the bottom panel of the figure), and continues as far as it does to the south-east because the groundwater flows laterally until it enters more permeable material, at which point it can then descend. The resistivity profiles reflect the general distribution of formations, with those to the north-west having lower true resistivities (10–130 Ω m) than those to the south-east ($> 130 \Omega$ m). The percentage-frequency-effect profile has a significant background of around 5%, which has been attributed to the presence of clayey materials and pyrite in the near-surface materials. The IP anomaly which is particularly prominent with $AO \approx 300$ m is due to an increase in disseminated pyrite deposited by circulating thermal waters as indicated from borehole data in the area.

Figure 9.23 (*opposite*) Percentage frequency effect and apparent resistivity data across a vapour-dominated geothermal system in the Yellowstone National Park (interpreted geological cross-section shown in the lowest panel) compared with a self-potential profile across the same feature. From Zohdy *et al.* (1973), by permission

9.4.3 Groundwater investigations

Vacquier *et al.* (1957), who were the first to use induced polarisation for hydrogeological investigations, have described two useful case histories. The first is for a site near Carrizozo, New Mexico, in a valley bounded to the south-east by the Sierra Blanca which comprise volcanic rocks from which potable water is derived. To the north and west, Cretaceous sediments of the Tularosa Valley contaminate the groundwater with chlorides and gypsum making it unusable.

Figure 9.24A shows a map of IP values 10 s after current shutoff; the contours are orientated approximately north–south and mirror the existing drainage pattern the boundaries of which correlate with contours of 2 mV/V in the east and 3 mV/V in the west. An irrigation well IR-1 produces about 4500 litres/minute of relatively poor quality water; hence the low IP overvoltage (≈ 3 mV/V). The closure with a value of 6 mV/V (L-1) was subsequently drilled and produced about 450 litres/minute of better quality water. The map shown in Figure 9.24B shows contours of the ratio of IP values 5 s after current shutoff to those after 10 s. Vacquier and colleagues interpreted these contours as highlighting a buried channel, with the higher ratios being associated with finer grained material.

Ogilvy and Kuzmina (1972) carried out a laboratory scale model experiment to examine the effectiveness of time-domain IP measurements in locating groundwater accumulations in sandy–clayey overburden. While standard constant-separation traversing over a hemispherical freshwater lens produced a broad apparent resistivity anomaly, the IP polarisability anomaly was both narrow and steep-sided (Figure 9.25A); a result which had also been obtained by Vacquier *et al.* (1957, p. 684). However, for a model with the same dimensions and geometry, but with a saline water lens, both the apparent resistivity and IP chargeability anomalies were broad in contrast with the ratio of IP overvoltages after 0.5 s and 5 s after current shutoff (Figure 9.25B).

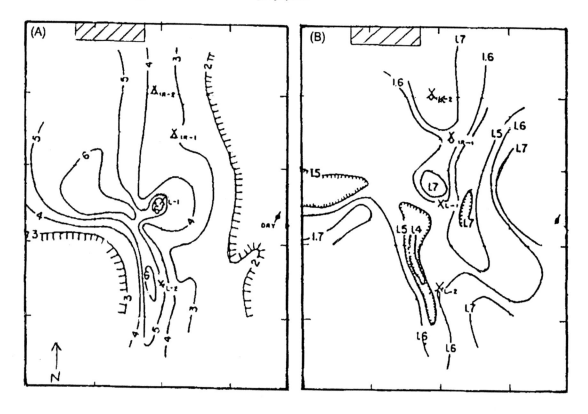

The examples from Vacquier and colleagues and from Ogilvy and Kuzmina demonstrate that for time-domain IP, maps of the ratio of the overvoltage for two different times can provide a sharper resolution of sub-surface water bodies than maps of the overvoltage for single cutoff times. While the IP method can provide a useful interpretation of groundwater bodies, the field method is less practical than modern electromagnetic induction methods, and this may account for the paucity of recent examples of the use of IP in groundwater investigations.

Figure 9.24 Maps of: (A) induced polarisation overvoltage after 10 s (contours at 1 mV/V intervals); and (B) the ratio of IP data after 5 s and 10 s after current switchoff, over a buried valley aquifer near Carrizozo, New Mexico, USA. The well sunk at L-1 produced about 450 litres/minute of potable water. From Vacquier *et al.* (1957), by permission

9.4.4   Environmental applications

Since the early 1980s, interest has been expressed in the possibility of using IP in the investigation of contaminated sites. Olhoeft (1985), for example, investigated the IP characteristics of rocks contaminated with organic pollutants. Organic chemicals can react with clay minerals so that the IP response of the clay mineral–electrolyte mixture changes. This work has been extended by Soininen and Vanhala (e.g. 1992), for example, who have investigated laboratory samples of glacial clays contaminated with ethylene glycol using spectral IP methods. They have found that both the phase and resistivity spectra

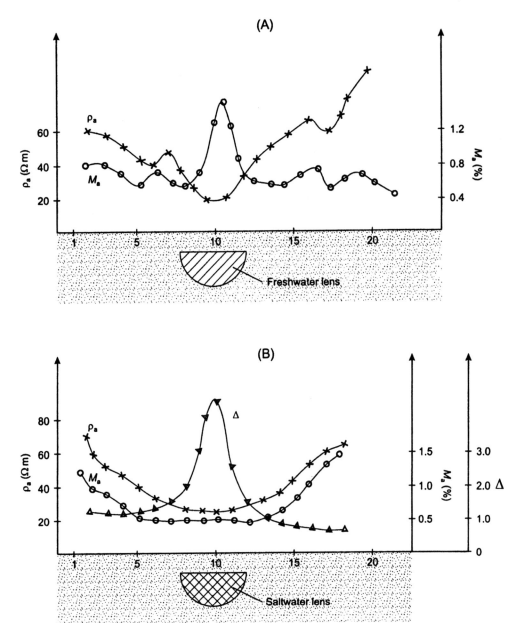

Figure 9.25 Scale model experimental results of apparent resistivity, chargeability, and ratio (Δ) of overvoltage measured after 0.5 minutes and 5 minutes after current switch obtained across a buried hemispherical lens of (A) freshwater and (B) saltwater. After. Ogilvy and Kuzmina (1972), by permission

(as functions of frequency) of contaminated samples differ significantly from those of uncontaminated samples. Whether this technique can be extended to achieve a fully commercial field survey technique has yet to be demonstrated.

Cahyna *et al.* (1990) have presented a case history where the field use of standard IP, in conjunction with appropriate laboratory measurements, was instrumental in mapping out an area seriously

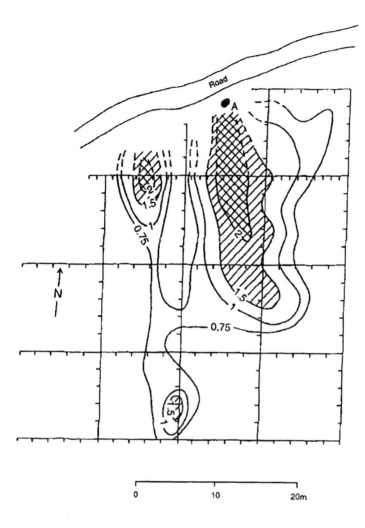

Figure 9.26 Chargeability map over a site contaminated with cyanide complexes. The location of a known outcrop of slag is indicated at A. Contours are in percentage chargeability. Shaded areas indicate the interpreted extent of contaminated land. From Cahyna *et al.* (1990), by permission

contaminated with cyanide complexes which originated from slag-type material from former plating works. They used both conventional resistivity sounding (using the Schlumberger array) and symmetrical resistivity profiling with IP (called SRP-IP by the authors). The resistivity survey failed to detect the contamination even though measured directly over a known slag deposit. In contrast, the SRP-IP survey, which was used to measure chargeability, was used successfully to identify not only the limit of the known slag, but also to detect a second and previously unknown area of contamination. A map of the SRP-IP chargeability was produced (Figure 9.26) on which the centres of contamination and the affected areas around them are clearly evident.

Section 4
ELECTROMAGNETIC METHODS

Chapter 10
Electromagnetic methods: introduction and principles

10.1 INTRODUCTION

10.1.1 Background

Among all the geophysical methods, the electromagnetic techniques must have the broadest range of different instrumental systems of any, matched by the remarkable range of applications to which these methods are being applied. These methods also show the greatest geographical diversity as some are used extensively and preferentially in the areas in which they were developed. For example, SIROTEM (see Section 11.3) is used predominantly in Australia where it was developed (named after the Australian Commonwealth Scientific Industrial Research Organisation, CSIRO), and Turam systems in

Sweden. The range of EM instruments manufactured by Geonics Ltd, Canada, has been used predominantly in North America, particularly eastern Canada, and now increasingly in Europe. There have been some major developments in the portability and ease of use of some instruments and their ensuing popularity has resulted in the techniques being used more widely. The interpretation methods available are largely dependent upon the instrumentation used for each survey and the information about the plethora of equipment available is widely scattered throughout the literature. However, the diversity of equipment provides a wide range of instruments to choose between in order to select the most appropriate tool for the task in hand. This, rather than being a disadvantage, is a major strength. Modern EM systems provide a very powerful suite of sophisticated instruments. Coupled with major advances in computer interpretation techniques, EM methods are set to become much more heavily used, especially for engineering and environmental applications.

Probably the first electromagnetic method to be used for mineral ore exploration was developed by Karl Sundberg in Sweden over two decades following the First World War (Sundberg 1931). What is now known as the Sundberg method was developed in 1925 and was also used in structural mapping in hydrocarbon exploration (Sundberg and Hedström 1934). Other pioneering work was done in the early 1930s by a Russian geophysicist V.R. Bursian, whose work is little known in the West. Other electromagnetic methods have been available commercially only since the Second World War and particularly since the mid-1960s. EM methods are especially important, not only in mineral and hydrocarbon exploration, but increasingly in environmental geophysics applications.

The different electromagnetic systems available are described briefly in Section 10.1.3 and in more detail in the next chapter. Chapter 12 is devoted to a discussion of 'ground penetrating radar' (GPR). A much more comprehensive and detailed discussion of the various electromagnetic methods, with the exception of ground penetrating radar, has been produced by Misac Nabighian (1987, 1991) and coauthors. Further discussions and descriptions of the various methods have been given in the three-volume treatise *Geotechnical and Environmental Geophysics* edited by Stan Ward (1990).

10.1.2 Applications

The range of applications of EM methods is large. It is dependent upon the type of equipment being used but can be broadly categorised as listed in Table 10.1. Not all EM methods are equally appropriate to the applications listed. For example, ground penetrating radar has very limited use in the direct investigation of landfills by virtue of the high ambient conductivity and the corresponding high attenuation of radiowaves with depth. Conversely, ground conductivity mapping

Table 10.1   The range of applications for EM surveying*

Mineral exploration
Mineral resource evaluation
Groundwater surveys
Mapping contaminant plumes
Geothermal resource investigations
Contaminated land mapping
Landfill surveys
Detection of natural and artificial cavities
Location of geological faults, etc.
Geological mapping
Permafrost mapping, etc.

* Independent of instrument type

does not have the required resolution in comparison with GPR in some archaeological investigations. Furthermore, GPR can be used with care inside buildings, while ground conductivity methods cannot by virtue of interference from ambient electrical noise from mains power lines, etc.

One of the main advantages of the EM methods is that the process of induction does not require direct contact with the ground, as in the case of electrical methods where electrodes have to be planted into the ground surface (see Chapter 7). Consequently, the speed with which EM surveys can be made is much greater than an equivalent survey using contacting electrical resistivity. Furthermore, the induction process also allows the method to be used from aircraft and ships, as well as down boreholes. Similar to electrical resistivity methods, scale model experiments can be undertaken to illustrate particular structures (e.g. Frischknecht 1987). However, numerical models may be used preferentially (Hohmann 1987) but still require large amounts of computing time, and are limited by the computational difficulties in defining especially two- and three-dimensional models. Aspects of the limitations of computer analysis are discussed in more detail in the next chapter.

10.1.3   Types of EM systems

Electromagnetic methods can be classified as either time-domain (TEM) or frequency-domain (FEM) systems. Frequency-domain instruments use either one or more frequencies whereas time-domain equipment makes measurements as a function of time. EM methods can be either passive, utilising natural ground signals (e.g. magnetotellurics) or active, where an artificial transmitter is used either in the near-field (as in ground conductivity meters) or in the far-field (using remote high-powered military transmitters as in the case of VLF mapping).

Table 10.2   A classification of electrical and electromagnetic systems

Transmitter type	Receiver type			
	Ground wire	Both wire and small coil	Small coil (ground)	Small coil (air)
Grounded wire				
Galvanic	Resistivity IP		Magnetometric resistivity (MMR) Magnetic IP (MIP)	
Inductive		CSAMT	Some TEM systems	
Small loop			Slingram Horizontal-loop EM Vertical-loop EM Tilt-angle method Ground conductivity meters (GCM) Some TEM systems Coincident loop Borehole systems	Airborne EM Time-domain towed-bird Helicopter rigid-boom
Large loop (long wire)			Large-loop systems Sundberg method Turam Many TEM systems Borehole systems	
Plane wave Vertical antenna Natural geomagnetic field		VLF-resistivity VLF		VLF
	Telluric currents			

Grounded wires measure potential difference per length, thus electric field. Coils (or fluxgate magnetometers or SQUIDS) measure magnetic field, or its time derivative. A small loop is a 3-D source (magnetic dipole). A long wire (or the long edge of a large loop) is a 2-D source. Natural EM sources are assumed to be 1-D sources. Receivers can be frequency-domain, time-domain (TEM), or both, CSAMT = controlled-oource audio magneto-telluric. This classification, which excludes the high-frequency techniques (radar, etc.), is based on Swift (1988, Table 1, p. 6)

A basic classification of EM systems is given in Table 10.2, which is based on Swift (1988). Each system is described briefly in this section; VLF, ground-conductivity, time-domain EM, telluric and magneto-telluric systems are described in more detail in Chapter 11. Ground penetrating radar is discussed comprehensively in Chapter 12.

Case histories are given where appropriate to illustrate the use of each of the main techniques. In most cases, the concept of each method is described rather than specific equipment systems which may change through continuing development work.

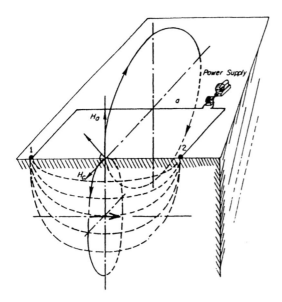

Figure 10.1 Basic concept of a magnetometric resistivity (MMR) field survey layout. Electrodes (1 and 2) are used to inject direct current into the ground. The secondary magnetic field arising from the current flow in the ground is measured at the mid-point by an extremely sensitive magnetometer. Reproduced with permission from Edwards and Nabighian (1991).

The term 'galvanic' used in Table 10.2 describes the injection of electrical current directly into the ground via electrodes; these methods are discussed in detail in Chapters 7 and 9.

10.1.3.1 Magnetometric resistivity (MMR)

Commutated direct current is injected into the ground through two widely separated electrodes. The anomalous conductivity contribution is determined at the midpoint by measuring the secondary magnetic field arising from the flow of current using an extremely sensitive low-noise magnetometer aligned perpendicular to the line between the electrodes (Figure 10.1). For further details, see Chapter 7 and, in particular, the review by Edwards and Nabighian (1991).

10.1.3.2 Small-loop systems

A frequency-domain EM system in which two small coils, one a transmitter and the other a receiver, separated by a constant distance of between 4 m and 100 m, are moved along a survey transect. The primary field is nulled so that the in-phase and quadrature components of the secondary field can be measured. The various combinations of coil orientation are shown in Figure 10.2. Slingram is synonymous with the *horizontal-loop method* (HLEM), Boliden, EM Gun, MaxMin and with Ronka EM methods. *Ground conductivity meters* (GCM) can be classified as being of this type of method. In this case, the quadrature component is normally taken to be a linear measure of the apparent conductivity of the ground; the coplanar

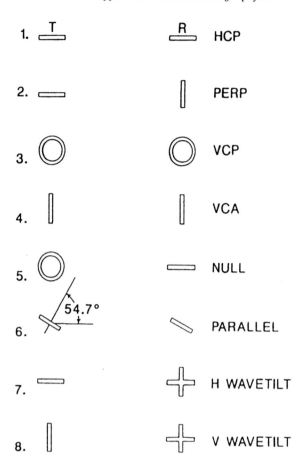

Figure 10.2 Eight common dipolar loop configurations (Tx = transmitter; Rx = receiver). Each rectangle represents the edge-on view of a coil; e.g. to move from configuration 1 to 3, both coils are rotated about a horizontal axis; to move from 3 to 4, each coil is rotated about a vertical axis. HCP = horizontal coplanar; VCP = vertical coplanar; VCA = vertical coaxial; PERP = perpendicular. Reproduced with permission from Frischnecht *et al.* (1991).

coils are deployed both horizontally and vertically; Chapter 11 has more details. For a detailed discussion of small-loop systems, see the review by Frischknecht *et al.* (1991).

10.1.3.3 *Large-loop systems*

There are two basic configurations in this classification, namely, the original method known as Sundberg's method, and the other, Turam.

Sundberg's method uses a long, grounded, insulated wire a few hundred metres to several kilometres long, or a rectangular loop with the long-side laid in the direction of geological strike (Figure 10.3A and B). Typical loop dimensions are 1200 m by 400 m. Measurements are made along profiles at right-angles to the cable or long side of the loop. Phase reference is taken by using a feeding coil located close to the source loop/cable using the compensator system shown in Figure 10.3C. Normally, only the vertical magnetic field is observed using the receiver coil. If the coil is deployed in three mutually perpendicular planes, then the EM field can be determined completely.

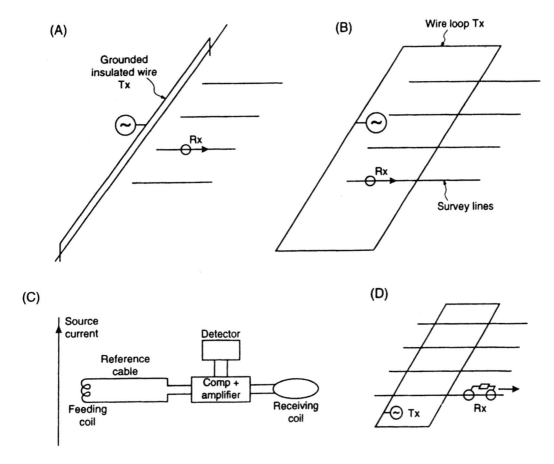

Figure 10.3 (A) Survey layout for the Sundberg method with a long grounded wire, or (B) a grounded wire loop with survey profile lines indicated. Phase reference is determined using a compensator (C) close to the source wire. In the Turam method (D), two separate receiver coils are deployed with a constant separation

The Turam technique overcomes a significant operational difficulty with the Sundberg method, i.e. the necessity to have a feeding coil close to the source cable/loop. In the Turam method, two separate receiver coils are used which are maintained at a constant separation, typically 10–20 m (Figure 10.3D). After each measurement, the coils are moved so that the rear coil then takes the position formerly occupied by the forward one, and so on along the transect. The two coils provide a means whereby a measurement is made of the ratio of the resultant vertical-field amplitudes and phase difference of the vertical fields at two neighbouring points. In effect, by having a constant coil separation and measuring parameters at each of the two locations, the horizontal gradient of phase of the resultant vertical field is determined. A more complete discussion of these two methods has been given by Parasnis (1991).

10.1.3.4 Time-domain systems

If a continuous EM field is produced by a transmitter, the secondary field is either determined by nulling the primary field so as to be able to detect the secondary field, or by measuring the resultant of both primary and secondary fields, and hence computing the secondary field parameters; those of the primary field are known by design. In time-domain or transient EM, the primary field is applied in pulses, typically 20–40 ms long, with the secondary field being measured once the primary field has been switched off over the following 100 ms, for example. One advantage of this is that the transmitter coil can also be used as the receiver. The basic field layouts are shown in Figure 10.4.

Typically a large ungrounded coil, through which a strong direct current is passed, is laid on the ground with the long axis parallel to

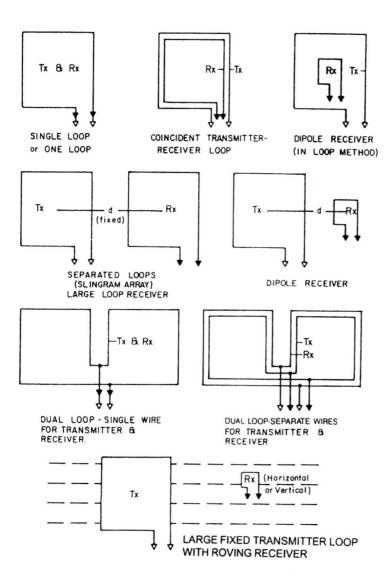

Figure 10.4 Field configurations for time-domain EM surveys. Reproduced with permission from Nabighian and Macnae (1991).

any geological strike. A small receiver coil is moved along transects perpendicular to the long axis of the ungrounded loop to obtain profiles of the measured parameters as a function of distance along the transect. Alternatively, instead of profiling, TEM systems can be used very effectively for depth soundings. Increased depth penetration is achieved by measuring the decay of the secondary field as a function of time. As the secondary field decays, the field parameters are measured at discrete time intervals (typically logarithmically arranged). It is analogous to the induced polarisation (IP) method in resistivity surveying. A specific system (INPUT) was developed in 1958 for airborne work (Barringer 1962). Following 1970, with improvements in technology and computing capabilities, a range of EM systems were developed by both academic institutions and commercial companies. By 1988, all instrument manufacturers had provided fully digital TEM systems.

As a guide, but depending upon the actual configurations and equipment being used, 50 TEM soundings per day is not unreasonable. With increasing pressure to use TEM in environmental applications where depths of penetration of less than 50 m are required, 'very-early TEM' (VETEM) systems are being developed. VETEM could be used in surveys where depths of penetration of less than 25 m but high vertical resolution are required, such as over closed landfills. For more detailed descriptions of TEM systems and of depth sounding, see the reviews by Nabighian and Macnae (1991) and by Spies and Frischknecht (1991) respectively.

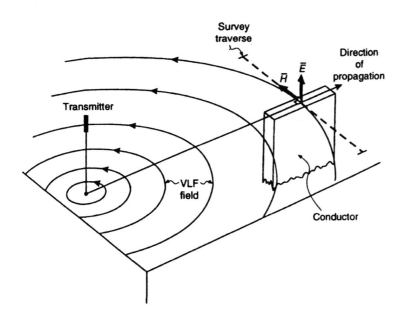

Figure 10.5 Artificial VLF source (e.g. military transmitter) provides a primary EM field which, at a sufficiently large distance, equates to a plane EM wave. Preferred survey directions over a linear conductor are tangential to the VLF field

10.1.3.5   *Very low frequency (V.L.F)*

High-powered military radio transmitters operating in the 15–24 kHz range (i.e. very low frequency in radio terms) are used to communicate with submarines even when submerged, and for long-range radio positioning. At very large distances from the transmitters, the EM field approximates to a plane wave which is used in geophysical exploration (Figure 10.5). The method can be used either on the ground or from aircraft. The method is discussed in more detail in the next chapter and has been reviewed by McNeill and Labson (1991).

10.2   PRINCIPLES OF EM SURVEYING

10.2.1· Electromagnetic waves

Electromagnetic methods use the response of the ground to the propagation of incident alternating electromagnetic waves which are made up of two orthogonal vector components, an electric intensity (E) and a magnetising force (H) (Figure 10.6), in a plane perpendicular to the direction of travel. An electromagnetic field can be generated by passing an alternating current through either a small coil comprising many turns of wire or a large loop of wire. The frequency range of electromagnetic radiation is very wide (Figure 10.7), from atmospheric micropulsations at a frequency less than 10 Hz, through the radar bands (10^8 to 10^{11} Hz) up to X-rays and gamma-rays at frequencies in excess of 10^{16} Hz. Of critical importance is the visible band ($\approx 10^{15}$ Hz).

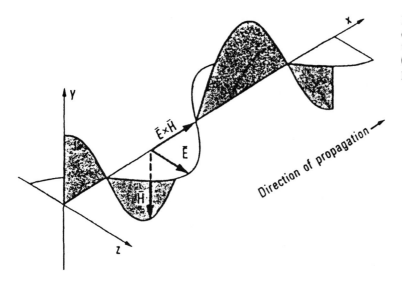

Figure 10.6 Basic elements of an electromagnetic wave, showing the two principal electric (E) and magnetic (H) components. Reproduced with permission from Beck (1981)

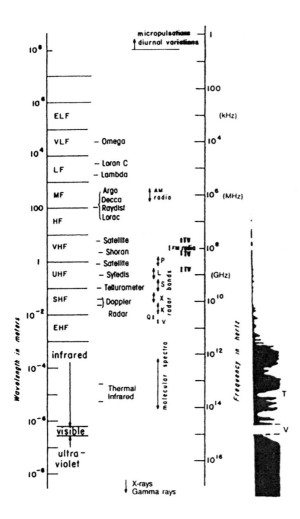

Figure 10.7 The electromagnetic spectrum. (A) designation of the various parts of the spectrum as a function of wavelength in metres. In (B), the dark portion of the graph shows zones of attenuation due to atmospheric absorption. Two windows are evident in the absorption spectrum at T (thermal infrared) and V (visible light). Reproduced with permission from Sheriff (1991)

For geophysical applications, frequencies of the primary alternating field are usually less than a few thousand hertz. The wavelength of the primary wave is of the order of 10–100 km while the typical source–receiver separation is much smaller (≈ 4–$100 + $ m). Consequently, the propagation of the primary wave and associated wave attenuation can be disregarded (Figure 10.8).

In general, a transmitter coil is used to generate the primary electromagnetic field which propagates above and below ground. When the EM radiation travels through sub-surface media it is modified slightly relative to that which travels through air. If a conductive medium is present within the ground, the magnetic component of the incident EM wave induces eddy currents (alternating currents) within the conductor. These eddy currents then generate their own, secondary, EM field which is detected by a receiver

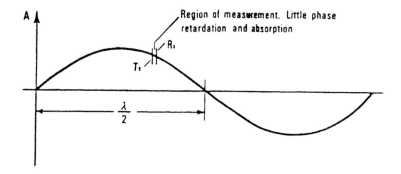

Figure 10.8 The physical separation of a transmitter (Tx) and receiver (Rx) is very small in relation to the wavelength of EM waves with frequencies greater than 3 kHz. Consequently, attenuation due to wave propagation can be ignored. Reproduced with permission from Beck (1991)

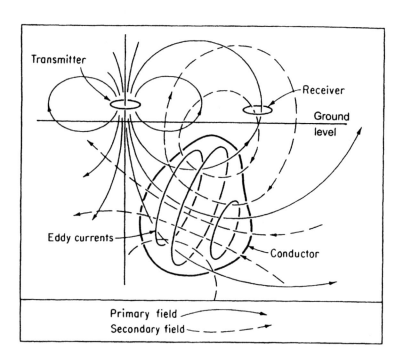

Figure 10.9 Generalized schematic of the EM surveying method. Reproduced with permission from Grant and West (1965)

(Figure 10.9). The receiver also detects the primary field which travels through the air, so the overall response of the receiver is the combined (resultant) effect of both the primary and the secondary fields. Consequently, the measured response will differ in both phase and amplitude relative to the unmodulated primary field. The degree to which these components differ reveals important information about the geometry, size, and electrical properties of any sub-surface conductor. Detailed discussions of electromagnetic theory, which are beyond the scope of this book, have been given by Grant and West (1965), Telford *et al.* (1990), and Ward and Hohmann (1991), among others.

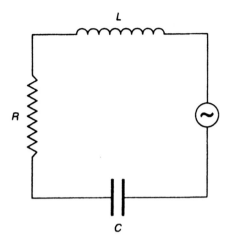

Figure 10.10   Basic electrical circuit containing capacitance (*C*), inductance (*L*) and resistance (*R*), the three electrical components that describe the equivalent behaviour of the ground.

It is useful to regard the ground under investigation as comprising three components: inductive (*L*), resistive (*R*) and capacitive (*C*); the electrical circuit equivalent is shown in Figure 10.10. The applied alternating voltage has the form of a sine wave with an angular frequency of $\omega\,(=2\pi f)$ and amplitude E_0 which varies as a function of time as described mathematically in Box 10.1. The current (*I*) which flows lags behind the applied voltage by an amount α, the phase lag. In EM exploration, a primary magnetic field is applied (*P*) which, in accordance with the properties of an EM wave, is in phase with its orthogonal electric component (*E*) (refer to Figure 10.6). Consequently, the form of the primary magnetic wave is $P = H_0 \sin \omega t$, where H_0 is the peak amplitude of the magnetic wave (Figure 10.11A). The voltage induced into a secondary perfect conductor as a result of the incident primary magnetic field lags behind the primary field by $\pi/2$.

According to Faraday's Law of EM induction, the magnitude of the induced voltage is directly proportional to the rate of change of the magnetic field. The induced voltage is directly proportional to the rate of change of the magnetic field. The induced voltage will be zero when the magnetic field is either at its maximum or minimum (Figure 10.11B). Eddy currents within a conductor take a finite time to generate, arising from an induced voltage. This generation time is manifest as the phase lag α (Figure 10.11C) which depends upon the electrical properties of the conductor. In good conductors this phase lag can be large, and conversely in poor conductors the phase lag is small. Once generated the secondary magnetic field interacts with the primary to form a resultant magnetic field (Figure 10.11D) which has a total phase lag (ϕ) behind the primary field.

Box 10.1   Time varying electrical field

The amplitude (E) of an alternating voltage is given by:

$$E = E_0 \sin \omega t.$$

The current (I) within the equivalent circuit (see Fig. 10.10) is described by:

$$I = E_0\{[\omega L - (1/\omega C)]^2 + R^2\}^{-1/2} \sin(\omega t - \alpha)$$

where

$$\alpha = \tan^{-1}[\omega L - (1/\omega C)]/R$$

and L is the inductance, C the capacitance and R the resistance.

The relationship between the primary, secondary and resultant fields can be represented in vector form (Figure 10.12A). The real (or in-phase) and imaginary (out-of-phase, or quadrature) components are shown on the vector diagram. The primary magnetic field is designated P and relates to the time-varying wave shown in Figure 10.11A. The induced voltage (cf. Figure 10.11B) lags $\pi/2$ (90°) behind the primary and the secondary current or magnetic field lags behind by α (cf. Figure 10.11C) and has a magnitude S. By normal conven-

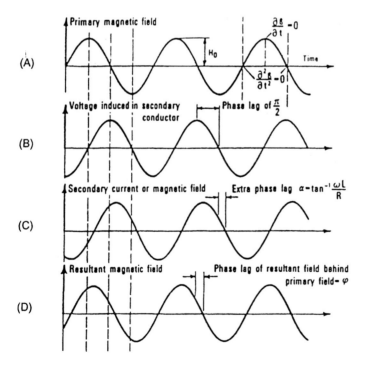

Figure 10.11 Relationships between induced voltages and associated phase lags between primary, secondary and resultant magnetic fields. Reproduced with permission from Beck (1981).

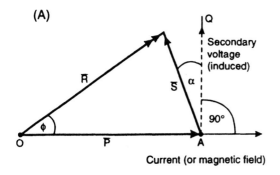

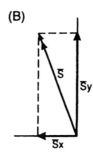

Figure 10.12 (A) Vector diagram defining the magnitudes and phase relationships of the primary and secondary fields. (B) The vectorial components of the secondary field in terms of the secondary voltage (Sy) and the current or primary magnetic field (Sx).

tions of vector diagrams, by completing the vector parallelogram, the resultant R of the primary and secondary fields (Figure 10.12A) is then defined with a total phase lag of ϕ (cf. Figure 10.11D). The secondary field S can be defined by the vectorial summation of its vertical and horizontal constituents (Figure 10.12B). Depending upon which equipment system is used, a number of these components can be measured from which an indication of the electrical properties of the sub-surface materials can be obtained.

10.2.2 Polarisation

It is important to consider two vectors P and S which differ in space by a spatial angle β (Figure 10.13A) and in phase by a phase angle ϕ. In order to calculate the resultant of these two vectors, it is necessary to resolve each into its horizontal and vertical components, denoted by suffices x and y, respectively. The mathematical summation is given in Box 10.2. The consequence of this summation process is that the resultant R always exists but varies continuously in magnitude and rotates in space. The tip of the resultant vector describes an ellipse in space, known as the 'ellipse of polarisation' (Figure 10.13C) which is inclined at an angle θ to the horizontal. The angle θ is known as the tilt or dip angle. Several EM methods (VLF and AFMAG) exploit this parameter and are known consequently as tilt-angle methods.

There are several special cases which should be mentioned. When the angle $\delta = 0$, equation (3) in Box 10.2 reverts to the equation of a straight line. This indicates that R is then a simple alternating vector and that the radiation comprises plane polarised waves. When $\delta = \pi/2$, the ellipse of polarisation is orientated such that axes are coincident with the x- and y-axes. The tilt angle θ becomes either zero or $n\pi/2$ when P and S are at right-angles and $\phi = \pi/2$. A further special case is when $\delta = \pi/2$ and $X = Y$, in which case equation (3) simplifies to the equation of a circle and the radiation is then circularly polarised.

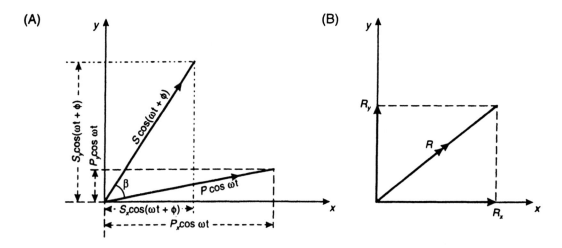

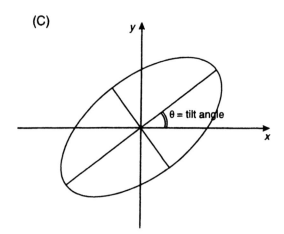

Box 10.2 Polarisation ellipse

The primary field $\boldsymbol{P}(t)$ is given by:

$$P(t) = P \sin \omega t. \tag{1}$$

The secondary field $\boldsymbol{S}(t)$ is given by:

$$S(t) = S \sin [\omega t - (\pi/2 + \phi)]. \tag{2}$$

The resultant $(\boldsymbol{R})$ can be resolved into its horizontal and vertical components, suffices x and y respectively, where $\boldsymbol{R} = i R_x + j R_y$ and $R^2 = R_x^2 + R_y^2$:

$$R_x = P_x \cos \omega t + S_x \cos(\omega t + \phi) = X \cos(\omega t + \phi_1)$$

$$R_y = P_y \cos \omega t + S_y \cos(\omega t + \phi) = Y \cos(\omega t + \phi_2).$$

continued

Figure 10.13 (A) Secondary field orientated in space at an angle β to the primary field. (B) The horizontal and vertical components R_x and R_y of the resultant of the summation of primary and secondary fields. (C) The ellipse of polarization inclined at the tilt-angle θ to the horizontal

— continued —

By solving the above equations and eliminating ωt, we obtain:

$$\frac{R_x^2}{X^2} + \frac{R_y^2}{Y^2} - \frac{2R_xR_y\cos\delta}{XY} = \sin^2\delta \qquad (3)$$

where

$$\delta = \phi_2 - \phi_1.$$

Equation (3) is the equation of an ellipse with its major axis inclined at an angle θ to the horizontal, where θ is defined by:

$$\tan 2\theta = \frac{XY\cos\delta}{X^2 - Y^2}.$$

10.2.3 Depth of penetration of EM radiation

Of prime importance in EM surveying are a consideration of the depth of penetrating of the EM radiation and the resolution as a function of depth. In an isotropic resistive medium, EM waves would travel virtually indefinitely. However, in the real world, where surface conductivities are significant, the depth of penetration is often very limited. The depth of penetration is largely a function of frequency and the conductivity of the media present through which the EM radiation is to travel. At the usual frequencies ($<5\,$kHz) used in EM exploration (excluding ground penetrating radar) attenuation effects are virtually negligible, but signal losses occur by diffusion.

A common guide to the depth of penetration is known as the *skin depth*, which is defined (Sheriff 1991) as the depth at which the amplitude of a plane wave has decreased to $1/e$ or 37% relative to its initial amplitude A_0. The mathematical definition of skin depth is given in Box 10.3. Given a known frequency for a particular equipment system, the unknown is the vertical variation of conductivity with depth. Different instrument manufacturers commonly cite effective depths of penetration for their instruments. For example, Geonics Ltd give the depth of penetration of their FEM systems (EM38/EM31/EM34) as a function of the inter-coil separation (see next chapter for details).

Box 10.3 Skin depth

Amplitude of EM radiation as a function of depth (z) relative to its original amplitude A_0 is given by:

$$A_z = A_0 e^{-1}.$$

The skin depth δ (in metres) is given by:

$$\delta = (2/\omega\sigma\mu)^{1/2} = 503(f\sigma)^{1/2}$$

where $\omega = 2\pi f$, and f is the frequency in Hz, σ is the conductivity in S/m, and μ is the magnetic permeability (usually ≈ 1). A realistic estimate of the depth to which a conductor would give rise to a detectable EM anomaly is $\approx \delta/5$.

10.3   AIRBORNE EM SURVEYING

10.3.1   Background

The earliest known airborne EM (AEM) system was developed by Hans Lundberg in 1946 and first used in eastern Canada. It consisted of two coils mounted inside the cabin of a helicopter which had to fly at only 5 m above the ground if any conductors were to be detected!

Figure 10.14 Input TEM system for airborne EM. (A) A primary magnetic field is generated that excites a receiver response (B). The measured response of the receiver is modified by an element (C) that is function of the ground properties. The decay curve (D) is sampled at designated time intervals (channels).

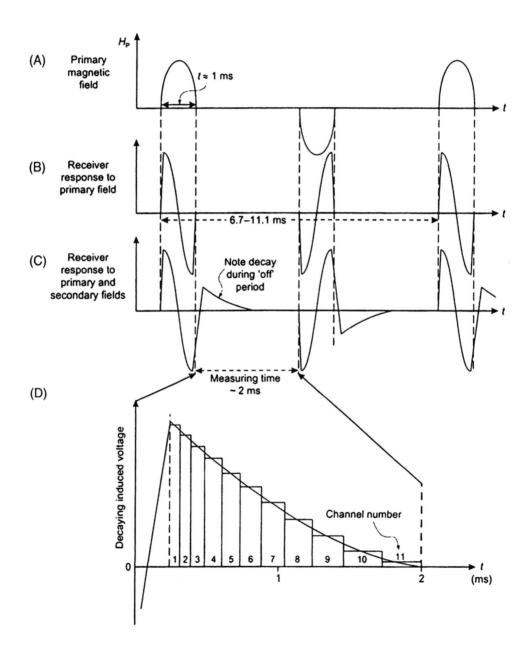

A detailed history of the development of airborne EM has been given by Pemberton (1962), Collett (1986), Becker *et al.* (1990), Palacky (1986) and Palacky and West (1991).

Following on from the early enthusiasm for AEM in the search for strategic base metals (such as copper, lead, zinc and nickel), many other airborne systems were developed. The most successful system was that developed in the late 1950s (Barringer 1962) and known as the INPUT system (INduced PUlse Transient), the principle of operation of which is shown in Figure 10.14. INPUT was developed further in the late 1970s to obtain greater depth penetration required in exploration for uranium and is now operated under the names QUESTEM (operated by Questor Surveys) and GEOTEM (Geoterrex Ltd). Further improvements to these systems were made in the mid-1980s with more powerful transmitters and modern computer technology. At this time an additional system known as PROSPECT became available along with its South African equivalent SPECTRUM; of the INPUT style AEM systems, only SPECTRUM, GEOTEM and QUESTEM are currently operational.

In the late 1970s, two clear styles of instrumentation deployment emerged, high-resolution helicopter surveying using towed instrument sondes and, particularly for deep penetration work, fixed-wing systems using rigid booms fitted to wingtips, mounted above the fuselage or on the nose and tail of the aircraft (Figure 10.15).

Applications of AEM surveying (excluding ground penetrating radar) to other than base metal exploration began in the 1960s with groundwater investigations, and later spread in the 1970s to include other forms of geological mapping, exploration for kimberlites in South Africa, and in the 1980s, mapping of Quaternary deposits in France, coal and lignite prospecting, detection of palaeochannels and salinity mapping in Australia, and shallow-water bathymetry and sea-ice thickness determinations in the USA. Further details of the range of applications of AEM can be found in the symposia edited by Palacky (1986) and Fitterman (1990), for example. Ground penetrating radar in the form of radio echosounding has been used from aircraft in polar regions to investigate major ice sheets since the late 1950s (see also Chapter 12). It is increasingly being used over glaciers at lower latitudes (e.g. Kennett *et al.* 1993).

10.3.2   AEM systems

The general principle of airborne EM surveying is shown in Figure 10.16. A powerful transmitter is mounted on the aircraft to generate the primary field in active systems, with a receiver either towed below in an instrument pod known as a 'bird' or on a separate part of the aircraft. The most commonly used systems are deployed from helicopters as they can be operated most easily at low flying heights and are much more manoeuvrable than fixed-wing aircraft.

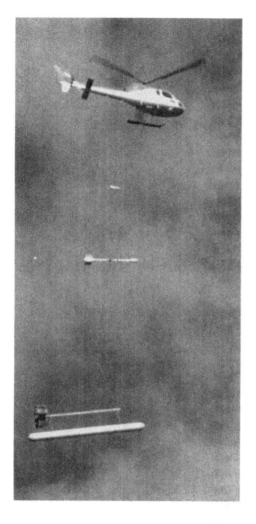

(A)

Figure 10.15   (A) Eurocopter AS350B towing three instrument pods. They are (from the top) 2-channel VLF-EM, total-field caesium magnetometer, and a 5-frequency electromagnetic induction system. (B) Cessna 404 fixed-wing aircraft with a total-field magnetometer mounted in a tail stringer. Photographs courtesy of Aerodat Inc., Canada

(B)

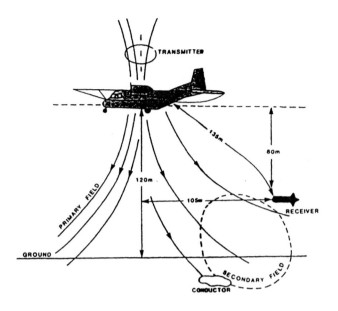

Figure 10.16 Principle of airborne electromagnetic surveying. The system shown deployed is of the towed-bird type. Reproduced with permission from Palacky and West (1991)

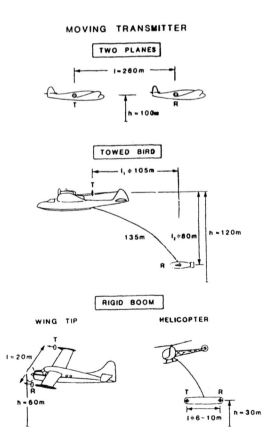

Figure 10.17 Transmitter–receiver geometry of five basic styles of active airborne EM systems. Reproduced with permission from Palacky and West (1991)

The various configurations of deployment of AEM systems are shown in Figure 10.17. There are two main types of transmitter–receiver geometries, towed bird and rigid boom systems. In some helicopter arrangements, the transmitter and receiver are both mounted in an instrument pod which is suspended below the aircraft. Further details of AEM systems have been given by Palacky and West (1991).

10.4   SEABORNE EM SURVEYING

10.4.1   Background

The principal difference between airborne and seaborne EM applications is largely one of scale. Whereas airborne surveys may involve flying heights of typically several tens to hundreds of metres, with transmitter–receiver distances of the order of 20–135 m, seaborne systems may have separations of tens of kilometres. Marine deployment of EM systems is usually for large-scale, crustal investigations and requires specialised instrument packages (Chave *et al.* 1991). There are a few examples of where land-based EM systems, such as a Geonics EM34, have been deployed in rubber inflatable boats and towed over shallow freshwater lakes and rivers in engineering investigations.

The main methods that have been adapted for use in the marine environment are magneto-telluric, magnetometric resistivity, and frequency- and time-domain systems. They have been reviewed in detail by Filloux (1987) and Chave *et al.* (1991).

The critical factor in all marine EM sounding is that the seawater is extremely conductive, and much more conductive than the geological materials at or below the seafloor. Seawater conductivity is strongly dependent upon salinity and temperature. The uppermost sediments under the ocean are usually water-saturated and have conductivities of the order of 0.1–1 S/m. This value decreases with increasing lithification and diagenesis which reduce the *in situ* porosity. Basaltic crust and upper mantle peridotite have conductivities ranging from 0.1 S/m at the base of the overlying sediments to three orders of magnitude less at a depth of about 10 km.

10.4.2   Details of marine EM systems

10.4.2.1   *Magneto-telluric (MT) methods*

All oceanic MT work has been specifically to probe the deep lithosphere and asthenosphere to obtain a vertical EM structural model to depths of hundreds of kilometres. The MT method is currently the only geophysical method capable of obtaining information about the electrical properties at depths greater than 30 km.

The only commercially available instruments for the measurements of a magnetic field are fluxgate magnetic sensors, with sensitivities of the order of 0.5–1 nT. These sensors are deployed directly on the seabed.

There are two types of device for the measurement of the electric component: (a) long-wire units, and (b) short-arm salt bridges. The long-wire system comprises an insulated wire typically 500 m to 1000 m long with Ag–AgCl electrodes connected to the ends of the wire and to a recording unit (Webb *et al.* 1985). The short-arm bridge apparatus utilises electrodes with spacings of only a few metres and salt bridges. Each salt bridge consists of a hollow tube attached to an Ag–AgCl electrode at one end and open to the sea at the other. The entire electrical unit has four arms (salt bridges and electrodes) which are spread out in the form of a horizontal cross which sits on top of a vertical cylinder housing the recording instruments. The base of this cylinder is located on a detachable tripod which, when lowered to the seafloor, is placed in contact with the seabed. Once measurements have been completed, the instrument module with salt bridges is released from the tripod and rises to the sea surface under slight positive buoyancy for subsequent recovery.

10.4.2.2 *Magnetometric resistivity (MMR)*

The marine version of MMR developed by Edwards *et al.* (1985), known as MOSES (Magnetometric Off-Shore Electrical Sounding), has been used for deep crustal sounding, mapping sulphide deposits near mid-ocean ridges (Wolfgram *et al.* 1986) and in the study of submarine permafrost below the Beaufort Sea. The general scheme of the MOSES method is shown in Figure 10.18.

The transmitter comprises a vertical long-wire bipole which extends from the sea surface to the seabed. A commutated current is fed

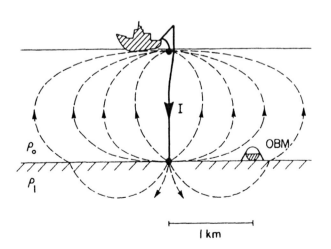

Figure 10.18 Schematic to illustrate the principle of the MOSES method. Current is passed via two electrodes, one at the sea surface and the other on the seafloor. The relatively small amount of current that enters into the resistive crust is proportional to the ratio of the conductivity of the crust to that of seawater. Only this small current contributes to the aximuthal magnetic field measured at a point on the seabed. OBM = ocean bottom monitor. Reproduced with permission from Chave *et al.* (1991).

to two large electrodes at each end of the vertical wire. The return electrical current passes through the seawater and the near-surface materials of the seafloor. A remote receiver located on the seafloor consists of two orthogonal horizontal component fluxgate magnetometers. Two orthogonal components of the magnetic field are measured as a function of frequency and source–receiver distance. The remote receiver consists of a concrete anchor shaped like an inverted cone into which a spherical instrument housing is located. The magnetometers are located within the detachable pressure case which can be released remotely from the concrete anchor for subsequent recovery.

A variation of the above system, called ICE-MOSES, was developed by Edwards *et al.* (1988) for use through sea ice. The sensor design is quite different from the original MOSES version. The sensor is deployed folded (and subsequently recovered) through a 25 cm diameter hole cut in the sea ice. Once through the hole, the unit unfolds to form a horizontal square which is lowered to the seabed. Along two sides at right-angles to each other are located the sensors which consist of coils wound on soft iron laminated cores and housed in stainless steel jackets.

ICE-MOSES is particularly important as it can be used to help define the physical properties of seafloor sediments to a depth of several hundreds of metres. Of particular interest in the Beaufort Sea, where ICE-MOSES was first used, is a seismically important permafrost layer between 100 and 600 m thick under seawater 10–100 m deep. This permafrost horizon is of importance for two reasons. First, a detailed knowledge of this layer is essential if reflection surveys undertaken in the same areas are to be interpreted accurately. Secondly, pockets of gas hydrate can be contained within the permafrost and these can be a possible resource as well as a hazard to drilling to deeper targets.

10.4.2.3   *Controlled-source EM methods*

Controlled-source EM systems use time-varying electric and magnetic dipole sources of known geometry to induce electric currents in the various conducting media present. The electric or magnetic character of the induced currents can be determined, from which estimates of the vertical electric conductivity structure of the geological materials present can be made. There are four basic source–receiver types but many combinations. The four are: *vertical and horizontal electric dipoles* (VED and HED) and *vertical and horizontal magnetic dipoles* (VMD and HMD).

In contrast to the land-based equivalent, marine controlled-source EM systems have both the source and receiver immersed in a conducting medium, and the electrical structure in both the seawater and the sub-seafloor materials affect the total induction achieved and thus

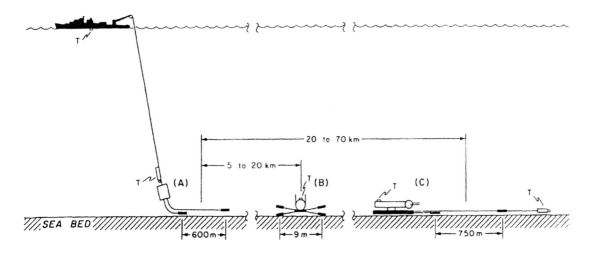

have to be taken into account in the interpretative modelling. In cases involving shallow water, for example over continental shelves, the position of the air/sea surface interface also has to be taken into consideration.

Three systems, two frequency-domain and one time-domain, will be described briefly to illustrate the diversity of systems currently being developed. The first is a submarine horizontal electric dipole (HED) frequency-domain system produced by Scripps Institution of Oceanography for deep sounding of the oceanic lithosphere. The source is a long (0.5–1 km) insulated cable terminating in stainless steel electrodes 15 m long. Receivers to detect the horizontal electric field are placed on the seabed between 1 and 200 km from the source. There are two types of electric receivers (Figure 10.19):

- The ELectric Field (ELF) free-fall recorder consists of a pair of rigid orthogonal antennae, each 9 m long, to the ends of which Ag–AgCl electrodes are fixed. ELF receivers are deployed between 5 km and 20 km from the transmitter.
- The Long antenna EM recorder (LEM) consists of 200–3000 m long insulated copper wire terminated by 0.5 m long Ag–AgCl electrodes. LEM recorders are placed up to 100 km or more away from the source.

The second basic system is also frequency-domain and is produced by Scripps for use over shallow continental shelves (Figure 10.20). The transmitter is made up of two 7 m long copper tubes 7 cm in diameter connected by 50 m of cable and powered directly by the survey vessel. The receiver array comprises a string of Ag–AgCl electrodes along a cable several hundred metres in length, all of which is in contact with the seabed. At the front end of the receiver array is a recording

Figure 10.19 Typical layout for a horizontal electric dipole (HED) deep-sounding experiment. Power is supplied from a surface source (e.g. ship) to the seafloor transmitter (A) through a single conductor with a sea-water return. The transmitter comprises an insulated antenna (with bare ends) of about 600 m length. Receivers are placed at ranges from 5 to over 70 km from the transmitter. Receivers may by either (B) an electric-field recorder (ELF) with a pair of rigid, orthogonal antennae of 9 m span, or (C) a long-antenna EM recorder (LEM), where the potential is measured between the ends of a 200–300 m insulated copper wire. Acoustics transponders (T) are used to locate all the seafloor components from a surface vessel. Reproduced with permission from Chave et al. (1991)

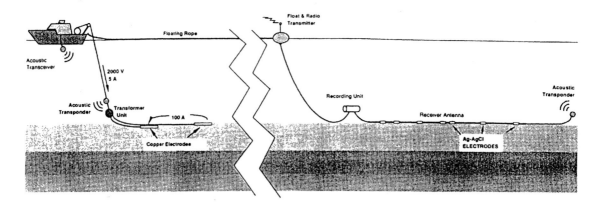

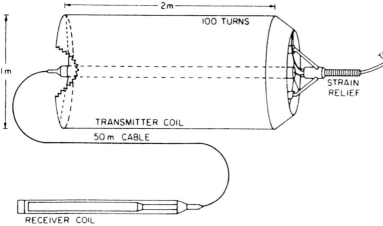

Figure 10.20 (*top*) Schematic to illustrate the components of a towed frequency-domain profiling system. The source antenna is towed immediately behind a research ship and is powered by the ship's generators. The receiving antenna is towed further behind from a radio-equipped buoy and consists of an array of Ag–AgCl electrodes. Acoustic transponders are used for location purposes. Reproduced with permission from Chave *et al.* (1991)

Figure 10.21 (*left*) Schematic of a horizontal magnetic dipole (HMD) transmitter that is connected to a surface vessel by an electric cable. The receiver is made up of a coil wound on an iron core and is encased in a protective plastic sleeve, all of which is streamed 50 m behind the transmitter. Reproduced with permission from Chave *et al.* (1991)

unit which is connected to a float and radio transmitter. The float is connected to the survey vessel by a floating rope, the length of which can be changed to alter the source–receiver separation. The point of having a surface radio transmitter is to allow the real-time relay of measured data from the submerged recorder unit, which also stores the data on to tape, directly to the survey vessel.

The third, time-domain, horizontal magnetic dipole (HMD) system (Figure 10.21) has been constructed by the University of Toronto, Canada. The transmitter comprises a 2 m long, 1 m diameter fibreglass cylinder in which 100 turns of wire are evenly embedded. Current is supplied to the transmitter from two car batteries located on the survey vessel. The polarity of the current is reversed every 5 ms to provide the transient EM signal. The receiver, which is made up of a modified iron core coil encased in a polycarbonate tube, is towed 50 m behind the transmitter. The entire source–receiver array is placed on to the seabed and is stationary during each measurement, which takes 90 s. The survey vessel is able to maintain headway by paying out additional cable during measurement periods and reeling in the extra cable between survey points. An advantage of this system is that the field source–receiver array is relatively small with a consequential improvement in the ease of deployment over some of the larger, and more unwieldy, frequency-domain systems. For any of these systems to become operational commercially, ease of operation is a major factor to be considered.

10.5   BOREHOLE EM SURVEYING

Whereas surface and airborne EM systems have regular geometries of sources and receivers, the addition of the third dimension via a borehole leads to an increased number of possibilities with associated complexities in interpretation. Borehole EM surveying differs from inductive well logging, which is used predominantly within the hydrocarbon industry, by virtue of the ability of being able to detect a conductive body at a significant distance away from the borehole. An induction logging device senses only those features through which the borehole has actually passed or within the near-field around the borehole (Figure 10.22).

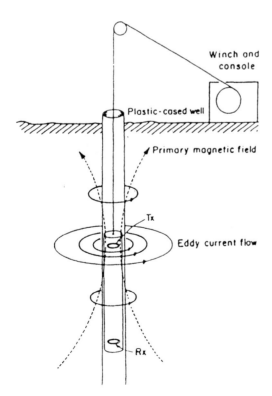

Figure 10.22 The basic principle behind an electromagnetic induction logger for use in boreholes. Reproduced with permission from McNeill (1990)

The principle of operation is the same as for ground conductivity meters. The system is able to measure the conductivity of the materials outside of a plastic cased borehole or well with diameters in the range 5–20 cm. The measurements are insensitive to the usually much more conductive borehole fluid within the casing (McNeill *et al.* 1988). Eddy currents are induced concentrically around the borehole using an intercoil separation of 0.5 m. This configuration provides a reasonable vertical resolution while at the same time maintains an adequate radial range of investigation (McNeill 1990). Drill-hole EM methods have been reviewed by Dyck (1991), and more details can be found therein.

There are three types of system in borehole surveying: dipole–dipole EM, rotatable-transmitter EM, and large-loop EM (LLEM) methods, of which the last is the most commonly used in mineral prospecting. The basic transmitter–receiver geometries are shown in Figure 10.23.

The dipole-dipole system has two coaxial coils separated by fibreglass rods with the transmitter preceding the receiver down the drill hole. Measurement points are taken as being the midpoint between transmitter and receiver. In-phase and quadrature components of the secondary magnetic field are measured as a percentage of the primary field. As the downhole system is deployed on a series of rods, the method can be used in near-horizontal and upwardly inclined holes,

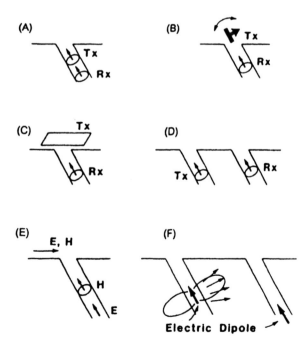

Figure 10.23 Drill-hole EM systems. (A) Dipole–Dipole EM. (B) Rotatable transmitter EM (with transmitter Tx shown side-on). (C) Large-loop EM. (D) Hole–hole dipole EM (variation of (A)). (E) remote transmitter (e.g. VLF radio source) for downhole measurement of electric and/or magnetic field. (F) Hole–hole wave propagation. Reproduced with permission from Dyck (1991)

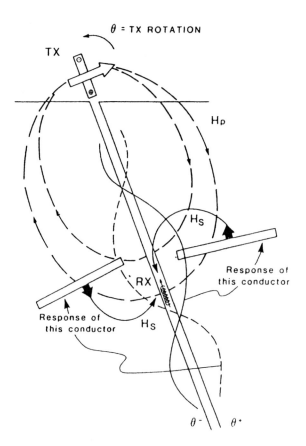

Figure 10.24 Magnetic fields produced by a rotatable transmitter and target conductor. The transmitter coil is rotated about an axis perpendicular to the plane of the diagram. The conductor to the right of the drill hole produces a negative (i.e. downward) component of secondary field H_s at the receiver position shown. A counterclockwise ($+\theta$) rotation of the transmitter is required to achieve a null by offsetting H_s with a component of H_p, assuming negligible change in transmitter–conductor coupling as the transmitter is tilted. Reproduced with permission from Dyck (1991)

and is then limited only by the ability to move the probes within the hole.

The rotatable-transmitter system is a version of the dipole–dipole method, but the transmitter remains at the drill hole collar throughout the survey while the receiver is moved up and down the hole (Figure 10.24). The receiver probe is moved down the hole in discrete intervals of several metres at a time. At each measurement point, the surface transmitter coil is rotated until a null point in the sensor is reached and the corresponding angle of tilt is recorded. The method is analogous to the surface tilt-angle technique.

The general layout for borehole LLEM surveying is shown in Figure 10.25. A loop transmitter is deployed at the ground surface adjacent to the borehole down which a detector is run to obtain a profile. Typical ground loop dimensions range from 100 m to 1000 m and are comparable to the depth of the drill hole being investigated. One ground loop, in conjunction with profiles down a number of drill holes from the surface and from within a mine gallery, are sufficient to resolve a sub-surface conducting target

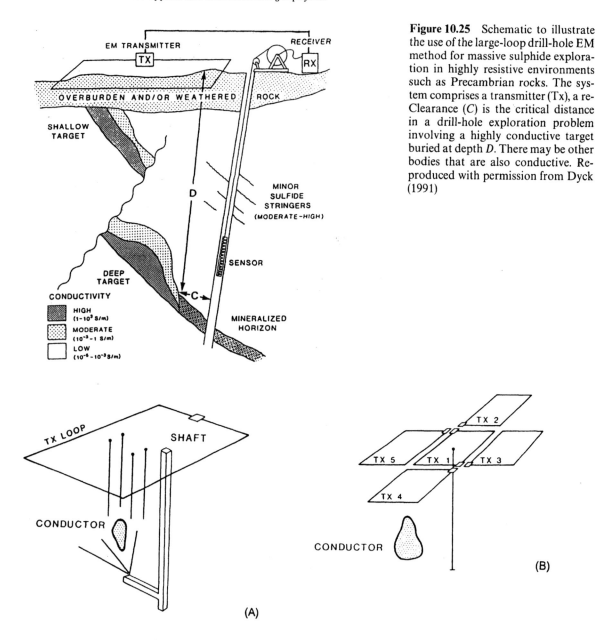

Figure 10.25 Schematic to illustrate the use of the large-loop drill-hole EM method for massive sulphide exploration in highly resistive environments such as Precambrian rocks. The system comprises a transmitter (Tx), a re- Clearance (C) is the critical distance in a drill-hole exploration problem involving a highly conductive target buried at depth D. There may be other bodies that are also conductive. Reproduced with permission from Dyck (1991)

Figure 10.26 Transmitter layouts for surveying (A) a group of drill holes collared underground; and (B) a single isolated drill hole. Tx 1–5 are successive locations of the transmitter loop. Reproduced with permission from Crone (1986).

(Figure 10.26A). In contrast, if only one drill hole is available, one loop on its own does not provide azimuthal information necessary to locate the target. Consequently, a number of loop positions located around a collared borehole (Figure 10.26B) can be used to provide the additional information required.

There are three types of LLEM system depending upon the received primary waveform of system function, namely impulse-type, step-function type (both of which are TEM systems), and multi-frequency (FEM) type. Further details of these systems are given by Dyck (1991). Other systems that are available include down-hole VLF, and inter-hole wave propagation (e.g. Newman 1994) which can include borehole tomographic techniques (see also 'borehole radar tomography' in Chapter 12). Three component (magnetic field) systems are currently under development, although one prototype has been successfully deployed by Boliden Mineral AB, in Sweden (Pantze *et al.* 1986). A 1×1 km ground loop was used in an FEM system which operated at two frequencies, 200 Hz and 2000 Hz. Three sensors were mounted in a 32 mm diameter probe with the y-axis always being horizontal, x parallel to the long axis of the probe, and z always at right-angles to x and y. In-phase residuals (computed after the removal of the primary field and the background response caused by the host rock) were plotted as a function of profile distance along the drill hole. The shape and size of the excursion of each component away from a normal value provided information about the location (depth and azimuth) of a sub-surface conductive target.

Chapter 11
Electromagnetic methods: systems and applications

11.1 INTRODUCTION

A basic introduction to the various types of electromagnetic systems and their respective operating principles has been given in Chapter 10. Here, five of the most commonly used types of method are discussed in more detail with case histories to illustrate their various applications. The five types of system can be classified into three groups:

- *near-field systems*, where the source is relatively close to the receiver;
- *far-field systems*, where the source is located at a very large distance from the receiver such that the EM wave can be treated as a plane wave (VLF surveying);
- *natural-source EM systems*, where no active generation of artificial electromagnetic radiation is necessary because naturally occurring ground currents provide the source.

The range of individual instruments is large, especially if airborne, seaborne and drillhole systems are included along with those deployed solely from the ground surface. It is not possible, therefore, to provide as in-depth account of every system that is available. The selection presented here is of those techniques most likely to be used in engineering, environmental and archaeological applications, although in some cases some techniques are used almost exclusively in mineral exploration. More detailed descriptions of specific instrument types have been given, for example, by Nabighian (1987, 1991) and coauthors; technical specifications of individual systems can be obtained from instrument manufacturers.

11.2 CONTINUOUS-WAVE (CW) SYSTEMS

11.2.1 Tilt-angle methods

Tilt-angle methods have been used extensively in both ground and airborne surveys, particularly for mineral exploration. The method obtains its name from the measurement of the angle (tilt) of the resultant of the applied primary field and the induced secondary fields arising from a buried conductive body (Figure 11.1), such as a buried massive sulphide orebody.

Given a horizontal primary field direction, induced eddy currents within a buried conductor generate a secondary magnetic field whose lines of force are concentric around the source of the currents, commonly taken as being the uppermost edge of the body. The secondary field is inclined upwards on the side nearest the transmitter and hence the resultant is also inclined upwards (positive tilt angle). Immediately above the conductive body, both primary and

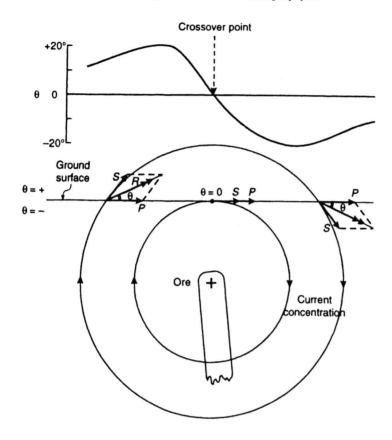

Figure 11.1 Tilt-angle (θ) profile over a conductive orebody arising from a plane EM wave from a remote vertical transmitter. *P* indicates the direction of the primary field vector, and *S* and *R* are the secondary and resultant field vectors, respectively. After Parasnis (1973), by permission

secondary fields are horizontal and in the same direction, hence the tilt-angle is zero. On the side furthest from the transmitter, the secondary field is inclined downwards and thus so too is the resultant (negative tilt angle). From a tilt-angle survey with a vertical coil axis (horizontal primary field), the conductive orebody is located at the point where the tilt angle passes from being positive to negative (the *crossover point*). When the primary field is orientated vertically downwards, the tilt angle passes through a minimum immediately over the conductive body (Figure 11.1).

When a conductive body is close to the ground surface, the steepness of tilt and the horizontal rate of change of tilt angle are greater than would be produced over a deep-seated conductive body. Furthermore, if the conductive body is vertical and the profile surveyed is at right-angles to the sub-surface target, the tilt-angle profile should be symmetrical about the crossover point. If the profile is asymmetrical, then the amount of asymmetry is indicative of the degree of dip of the sub-surface target. As the dip of the conductive target decreases (becomes less vertical relative to the ground), so the

amount of asymmetry increases. Principal tilt-angle methods currently in use are the VLF and AFMAG techniques which are described in
more detail in Sections 11.4 and 11.6 respectively.

11.2.2 Fixed-source systems (Sundberg, Turam)

Where a fixed source such as a large loop of wire or a long grounded
cable is used, as in the Sundberg method, the primary magnetic field is
inclined towards the ground (Figure 11.2). In the presence of a vertical, thin, and laterally continuous conductor, eddy currents induced
at the top edge of the sub-surface body generate a secondary magnetic
field which interferes with the primary field. The resultant or secondary vertical component of the magnetic field is measured by the
receiver. In addition, the gradient of the secondary magnetic field can
be measured and displayed as a profile.

  In the case of a vertical thin conductor, the secondary magnetic
field is inversely symmetrical about the crossover point immediately
above the top of the sub-surface target (Figure 11.2). At the same
location, the gradient of the secondary field reaches a minimum value.
If the conductive sheet is not vertical, the anomalies produced
increase their degree of asymmetry with decreasing dip angle.

  Care has to be exercised when using the Sundberg method because,
as with many other EM techniques, topography affects the quality of
data acquired. If the ground surface over which a survey is being
undertaken is rough, such that the receiver as at an elevation significantly different from that of the source, appropriate topographic

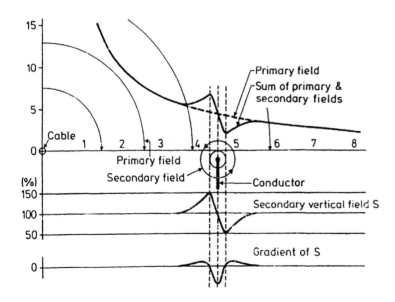

Figure 11.2 In the Sundberg method, the source is a long wire which
generates the primary field that produces a secondary field over a sub-
surface conductor. The anomaly due
to the sub-surface conductor is superimposed on the primary field but can
be residualised by the removal of the
primary field. The horizontal gradient
of the residualised secondary field
yields a minimum at a point directly
over the top of the conductor. From
Parasnis (1986), by permission

corrections have to be applied to the real component data of the vertical primary field in order to compensate.

In the case of the Turam method, two receiver coils (usually placed horizontally) are separated by a fixed distance (c) and are moved along profiles at right-angles to the source wire. Given a number of stations (1, 2, 3, 4,...) at which each coil is located in turn, the amplitude (V) and phase (α) of the vertical electromagnetic field are measured at each location. The ratio of the amplitudes at each successive pair of stations (e.g. V_1/V_2, V_2/V_3, V_3/V_4,...) and the horizontal gradient of the phases $((\alpha_2 - \alpha_1)/c$, $(\alpha_3 - \alpha_2)/c$, $(\alpha_4 - \alpha_3)/c$,...) are plotted at the location of the midpoint between the coils along the profile. In addition, as the primary field (P) decreases in amplitude away from the source, the measured amplitude ratios are divided by the normalised amplitude ratios $(P_1/P_2, P_3/P_2, P_4/P_3,...)$ to give $V_1 P_2/V_2 P_1$, $V_2 P_3/V_3 P_2$, $V_3 P_2/V_4 P_3$,..., which are known as *reduced ratios* (RR). If either or both receiver coils are located at an elevation different from that of the source cable, then a topographic correction has to be applied to the reduced ratios.

It is usual to plot a Turam profile in terms of the reduced ratios and horizontal gradient of phase (e.g. Figure 11.3). As in the case of the Sundberg anomalies, symmetrical anomalies of both measured parameters should be obtained over a vertical conductor. If dip decreases from the vertical, the degree of asymmetry increases; higher values of the reduced ratio and smaller values of horizontal gradient of phase are measured over the downdip side of the conductive target.

Figure 11.3   Reduced-ratio and phase-difference gradient profiles over an inclined conductor obtained using a Turam survey

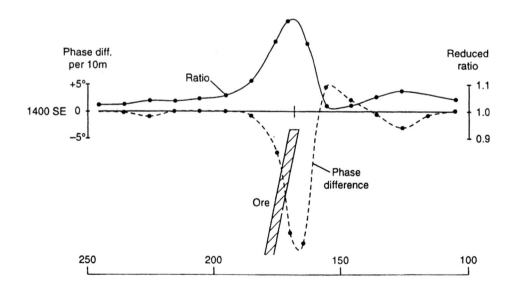

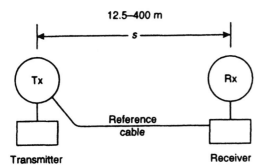

Figure 11.4 Moving dual-coil EM system; circles indicate the transmitter (Tx) and receiver (Rx) coils

11.2.3 Moving-source systems

The most commonly-used EM surveying method in environmental and engineering geophysics is the *moving-source dual-coil method* (McNeill 1990). Two separate coils connected by a reference cable provide the basis of the system (Figure 11.4); one coil serves as a transmitter to generate the primary field and the other acts as a receiver. The inter-coil separation is maintained at a fixed distance and the dual-coil pair is moved along the survey transect in discrete intervals. The point of reference for the measurement is usually the mid-coil position. Typically, dual-coil systems measure the quadrature component only or both the quadrature and in-phase components.

In the case of *ground conductivity meters* (GCM), as manufactured by Geonics Ltd, for example, the instrument provides a direct reading of the quadrature component as the apparent conductivity (σ_a) of the ground in units of millimhos per metre (SI equivalent units are millisiemens per metre (mS/m)). The in-phase component is measured in parts per thousand.

The ratio of the inter-coil spacing (s) divided by the skin depth (δ) is known as the *induction number* (B) (see Box 11.1). Where the induction number is much less than one, then the ratio of the secondary to the primary magnetic fields at the receiver is directly proportional to apparent conductivity (σ_a). If the ground is entirely homogeneous and isotropic, the instrument should give a measure of the true conductivity of the ground. However, real geological materials comprise a mixture of constituents, most notably a solid matrix with pore spaces that may be partially or fully saturated with pore fluids which, in some cases, can be highly conductive. Furthermore, the ground usually is made up of layers (soil over weathered material over bedrock, for example). Each material has its own diagnostic true conductivity and will contribute to the measured apparent conductivity value.

Box 11.1　Apparent conductivity at low induction numbers
(McNeill 1980)

The skin depth (δ) is given by:

$$\delta = (2/\omega\mu_0\sigma)^{1/2} = (2i)^{1/2}/\Gamma$$

and

$$\Gamma s = (2i)^{1/2}s/\delta = (2i)^{1/2}B \quad [\text{as } B = s/\delta]$$

where:

$\omega = 2\pi f$ and f is the frequency (Hz)

$\mu_0 = $ permeability of free space

$i = \sqrt{-1}$

$\Gamma = (i\omega\mu_0\sigma)^{1/2}$

$\sigma = $ conductivity

$s = $ inter-coil separation (m).

The ratio of the secondary (H_s) to primary (H_p) magnetic fields at the receiver at low induction numbers (i.e. $B \ll 1$) is given by:

$$H_s/H_p \approx iB^2/2 = i\omega\mu_0\sigma s^2/4.$$

The measuring instrument is designed to ensure that with the selected frequency (f), a given inter-coil separation (s), a designed response of H_p for a given transmitter, the only unknowns are H_s, which is measured by the instrument, and the ground conductivity (σ).

Put another way:

$$\sigma_a = (4/\omega\mu_0 s^2)(H_s/H_p)_q$$

where the subscript q denotes the quadrature phase.

A ground conductivity meter responds to the conductivity composition of the ground, depending upon the orientation of the coils. Typically, there are two modes: horizontal coils with a vertical magnetic dipole (VMD), and vertical coils with a horizontal magnetic dipole (HMD). If a thin semi-infinite horizontal layer is located at a normalised depth z (where z is the actual depth divided by the inter-coil separation), then the relative contribution of that thin layer to the secondary magnetic field (H_s) at the receiver is denoted by the impulse response function (ϕ). The form of this response for both the vertical and horizontal magnetic dipoles (Figure 11.5) is very important. In the case of the vertical magnetic dipole, there is little relative

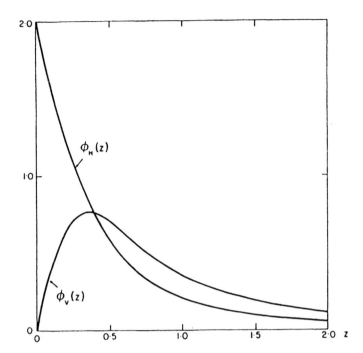

Figure 11.5 Impulse response functions (ϕ) for horizontal and vertical magnetic dipoles as a function of normalised depth (McNeill 1980)

contribution from the very near surface. Consequently, in this configuration, the technique is relatively insensitive to features very near to the surface. The maximum relative contribution arises from a normalised depth of $z = 0.4$. In contrast, the relative response for a horizontal magnetic dipole decreases with depth from a maximum at the surface. In this case, the dipole configuration makes the system very sensitive to near-surface features.

Rather than considering the relative contribution of a single thin layer at a depth z, the relative contributions of all materials within the depth z to the secondary magnetic field (or apparent conductivity) can be calculated. The sum of all the relative impulse responses for all the depths to z is expressed mathematically as the integral of all the impulse response functions. The total contribution so calculated is called the *cumulative response function*, $R(z)$, and has different forms for the VMD and HMD configurations (Figure 11.6).

The graphs of cumulative response function for each dipole orientation can be used in the calculation of true conductivities for simple 2- or 3-layer models. An example of such a calculation is given in Box 11.2.

Consider the case of a model involving two semi-infinite horizontal layers where the inter-coil separation is much less than the skin depth for all layers. The measured apparent conductivity is made up of the contribution of the first layer plus a contribution from the underlying

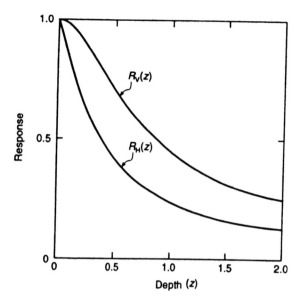

Figure 11.6 Cumulative response functions (R) for horizontal and vertical magnetic dipoles as a function of normalised depth (McNeill 1980)

Box 11.2   Use of the cumulative response function to calculate a layered-earth model from a measured apparent conductivity

In a 2-layer model, the contribution from the upper layer to the measured apparent conductivity σ_a is given by:

$$\sigma_a = \sigma_1 (1 - R)$$

where σ_1 is the true conductivity of the first layer and R is the cumulative response function for the appropriate dipole orientation (VMD or HMD). R is a function of normalised depth $(z = d/s)$, where d is the actual depth and s is the inter-coil separation.

The contribution arising from the underlying layer is given by:

$$\sigma_a = \sigma_2 R$$

where σ_2 is the true conductivity of the underlying material.

The total contribution to the apparent conductivity is the sum of these two contributions, such that:

$$\sigma_a = \sigma_1 (1 - R) + \sigma_2 R.$$

For a 3-layer case:

$$\sigma_a = \sigma_1 (1 - R_1) + \sigma_2 (R_1 - R_2) + \sigma_3 R_2$$

continued

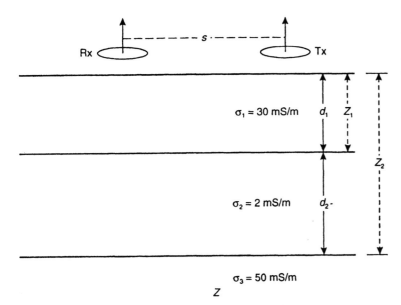

Figure 11.7 Example of the use of cumulative impulse response functions (R) and true conductivities (σ) to calculate a theoretical value of apparent conductivity (σ_a) for a given 3-layer earth model. Given $R_1 = 0.71$ and $R_2 = 0.31$, and using Figure 11.6 for $d_1 = 5$ m and $d_2 = 10$ m, for $s = 10$ m, then: $\sigma_a = 30(1 - 0.71) + 2(0.71 - 0.31) + 50 \times 0.31 = 25.0$ mS/m, using the formula for a 3-layer case given in Box 11.2

---
continued
---

where $\sigma_{(1,2,3)}$ are the true conductivities and $R_{(1,2)}$ are the cumulative response function values of the first, second and third layers, respectively.

Examples of the calculation are given in the caption to Figure 11.7.

---

material. The weighting of the conductivity contribution is provided by the cumulative response function. A similar logic follows for a 3-layer case.

This type of calculation is useful if the likely value of apparent conductivity is required over a model where layer conductivities have been estimated along with layer thicknesses. The values of apparent conductivity so obtained are estimates and will be only as accurate as the validity of the assumptions for a given situation. If the ground is not approximated by semi-infinite horizontal planar layers, there is some lateral as well as vertical variation in conductivity and thickness, then the calculation will only be at best a rough guide. If there is a three-dimensional object within the sphere of influence of the EM measuring system, or if the interfaces between layers is sloping or non-planar, then the validity of this calculation will be substantially reduced.

In the case of an APEX MaxMin I-10 dual-coil system, up to 10 different frequencies can be selected (110, 220, 440, 880, 1760, 3250, 7040, 14 080, 28 160 and 56 320 Hz), using one inter-coil separation

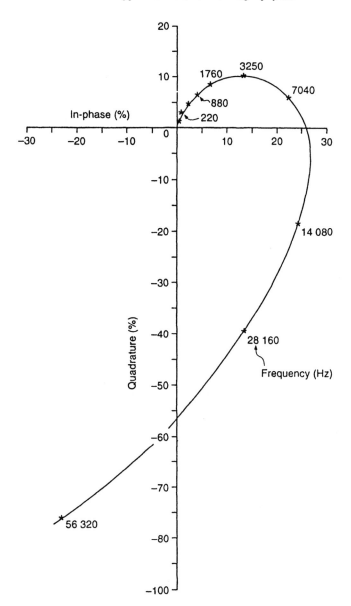

Figure 11.8 Example of a phasor diagram where the in-phase and quadrature data for each frequency (shown by asterisks) are plotted and compared with a best-fit model (solid line) for a horizontal loop EM sounding (Palacky 1991)

which can be selected over the range 20 m to 500 m. The normal mode of operation is referred to as *horizontal loop EM* (HLEM), although measurements can also be made in the perpendicular dipole position and in the tilt-angle configuration (Frischknecht *et al.* 1991). Commonly for mapping Quaternary sediments, an inter-coil separation of 100 m is used. Both the quadrature and in-phase components are measured at each frequency at a given station location and are displayed in a phasor diagram (Figure 11.8). The measured HLEM values depend on a response parameter α (Box 11.3) that is directly proportional to the conductivity of the ground it is desired to measure, and the frequency of measurement (one of the 10 available frequencies) and the inter-coil separation, both of which are chosen by the system operators; other parameters are constants (Palacky 1991). Master-curve phasor diagrams are used in the interpretation of the measured data (see next section).

Box 11.3 Horizontal loop EM response parameter α

The HLEM response parameter α is given by:

$$\alpha = \mu\sigma\omega L^2$$

where μ is the magnetic permeability of free space $(4\pi \times 10^{-7}\,\text{H/m})$, σ is the overburden conductivity, ω is the angular frequency where $\omega = 2\pi f$, and f is the frequency of measurement, and L is the inter-coil separation (m).

11.2.4 Interpretation methods

11.2.4.1 *Profiling and depth sounding*

Electromagnetic data can be analysed in a number of different ways, according to the manner in which they have been acquired. Measured parameters may be plotted as profiles or as gridded and contoured maps on which anomalous zones can be identified. These approaches tend to be qualitative and first-order interpretations.

For reconnaissance mapping or 'anomaly spotting', qualitative interpretation may suffice. However, there are certain pitfalls that can befall the unwary if the characteristic responses of certain features are not recognised. For example, it is a misconception when using a dual-coil system, such as Geonics EM31, that a target produces only a single peak over a thin conductive target; spotting anomaly 'highs' is fraught with danger!

Consider a typical apparent conductivity profile produced over a 10 cm diameter metal gas pipe buried at around 1 m (Figure 11.9A). Two peaks are evident with a strong low or even negative occurring immediately over the target. Note that the distance between the anomaly peaks is the same as the dipole length. Always check the inter-peak distance; if it is curiously similar to the inter-coil separation being used, then the target causing the anomaly is at the midpoint between coils. Depending upon the spatial sampling interval relative to the position of the target, the anomaly peaks and low may be slightly broader or narrower than one dipole length by the sampling being skewed relative to the target (see the sub-sampled profile in Figure 11.9B). The same effects may be more subdued with larger inter-coil separations.

(A)

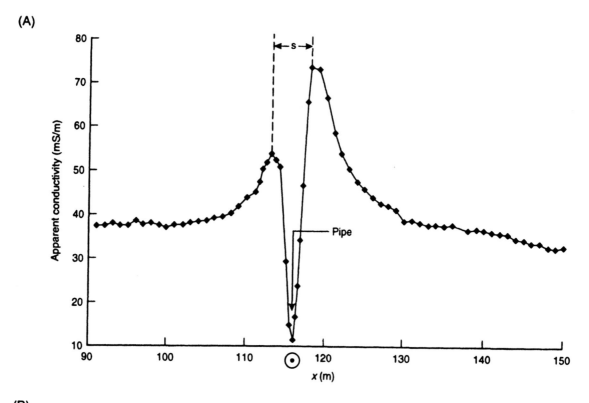

(B)

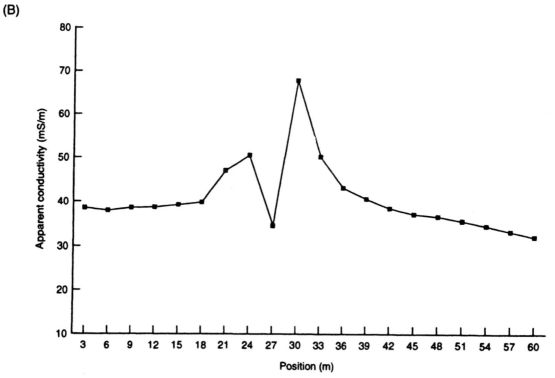

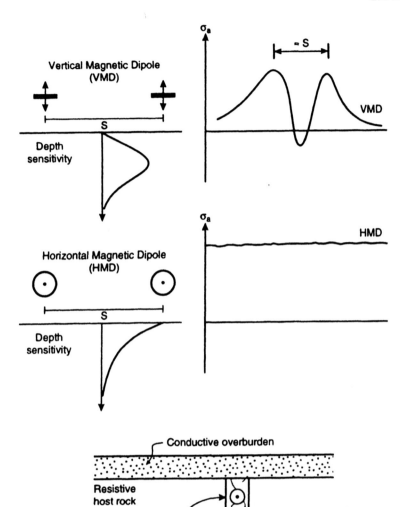

Figure 11.9 (*opposite*) Example of the effect of (A) a metal gas pipe on apparent conductivity data, and (B) the horizontal smearing of the anomaly caused by inadequate spatial sampling. Note that the peak-to-peak distance in (A) is equal to the inter-coil separation

Figure 11.10 Example of the difference in output from both vertical and horizontal magnetic dipoles over a vertical conductor for the same intercoil separation

Similarly, the shape of the anomaly will also vary depending upon which dipole orientation is used. An example of the difference of the output from both dipole orientations over a vertical conductor for the same inter-coil separation is shown in Figure 11.10.

The spatial smearing effect is also noticeable in gridded data. If the spatial sampling is too coarse, the anomalies arising from small (particularly 3-D) targets may be effectively smoothed (aliased) and the targets may be missed. The resolution of ground conductivity mapping is discussed in Section 11.2.4.3.

11.2.4.2   *Computer analysis*

More quantitative analysis can be undertaken using specialist software. Two approaches are possible: (a) using EM data to form a depth-sounding at a single location, and (b) undertaking EM profiling along a line or over a grid to produce a two-dimensional geoelectric cross-section.

In the case of depth-sounding, a limited number of data are obtained as a function of inter-coil separation (which is in itself a function of frequency; greater depth penetration is achieved using the lowest frequencies and greatest inter-coil separations) and dipole orientation. For this discussion, the range of ground conductivity meters made by Geonics Ltd will be used as these instruments are employed commonly in environmental and engineering surveys. Data obtained with other makes of instruments can be interpreted in similar ways.

If a Geonics Ltd EM34-3 ground conductivity meter is used, three inter-coil separations and two dipole orientations are available, giving a maximum number of six data points obtainable using the one instrument at a given sampling point. The apparent conductivity values measured at each inter-coil separation and with each dipole orientation are entered into a program such as EMIX-MM (Interpex Ltd, USA). If an APEX MaxMin I-10 has been used, the 10 pairs of in-phase and quadrature data can be entered into the same software. The program is used to invert the data to produce a layered earth model in which the true conductivity of each layer and its associated thickness are estimated and entered into the program. Synthetic values of apparent conductivity are calculated for the selected model and compared with those actually observed. The computer model is automatically adjusted until the difference between the measured and observed apparent conductivities satisfies some statistical criterion – e.g. an RMS error of less than 2%. The final output of the program is a vertical depth–true conductivity profile. Some versions of the software allow equivalence testing (see also Chapter 7). Correlation with borehole data can help to constrain the model layer thicknesses in order to obtain more realistic values for the layer true conductivities.

While the model so obtained may be statistically adequate, the question of geological reasonableness still has to be asked. For APEX MaxMin data, a series of master-curve phasor diagrams can be produced where the number of layers and their respective conductivity values and thicknesses can be estimated from the fit with the observed phasor diagram. An example of an interpretation is shown in Figure 11.11.

An extension of the depth-sounding interpretation is the profile inversion. Instead of having one set of sounding data, a series of values of apparent conductivity for each inter-coil separation and dipole

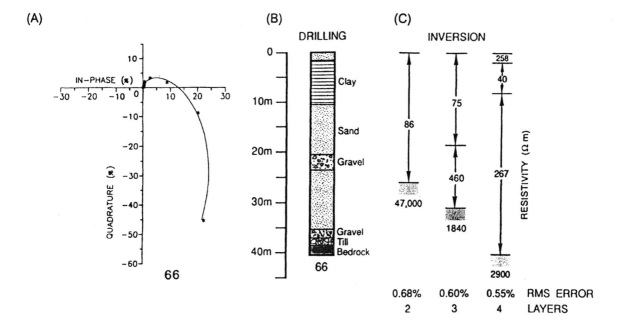

Figure 11.11 Use of model phasor diagrams in the interpretation of HLEM data. (A) Eight frequency HLEM data acquired using an APEX MaxMin instrument; (B) the drilling log; and (C) the various inversions produced using 2-, 3- and 4-layer models. From Palacky (1991), by permission

orientation are obtained along a survey line at discrete intervals. The ensuing apparent conductivity profile can be inverted using a sister programme to EMIX-MM called EMIX34PLUS (Interpex Ltd). The final output is a pseudo-geological two-dimensional section displaying true conductivities and layer thicknesses along the profile line. It is important that sufficient data be collected to provide adequate sampling both spatially and as a function of depth. At least three different inter-coil separations are required for either depth-sounding or profile-inversion to be achieved. Additional data can be obtained by supplementing the Geonics EM34-3 with the EM31 and the EM38, where appropriate. Care has to be taken in the calibration of each instrument when using more than one type on the same survey line, to ensure consistency.

11.2.4.3 Resolution

As with any geophysical technique, consideration has to be given as to the resolution achievable. In the case of dual-coil systems, for example, a number of factors have to be taken into account on any survey if it is to be completed successfully.

Ground conductivity depth-sounding and profiling are best suited to horizontal or sub-horizontal layered structures where the vertical conductivity contrast between horizons is significant. However, in environmental and engineering geophysics, EM techniques are increasingly being used at less than ideal sites. The criterion stated above is seldom applicable and, to complicate matters, the sites may often be cluttered with above-ground structures, old pipes and cables, buildings, tanks, metal fences, metal signposts, etc., each of which may degrade the quality of data and reduce the reliability of, if not negate, any quantitative analysis. In some cases it may be possible to filter out the effects of a power line or pipe if the anomaly due solely to that feature can be identified clearly.

Edge effects associated with changes of slope, or margins of sites contained in old quarries, for example, also affect the data quality. Some of these effects can be reduced in the survey design stage if the presence of these features is known in advance.

A further factor to be considered is the likely resolution of the method for a given type of sub-surface target. The more a target deviates from being a semi-infinite homogeneous horizontal half-space (i.e. one that exists only in textbooks!) the more difficult it is to detect it. For example, consider the case of a steel drum buried at 2 m in an otherwise homogeneous ground. This target should be located readily using an EM31 given a fine enough survey grid. If, however, an EM34-3 is used with 40 m coil separations, the volume of ground being sampled is increased enormously compared with that being sampled by an EM31, and consequently, the drum may be missed. Similarly, a conductive target of 1 m diameter buried at 10 m is unlikely to be resolvable with any EM method unless the utmost care is used and the site is virtually noise-free. Furthermore, a station interval of 1 m would have to be used to stand any chance of providing adequate spatial sampling.

A further consideration in the interpretation of EM data is the position of a sub-surface target with respect to the dual-coil dipoles. If a survey is being conducted with both coils aligned along the survey transect (an 'in-line' configuration) with the transmitter preceding the receiver, and an apparent conductivity anomaly is observed, the source of that anomaly may not actually lie in the line of the transect but to one side (Figure 11.12). This makes the precise identification of the location of a sub-surface target, such as a mineshaft, quite difficult. It is for this reason that one often finds the location of an apparent conductivity anomaly does not necessarily coincide with that of a magnetic anomaly arising from the same metallic conductive target. This apparent lack of coincidence can lead to misinterpretation and the feature being missed. To assist with determining lateral variability in ground conditions, it may be sensible on some sites to rotate the dual-coil system from being 'in-line' to being 'broadside'.

In the case of an EM31, the operator should remember that the instrument has a finite response time if the apparent conductivities are different between the two orientations, and allow the instrument to settle to the correct value before proceeding to the next measurement station. If measurements of apparent conductivity are made too quickly and while the instrument is settling down, inaccurate values

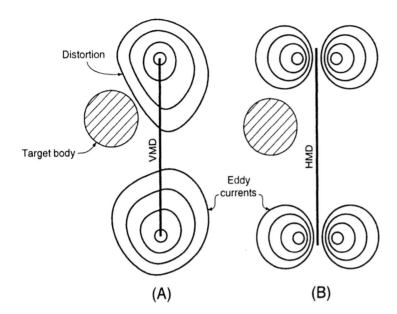

Distortion

Target body

VMD

HMD

Eddy
currents

(A)   (B)

Figure 11.12 Cartoon to illustrate the difference in eddy currents (plan view) produced by (A) the vertical and (B) the horizontal magnetic dipoles (Stoyer 1989). A conductive (or resistive) target in a resistive (or conductive) medium will distort the eddy currents even when off the line of the survey section

will be obtained, thus reducing the value of the entire EM survey. Other aspects of interpretation will be evident by reference to the various case histories presented in the next section.

11.2.5 Applications and case histories

The range of mineral exploration case histories available in the literature is quite large but there are relatively few well-documented environmental examples. While there are a number of different types of EM instruments available, the majority of environmental case histories make reference to EM systems made by Geonics Ltd.

11.2.5.1 Location of orebodies

The most common use of EM profiling is undoubtedly exploration for mineral deposits, mostly as a means of locating possible targets. EM methods have been instrumental in the location of many significant economic orebodies (Frischknecht *et al.* 1991): examples are the Temegami Mine, Ontario; the Poirier deposit, Quebec; the Faro Deposit, Yukon; the Caribou Deposit, New Brunswick; and the Kidd Creek Mine, Ontario, among many others.

EM results on their own are not usually diagnostic of whether a conductive body is economic or not. For example, it may be difficult to distinguish between a carbonaceous, a graphitic or a sulphide body purely on conductivity values alone. Consequently, other geophysical techniques, including other EM methods, are used in conjunction with each other to aid the interpretation.

There are many case histories describing the wide range of EM methods in mineral exploration and a good number have been described by Frischknecht *et al.* (1991). It is important to note that the geophysical responses observed in one geological environment are not necessarily exactly the same in other areas. The specific geophysical responses are determined by the individual blend of mineral and structural associations present at a given site and, in some cases, may be unique to individual geographical locations.

One example is given here to illustrate the combined approach needed to differentiate between target types. Slingram and VLF EM methods were used by Barbour and Thurlow (1982) in exploration where long graphitic zones are common in Newfoundland. In addition, gravity data were used to help differentiate between probable sulphide occurrences and massive sulphide deposits. At the Tulks East deposit, black graphitic shales and mudstones occur within 50 m of more-conductive massive sulphides (Figure 11.13). Both quadrature and in-phase EM components (expressed as a percentage of the primary field) were measured using a MaxMin horizontal loop system. Two frequencies were used (222 Hz and 3555 Hz) but the greatest difference between target types was observed on the higher-frequency dataset (Figure 11.13). The massive sulphide deposit also gives rise to a peak in both the Bouguer gravity profile and the Fraser-filtered VLF transect.

A second example is presented which shows how important adequate signal processing may be to locating a mineral target accurately. A Turam profile across the Kimheden orebody in north Sweden (Parasnis 1991) is shown in Figure 11.14A. Two components (reduced ratio and successive phase difference) are plotted as a function of distance and both show very distinctive positive (RR) and negative (phase difference) anomalies over a steeply dipping pyrite orebody. The point of occurrence of an orebody is usually taken as corresponding with the reduced ratio maximum and phase difference minimum. The secondary fields calculated from the Turam profile in Figure 11.14A and normalised to the local primary field are shown in Figure 11.14B. The reduced ratio maximum is displaced by about 4 m relative to the true position of the current as revealed by the secondary fields. While a 4 m discrepancy in position may have little importance for shallow drilling, in cases where deeper drilling is required, this lateral shift could result in the sub-surface target being missed or inadequately sampled.

An example of the use of HLEM has been presented by Palacky (1991) in an investigation over a conductor buried beneath Quaternary sediments in north-eastern Ontario, Canada. An Apex Max-Min-I (eight-frequency) system with an inter-coil separation of 100 m was used along a 1 km profile south of Fraserdale. In-phase and quadrature components measured at each of eight frequencies are shown in separate families of graphs in Figure 11.15. It is clear that

Figure 11.13 (*opposite*) HCP slingram, VLF, and Bouguer gravity profiles across the Tulks East Prospect, Newfoundland. The massive sulphide can be distinguished from 'graphitic' shale at 222 Hz by the larger conductance of the sulphide. From Barbour and Thurlow (1982), by permission

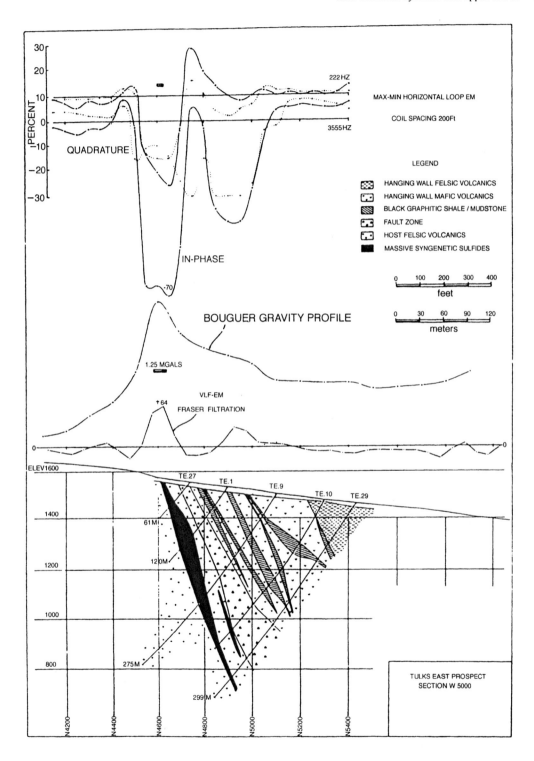

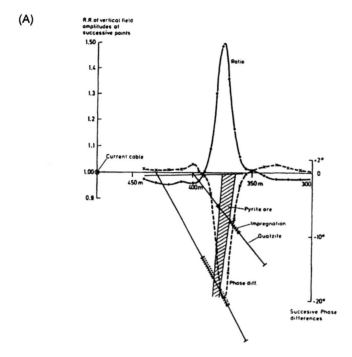

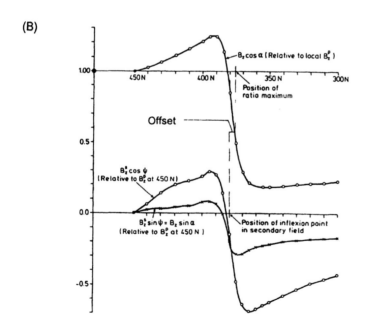

Figure 11.14 (A) Turam profile across the Kimheden orebody in north Sweden, showing reduced ratio (RR) and successive phase differences. (B) The secondary field calculated from the Turam profiles in (A). Note the slight offset in location of the position of the RR maximum and that of the point of inflection in the secondary field. From Parasnis (1991), by permission

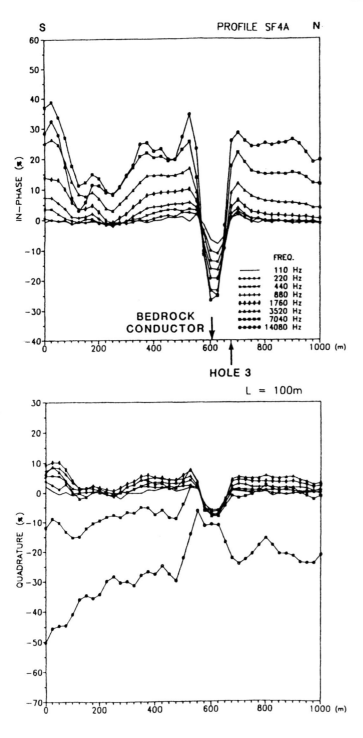

Figure 11.15 HLEM profile over a 1000 m line, 20 km south of Fraserdale, Ontario, over a bedrock conductor (at station 600) covered by Quaternary sediments. In-phase (top) and quadrature (bottom) data were acquired at the frequencies specified. Coil separation is 100 m. The location of borehole 3 is indicated. From Palacky (1991), by permission

around station 600, the in-phase component goes negative and a corresponding trough-shaped anomaly is evident in the quadrature component data. The width of this anomaly at the zero level is equivalent to the inter-coil separation plus the width of the conductor. The character of this particular anomaly is typical of a narrow sub-vertical bedrock conductor of high conductance. In the case of a mineral exploration survey, such an anomaly would be indicative of a possible target zone. The reversal of the trough-shaped anomaly on the quadrature data at 14 080 Hz around station 600 (upside-down with respect to the other graphs) is a result of the presence of moderately conductive overburden. The Quaternary sediments present locally comprised alternating clay–sand units 37 m thick with 2 m of clay over 8 m of sand, 22 m of glacial till and 5 m of sand.

In contrast to the above example, Palacky (1991) also presented a HLEM profile over shear zones covered with thick clay (Figure 11.16) along a 12 km profile south of Kapuskasing, about 80 km south-west of Fraserdale, Ontario. Two shear zones are indicated around stations -375 and -635. A drillhole located at station -500 passed through 35 m of massive clays. The measured amplitudes of both components are much larger than in the previous example and are due to the presence of the very conductive clay. The quadrature data are negative at the four highest frequencies while the in-phase component is negative at 14 080 Hz.

Phasor diagrams obtained for HLEM data at three locations with different sediment types and at which borehole control was available are shown in Figure 11.17, with the simple borehole results. The phasor diagram for borehole 42 (which lies on the profile shown) shows that the fit between the measured data (as indicated by asterisks in Figure 11.17A) and the calculated response of the model (solid line) is imperfect. The overburden thickness was constrained at 35 m (the depth to bedrock as determined by drilling). However, an unconstrained inversion for a 2-layer model produced an interpreted thickness of 42 m and resistivity values of 51 Ωm for the upper layer and 8000 Ωm for the bedrock. The discrepancy of 7 m in depth estimates can be attributed to the effect on the HLEM response of the shear zones nearby.

The phasor diagram associated with the location of borehole 67 (on the profile shown in Figure 11.15) indicated a depth of overburden of 40 m with a resistivity of 210 Ωm from an unconstrained model (Figure 11.17B). The drilled depth to bedrock was 39 m, which is in very close agreement. However, the bedrock resistivity was poorly resolved.

The third phasor diagram (Figure 11.17C) was obtained over thick sand. The best-fit model obtained produced an overburden 41 m thick with a resistivity of 350 Ωm over bedrock with resistivity 500 Ωm. However, when the resistivity values are large and the contrast between layers is small, the determination of layer thicknesses be-

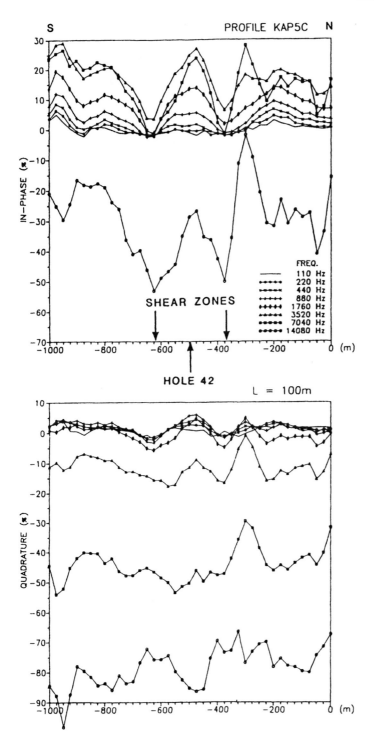

Figure 11.16 HLEM profile over a 1000 m line, 12 km south of Kapuskasing, Ontario, with shear zones concealed by thick clay. From Palacky (1991), by permission

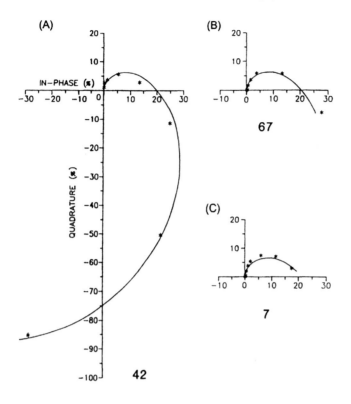

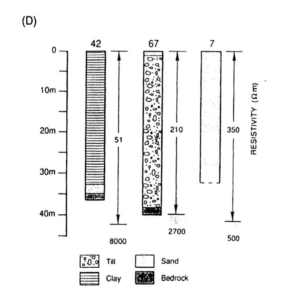

Figure 11.17 (A)–(C) Phasor diagrams, and (D) corresponding borehole results, with HLEM-derived models over three sites with contrasting sediment types. (A) = thick clay adjacent to a shear zone (hence the discrepancy in depth estimates between borehole and HLEM model); (B) = glacial till; (C) = thick sand within a esker. From Palacky (1991), by permission

comes unreliable. Drilling in this case only went through 32 m of sand when the hole was terminated as there was little point in going deeper through an esker where no basal till was expected.

These three examples provided by Palacky demonstrate the effectiveness of the use of HLEM profiling to locate shear zones and conductive targets beneath Quaternary sediments. The method produces phasor diagrams characteristic of the types of sediments present with different associated resistivities. The types of phasor diagrams can be used to help discriminate between different material types and has obvious benefits in geological mapping.

11.2.5.2 Groundwater investigations

Electromagnetic methods have been demonstrated to provide a powerful suite of tools in hydrogeological investigations since the late 1970s. The methods lend themselves to environments where conductivity contrasts are high but the ground surface precludes the simple deployment of DC resistivity methods owing to high surface resistances such as are found in very arid regions.

Two approaches tend to be taken. One is for the general investigation of a groundwater regime, where the groundwater is prevalent within aquifers. The second approach is to search within the local bedrock for fractures which may contain small but usable reservoirs of potable water. The frequency-domain EM methods are generally used for near-surface mapping investigations within a hydrogeological investigation. For deeper investigations, TEM methods are used (see Section 11.3.3.1).

Goldstein *et al.* (1990) have presented a case history where ground conductivity measurements were used to delineate contaminated water emanating from a series of water storage lagoons at Kesterton, Merced County, California. Agricultural drainwater with trace amounts of selenium and other toxic elements had been discharged into 12 unlined storage ponds totalling 5.2 km^2. These ponds provided both a year-round and seasonal habitat to migratory and local waterfowl. The high levels of selenium within the local food chain had resulted in physical deformities within the waterfowl population. A large-scale environmental study to determine the extent and degree of the contamination was initiated so that a plan could be developed for the remediation of the reservoir.

Conductivities of the local unpolluted groundwater were found to be between 300 and 400 mS/m, in contrast to those of the contaminated groundwater which reached 1000–1700 mS/m. From a series of local boreholes it was known that the contamination appeared to be restricted to a depth of less than 40 m, and typically around 20 m.

Ground conductivity measurements were made in 1987 along a series of transects (Figure 11.18A) with GCM instruments made by

Geonics Ltd (EM31 and EM34). Maps of apparent conductivity were produced for each instrument and coil separation (horizontal magnetic dipole configuration) (Figure 11.18B–D). In order to remove the effects of near-surface conductive features, the apparent conductivity values obtained using the 20 and 40 m coil separations were processed using a technique described by McNeill (1985). The conductivity value measured using the 40 m coil separation was doubled and the apparent conductivity value obtained using the 20 m coil separation was subtracted from it to give a new apparent conductivity value which was also displayed in map form (Figure 11.18E). It can be seen that the anomalies shown in the map of the processed data are much more intense than in either of the apparent conductivity maps for each of the two coil separations individually.

The final interpretation of the data is shown in Figure 11.18F. Zone A indicates normal background conductivity values; these values (150–200 mS/m) are high but are consistent with the type of local regional soils known to have a high ambient salinity. The highest apparent conductivity values (>400 mS/m) are found adjacent to Ponds 1 and 2 (area B in Figure 11.18F). The leading edge of the contaminant plume was interpreted to be where the apparent conductivity values measured using the 40 m coil separation declined to background values. This indicated that the plume front was about

Figure 11.18 (A) Location map, (B)–(D) a series of isoconductivity maps, (E) processed data isoconductivity map, and (F) interpretation map at Kesterton, California. From Goldstein *et al.* (1990), by permission

(A)

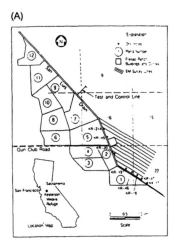

(B)

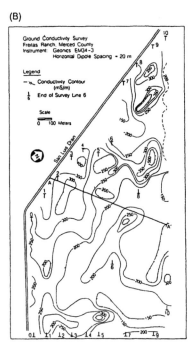

(C)

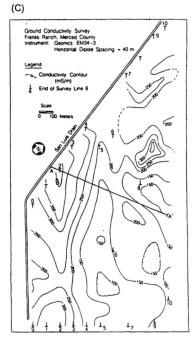

(D)

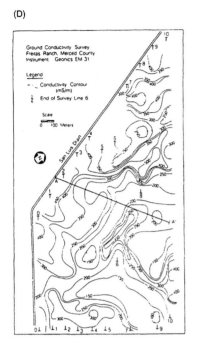

(E)

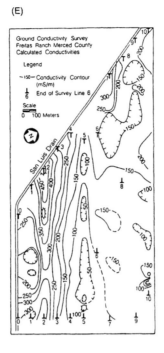

(F)

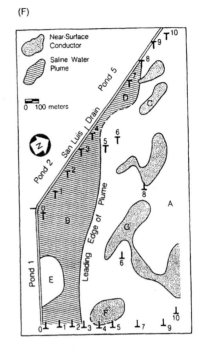

Figure 11.18 (*continued*)

350 m to the east of the San Luis Drain. Anomaly C appears to be related to high-salinity soil and correlated with an area evident on aerial photographs throughout the rainy season as a bleached zone. Area D is part of the saline plume but is confined to the area immediately adjacent to the land drain. A low-conductivity zone (area E) was identified adjacent to Pond 1. This area is thought to have had the effects of the saline plume diluted by local rainwater as drainage water had not been added to Pond 1 during the two years prior to the survey being undertaken. Area F is a deeper-penetrating higher-conductivity zone and is thought to be a plume of contaminant water which is a residual from saline water drainage prior to 1985. Zone G, which has abnormal vegetation cover, appears to be due to a near-surface conductivity feature unrelated to the contaminant plume but probably a function of local soil salinisation processes.

Monier-Williams *et al.* (1990) have described an investigation of a leachate plume emanating from a municipal landfill near Novo Horizonte, a town located 484 km north-west of São Paulo in Brazil. The location of the site is shown in Figure 11.19. The landfill had been in operation for about 5 years prior to the investigation. A well in a local farmhouse located to the north-east of the landfill was contaminated. The local geology comprises unconsolidated

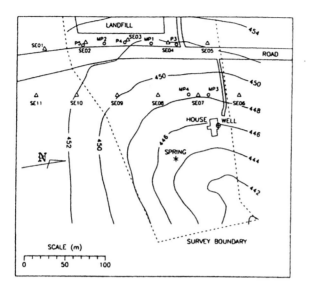

Figure 11.19 Location map of Novo Horizonte municipal landfill and adjacent farm and well. From Monier-Williams *et al.* (1990), by permission

Figure 11.20 (*opposite*) Schlumberger resistivity sounding interpretations and EM ground conductivity profiles, Novo Horizonte, Brazil. From Monier-Williams *et al.* (1990), by permission

Figure 11.21 (*opposite*) (A) Measured EM 34-10 H data, and (B) the same data topographically corrected, at Novo Horizonte, Brazil. From Monier-Williams *et al.* (1990), by permission

Cretaceous to argillaceous sediments which were thought to have been deposited in a fluvio-lacustrine environment. These sediments lie unconformably on basalts. From the local topography and information from a series of piezometers, the local groundwater gradient was found to be from the landfill towards the farmhouse. Uncontaminated groundwater in the area was known from other borehole information to have conductivities in the range 20–50 μS/cm. Resistivity depth-soundings were also undertaken in the area (at locations depicted by triangles in Figure 11.19).

The results of these soundings and of the EM conductivity surveys along the same lines are shown in Figure 11.20A and B. From these it is clear that, below the soil layer and away from the contamination, high resistivities are found within clay-free sands. A clay horizon is found on each section below an elevation of 432 m. Contamination results in resistivities of less than 40 Ω m in the upper sand, increasing to 200–250 Ω m at the sand–clay interface. The apparent conductivity highs evident on the two EM transects is entirely consistent with a conductive plume emanating from the landfill. Maps showing the observed and topographically corrected apparent conductivity values measured using the EM34-10H (displayed in units of decibels) are shown in Figure 11.21A and B, respectively. The topographically corrected data clearly reveal a more conductive element between the north-eastern corner of the landfill and the farmhouse.

Other examples of the use of ground conductivity meters in the investigation of groundwater contaminant plumes have been given by Monier-Williams *et al.* (1990), Cartwright and McComas (1968), Slaine and Greenhouse (1982), and by Greenhouse and Slaine (1983), among others.

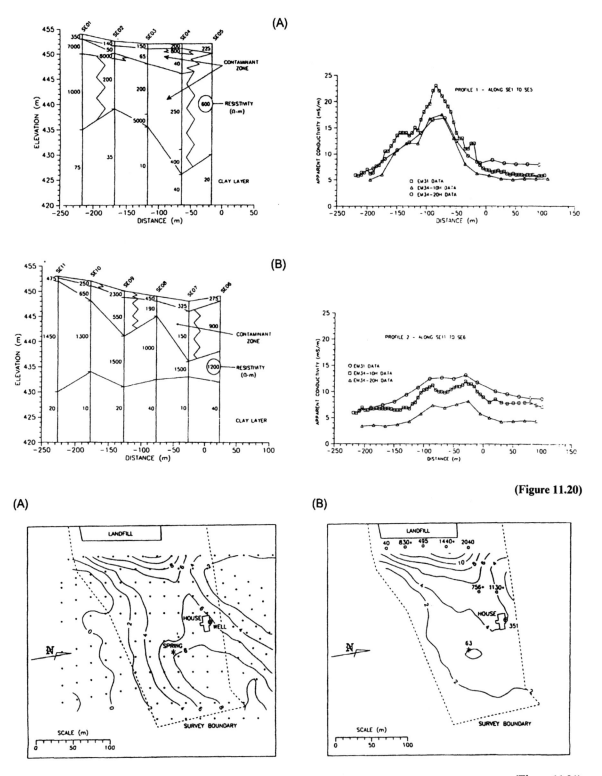

(Figure 11.20)

(Figure 11.21)

11.2.5.3   Detection of underground cavities

During a major investigation into the human biological history of the
Near East, a Geonics EM31 ground conductivity meter was used as
part of an archaeological study at Bab edh-Dhra in Jordan (Frohlich
and Lancaster 1986). The aim of the work was to locate and indicate
the condition of shafts and tomb chambers. An example of an
apparent conductivity profile over what was later found to be a shaft
and two burial chambers is shown in Figure 11.22. Where a chamber
was silted up, a slightly higher apparent conductivity was observed
relative to a background value; where a chamber was intact (air-
filled), a low apparent conductivity was observed. By mapping the
excavation site with the EM31, various anomalous zones were identified. Of seven examined by direct excavation, all were found to be
infilled tomb shafts about 1.5 m wide and about 2 m deep.

Where cavities are large relative to their depth, and there is good
electrical contrast between the cavity and the host material, then such
a feature should be readily detectable. However, small cavities with
low contrasts in conductivity buried at depth are unlikely to be
resolvable.

11.2.5.4   Location of frozen ground

In areas affected by extensive permafrost, it is of significant engineering importance to be able to differentiate between frozen and unfrozen ground. In the Arctic, the location of ice-bonded permafrost is
vital when planning major projects such as pipelines. Ground frozen

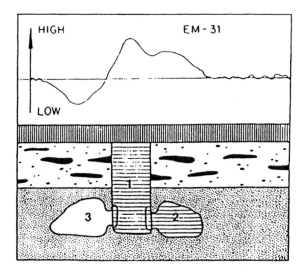

Figure 11.22 An apparent conductivity profile over a shaft tomb at Bab
edh-Dhra. The shaft (1) which leads
down to the surrounding burial chambers (2 and 3) gives rise to an apparent
conductivity high; the silted chamber
(2) and the air-filled chamber (3) are
associated with intermediate and low
apparent conductivities, respectively.
From Frohlich and Lancaster (1986),
by permission

solid restricts the use of resistivity surveys because of the difficulty in implanting electrodes. Consequently, non-contacting inductive methods are very practical alternatives. This has been demonstrated in the first two case histories. While seismic reflection can be used in overwater surveys, EM dual-coil profiling can be much more rapid in covering the ground. An example of this is given in the third case history. A review of the use of geophysical methods in the investigation of permafrost has been given by Scott *et al.* (1990).

The first case demonstrates the use of a Geonics EM31 in providing a very rapid method of mapping the extent of frozen and unfrozen ground in the Mackenzie River Delta, Northwest Territories (Todd *et al.* 1991). A 5 km traverse was undertaken across the ice of a small channel in the Delta, over frozen tundra and across lake ice over Big Lake. The apparent conductivity profile obtained is shown in Figure 11.23. Apparent conductivity values were found to be around 1.5 mS/m over solidly frozen ground but over 10 mS/m over unfrozen sediments and water. The lateral extents of the frozen/unfrozen portions at and near the surface correlate exactly with the marked changes in the values of the measured apparent conductivity. An example of the use of an EM31 to locate massive ice within frozen silt

Figure 11.23 Apparent conductivity profiles obtained with a Geonics EM 31 ground conductivity meter over part of the Mackenzie Delta, Canada, showing areas affected by permafrost. From Todd *et al.* (1991), by permission

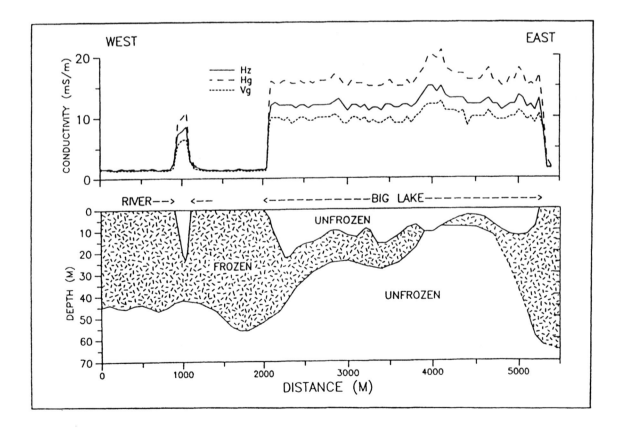

is shown in Figure 7.44 in Chapter 7 in a comparison with electrical resistivity profiling. Massive ice has a lower conductivity than frozen soil and so gives rise to readily identifiable zones along a transect.

The second example, provided by Sinha and Stephens (1983), describes an EM ground conductivity survey over a drained lake at Illisarvik, Northwest Territories. Both an EM31 and an EM34 (with all three coil separations) were used over a 300 × 600 m grid at 25 m centres. The apparent conductivity data from both vertical and horizontal magnetic dipoles for all three coil separations were contoured to give six apparent conductivity maps. Examples of the maps of EM31 and EM34(40) VMD data are shown in Figure 11.24. It is clear that there is an apparent conductivity anomaly centred around (25 mW, 100 mS). This relates in part to a residual pond of water at the surface. Note, however, that the centre of the conductivity anomaly on the 40 m dataset is at (25 mE, 75 mS). This is thought to be related to a zone of partially frozen sediments within the lake bed which have been gradually freezing up after the artificial draining of the lake three years before the geophysical survey was undertaken. Sinha and Stephens undertook some simple modelling of the conductivity data and estimated that the frozen layer was between 11 and 23 m thick in the central part of the former lake area but thicker towards the former shorelines. The draining of the lake has permitted the gradual freezing of the sub-lake sediments.

Figure 11.24 Maps of apparent conductivity at Illisarvik, Northwest Territories, Canada, obtained with (A) an EM 31 and (B) an EM 34-3 with a 40 m inter-coil separation. From Sinha and Stephans (1983), by permission

(A)   (B)

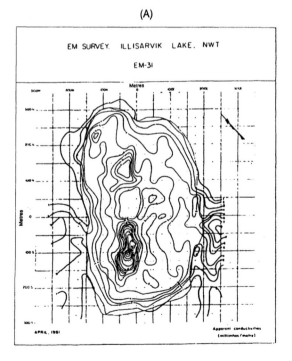
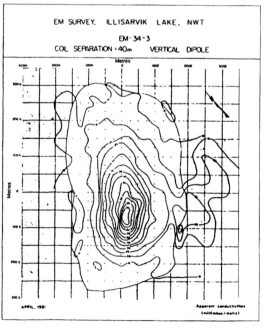

In areas where lake or sea ice is present to provide a safe working platform, ground EM measurements can be made over seawater to determine sub-bottom geological information. An example of where this approach has been taken is in a case history described by Palacky and Stephens (1992). Field experiments using horizontal loop EM (HLEM) were undertaken by the Geological Survey of Canada over sea ice on the Beaufort Shelf. Between 60 and 80 soundings were achieved in a day. To compare results, seismic refraction and reflection measurements were also made along with a number of drillholes. The ice and water layers are readily defined in terms of thickness and conductivity. Sea ice is resistive ($5000 \, \Omega \, m$) in contrast to seawater and seawater-saturated sediments ($1 \, \Omega \, m$); sub-bottom ice-bonded permafrost is also resistive ($5000 \, \Omega \, m$) whereas unfrozen sediments were found to have an average resistivity of $2 \, \Omega \, m$.

An example of one profile is given in Figure 11.25. The seawater and seawater-saturated sediments have not been differentiated as there is insufficient contrast in resistivities for this to be possible. Depths to the top of ice-bonded permafrost are indicated by circled crosses. Seismic reflection data were only of limited use in this area because of shallow accumulations of gas within the near-surface sediments producing acoustic blanking. At one location a drillhole was constructed through 10 m of water to 75 m below seabed. The top 10 m of sediments were found to be water-saturated sandy clays. Ice-bonded permafrost was identified in fine to medium sands between 10 and 36 m below the seabed and was underlain by silty clays and a sandy unit. The borehole was logged using seismic, gamma-ray and EM probes. The borehole EM probe was a Geonics EM39 with two coaxial coils 0.5 m apart.

Figure 11.25 Conductivity section of a 5 km profile over the Beaufort Shelf, Northwest Territories, Canada, obtained using an Apex MaxMin I horizontal loop EM (HLEM) system. Encircled crosses indicate depths interpreted from seismic refraction surveys to the top of the permafrost layer; triangles indicate seismic refraction surveys where no permafrost was found. From Palacky and Stephens (1992), by permission

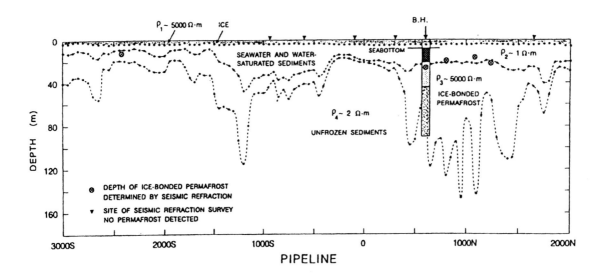

The HLEM interpretation is compared to the results of the EM logging and borehole lithostratigraphy in Figure 11.26 (A to C) respectively. Within the sub-permafrost horizon, the low conductivity values are associated with the sand-rich units while the conductivity highs are associated with silty clay horizons. It is clear that the HLEM interpretation and the drillhole results correlate very well, with all three showing the base of the permafrost horizon to be at a depth of 36 m below the seabed. The seismic refraction depths to the top of the permafrost horizon were in reasonable agreement with the interpretation of the HLEM data. This case history provides an excellent example of the efficacy of the EM method and the reliability of its results over this type of geological environment. From a practical point of view, the HLEM proved to be more rapid and easier to deploy in the extreme cold (down to −30°C), although battery power at such low temperatures is always problematic.

11.2.5.5 *Contaminated-land mapping*

Dual-coil EM mapping provides a rapid means of surveying possibly heavily contaminated sites in a cost-effective and environmentally benign manner. As a trial survey, British Rail Research commissioned a combined EM and magnetometer survey over a trial 50 × 50 m area

Figure 11.26 (A) HLEM unconstrained conductivity model of the borehole shown in Figure 11.25. (B) Conductivity profile obtained by logging the borehole with Geonics EM 39 dual-coil probes. (C) Composite lithostratigraphic results of the drilling. After Palacky and Stephens (1992), by permission

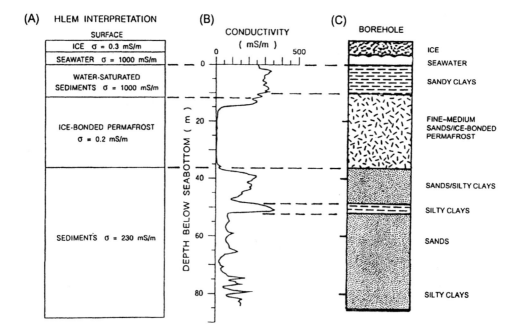

at a former railway welding yard at Dinsdale, Teeside, UK (Reynolds 1994). It was known that part of the yard had previously been the site of an old iron foundry but the only plans of the works were unreliable. Geophysical methods were deployed to demonstrate what they could find and, in particular, to locate evidence of the iron foundry. The ground was featureless apart from the obvious lines where the railway tracks had once been, but was covered all over in railway ballast made of crushed limestone and dolerite. The two material types were laid in defined lines.

An EM31 was deployed on a 1×1 m and 2×1 m grid over the area, and both the quadrature and in-phase components were recorded. In addition, the Earth's magnetic field intensity was measured at two different heights above the ground and the difference between the two computed as a vertical gradient. Isometric projections of the apparent conductivity, in-phase component and magnetic gradient are shown in Figure 11.28A–C, respectively. On each, there is clearly an effect due to a major overhead metallic gantry used for moving welded railway track around. It is clear that, in addition to the effects of the gantry, different anomalies are present on each display. For example, in Figure 11.27A (conductivity), six conductive zones were found (lettered A to F). However, only anomaly A appears on the in-phase data, suggesting that the feature causing it is both conductive and metallic (as the in-phase component is particularly sensitive to metallic objects). The other anomalies (B–F) are not evident, suggesting that their cause is due to sub-surface conductive material but which is not metallic. This interpretation is consistent with the results of the magnetic gradiometer survey (Figure 11.27C) on which anomaly A is again clearly evident.

Direct excavation of each of these zones revealed anomaly A to be a series of I-shaped steel girders within reinforced concrete buried at a depth of 1 m. The anomaly width was 12 m and the concrete foundations (with sub-slab void!) was 11 m wide and was contained within the extent of the anomaly. Furthermore, where anomaly A passes into anomaly D, it was found that the concrete foundations stopped and vitrified sulphurous slag was present. Sulphur had leached out from the slag to form yellow crystals on the surface of the slag and presumably giving rise to the elevated conductivities. Anomalies E and F were also found to be due to vitrified but broken-up pieces of sulphurous slag. Anomalies B and C were found to be buried tips of fine ash.

The linear anomalies evident on all three datasets relate to the surface ballast material. The orientation of the anomalies is exactly parallel to the lines of the former railway tracks. One linear feature (L1) was found to correlate with a cable duct which terminated in a metal box evident at the ground surface. On the conductivity data, the effect of this duct diminishes in amplitude towards the metal box. This effect is thought to be due to the increased depth of burial of the

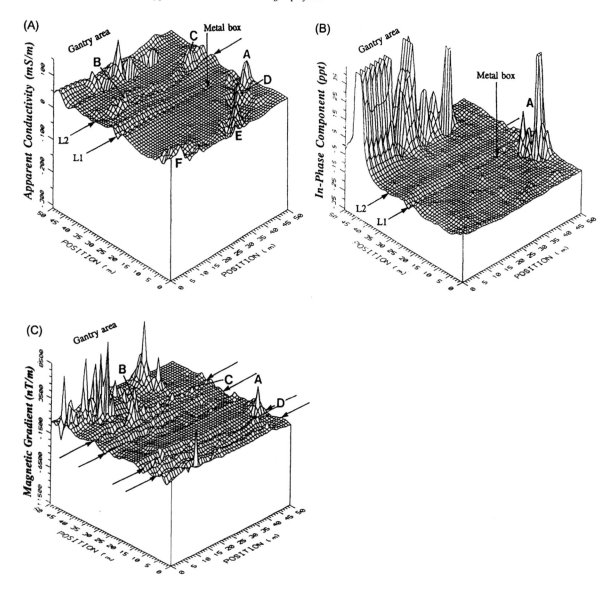

ducting below the ballast towards the box. No evidence of the ducting, on either the ground or within the three datasets, was found between the box and the location of anomaly A. In Figure 11.27A, it can be seen that there is no conductivity anomaly over this zone either. Despite the site being so potentially noisy, and with railway ballast covering the area, the EM and magnetic surveys demonstrated that they could be used not only to locate anomalous material within the sub-surface, but to differentiate between material types.

Figure 11.27 Isometric projections of (A) apparent conductivity and (B) in-phase component, both measured using a Geonics EM 31, and (C) vertical magnetic gradient, at a former railway welding yard at Dinsdale, Teeside, UK. Courtesy of British Rail Research (Reynolds, 1994)

(A)

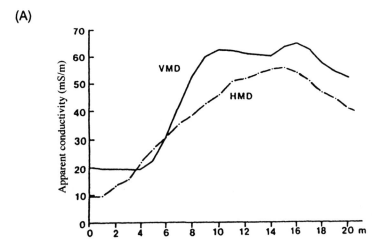

(B)

(C)

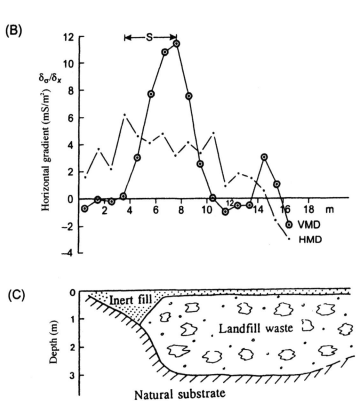

Figure 11.28 (A) EM 31 apparent conductivity profiles with both vertical and horizontal magnetic dipoles (VMD and HMD, respectively) over the edge of a shallow landfill in North Wales. (B) Schematic section of a trenched excavation. (C) Horizontal apparent conductivity gradients for the data in (A). After Reynolds and McCann (1992), by permission

11.2.5.6   Landfill investigations

Not only have FEM methods been used to detect contaminant plumes arising from landfills, they have also been used successfully at three stages in the life of a landfill. These stages are (a) site appraisal prior to the development of the waste facility; (b) investigation of the base of the site during excavation and construction; and, most commonly, (c) over closed landfills.

It is pertinent to consider the factors that need to be taken into account when contemplating undertaking a geophysical survey over any landfill site, at whatever its stage of development. These factors are listed in Table 11.1 (from Reynolds and McCann 1992). The types of waste material, the nature of the base and lining materials, the style of cover, etc., all affect the deployment of geophysical methods. Of particular interest to waste regulatory authorities and those responsible for closed landfill sites, are the location of the edges of the waste material, the depth of waste present, and any information concerning leachate. Three case histories are provided here, one to illustrate a simple method of determining the edge of a landfill, the second to show how EM methods have been used in leachate studies, and the third to demonstrate how ground conductivity meters can be used to provide a wide range of information about the type of fill material.

Of importance to those responsible for old closed landfills is a knowledge of the boundary of the waste. This is often required in relation to building new properties which are not allowed in the UK to be within 250 m of the edge of a waste site for fear of gas and leachate migration. The other problem is that the responsible authority very often has many hundreds of sites within its jurisdiction but insufficient money to investigate them all. What is required is a simple, cost-effective method that can be deployed without interfering with the site in any way.

To detect the edge of a landfill, a Geonics EM31 has been demonstrated to be a useful tool (Reynolds and McCann 1992). A former North Wales landfill containing domestic waste up to 12 m thick and capped with about 1 m of inert soil was investigated using a variety of geophysical methods. The site was about 150 m long by 80 m wide. Trenches through the edges of the landfill had been constructed previously as part of a remediation programme and these provided corroborative evidence against which to check the geophysical interpretation. Short (< 25 m) EM31 profiles were undertaken over the landfill site at right-angles to the perimeter hedge. Both vertical and horizontal dipoles were used but only the quadrature (conductivity) component was measured.

One such profile is shown in Figure 11.28A, with the corresponding trench information (Figure 11.28B). It is clear that the apparent conductivity values increase sharply across the landfill boundary. It is also noticeable that the shallower penetrating EM31H has a less rapid change compared with the data obtained using the vertical dipole. This reflects the slanted edge of the landfill. The EM31H responds to the gradually increasing thickness of the inert fill before finally responding to the refuse itself. The EM31V, however, responds more to the deeper material which changes more abruptly from inert

Table 11.1 Potential unknowns for any landfill (Reynolds and McCann 1992)

Type of void space:
  Hard-rock quarry
  Sand/gravel quarry
  Brickearth quarry
  Shallow valley
  Estuarine creek
  Engineered site
  Previous industrial site

Type of lining:
  None
  Mineral linear (compacted clay)
  Artificial linear (e.g. HDPE geomembrane)
  Combination linear (e.g. mineral + geomembrane, double membrane, etc.)

Type of capping:
  None (natural venting)
  Clay
  Artificial (e.g. HDPE geomembrane)

Site dimensions:
  Areal size
  Depth
  Shape, particularly of margins

Tipping history:
  Types and mixtures of wastes
  Duration of tipping (likely volumes)
  Style and degree of compaction and cover during tipping operations
  Age

Geological factors:
  Type(s) of substrate
  Local hydrogeology
  Sub-site faulting
  Previous resource activity at or beneath site (e.g.
    coal mining, quarrying, etc.)
  Sub-site natural cavities
  Site (slope) stability

Factors related to infill material:
  Degree of saturation
  Gas generation
  Internal temperature and variability
  Liquor/leachate generation
  Mobility and conductivity of lechate
  Compaction density and variability
  Material composition (e.g. inert builder's rubble, putrescible material,
    industrial refuse, etc.)

to refuse material. In order to highlight the effect, the horizontal gradient of each component has also been computed (Figure 11.28C). This demonstrates that the edge of the landfill as defined by the EM31H results lies outside that determined by the EM31V. This is clearly consistent with the observations within the trench.

By carrying out a series of short EM31 transects at right-angles to the site boundary, and using the criteria described above, the edge of this landfill site was determined accurately (to within 1–2 m) using a survey which took one person about one day to execute. The results could be interpreted on site and the landfill boundary marked with reasonable confidence. The approach taken here is very similar to that demonstrated by Zalasiewicz *et al.* (1985) for the interpretation of EM31 data in mapping geological boundaries at shallow depth.

A major problem with many landfills is the formation of leachate. As long as the liquor remains on site and is treated it poses no significant problems. However, should leachate begin to migrate away from a site it may cause potential contamination of local potable water supplies. A further difficulty is in the treatment of wastes from some coal workings. Oxidation of pyrite associated with coal and coal-bearing strata causes increased acidity of minewater which also typically contains high concentrations of iron, sulphates and trace metals, especially manganese and aluminium. Ladwig (1983) has described several case histories associated with acid mine drainage (AMD). One of these examples relates to an abandoned strip mine in Butler County, Pennsylvania. The 8-hectare site was thought to have been mined in the 1950s but detailed records were no longer available.

In 1981, the mine was reclaimed and a lake previously filling the mine was drained and the void space backfilled with mine spoil and regraded. Up to 7 m of soil was placed over the high-wall of the mine grading down to less than 2 m in the central and eastern parts of the site. A limestone-lined sub-surface drain was built along the base of the high-wall to take away seepage. Following completion of the reclamation, discharge from the sub-surface drain had been slightly acidic (acidity 20 mg/litre with iron 18 mg/L and sulphate 200 mg/L). Furthermore, an intermittent seep of acid water had occurred on the south-eastern side of the site.

Over a 1.6-hectare part of the site near to where the seep was observed, a series of EM ground conductivity profiles was undertaken using a Geonics EM34 with a 10 m coil separation and 10 m station interval. The field survey took $1\frac{1}{2}$ days. The resulting map of apparent conductivity values is shown in Figure 11.29. The closely spaced contours (14–24 mS/m) to the west and east of the site correspond to the edges of the mine. The rapid increase in apparent conductivity observed over these areas is indicative of passing from resistive bedrock into conductive regarded spoil. The apparent conductivity

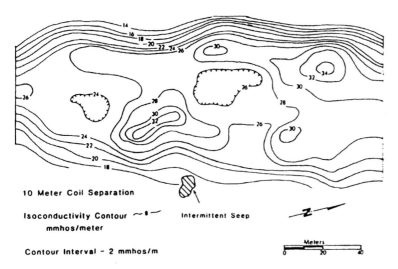

Figure 11.29 Apparent conductivity map over a backfilled former strip mine, Butler County, Pennsylvania, USA. From Ladwig (1983), by permission

10 Meter Coil Separation

Isoconductivity Contour ~•~   Intermittent Seep
    mmhos/meter

Contour Interval - 2 mmhos/m

highs (30 and 34 mS/m) in the north-west part of the site are associated with the sub-surface drain. The distribution of apparent conductivities (apart from the features already described) is related to the infiltration of conductive acid minewater moving downgradient towards the lower eastern part of the site and ponding within the spoil. This is thought to be the cause of the conductivity high of 32 mS/m in the southern part of the site. Further migration of contaminated groundwater is restricted by the presence of the low-wall of the mine and what is interpreted to be a more-resistive bedrock ridge of unmined material separating the southern conductivity anomalous zone from a smaller one to the north. The location of the intermittent seep occurs downslope of the main conductivity anomaly, giving credence to the ponded groundwater model. A summary of the interpretation of the EM survey is shown in Figure 11.30. As a consequence of this survey, the physical model was used as a basis for siting injection and observation wells installed as part of a remediation test using AMD inhibiting agents.

While the station interval (10 m) chosen was quite coarse, it provided sufficient spatial detail to meet the objective of the survey. However, if the edges of the mine were to have been mapped, a closer station interval would have been necessary. In this case, the mine high-wall could possibly have been located to better than ± 2 m. Given the speed of the EM survey, it demonstrates how much useful information can be obtained in such a short time.

The third example is of a combined EM31 and EM34 survey over a site near Manchester at which a landfill facility was being constructed. The objective of the survey was to determine and map the

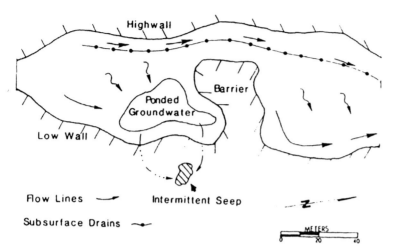

Flow Lines ———➤   Intermittent Seep

Subsurface Drains ——➤

Figure 11.30 Schematic of the interpretation of the EM data shown in Figure 11.29. From Ladwig (1983), by permission

thickness of the boulder clay underlying the site, because a requirement for the site licence was a certain thickness of boulder clay to act as an impermeable mineral liner. The survey was undertaken over a series of parallel traverses using both magnetic dipole orientations and 10 and 20 m coil separations with the EM34. The resulting six values of apparent conductivity per station were entered into EMIX34PLUS (Interpex Ltd) to produce two-dimensional interpreted cross-sections through the site. The general stratigraphic sequence was boulder clay over a saturated fine sand over bedrock. There was a reasonable conductivity contrast between the various layers which were delimited on the interpreted models. One such profile (reproduced courtesy of RUST Environmental) is shown in Figure 11.31.

Borehole information available at the edge of the site was used to constrain the EM models. The interpreted thicknesses along each survey line were correlated between adjacent lines and an isopachyte map of clay thickness and a map of the elevation of the base of the clay layer were produced. As the landfill design required that a certain void volume be available to accommodate waste for the site to be commercially viable, a knowledge of the thickness of the clay and the elevation of its base were essential. As a consequence of the EM survey, various cell walls within the landfill were relocated. The geophysical survey provided the site operator with information of sufficient quality and reliability that he was able to modify his landfill design and construction appropriately, thereby avoiding potential problems later had the clay liner been too thin.

These three case histories are but a small representation of what ground conductivity meters can be used for. It is worth reiterating, however, that in all cases, careful survey design is essential if the

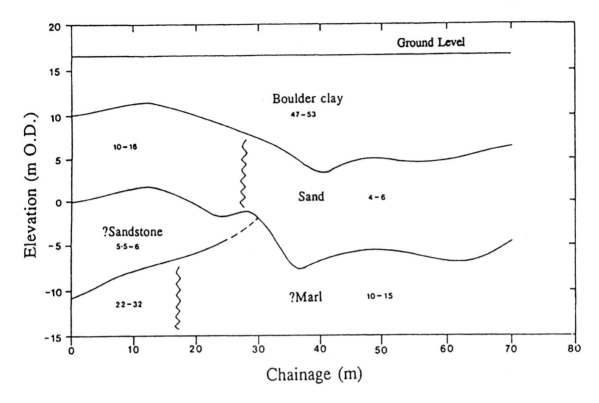

Figure 11.31 Pseudo-geological cross-section derived from electromagnetic ground conductivity data derived from an EM 31/34 survey over a base of a landfill being engineered with a natural clay liner. Values cited for geological materials are true conductivities in units of mS/m. From Raynolds and McCann (1992), by permission

objectives of a survey are to be met. In addition, while isoconductivity maps can provide excellent qualitative spatial information about a site, detailed quantitative analysis using specialist software can give highly reliable two-dimensional models. When used as part of a grid of survey lines, the models along each line can be correlated between lines to produce a pseudo-three-dimensional model of the subsurface, under the right conditions. There are situations where no satisfactory model can be computed which then suggests that there may be three-dimensional targets present within the ground. This in itself is useful information. For example, when modelling some EM data over an old colliery in the north-east of England, consistent layered models were successfully produced over much of the site. However, along one line, the data became erratic over a very short distance and no satisfactory models were producible. This effect was thought to be due to the presence of pillar and stall mine workings at shallow depth (< 10 m). When the location of the anomalous zone was drilled, soft ground indicative of partially collapsed mine workings was found, thus confirming the inference derived from the modelling.

11.3   PULSE-TRANSIENT EM (TEM) OR TIME-DOMAIN EM (TDEM) SYSTEMS

11.3.1   TDEM/TEM surveys

In a typical TDEM survey, a large direct current is passed through a large ungrounded loop transmitter to energise it. After a discrete period of time (a few tens of milliseconds), during which any effects due to switching the current on would have died away (known as 'turn-on transients'), the applied current is interrupted abruptly. If a conductor is present within the vicinity, the sharp change in the primary field will induce eddy currents within the conductor, initially at its surface only. This is known as the 'early-time' stage of the transient process. These surface currents then start dissipating through ohmic losses. The zone immediately within the conductor then experiences a decreasing magnetic field with a consequential flow of eddy currents through it. Effectively, this is the start of the inward diffusion of the current pattern caused by the eddy currents towards the interior of the conductor. This is the 'intermediate-time' stage of the transient process. The final or 'late-time' stage of this process is reached when the induced current distribution is invariant with time. The only change observed is a decrease in the overall amplitude with time. If the conductor present is very large relative to the dipole source being used, the eddy currents may spread out laterally as well as diffuse into the interior of the conductor. The rate of change of these currents and of their respective magnetic field depends on the size and shape of the conductor and on its conductivity. In contrast, the initial distribution of surface current is dependent only on the size and shape of the conductor as this is a geometrical phenomenon, not one due to the conductivity of the body. The whole process of the step-wise excitation of the current loop is repeated many times (Figure 11.32) and the data stacked for a given location. A detailed description of TEM prospecting methods has been given by Nabighian and Macnae (1991) which has been used as the basis for this section.

The transient electric field reaches a maximum at a distance known as the diffusion depth (d), which is to TEM what the skin depth δ is to frequency-domain EM (see Box 11.4). In the time domain the diffusion depth is directly proportional to $\sqrt{t}$, whereas in the frequency domain it is inversely proportional to $\sqrt{\omega}$ (where $\omega = 2\pi f$). This local maximum propagates downwards with a finite velocity (v).

In the case of a semi-infinite half-space, i.e. uniform horizontally layered media, the 'early-stage' surface currents are located primarily in the vicinity of the transmitter loop. With the passage of time, diffusion occurs by the downward and outward spreading of the induced current loop, much like the downward movement of a system of smoke rings (Figure 11.33), with a consequential decay of the

(A)

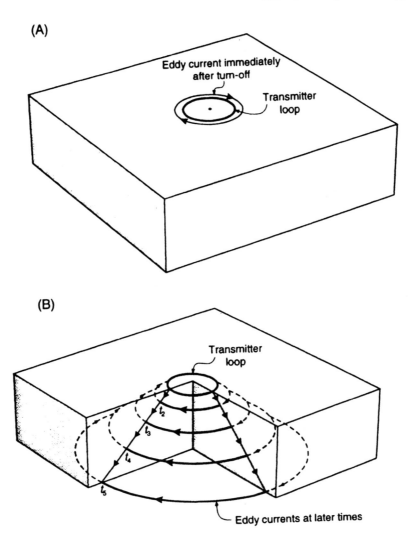

(B)

Figure 11.32 (A) The form of an eddy current immediately after turn-off of the primary field. (B) Downward and outward propagation of the eddy current filament at successive intervals of time $(t_2, \cdots)$ over homogeneous ground, rather like smoke rings

amplitude with time. The same principle applies in the case of a horizontally layered earth. Normally, the ground materials are assumed to be non-polarisable and conductivity is taken to be independent of frequency or delay time. These induced polarisation effects affect the reliability of interpretation.

Where materials within the ground have slight variations in magnetic permeability, typically of the order of 1% of the Earth's magnetic field intensity (e.g. 550 nT in a field of 55 000 nT), small TEM effects may be detectable. The TEM response is likely to be enhanced by about 1% in such cases. Where lateritic soils are present, superparamagnetic effects may cause anomalous transient recordings with the SIROTEM system. It manifests as a $1/t$ dependence which results

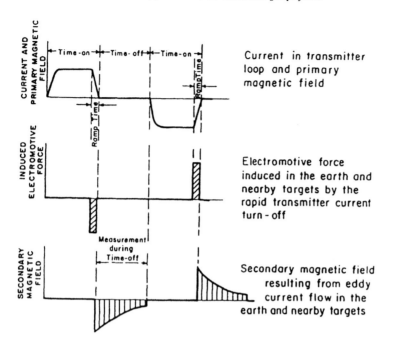

Current in transmitter
loop and primary
magnetic field

Electromotive force
induced in the earth and
nearby targets by the
rapid transmitter current
turn - off

Secondary magnetic field
resulting from eddy
current flow in the
earth and nearby targets

Figure 11.33 Time-domain EM
waveforms. From McNeill (1990), by
permission

Box 11.4   Diffusion depth (*d*) and velocity (*v*)

In a uniform conducting medium, the transient electric field achieves a maximum at the diffusion depth (*d*) such that:

$$z = (2t/\sigma\mu)^{1/2} = d$$

where σ and μ are the conductivity and the magnetic permeability of the medium. The maximum travels downwards with a velocity (*v*) such that:

$$v = (2\sigma\mu t)^{-1/2}.$$

In a conducting half-space, the downward velocity is given by:

$$v = 2(\pi\sigma\mu t)^{-1/2}.$$

in erroneous determinations of apparent resistivity with time. Similar effects are also likely to occur where either the conductivity or the magnetic permeability of the ground varies as a function of frequency. Removal of superparamagnetic effects can be achieved by displacing the receiving loop by 2–3 m relative to the transmitting loop where deployed in a coincident loop configuration.

There are three principal sources of error in TEM measurements: (a) geometric errors in transmitter–receiver positions and topographic effects; (b) static cultural noise; and (c) dynamic cultural noise.

Most TEM methods are largely insensitive to geometrical errors and, in the case of a resistive ground, are also relatively insensitive to topographic effects. However, where a conductive overburden is present, topography can produce severe coupling errors and deliberate procedures need to be followed to correct for such effects (Nabighian and Macnae 1991).

Static cultural noise arises from the presence of pipes and cables, metal fences or other utilities present in the survey area. Some metallic utilities serve as current channellers which can cause distortions in the TEM data. Live electric cables have distinctive effects at particular frequencies and their harmonics, but these can be readily removed by notch filtering. The effect of channelling can be reduced by laying the transmitter loop symmetrically over the utility.

Dynamic cultural noise is caused by a number of sources. At frequencies less than 1 Hz, the source is geomagnetic signals from within and above the Earth's ionosphere. At frequencies above 1 Hz, typically in the 6–10 Hz range, signals generated by distant lightning discharges produce sferics which are natural EM transients. Higher frequency sources of noise are AC power lines (50–60 Hz) and VLF transmitters (10–25 kHz). Of particular importance in airborne EM and in surveys undertaken in wide open spaces is wind noise which causes motion of magnetic field sensors within the Earth's magnetic field. The fields used in TEM work are typically five orders of magnitude smaller than the Earth's magnetic field.

11.3.2   Data processing and interpretation of TEM surveys

There are many ways in which TEM data can be processed and these are largely dependent upon which instrument system is used to acquire the original data. Most TEM systems record the transient voltage at a number of discrete intervals during the voltage decay after the applied current is switched off. Each time the current is applied and then stopped, measurements are taken; when the current is applied again and switched off, a repeat set of measurements is taken. This process may be repeated many tens of times at a given location with all the data being logged automatically. Consequently, these many data can be processed to improve the signal-to-noise ratio. At the same time, the field data are checked for repeatability. Commonly, the data are normalised with respect to the transmitter current or other system parameter, and the effects of the time decay may be amplified in compensation by normalising the observed field at each point with the respective primary field value at the same point.

As the field measuring systems become more sophisticated and the amount of data increases, more careful thought needs to be given to the often quite involved data processing sequences now available. For example, Stephan *et al.* (1991) describe a data processing sequence for 'long offset transient EM' (LOTEM) sounding undertaken in Germany. Three data processing stages were formulated: (a) prestack processing; (b) selective stacking; and (c) post-stack processing. Prestack processing was used to remove unwanted periodic noise using filtering such as a notch filter to remove noise associated with AC power lines and the German electric railway grid. A selective stacking algorithm was applied to average only a percentage of the data around the median of the individual time samples. The consequence of this was to reduce the noise content, thereby improving the signal-to-noise ratio. The final stage was to apply a slight time-variable smoothing filter. The culmination of this processing was the production of logarithmic plots of apparent resistivity as a function of decay time.

A variety of plots of processed data can be produced, such as transient decay (logarithmic) plots of voltage (in mV) versus decay time (in milliseconds); response profiles (graphs of measured voltage at a selected decay time at all stations in a survey area); response contours (the response profile data plotted in map form); apparent resistivity plots, either as profiles or maps; and vector plots, display components of the data in different orthogonal planes (*xz* or *yz* planes).

Interpretation methods are as varied as the different types of data plots and systems used to acquire the data. Typically, the interpretation is undertaken in two stages. The first is to locate a possible sub-surface target on the basis of the shape, size and location of anomalies evident on profiles and maps of relevant parameters. The second, more quantitative, stage is to determine the 'quality' of the conductor using time constants determined from decay plots of the field intensity at one or more locations.

Various types of display parameter are useful for different applications. For example, apparent resistivity soundings can be extremely useful in hydrogeological investigations and in geological mapping but provide very little information appropriate for mineral exploration. In the latter application, time-decay rates are more valuable as the curves produced can be characteristic of specific types of conductor. For example, the decay curve of an isolated conductor in a resistive medium shows a rapid decrease in amplitude in the 'early-time' stage but this changes to a straight-line segment at late delay times. The gradient of this straight-line segment is used to derive a characteristic time constant (τ) when plotted logarithmically. These time constants are indicative of different types of causative bodies; examples of time constants for four target types are listed in Table 11.2. An analytical approach to the calculation of time constants is given in Box 11.5. For mineral exploration, time constants in the range 0.5–20 ms are of particular interest. Pyrrhotitic bodies often have very

Table 11.2 Time constants for four common target types

Conductor type	Time constant τ
Sphere of radius a	$\sigma\mu a^2/\pi^2$
Cylinder of radius a	$1.71\,\sigma\mu a^2/\pi^2$
2-D conducting plate of finite depth extent (l)	$2t\sigma\mu l/\pi^2$
Thin prism of thickness t and average dimesion L	$\sigma\mu Lt/10$

Box 11.5   Analytical approach to the calculation of time constants

The general expression for a time constant τ of a conducting target is given by (Nabighian and Macnae 1990):

$$\tau = K\sigma\mu A$$

where K is a numerical coefficient and A is proportional to the effective cross-sectional area of the conductor (see Table 11.2).

The time constant of a conducting target can be obtained from the straight-line segment of a TEM decay graph plotted on semi-log axes. The cotangent of the slope angle (in degrees) gives directly the value of the conductor's time constant.

Analytically, given an initial amplitude A_0, then the decayed amplitude at a time t (A_t) is given by:

$$A_t = A_0 \exp(-t/\tau).$$

Taking the logarithm of both sides, this can be rewritten as:

$$\ln(A_t) = \ln(A_0) - t/\tau$$

which is the equation of a straight line with negative gradient of $1/\tau$.

If the amplitudes at two times t_1 and t_2 are measured, the expression for the time constant becomes:

$$\tau = (t_2 - t_1)/\ln(A_1/A_2).$$

large time constants (several tens of milliseconds). With the exception of nickel associations, such targets are often of little economic interest. However, time constants alone should not be used to attach either geological or economic significance to any particular target.

The sense and degree of dip can be gauged from the asymmetry in measured components when plotted as functions of lateral distance along survey traverse lines.

While computer modelling is being used increasingly in the interpretation of TEM data, classical inversion is notoriously difficult to apply to three-dimensional models. Software for such processing is largely still research-based and requires inordinately long computer execution times on mainframe computers. Consequently, such processing has yet to be applied routinely to commercial TEM projects when three-dimensional modelling is required.

However, for engineering and environmental applications, TEM-sounding interpretation is carried out routinely using personal computers; appropriate software is available commercially. Very detailed discussions of data processing and modern interpretational methods for a wide variety of EM systems have been given by Spies and Frischknecht (1991) and Nabighian and Macnae (1991), among others.

11.3.3 Applications and case histories

11.3.3.1 Groundwater investigations

Electromagnetic methods have long been used in hydrogeological investigations. Most commonly, frequency-domain systems have been used for shallow investigations and electrical resistivity depth-sounding for greater depth penetration. However, development during the 1980s of TEM systems with faster shutoff rates and earlier time sampling resulted in an increased use of TEM in hydrogeology. A theoretical approach has been described by Fitterman and Stewart (1986) which provides a range of hypothetical TEM responses for a range of commonly found hydrogeological problems. One advantage with TEM is that with the relatively small loop sizes available, measurements can be made on sites, such as public open space and sports fields, whose size may preclude the use of resistivity soundings, for example. Furthermore, in urban areas with high ambient dynamic electrical noise, the signal stacking capability of modern digital systems helps to improve the signal-to-noise ratio and recovery of the all-important signals from background noise.

TEM measurements have found increasing application in the mapping of saline–freshwater interfaces in coastal regions. A good case history describing such an application has been published by Mills *et al.* (1988) and Hoekstra and Blohm (1990). It describes the mapping of four overlapping aquifers shown schematically in Figure 11.34. These comprise a perched aquifer in which the groundwater has been heavily contaminated by fertilisers; a 60 m thick aquifer (known as the '180 ft') into which saltwater has intruded a significant distance inland; a 123 m thick aquifer (known as the '400 ft') into

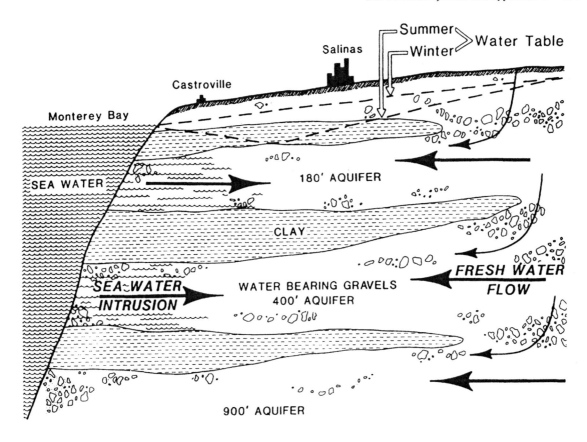

SALINAS

Castroville

Monterey Bay

Summer
Winter
Water Table

SEA WATER

180' AQUIFER

CLAY

SEA WATER
INTRUSION

WATER BEARING GRAVELS
400' AQUIFER

FRESH WATER
FLOW

900' AQUIFER

which saltwater has penetrated less far; and the deepest ('900 ft') aquifer which was untainted by saltwater at the time of the surveys. The layer-cake sequence of contaminated aquifers provided a difficulty in penetrating through the upper, more extensive, saline intrusions to detect the lower ones. To this end, a TEM system employing 100×100 m transmitting loops was used to map the '180 ft' aquifer; for the deeper aquifers, a 200×200 m transmitter loop was used.

Four late-stage apparent resistivity sounding curves are shown in Figure 11.35, with the corresponding one-dimensional inversions. These soundings relate to the positions on the interpreted geoelectric section B–B' shown in Figure 11.36. As an aid to interpretation, information from monitoring wells was used to constrain the number of layers used in the TDEM data inversion process. This information was also used to correlate derived true resistivities with equivalent chloride concentrations. It was found that a resistivity of approximately $8 \, \Omega$m correlated with a 500 ppm chloride concentration. Using this information, and the spatial information provided from the TDEM surveys, interpreted positions of the 500 ppm isochlor contours were plotted in map form for both the '180 ft' and '400 ft' aquifers and are shown in Figure 11.37. Also shown are the locations

Figure 11.34 Schematic hydrogeological section in the Salinas Valley, California. From Hoekstra and Blohm (1990), by permission

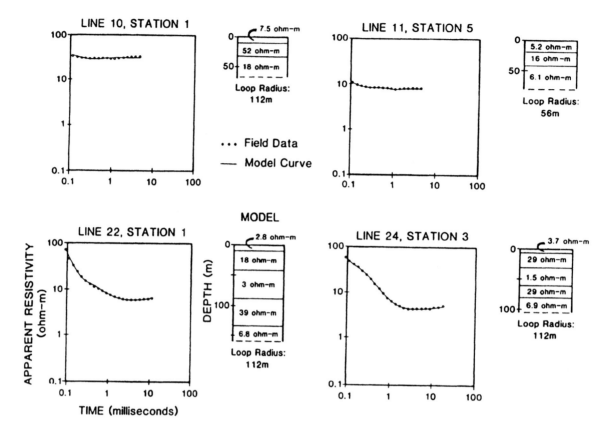

of the 500 ppm isochlor contours derived from monitoring wells. The greater detail on the TDEM-derived contours is a consequence of the greater spatial sampling provided by the TDEM survey compared with that of the monitoring wells.

A very similar application has been described by Goldman *et al.* (1994) to map saline intrusion within the coastal strip immediately to the west of the Dead Sea in Israel. They carried out TDEM sounding using a Geonics PROTEM 37 with inversion and equivalence testing with TEMIXGL software (Interpex Ltd, USA). A feature of this example was the use of TDEM to image the transition zone between the fully saline intrusion and the freshwater above.

An example of the use of TDEM mapping in conjunction with resistivity, magnetic and gravity mapping in the Murray Basin, New South Wales, Australia, was given by Dodds and Ivic (1990). In addition to a Geonics EM34 (FEM) system, a SIROTEM time-domain system was used with 100 × 100 m loop transmitter with a multiturn receiver at its centre. Delay times used were in the range 0.4 ms to 52 ms. An example of the SIROTEM pseudosection along one traverse is shown in Figure 11.38, with the corresponding interpreted geoelectric section. The results of the inversion of four electri-

Figure 11.35   Four late-stage apparent resistivity curves and their corresponding one-dimensional inversions along section B–B' (see Figure 11.36). From Hoekstra and Blohm (1990), by permission

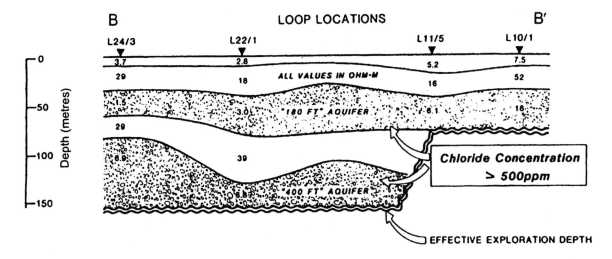

LOOP LOCATIONS

ALL VALUES IN OHM-M

Chloride Concentration
> 500ppm

EFFECTIVE EXPLORATION DEPTH

cal resistivity depth-soundings are also shown on this figure for comparison. A zone (shaded in Figure 11.38B) with a true resistivity in the range 0.6–2.5 Ω m is interpreted as being due to saline groundwater. There is a clear zone where saline groundwater was not found and this zone was interpreted to be a basement high. This interpretation, which was also consistent with the results of the gravity and magnetic surveys, was subsequently proven by drilling. One aspect of this investigation is that the VES inversions were then used to fix the layer parameters for the top two layers in the TEM inversion routine. Examples of this are shown in Figure 11.39.

These case histories demonstrate the usefulness of TEM soundings and profiles constructed with closely spaced soundings in hydrogeology. With the right combination of transmitter sizes and closely spaced soundings, profiles can be produced with a high degree of vertical and lateral resolution. Such spatial sampling, and the

Figure 11.36 Interpreted geoelectric section B–B′ derived from TDEM soundings illustrated in Figure 11.35. From Hoekstra and Blohm (1990), by permission

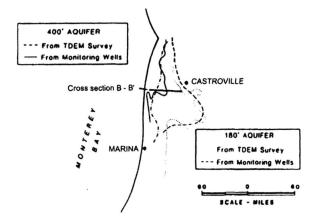

Figure 11.37 Comparison of the position of the 500 ppm isochlor within the '180 ft' and '400 ft' aquifers derived from monitoring wells and TDEM soundings. From Hoekstra and Blohm (1990), by permission

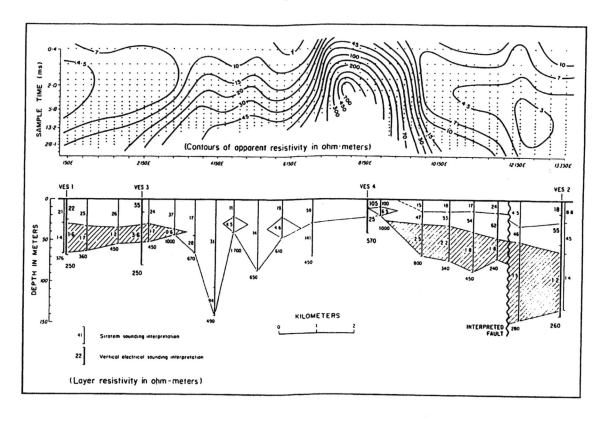

Figure 11.38 (A) SIROTEM apparent resistivity pseudosection in New South Wales, Australia, with (B) the corresponding geoelectric section. Saline groundwater (shaded zone) is evident each side of the bedrock high. From Dodds and Ivic (1990), by permission

increasing reliability of the inversion process, enables detailed two-dimensional geoelectric sections to be produced much more cost-effectively than could have been achieved with boreholes alone.

11.3.3.2 Contaminant plume mapping

Hoekstra *et al.* (1992) have provided a case history which demonstrates the benefit of a combined deployment of both FEM and TEM soundings. The survey was conducted in the vicinity of a brine lagoon in south-west Texas, USA. A Geonics EM34-20H FEM system was used to map variations in apparent conductivity around the brine pit. It is obvious from the resulting contour map (Figure 11.40) that elevated apparent conductivities are evident a considerable distance away from the brine pit. However, a single dipole configuration with a single inter-coil separation can provide only a qualitative impression of the distribution of this conductive contaminant plume. If both dipole orientations had been used with the 10 m and 40 m inter-coil separations, quantitative inversion would have been possible to determine the depth of the plume.

Instead of increased number of measurements with the EM34, TEM soundings were undertaken using a Geonics EM-37 with

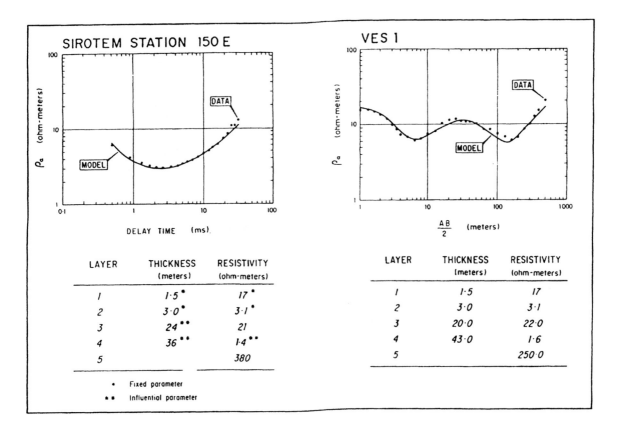

LAYER	THICKNESS (meters)	RESISTIVITY (ohm·meters)
1	1·5 *	17 *
2	3·0 *	3·1 *
3	24 **	21
4	36 **	1·4 **
5		380

* Fixed parameter
** Influential parameter

LAYER	THICKNESS (meters)	RESISTIVITY (ohm·meters)
1	1·5	17
2	3·0	3·1
3	20·0	22·0
4	43·0	1·6
5		250·0

Figure 11.39 (*top*) Combined TEM and resistivity inversion using vertical electrical soundings (VES) to fix the layer parameters for the top two layers in the TEM inversion. From Dodds and Ivic (1990), by permission

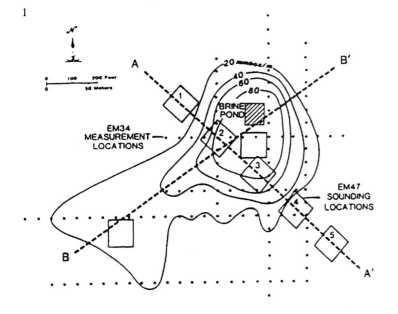

Figure 11.40 Location map and apparent conductivity contour map derived from measurements made with a Geonics EM34 at 20 m coil separation, horizontal magnetic dipole, around a brine evaporation pit. Numbered squares indicate the locations of Geonics EM47 loop soundings. From Hoekstra *et al.* (1992), by permission

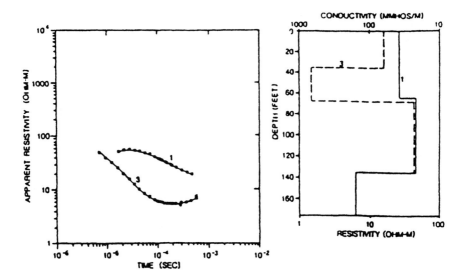

a transmitter loop side length of 100 ft (30 m). Examples of two TEM soundings are shown in Figure 11.41 with their respective one-dimensional inversion results. It is clear that a layer about 10 m thick with a high conductivity is present on sounding 3 but is not seen on sounding 1, although the conductivities for deeper layers are the same for both. This high-conductivity zone is attributed to the saline contaminant plume. Using a series of soundings, a geoelectric section has been produced (Figure 11.42) on which the position of the contaminant plume is obvious (shown as the shaded area). The advantage of the TEM soundings in this case history is that a model of the probable migration pathways of the contamination can be developed using the TEM models (Figure 11.43). Not only can the EM data be used to map the contaminant plume spatially, but they have been

Figure 11.41   TDEM apparent resistivity sounding curves and their corresponding one-dimensional inversions along cross-section A –A' (see Figure 11.40). From Hoekstra *et al.* (1992), by permission

Figure 11.42   Geoelectric section derived from one-dimensional inversions of TDEM soundings along A–A' (see Figure 11.40). From Hoekstra *et al.* (1992), by permission

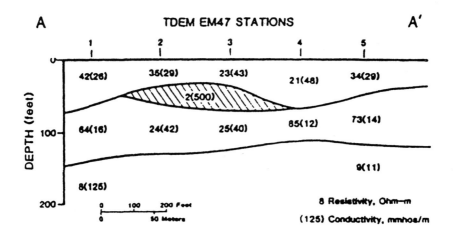

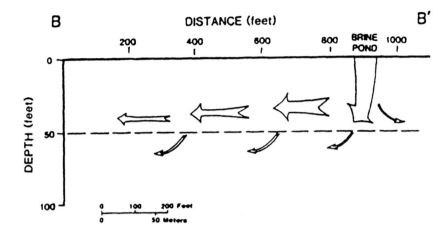

Figure 11.43 Model of probable pathways of brine migration along cross-section B –B' (see Figure 11.40) constructed from geophysical surveys. From Hoekstra *et al.* (1992), by permission

demonstrated to provide an indication of the thickness too. Furthermore, and perhaps more fundamentally, they provide important information for the development of a model of migration pathways. This knowledge is vital if any remediation is to be undertaken successfully.

Buselli *et al.* (1990) have provided examples of the use of TEM in mapping contaminant plumes near Perth in Western Australia. One of these has been selected for discussion here.

A waste disposal facility had been established at Morley, near Perth. Sand was quarried from the Pliocene–Holocene sequence of surficial sand formations and the resulting pit was then filled with domestic solid waste. The area underlain by the predominantly clay Osborne Formation (Figure 11.44) was found to have a resistivity of $10\,\Omega\,m$. The unsaturated zone within the sand was found to have a resistivity of $3600\,\Omega\,m$ and to be up to 50 m thick. The saturated zone of the surficial sand formation was between 20 and 30 m thick with a resistivity of $50\,\Omega\,m$. The sand formations consist of unconsolidated very fine to medium sand which coarsens progressively with depth to become gravely at the contact with the Osborne Formation. The general direction of groundwater flow is towards the south-west.

TEM measurements were made using a SIROTEM system with a transmitter loop of either 25 m or 50 m side length with an in-loop geometry with a dipole receiver of $10^4\,m^2$ effective area placed in the centre of the transmitter loop. The instrument had a capability of measuring the response in a delay time range of $49\,\mu s$ to 160 ms. The locations of all TEM loop centres are shown in Figure 11.44 and borehole locations in cross-section in Figure 11.45.

The results of the TEM survey are shown in Figure 11.46, in which values of resistivity of the sand formation derived from the TEM inversion are contoured. The contaminant plume arising from landfill

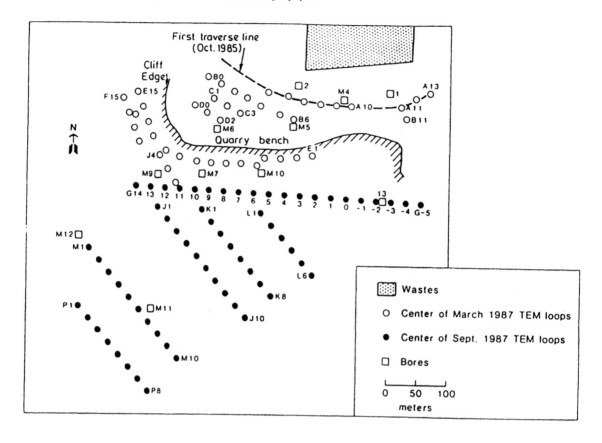

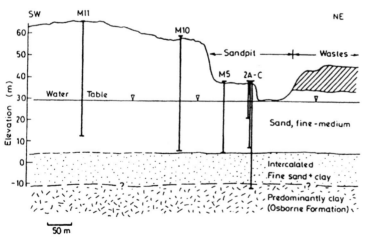

Figure 11.45 Cross-section through the sequence of boreholes M11, M10, M5 and 2A–C and the landfill at Morely, Perth, Western Australia. From Buselli *et al.* (1990), by permission

Figure 11.44 Plan of the Morley landfill and survey area with centres of the 25 m TEM transmitter loops and borehole locations indicated. From Buselli *et al.* (1990), by permission

has been interpreted to be where resistivity values less than or equal to 75 Ω m were determined (tinted area in Figure 11.46). The general orientation of this anomalous zone is the same as the direction of general groundwater flow. This is further confirmed by superimposing

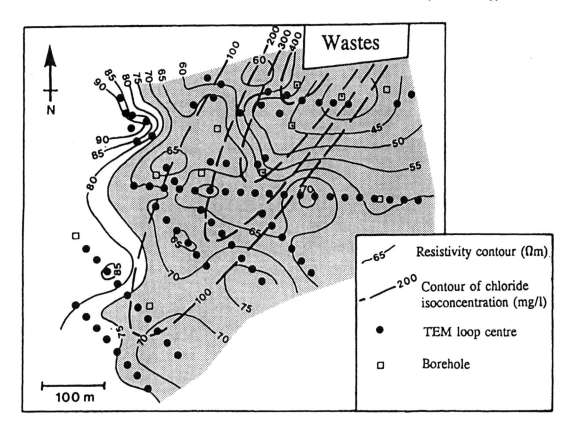

Figure 11.46 Isoresistivity map derived from TEM sounding data with contours of isochlor concentrations determined from chemical analysis of water samples from the boreholes. The contaminant plume is indicated by a tint, as determined by resistivity values $\leqslant 75\,\Omega$ m. After Buselli *et al.* (1990), by permission

the isochlor concentrations as determined from chemical analyses of water samples from the various boreholes over the resistivity map.

The ease with which TEM measurements can be made makes it very simple to re-occupy the same survey locations after discrete intervals of time, such as every 3–6 months, in order to observe changes in the resistivity values. This provides a means of monitoring the flow of a contaminant plume not only with respect to space but also in time. The natural extension of this is the use of TEM to monitor any remediation undertaken.

11.3.3.3   Mineral exploration

A surface DEEPEM survey was undertaken by Crone Geophysics Ltd for Cogema Canada Ltd in the Athabasca Basin, Saskatchewan, Canada. As a consequence of this survey, in conjunction with other techniques, a high-grade uranium deposit was discovered (Crone 1991). The uranium deposit, which is a typical Athabasca type, comprises a long horizontal 'tube' of cross-sectional width less than 100 m of high-grade uranium mineralisation which occurs at the base

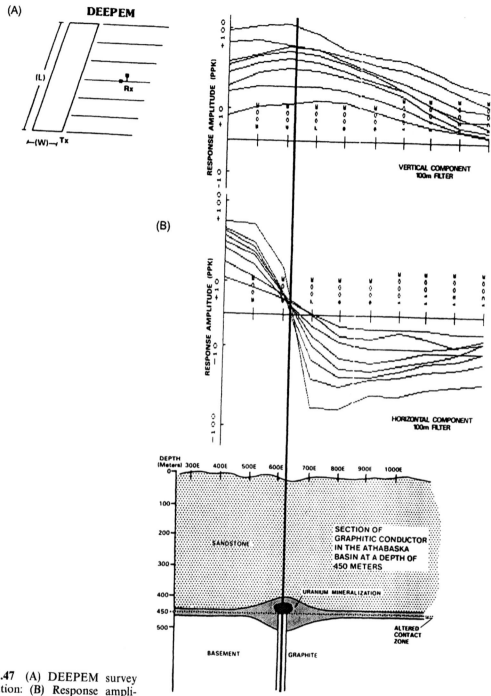

Figure 11.47 (A) DEEPEM survey configuration: (B) Response amplitudes for both horizontal and vertical components (Fraser filtered with a 100 m station interval filter gate). From Crone (1991), by permission

of flat-lying sandstone. The mineralised 'tube' is commonly associated with the upper interface of a large near-vertical graphitic
structure that occurs in Archaean basement rocks. The EM surveys
may be employed to locate the graphitic conductors. Between the
basement and the sandstone is an altered contact zone up to 100 m
thick which is made up largely of clay and which is commonly found
in areas of uranium mineralisation.

In this survey, a large-loop (400 × 800 m) pulse EM system was
used, powered by a 2 kW waveform generator. Horizontal and vertical component measurements were made by the receiver coil at each
survey location (Figure 11.47A). The data were filtered using a Fraser
filter to emphasise selectively the response from currents at up to
a depth of five times the filter station interval. A 100 m station interval
was used in the filter in order to pinpoint the location of the graphitic
conductor. The filtered results of the survey are shown in Figure
11.47B. The data for the horizontal component are particularly
revealing, with the location of the graphitic conductor being indicated
by the node of the family of graphs as indicated in the figure.

Not all massive sulphide mineral deposits are necessarily good EM
targets. For example, Cyprus-type massive sulphide deposits, which
occur as silicically altered mineralisation in altered conductive pillow
lavas, present very small physical parameter contrasts with the host
rocks. Experiments using TEM over known mineral deposits in
Cyprus failed to detect the orebodies (Cooper and Swift 1994). Even
when induced polarisation was used, only the true chargeability could
discriminate between highly silicified structures and mineralised targets. Even chargeability was found to be an unreliable exploration
parameter. The occurrence of extensive local disseminated pyrite
masks electrical mineralisation anomalies at a survey scale typical of
that used for mineral exploration. However, despite the lack of direct
success with TEM in locating orebodies, it was used to help to
delineate geological detail (faults, determination of lithology thicknesses, trends, etc.) which in turn can be used to build up a three-
dimensional structure of a given area (Cooper and Swift 1994).

11.3.3.4 Mapping sub-surface voids

A SIROTEM system was used in the investigation of a series of sink
holes along the route of the proposed Alice Springs to Darwin railway
across the Wiso Basin, Northern Territories, Australia (Nelson and
Haigh 1990). The occurrence of sinkholes in this area was well known,
but the sudden collapse of part of the Buchanan Highway into
a sinkhole in 1982 highlighted the engineering hazard which then
prompted a full-scale investigation.

Various geophysical methods were tested as to their effectiveness
over terrain which was undulating and irregular, and which was
covered with lateritic soils. Intuitively, air-filled cavities should give

rise to resistive anomalies, but the contrasts found were sufficiently small that the method was not considered practicable for the required coverage of large tracts of land along the proposed railway route. It was found, however, that the walls and floors of known dolines tended to contain electrically conductive clay-rich material. Furthermore, the process of doline (sinkhole) formation was associated with leaching of silica, which produced pipes or structures with leached, weathered, transported or altered detritus that were also found to be charactristically conductive. Consequently, the target type for which the geophysics was selected was not the more intuitively obvious 'hole-in-the-ground' but the conductive secondary targets. Indeed, TEM would not have been suitable for air-filled voids owing to the diffusion velocity through the void being of the same order as the velocity of light. Similarly, these same silica-leached features also caused significant and detectable anomalies on time delays, relative amplitude attenuation and spectral content of seismic waves.

Following a considerable amount of trial work with a variety of methods, it was established that coincident-loop SIROTEM was the best method for rapid reconnaissance with the seismic surveys being used over specific features. The seismic work, in addition to the SIROTEM investigation, has been described in considerable detail by Nelson and Haigh (1990).

Of several test areas, the Buchanan Highway Test Site, located 60 km west of the Stuart Highway, was known to be a major zone of doline formation. As part of this work, 25 m coincident-loop SIROTEM surveys were undertaken, with four turns of wire per loop to improve the signal strength. The rate of surveying with this system was up to 4 km per day. Two lines parallel to each other but 25 m apart were surveyed, with additional parallel lines with greater offset being investigated as required to help delimit the extent of features identified.

The general findings of the SIROTEM investigation were that the local geology was characterised by a generally uniform resistivity distribution, with apparent resistivities in the range $50-100\,\Omega$ m, and with a gently dipping and consistently layered structure. In contrast, areas affected by dolines were found to have lateral resistivity fluctuations of up to $20\,\Omega$ m with a complex heterogeneous geometrical appearance. A map of apparent resistivity values obtained using the SIROTEM system for part of the test site, where two dolines were known to occur, is shown in Figure 11.48. Sinkholes HV4 and HV5 were the two that occurred suddenly overnight in 1982. A car that drove into the newly formed hole was totally wrecked but the driver escaped with only minor injuries! As a consequence of this incident, the highway was diverted along the route shown in Figure 11.48. However, a prominent apparent resistivity low (to $10\,\Omega$ m) was found at coordinates (90 m E, 40 m N) almost on the diverted highway. This same feature correlated with seismic arrival time anomalies and is interpreted to be an imminent doline collapse feature!

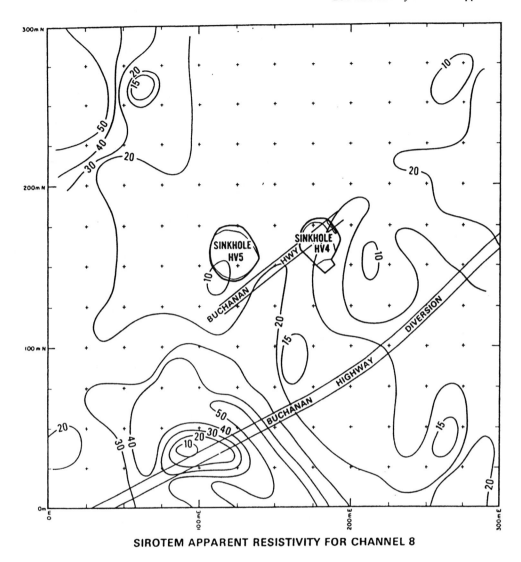

SIROTEM APPARENT RESISTIVITY FOR CHANNEL 8

Figure 11.49 shows composite results of the two parallel SIROTEM traverses 25 m apart as contoured apparent resistivity values. A complex anomalous zone is evident between 89 900 N and 90 200 N on both traverses; in comparison with anomaly shapes over known dolines, this anomaly is characteristic of sinkhole-prone terrain. The SIROTEM anomaly is also consistent with the interpretation of seismic refraction results in terms of amplitude decay and travel-time delay over the anomalous feature. This combined correlation of the anomalous zone provides increased confidence in the overall interpretation. The full scope of the investigation carried out by Nelson and Haigh (1990) and co-workers deserves being read in

Figure 11.48 SIROTEM apparent resistivity anomaly map over part of the Buchanan Highway test site, Northern Territory, Australia. From Nelson and Haigh (1990), by permission

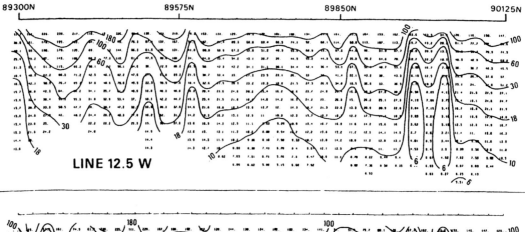

LINE 12.5 W

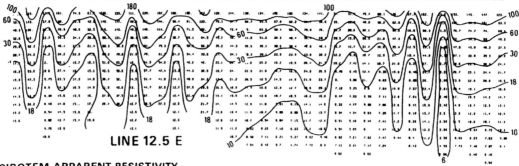

LINE 12.5 E

SIROTEM APPARENT RESISTIVITY
– TIME SECTIONS

N.T. RAIL INVESTIGATION - BUCHANAN NORTH AREA

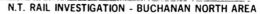

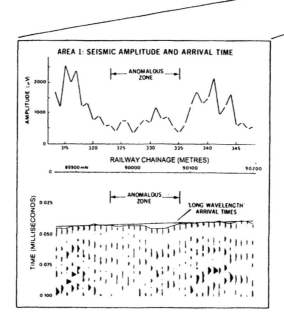

detail as it is an excellent example of a well-thought-out investigation that was well executed. As with so many projects like this, lack of finance at a critical stage denied the final conclusive direct investigative work. A series of appropriately sited boreholes could have helped test the geomorphological model for the formation of these features. Considering the major engineering significance of these features over a very large tract of land in the Northern Territory of Australia, the incomplete funding of such work is a major oversight.

11.3.3.5 *Geological mapping using airborne EM*

Huang and Palacky (1991) have described the use in 1983 of airborne EM surveying for geological mapping in Dongling, Anhui Province, China. The survey was undertaken using an INPUT-type system made in China (the M-1 instrument) which uses a transmitter made up of a seven-turn vertical axis loop mounted on a fixed-wing aircraft. The receiver used was a towed bird, nominally 90 m behind and 63 m below the aircraft. Although designed for an optimum flying height of 120 m, the system was flown at significantly greater heights. The channels at which the signals were measured were 0.3, 0.5, 0.8, 1.2, 1.7 and 2.3 ms after the transmitter switch-off.

The survey area was known to contain lead–zinc mineralisation of the replacement type in Mesozoic limestones. The objective of the AEM survey was to locate previously unknown mineralised bodies and to determine the thickness of fluvial clays in the Yangtze River plain and the extent of the weathering in areas underlain by Mesozoic rocks. The line spacing was 500 m with a NW–SE flight line direction.

Figure 11.50 shows the AEM data and the corresponding geological section with the altimeter trace (top), and measured amplitudes at six channels with indicated base levels for each. The fiducial markers are at approximately 500 m intervals. The geological section was derived from the interpolation of drilling data. The peak observed on the AEM data at fiducial 54 for channels 1 and 2 (arrowed) is thought to be due to the change in flying height. Strong AEM anomalies evident on the early-time channels (1–3) are due to conductive fluvial clays. Resistivities of the local geological materials were obtained from ground-based electrical soundings with the fluvial clays having a resistivity in the range 5–15 Ω m and the weathered material, up to 50 Ω m with the underlying Mesozoic strata having resistivities of hundreds to thousands of ohm-meters. The inversion results of the same data are shown in Figure 11.51. The apparent resistivity (ρ_a) was derived from the inversion of the AEM data from all six channels, assuming a homogeneous half-space model. The marked increase in apparent resistivity near fiducial 51 is associated with the transition from the fluvial clays to Mesozoic sediments. The segments of the resistivity graphs shown dotted are indicative of less reliable results owing to the higher noise levels within the system.

Figure 11.49 (*opposite*) SIROTEM 25 m coincident-loop apparent resistivity pseudosections along two lines 25 m apart, either side of the centreline of a proposed railway route at Buchanan Highway. Seismic amplitude and waveform sections over the SIROTEM anomalous zones confirm the probability of this region being indicative of sinkhole-prone terrain. From Nelson and Haigh (1990), by permission

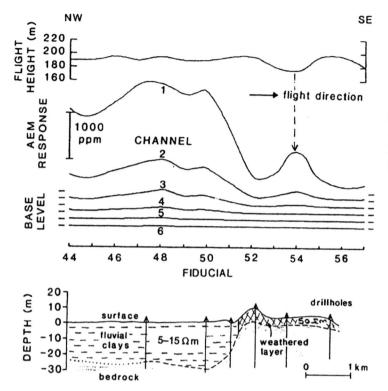

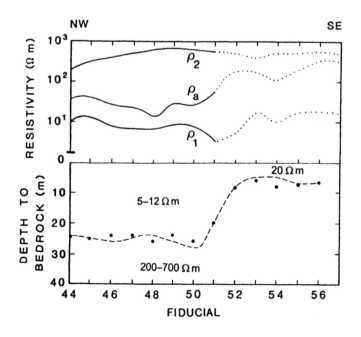

Figure 11.50 (A) Altimeter trace for an airborne EM (AEM) survey at Dongling, Anhui Province, China. (B) AEM responses at six channels with respective base levels for each channel with fiducial fix numbers. (C) Interpreted geological section derived from AEM data and borehole information. The true resistivity ranges (in Ω m) of the superficial materials are indicated. From Huang and Palacky (1991), by permission

Figurer 11.51 Apparent resistivity (ρ_a) and first- and second-layer resistivities (ρ_1 and ρ_2) obtained by inversion of the AEM data shown in Figure 11.50. The interpreted depth to bedrock at each fiducial position derived from the AEM inverion (dots) and from drilling (dashed lines) are shown for comparison. From Huang and Palacky (1991), by permission

The inversion was carried out using a 2-layer model for which resistivities were found to be 200–700 Ω m for the Mesozoic strata, 5–12 Ω m for the Cenozoic fluvial clays and about 20 Ω m for the weathered Mesozoic sediments, all of which are in good accord with the results of the ground-based investigations. Furthermore, the depths derived from the inversion (dots) are in very good agreement with those found from drilling (dashed line) as shown in Figure 11.51. This example demonstrates the effectiveness of the AEM method coupled with appropriate inversion processing in mapping significant geological boundaries. The method shows that it is particularly the early-time responses that are sensitive to the conductivity contrasts in the near-surface.

11.4 VERY-LOW-FREQUENCY (VLF) METHODS

11.4.1 Introduction

The first commercially available ground VLF instrument was made in 1964 by Ronka (Paterson and Ronka 1971) with others being manufactured within the following few years. The most widely accepted ground survey instrument has undoubtedly been the Geonics EM16. Its resistivity mapping mode is also very well used in the EM16R configuration. In recent years, new VLF instruments have been produced such as the WADI from ABEM, EDA's integrated magnetometer/VLF system called the OMNI IV, the 'VLF-3' and 'VLF-4' from Scintrex, and the 'VLF-2' from Phoenix, among others. Although VLF measurements can also be made from the air, only ground-based systems wlll be considered here. A comprehensive review of VLF methods has been given by McNeill and Labson (1991).

The VLF method has remained an excellent, cheap and rapid tool for reconnaissance mapping of conductive mineralised bodies and water-bearing fractures. Its use in engineering and environmental work has as yet remained small. However, with the advent of modern VLF systems with integrated data-loggers, auto-selection of appropriate transmitters and enhanced display and interpretation methods, the technique is gradually being tried for non-exploration purposes such as in cavity detection and in mapping landfills.

11.4.2 Principles of operation

There are eleven major VLF transmitters distributed around the world used primarily for military communications. These provide very powerful EM waves which, when sensed from a distance greater than a few tens of kilometers, behave as plane waves propagating outwards horizontally. Signal level contours for two of the major VLF transmitters are shown in Figure 11.52. Areas enclosed within the 54 dB contours have good signal strengths while those within the 48 dB contour have only marginal signal strength. Areas left unshaded have signal strengths too weak for the method to be used effectively or without special techniques being used.

If a vertical sheet conductor is orientated such that its long axis lies on a radial direction from an active transmitter, the magnetic vector acts tangentially (as depicted in Figure 10.5 in the previous chapter) across the conductor. Eddy currents are induced within the conductor to produce a secondary electromagnetic field. For a conductor not so aligned, the production of eddy currents is much less efficient and the strength of the induced secondary field is much reduced.

For a vertical sheet conductor in a resistive medium, for example, and a profile direction aligned along the magnetic vector, then the tilt-angle response obtained is derived from the vector summation of the primary and secondary components, as shown in Figure 11.53. The primary magnetic vector is horizontal. The induced electromagnetic field varies in amplitude and direction with position relative to the target. On one side of the target, the angle between the two vectors reaches a maximum and then passes to a minimum on the other side before returning to zero beyond the influence of the target. The point at which the tilt angle passes through zero from the positions of the maximum and the minimum, known as the 'crossover' point, lies immediately above the conductive target. If the target dips, then the anomaly shape is distorted with either the positive or negative element being emphasised at the expense of the other component.

The largest amplitudes of the various electric and magnetic components lie at the ground surface and diminish with depth (Figure 11.54). VLF surveys are used largely for spatial mapping of sub-surface targets, not for depth determination as the depth penetration is generally restricted to the near-surface only.

The VLF instrument contains one or more aerials whose characteristics are appropriate for specific transmitters. The VLF EM16 requires the correct crystal aerial for a given transmitter to be physically plugged into the device. To use the instrument, the operator holds the device out horizontally and then rotates it around a vertical axis until a null position is sensed using an audio signal. The direction in which the device is then pointing is along the line of the horizontal component of the electric vector. Survey profiles are conducted at right-angles to this direction, i.e. along the line of the magnetic vector for the given transmitter. Once aligned along the profile direction, at each station, the device is rotated about a horizontal axis. The operator holds the device to his/her eye and views the tilt of the instrument through an eyepiece. The operator rocks forwards and backwards to sense a null in the audio signal, at which

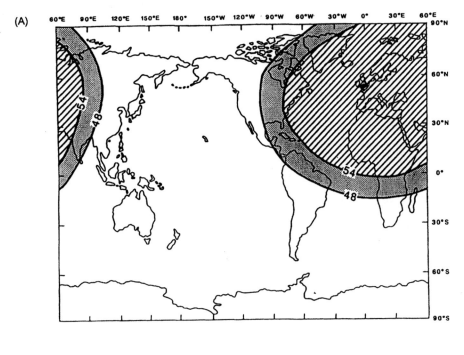

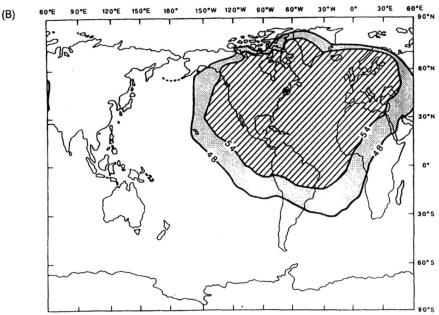

point the tilt angle is then read from the instrument. VLF devices can also be used to measure the electric component by using a grounded dipole up to 5 m long made up of a piece of wire connected to the ground at each end. In this configuration (VLF-R) the apparent

Figure 11.52 Signal level contours for VLF transmitters at (A) GBR, Rugby, UK, and (B) NOAA, Cutler, USA. From McNeill and Labson (1991), by permission

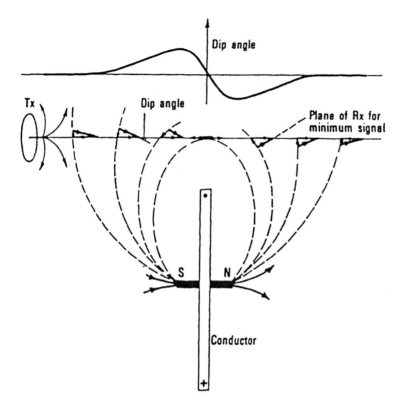

Figure 11.53 Tilt-angle profile over a vertical sheet conductor. From Beck (1981), by permission

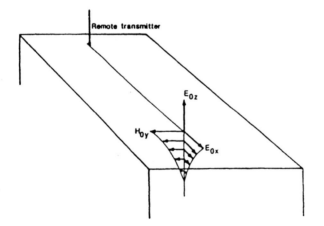

Figure 11.54 Schematic of the diminution of the amplitude of field components with depth. From McNeill and Labson (1991), by permission

resistivity and phase of the horizontal electric field in the ground can also be determined.

In a modern instrument, such as the ABEM WADI, three different frequencies are used simultaneously and the device is able to sense from which transmitter the strongest signal is emanating. Without having to orientate the instrument, the operator walks along the chosen profile line and takes readings. While this has obvious benefits in reducing setup time, there is one major disadvantage with auto-tuning devices. Repeatability of measurements is sometimes difficult along the same profile when surveyed on different occasions as the transmitter with the strongest signal may be different each time. Thus survey parameters are difficult to keep constant.

Another drawback with any VLF system is that the method is totally dependent on there being an appropriate transmitter operational. There are occasions when transmitters are off-air and no source signal from a transmitter with an appropriate azimuth to the target is available. Sometimes, a given transmitter can be switched off while an operator is in mid-survey. There is nothing within the control of that operator that can be done to rectify the situation unless another transmitter is available with a suitable azimuth and the instrument being used has the appropriate aerial.

With modern instruments it is now possible to measure the respective fields from two orthogonal transmitters with one instrument in the same pass along the survey line. The direction to one transmitter is parallel to the strike of the target with the other at right-angles. These two orientations are then referred to as the E- and H-polarisation modes. Thus with modern instruments in both VLF-EM and VLF-R modes using two orthogonal transmitters, eight parameters can be obtained at each station (in-phase and quadrature components in VLF-EM mode; resistivity and phase in the VLF-R mode for each transmitter).

11.4.3   Effect of topography on VLF observations

VLF measurements can be adversely affected by topography, and so some sort of topographic correction may have to be applied before a target anomaly becomes obvious. If a topographic high, such as a ridge, is parallel to the strike direction of the target body, and thus is at right-angles to the survey direction, the topographic effect is a subdued version of the actual topography. That is, on rising up the slope, the tilt angle is increased slightly, and on descending down the slope the other side, it decreases. If the survey is conducted parallel to the survey direction along the ridge, then no association with topography may be evident. The actual topographic response can be complex and depends on the resistivity of the local materials. If the tilt-angle profile mirrors the actual topography, then this is strongly suggestive that topographic effects are present. Similarly, if positive polarity anomalies correlate with topographic highs and negative polarity anomalies are located at the bottoms of valleys, this also indicates the presence of topographic effects.

If the survey has been conducted over topography with a wavelength much longer than the width of the target along the survey line, topographic effects may be removed by the application of an appropriate filter.

11.4.4   Filtering and interpretation of VLF data

VLF tilt-angle data are commonly interpreted only qualitatively. The point where the tilt angle crosses over from being positive to negative polarity is usually interpreted as being immediately above the top of the conductor causing the anomaly. In profile, this crossover is usually quite clear. When plotted spatially in map form, however, the locus of all zero-points (a line joining the crossover points from each profile) is not as easy to identify.

One way by which this problem has been resolved is the use of a filter, such as that devised by Fraser (1969). His filter was designed to shift the tilt-angle data by 90 degrees so that crossover and inflection points become peaks. The filter also attenuated long spatial wavelengths to help overcome some aspects of topographic effects and also to reduce the slow temporal variations in signal strength of the transmitter. Furthermore, his filter was designed not to increase the noise content of the data and, importantly, to be simple to apply.

The Fraser filter uses four consecutive data points, where the data have been acquired at a regular interval and can be applied very simply using a hand calculator or spreadsheet. The sum of the first and second data points is subtracted from the sum of the third and fourth values and plotted at the midpoint between the second and third tilt-angle stations (see Box 11.6). It is important to remember to take account of the polarity of the tilt angle, whether positive or negative, in the calculation of the filtered values.

Box 11.6   Fraser filtering VLF tilt-angle data

Given a sequence of tilt-angle data, $M_1, M_2, M_3, \ldots, M_n$, measured at a regular interval, then the Fraser filter F_i is applied as follows:

- The first filtered value, $F_1 = (M_3 + M_4) - (M_1 + M_2)$ and is plotted half-way between stations 2 and 3.
- The second filtered value, $F_2 = (M_4 + M_5) - (M_2 + M_3)$ and is plotted half-way between stations 3 and 4, and so on along the profile.

For interpretation, it is arguably better to use both the raw and filtered data. The effect of filtering the data can result in the anomaly peak being displaced laterally along the survey line. Thus for more accurate target location, the raw data should be used.

For more quantitative interpretation, two approaches can be taken. One is to use a set of *master curves* and associated interpretational aids published by Madden and Vozoff (1971). However, these curves have not been widely used as most interpreters appear to favour their own empirical approach based on experience or on nomograms calculated for simple dipping sheet-like conductors (Sinha 1990). For more detailed interpretation, especially in connection with two-dimensional and three-dimensional targets, numerical modelling can be used (e.g. Nissen 1986; Ogilvy and Lee 1989), although this tends to be undertaken more in research work rather than mainstream commercial surveys. A detailed discussion of the quantitative analysis of VLF data has been given by Telford *et al.* (1990) and by McNeill and Labson (1991). The main value in VLF surveying lies in its simplicity of use in the field for reconnaissance work for which qualitative interpretation is usually adequate.

11.4.5 Applications and case histories

11.4.5.1 Detection of orebodies

A VLF profile acquired across mineralised metasediments at Sourton Tors adjacent to the Dartmoor granitic intrusion, Devon, is shown in Figure 11.55. The VLF profile shows a marked change in polarity at about 210 m along the survey line. This location correlates with a known area of stratabound mineralisation within the chert/shale zone marked 'MZ'. For comparison, profiles obtained using SP and magnetometry are also shown. Both of these reveal anomalies around the same location and with the shapes of anomalies consistent with a conductive magnetic target dipping towards the north-west. Another section across Sourton Tors is shown in Chapter 2 (Figure 2.43). The VLF data were acquired using a Geonics EM16 with GBR, Rugby, as the transmitter.

11.4.5.2 Location of sub-surface cavities

Ogilvy *et al.* (1991) have described a VLF survey undertaken near Alcala de Henares, about 20 km east of Madrid, Spain. In medieval times, artificial galleries were constructed at a depth of 2–4 m to drain superficial Quaternary gravel terrace deposits which overlie an impermeable Tertiary clay formation in order to provide the town of Alcala with fresh water. The location and the lateral extent of these galleries were uncertain. The shallowness of the galleries meant that they posed a potential for collapse. Indeed, one road subsided as a direct consequence of being constructed over one of these galleries.

An example of a geological cross-section through a typical gallery is shown in Figure 11.56A. The galleries are on average 1 m wide and 2 m high and are largely free-standing and straddle the gravel–clay

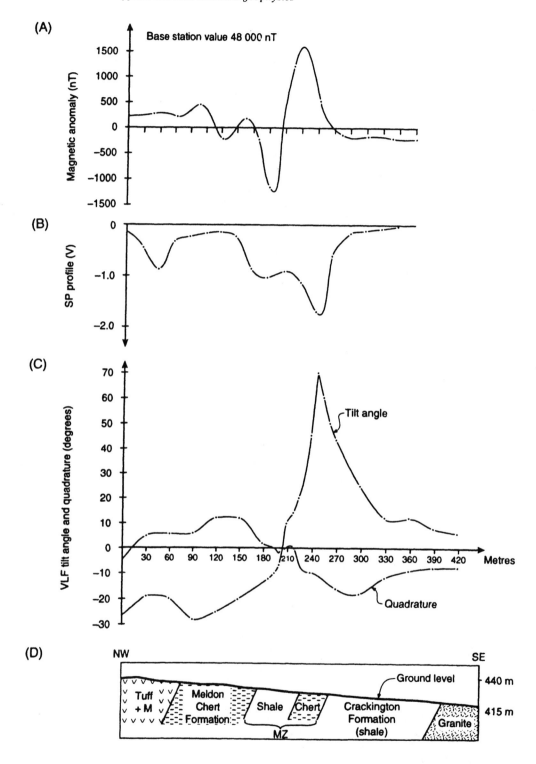

(A)

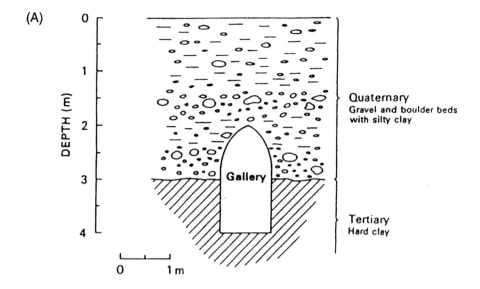

(B)

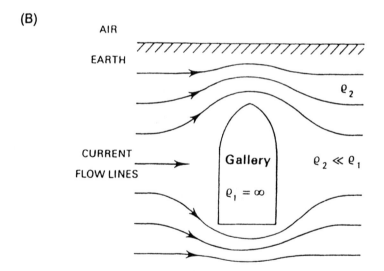

Figure 11.56 (A) Typical cross-section, and (B) schematic representation of primary current flowlines around an air-filled (infinitely resistive) drainage gallery at Alcala, near Madrid, Spain. From Ogilvy *et al.* (1991), by permission

interface. At the time of the survey, the galleries were dry and thus constituted air-filled tunnels.

A $100 \times 100\,m$ grid was established over one part of the known gallery system with lines orientated at right-angles to the direction of the galleries with an inter-line separation of $10\,m$ or less. Station intervals along the lines varied between $5\,m$ and $1\,m$, as required to sample the observed anomalies spatially. The position of the gallery system was also mapped underground by theodolite and compass so as to provide information with which to compare the VLF results.

Figure 11.55 (*opposite*) A variety of geophysical profiles acquired using different geophysical methods over mineralised metasediments at Sourton Tors, Devon, UK: (A) magnetic total field; (B) SP; (C) VLF, and (D) the geological cross-section

The VLF instrument used was a Scintrex VLF-3 in both VLF-EM and VLF-R modes using two orthogonal VLF transmitters: NAA, USA (24 kHz) and Ste Assise, France (16.8 kHz) providing the H-polarisation and E-polarisation data, respectively. The type of current flow around an infinitely-resistive two-dimensional void, such as provided by the galleries, over a conductive substrate is shown in Figure 11.56B. There is obvious current channelling occurring as indicated by the closeness of the current flowlines around the gallery.

Figure 11.57 (A) VLF-R apparent resistivity, and (B) phase profiles over a known drainage gallery: H-polarisation mode using the NAA, USA, transmitter (24 kHz). From Ogilvy *et al.* (1991), by permission

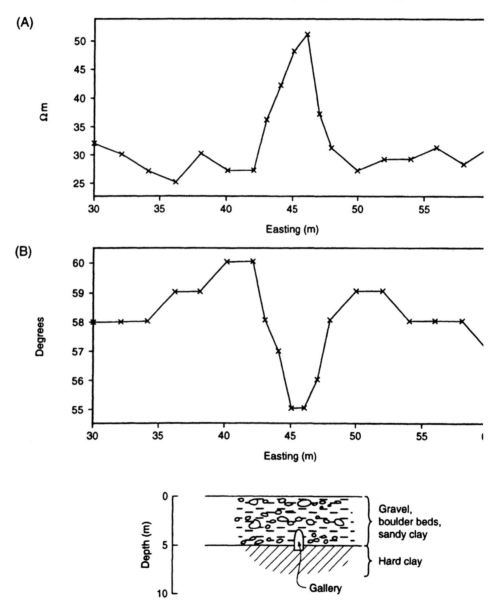

The VLF-EM survey produced absolutely no response from the gallery system. Given the true resistivities of the gravel beds ($200 \, \Omega \, m$) and the underlying clays ($10 \, \Omega \, m$), and the geometry of the gallery with respect to the H-polarisation, no vertical magnetic gradient is generated by such a structure.

In contrast, however, the VLF-R H-polarisation results produced anomalies which could be directly correlated with the gallery system and which could be modelled using two-dimensional inversion routines. Pronounced apparent resistivity ($52 \, \Omega \, m$) and phase ($5°$ peak-to-peak) anomalies were observed directly over a known gallery position (Figure 11.57). The decrease in phase above the gallery is indicative of a resistive structure below the uppermost layer. As it is known that the substrate is conductive clay, the only resistive target available is the gallery. This demonstrates the effectiveness of H-polarisation VLF-R measurements in the detection of air-filled voids over conductive media.

This example also demonstrates the importance of using orthogonal transmitters. The absence of any detectable VLF-EM magnetic field response for H-polarisation suggests that the observed anomaly must be caused by the electric field component E_x of the complex impedance ratio E_x/H_y. It is also a good example of the dominance of galvanic currents over vortex current flow arising from the deviation and concentration of current flow lines around the air-filled gallery. Such current flow results in larger-than-normal primary electric fields, with a consequent increase in apparent resistivities; hence the observed apparent resistivity high immediately over the target.

11.5 THE TELLURIC METHOD

11.5.1 Principles of operation

As a consequence of the presence and fluctuation of the Earth's magnetosphere (see Chapter 1), natural low-frequency magnetotelluric fields occur which induce alternating currents within the ground. These currents flow parallel to the ground surface and cover huge areas and are known as 'telluric' currents – named after Tellus, the Earth Goddess in Roman mythology. The electric current fields fluctuate continuously in direction and magnitude at any point in response to the temporal variations in the ionosphere and magnetosphere caused by extraneous influences (solar wind, etc.). Distant lightning gives rise to frequencies in the range $1–400 \, Hz$, and changes in density of conductive plasma (solar wind) impinging on the Earth's magnetic field generate frequencies between 0.0005 and $1 \, Hz$. The magnitude of the electric field gradient is of the order of $10 \, mV/km$. Electrical noise is also present, generated by electric storms, seawater currents, sferics, and man-made sources, such as electrified railway and tram lines.

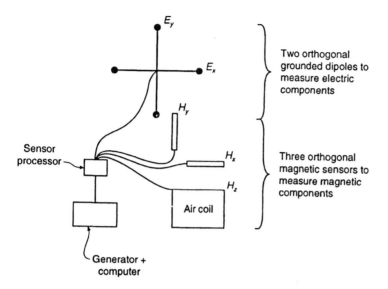

Figure 11.58 Telluric current flow around a sub-surface structure

In a uniform conductive earth, the telluric currents flow parallel to the ground surface with decreasing amplitude with depth. If there is a large-scale resistive sub-surface structure present, such as a salt dome or anticline, then the telluric current flow will be distorted and directed into the more conductive material above; similarly, if a large conductive target is present, such as a massive orebody, then the current will flow towards the more conductive material (Figure 11.58). The exploration method is designed to locate the distortions of the telluric fields. Historically, the telluric method has been used in the location of salt domes in exploration for hydrocarbons, particularly in Russia, Europe and North Africa. It has not been used widely in the USA as the salt domes there are generally too small to give rise to significant telluric anomalies. The method has also been used in the search for massive sulphide deposits and in geothermal resource evaluation.

11.5.2 Field measurements

The potential gradient is measured across two pairs of orthogonal non-polarisable electrodes, at a base station located over ground that is thought to be electrically uniform or remote from the type of target being sought. A second pair of orthogonal non-polarisable electrodes are used as mobile search dipoles. The potential gradient across each dipole within each pair is measured at both the mobile and base stations simultaneously over a period of several minutes. If the ground is electrically uniform below the base station, the two horizontal components of the electric field measured at right-angles should be the same, irrespective of azimuth. The locus of the electric vector at the base station should, therefore, describe a circle.

In actuality, with the presence of various components of electric noise in addition to the telluric currents, the base-station field vector does not describe a circle. However, a mathematical function can be applied to the data to constrain the results to conform to a circle with unit radius. This same mathematical function is applied to the dataset measured simultaneously at the mobile pair of dipoles in order to correct for the electric noise and time-variant field. By referring the measured signals at the mobile dipoles to those at the base station, the data from the mobile station are normalised. If there is no perturbation to the flow of telluric currents, the electric vector at the mobile station also describes a circle. If, however, there is some distortion due to the presence of a sub-surface target, the field vector at the mobile station describes an ellipse. The orientation of the major axis of the ellipse at the mobile station is aligned to the direction of maximum current flow at that point. The ratio of the area of the ellipse to that of the base-station circle gives a relative indication of the amplitude of the telluric anomaly.

Interpretation of telluric results can be qualitative, especially when the method is used for reconnaissance purposes. For more quantitative analysis, the effect of certain simple two-dimensional geological structures, such as an anticline, fault or horizontal cylinder, and three-dimensional shapes, such as a sphere or ellipsoid, can be calculated theoretically. Model curves can be produced against which measured anomalies can be matched. Detailed discussions of such procedures have been given by Keller and Frischknecht (1966) and by Telford *et al.* (1990).

11.6   THE MAGNETO-TELLURIC (MT) METHOD

11.6.1   Principles of operation

The magneto-telluric method uses measurements of both the electric and magnetic components of the natural time-variant fields generated as described in Section 11.5.1. The major advantage of this method is its unique capability for exploration to very great depths (hundreds of kilometres), as well as in shallow investigations, all without the use of an artificial power source (with the exception of 'controlled-source' versions of the method).

The *natural-source magneto-telluric* method uses the frequency range 10^{-3} to 10 Hz, while the *audio-frequency MT* (AMT or

AFMAG) method operates within the higher range 10–10^4 Hz using sferics as the main energy source. The main disadvantage with the natural-source MT methods is the erratic signal strength. The variability in source strength and direction requires substantial amounts of stacking time (5–10 h) per site, thus making MT soundings expensive and production rates slow. AMT measurements can be made faster owing to the slightly higher frequencies, but variability of local thunderstorm sources and signal attenuation around 1 Hz and 2 kHz can degrade data quality. For an example of the relative speed of measurements, three AMT soundings were made in six hours within the crater region of White Island, an active volcano 50 km to the north of North Island, New Zealand (Ingham 1992).

In the early 1970s, David Strangway and Myron Goldstein (Goldstein and Strangway 1975), at the University of Toronto, introduced the principle of an artificial signal source which was dependable and strong enough to speed up data acquisition and improve the reliability of results. The *controlled-source MT* methods typically operate within the frequency band 0.1 Hz to 10 kHz. The first commercial systems were produced by Zonge Engineering and Research Organization Inc., from 1978. Excellent overviews of the range of methods included within the MT family have been given by Vozoff (1986, 1991), and for controlled-source audio-frequency MT (CSAMT) by Zonge and Hughes (1991).

Table 11.3 Applications of controlled-source audio magneto-telluric (CSAMT) surveying

---

Exploration for:
 Hydrocarbons
 Massive sulphides
 Base and precious metals
 Geothermal resources

Geological mapping:
 Structure
 Lithology

Environmental applications:
 Mapping brine leakage from wells
 Mappings brine plumes from leaking tanks, etc.
 Mapping spilled petroleum products
 Monitoring leachate solution in *in-situ* copper recovery projects

Geotechnical applications:
 Structural analysis in mine planning
 Void detection in underground mines
 Mapping burn fronts in underground coal mine fires
 Monitoring enhanced oil recovery

References to all published sources have been given by Zonge and Hughes (1991)

AMT has been used in groundwater/geothermal resource investigations and in the exploration for major base metal deposits over the depth range from 50–100 m to several kilometres. The main application of the MT method, however, has been in hydrocarbon exploration, particularly in extreme terrain and to penetrate below volcanic materials, both types of areas where reflection seismology is either extremely expensive or ineffective. It has also been used recently to investigate a meteoric impact structure in Brazil (Masero *et al.* 1994).

Since the mid-1970s, CSAMT has been used in an increasing range of applications, and especially since the early 1980s within geotechnical and environmental investigations (Table 11.3). It is considered that CSAMT is an under-utilised method with many potential applications in the future. For this reason, CSAMT is described in some detail here.

11.6.2   Field measurements

The general field layout for a magneto-telluric survey is shown in Figure 11.59. It comprises two orthogonal electric dipoles to measure the two horizontal electric components, and two magnetic sensors parallel to the electric dipoles to measure the corresponding magnetic components. The magnetic sensors are made up of coils with several tens of thousands of turns around highly permeable iron cores with a total sensor length typically 2 m long. A third sensor measures the vertical magnetic component. Thus at each location, five parameters are measured simultaneously as a function of frequency. By measuring the changes in the magnetic (H) and electric (E) fields over a range

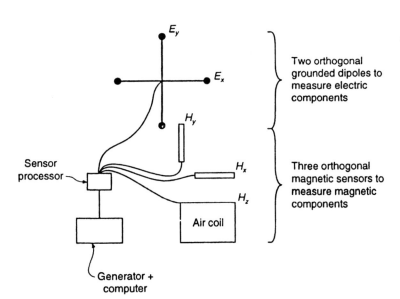

Figure 11.59 Generalised field layout for a magneto-telluric survey

of frequencies, an apparent resistivity sounding curve can be produced, analogous to that produced for electrical resistivity sounding but measured as a function of frequency rather than inter-electrode separation (Box 11.7). The lower the frequency, the greater is the depth penetration.

The data are displayed on log–log plots as apparent resistivity versus either frequency (f) or period ($1/2\pi f$). Over a uniform earth, the phases of the two orthogonal components differ by $\pi/4$, with the magnetic component lagging behind the electric component. If, however, the measured phase difference (θ) is not $\pi/4$, this is indicative of the ground being non-uniform. The basic definitions of apparent resistivity using magneto-telluric parameters were formulated by Cagniard (1953) whose name is given to the apparent resistivity and the impedance term which is the ratio of the orthogonal electric and magnetic horizontal components. The definitions apply to a layered earth; for more complex structures the full tensor impedance must be used (see later in this section).

Box 11.7   Determination of apparent resistivity from magneto-telluric data

Apparent resistivity (ρ_a) is approximated by:

$$\rho_a = \frac{0.2}{f}\left|\frac{E_x}{B_y}\right|^2 \equiv \frac{0.2}{f}\left|\frac{E_x}{H_y}\right|^2 = \frac{0.2}{f}|Z|^2$$

where E_x (nV/km) and B_y are the orthogonal electric and magnetic components, respectively. B_y is the magnetic flux density in nT, which is numerically equal in these units to the magnetising force H_y (A/m). The term Z is the *Cagniard impedance*.

For controlled-source MT surveys, either a loop or grounded dipole is used as a transmitter (the controlled source) with the same measurement configuration as described above. The grounded dipole is typically between 1 km and 3 km long; commonly two orthogonal grounded dipoles are used to provide two different source polarisations (Figure 11.60). The source may be several kilometres away from the receiver sensors. The location and orientation of the bipole source are important in determining the response of the ground at the receiver and have implications for the style of interpretations appropriate. Particular aspects of source polarisation effects have been discussed by Kellett *et al.* (1994) with respect to two massive sulphide deposits in Australia.

A variety of different field configurations are available, ranging from the simple scalar CSAMT, which provides a measure of the two

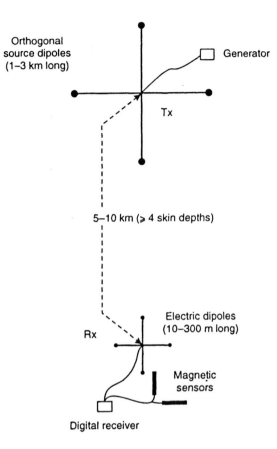

Orthogonal
source dipoles
(1–3 km long)

Generator

Tx

5–10 km (⩾ 4 skin depths)

Rx

Electric dipoles
(10–300 m long)

Magnetic
sensors

Digital receiver

Figure 11.60 Generalised field layout for a controlled-source magnetotelluric survey

orthogonal electric and magnetic components as shown in Figure 11.61A, to vector CSAMT with one source and two sets of orthogonal components (Figure 11.61B), and full tensor CSAMT with either coincident or separated sources (Figures 11.61C and D, respectively).

11.6.3   Interpretation methods

Since the early 1980s there has been a tremendous advance in the interpretation of both magneto-telluric and controlled-source MT soundings, particularly in relation to the computer inversion of sounding curves. Detailed discussions of interpretational methods have been given by Telford *et al.* (1990), Vozoff (1991), and Zonge and Hughes (1991).

Two examples of magneto-telluric soundings are shown in Figure 11.62. It is evident that the combined interpretation of phase difference as well as apparent conductivity can yield important information. As frequency decreases, phase anticipates the resistivity behaviour. For example, Model A in Figure 11.62 is for a resistive basement. With decreasing frequency, the phase difference increases

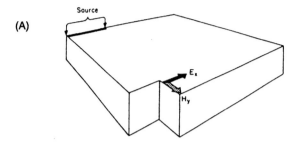

(A)

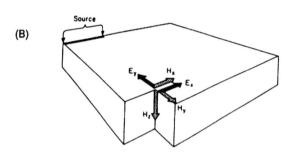

(B)

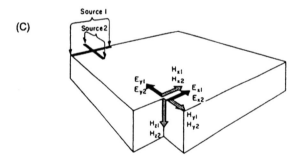

(C)

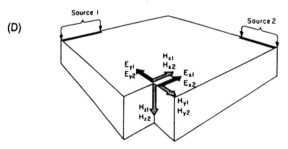

(D)

Figure 11.61 Schematics of: (A) scalar controlled-source audio-magneto-telluric sounding (CSAMT); (B) vector CSAMT; (C) tensor CSAMT with coincident sources; and (D) tensor CSAMT with separated sources. From Zonge and Hughes (1991), by permission

to achieve a peak at mid frequencies while the apparent resistivity increases to reflect the basement value. Conversely, in the case of a conductive basement, the phase passes through a minimum while the apparent resistivity decreases with increasing depth penetration.

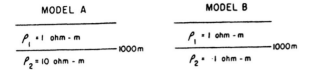

MODEL A   MODEL B

$\rho_1 = 1$ ohm - m $\rho_1 = 1$ ohm - m

———————— 1000m   ———————— 1000m

$\rho_2 = 10$ ohm - m $\rho_2 = \cdot1$ ohm - m

Figure 11.62 Magneto-telluric apparent resistivity and phase responses of 2-layer models. Model A = resistive basement; Model B = conductive basement. From Vozoff (1991), by permission

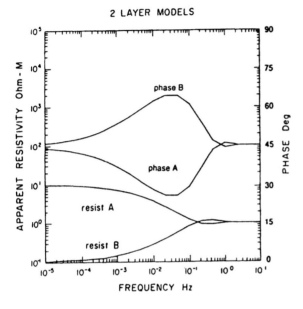

2 LAYER MODELS

As long as the apparent resistivity is asymptotic to a constant value at low frequency, then phase difference should revert to 45° ($= \pi/4$). The apparent resistivity profile, however, is sensitive to near-surface inhomogeneities which may reduce the reliability of a resistivity-only inversion. Furthermore, at middle and higher frequencies the phase difference is more sensitive to deeper structures than apparent resistivity. Very shallow features may not be evident on the phase difference sounding while they may be seen on the apparent resistivity data.

In CSAMT surveys, considerable amounts of data processing may be necessary prior to inversion (see Figure 11.63). The processing is usually undertaken in two stages: pre-processing and interpretative processing.

Pre-processing conditions the acquired data by the removal of errors and noise. Interpretative processing includes optimising plotting conventions for the particular measured or derived parameters. In addition, certain data enhancement processes may be applied such as normalising, static correction, filtering and derivative calculations. As explained by Zonge and Hughes (1991), normalisation removes the effects of layering by subtracting, dividing or other means of deconvolving equal-frequency or equal-depth average values from a set of data. Regional effects can be removed by deconvolution,

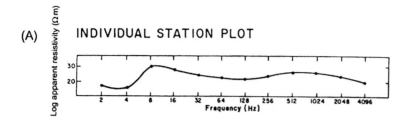

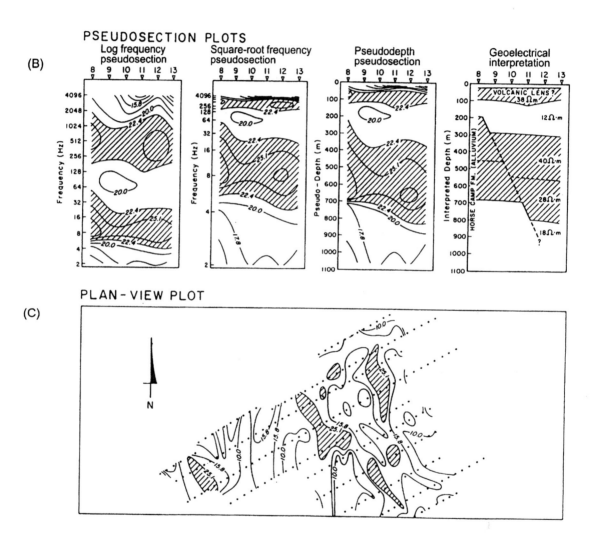

which can be used as well to enhance subtle lateral effects in survey areas with complex layering.

With multistep processing it is important to compare the processed results with the original resistivity and phase information in order to maintain a sense of reality in the processing. If the basic trends evident in the original data cannot be observed in the processed results, then this suggests that something is wrong with the processing. The interpretative processing stages should emphasise particular trends in the original dataset, not create new ones.

11.6.4   Applications and case histories

Zonge and Hughes (1991) have provided a selection of excellent case histories, several of which are abstracted here.

11.6.4.1   *Mineral exploration*

In central Washington, USA, gold mineralisation occurs in hydrothermally altered, silicified zones (reefs) within arkosic sandstones and conglomerates. The silicified zones occur in linear, steeply dipping bodies and are mostly concealed beneath surface cover.

A CSAMT apparent resistivity pseudosection (shown in Figure 11.64) was obtained across a known silicified reef several kilometres from Cannon mine, which is a major gold producer. A strong electrically resistive feature is associated directly with the silicified reef which outcrops at the surface along this transect. Each side of this feature are two conductive zones thought to be due to conductive clay alteration surrounding the silicification. Similar CSAMT lines were acquired over areas with surface cover and the resistive feature was mapped along strike. Such information was used to identify specific drill targets, thereby reducing the overall cost of the exploration drilling programme.

Another example of the use of CSAMT in gold exploration was that undertaken by Phoenix Geoscience in north-central Nevada, USA. The geology (Figure 11.65A) consists of alluvium over 60 m

Figure 11.63 (*opposite*) Display types for CSAMT data. (A) Plot of log apparent resistivity versus frequency. (B) Pseudosection displays with the corresponding geoelectric interpretation. (C) Plan-view plots. Other parameters, such as phase, may be plotted in similar ways. After Zonge and Hudges (1991), by permission

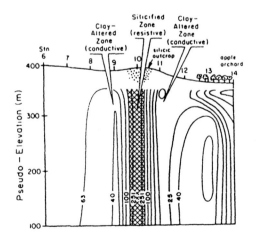

Figure 11.64 CSAMT apparent resistivity pseudosection over a silicified reef several kilometres from a gold mine which exploits this same structure. From Zonge and Hudges (1991), by permission

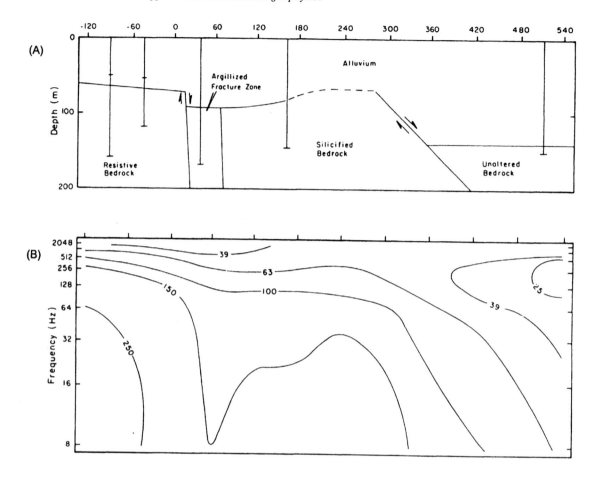

thick overlying silicified Palaeozoic sediments. Graben faulting has resulted in the bedrock interface varying considerably in depth. The gold deposit occurs in a steeply dipping fracture zone which exhibits argillic alteration. The gold deposit has been proved by drilling but has not been mined since 1988.

An example of a CSAMT Cagniard resistivity pseudosection, obtained using 60 m dipoles, across the above structure is shown in Figure 11.65B. The general level of the bedrock interface is well imaged by the 63 Ω m and 100 Ω m contours and shows a dipping trend beyond position 300 with low-resistivity material at greater depth beyond this point. The 150 Ω m contour shows a prominent cusp around position 60, indicating the presence of a conductive feature. This correlates with the clay-rich fracture zone known to contain gold. The results of one-dimensional inversion of the resistivity data agree closely with basement depths obtained by drilling.

Figure 11.65 CSAMT results of a survey over a buried basement structure associated with a hydrothermal gold deposit: (A) geology and CSAMT results, and (B) CSAMT Cagniard resistivity data, courtesy of Phoenix Geophysics. From Zonge and Hughes (1991), by permission

11.6.4.2   *Hydrocarbon investigations*

CSAMT surveys have been used predominantly for hydrocarbon exploration where either the terrain is unsuitable for seismic reflection surveys or where volcanic sequences obscure deeper structure from being imaged seismically. One such example is the Trap Spring Field in Railroad Valley, east-central Nevada, USA. Within the area, there are many graben-type sub-basins, each with its own unique lithology, structure, reservoir and source rocks, thermal history, etc. Fractured Oligocene ignimbrites source the oil which is trapped by the truncation of volcanics in an up-dip direction by a high-angle, basin-margin fault. The volcanics are overlain by 300 m to 3000 m of unconsolidated valley fill material. The material below the volcanics consists of Palaeozoic shales, dolomites and limestones. The top seal to the oil trap is a heavily argillized, unwelded zone at the top of the volcanics. There is only limited drilling information. Seismic reflection surveys are very expensive to procure, and the results less than satisfactory given the cost. Volcanic lenses produce unwanted multiples, unresolvable complex structures, a lack of predictability of target type and, occasionally, poor signal coupling.

A single line of scalar CSAMT was surveyed across the southern portion of the field and the Cagniard resistivity and phase data obtained (shown in Figure 11.66A and B, respectively). The resistivity versus frequency data reveal a characteristic shape with a cusp occurring below a frequency of 8 Hz. The steepness of the cusp sides increases as a result of a tuning effect associated with an extreme 2-layer low-over-high resistivity contrast which occurs at the top of the volcanics where a conductive argillized zone overlies resistive ignimbrites. This electrical marker is particularly sensitive to changes in layering and fault-related displacements, making the CSAMT method an effective structural tool with which to map the resistive basement.

In detail, a significant lateral discontinuity can be seen, especially on the 8 Hz data in Figure 11.66A (Cagniard resistivity plot), at position −2. This termination is associated with the basin-margin fault which bounds the western margin of the field. The overall interpretation utilising both CSAMT and drillhole information is shown in Figure 11.66C. The conclusions arising from the combined interpretation are: (1) the basin-margin fault dips moderately steeply to the east and (2) probably extends to the ground surface; (3) the eastern block is downthrown approximately 200 m with respect to the western side; and (4) the eastern block deepens consistently to the east as a result of successive step-faulting.

The trial CSAMT survey across the Trap Spring Field demonstrated that the method is an effective reconnaissance approach at only a quarter of the cost of reflection seismology. More detailed seismic surveys could be targeted on the basis of further CSAMT surveys.

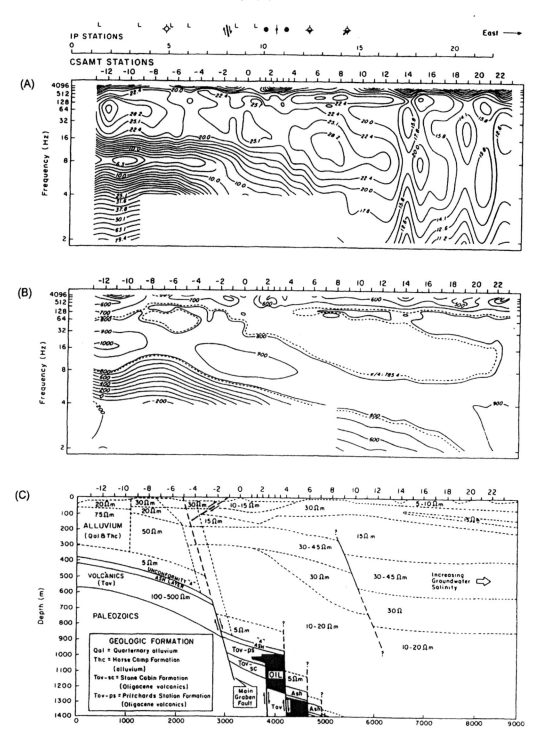

11.6.4.3 Environmental applications

Zonge and Hughes (1991) have provided several case histories of environmental applications of the use of CSAMT surveys. Space constraints at typical sites, high ambient cultural noise, lack of experienced CSAMT operators, and significant survey costs have precluded the use of the CSAMT method for environmental applications within the UK particularly, and to a lesser extent within western Europe. However, with a greater awareness of the effectiveness of the CSAMT method, applications at suitable sites may be forthcoming in the future.

An engineering company was commissioned to undertake a hydro-geological investigation over a 2.5 km² site of a potential commercial development of a desert area north of Phoenix, Arizona, USA. The supply of water to the development was of major concern; to have imported water via a surface pipeline would have resulted in enormous cost and potential legal problems. To have a water supply at the site would have significant benefits.

The local geology, knowledge of which was very limited at the start of this project, comprises granitic outcrops trending N50°W with basaltic lava flows, tuffs, agglomerates, and cinders, and separate schistose units, all laterally discontinuous suggestive of some structural control in the area. Apart from some basic regional gravity data and limited surface geological information, little else was known.

Two boreholes were constructed at the southern part of the property to a depth of about 150 m and approximately 450 m apart. In one hole, water was found at a depth of 91 m and at 43 m in the other, but pressure heads were minimal and the saturated zones were thin.

At this stage, three survey lines of CSAMT were acquired to provide additional information about the sub-surface hydrogeological environment. All three lines showed similar features on the apparent resistivity pseudosections, one of which is shown in Figure 11.67. A significant near-vertical feature is evident around station 8 which is thought to be due to a fault.

The apparent resistivity versus frequency data display the presence of a cusp at low frequencies (less than 128 Hz) as shown in Figure 11.68. The data show a spatial dispersion with the resistivity profile away from the supposed fault as indicated by line A in Figure 11.68 with a cusp at 64 Hz. Survey stations near the fault exhibit type-B behaviour with a much steeper cusp, the side slopes of which are increased owing to the focusing effect of decreased resistivity at depth near the fault. (This same type of behaviour was described in Section 11.6.4.2.) Computer modelling indicated the presence of a conductive layer at a depth of around 300 m, which could have been caused by a previously unknown aquifer at depth. Consequently, it was decided to deepen one borehole to investigate the interpreted conductive target. A significant water-saturated zone was encountered by the

Figure 11.66 (*opposite*) (A) Cagniard resistivity (Ω m), (B) phase difference (mrad) and (C) CSAMT/drillhole interpretation over the Trap Spring Field, Nevada, USA. From Zonge and Hughes (1991), by permission

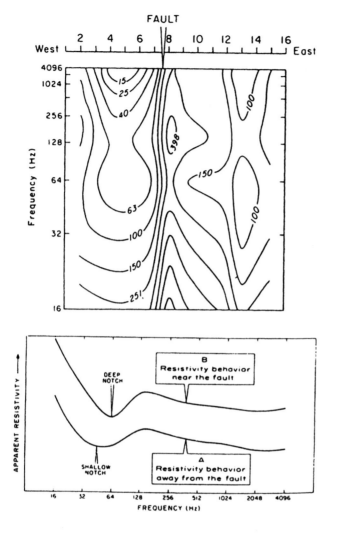

Figure 11.67 Apparent resistivity pseudosection across a fault zone; lower frequencies correspond to greater depth penetration. From Zonge and Hughes (1991), by permission

Figure 11.68 Comparision of resistivity responses adjacent to, and removed from, the fault. From Zonge and Hughes (1991), by permission

extended drilling at a depth of 317 m. The saturated zone was at least 61 m thick and had a pressure head of 360 psi on a 6-inch borehole. With a 17-inch diameter well, it was thought that the hole would have yielded between 600 and 1000 gallons per minute.

The successful use of CSAMT at this site helped avoid the premature abandonment of the property or the instigation of expensive and legally complicated surface pipeline option.

In the USA, there are thousands of abandoned, improperly plugged oil and gas wells. Where the producing horizon is overpressurised following secondary hydrocarbon recovery techniques, these wells can allow injected oilfield brines to migrate up the borehole and into shallower potable-water aquifers. For example, at one oilfield at Sac and Fox tribal lands of Lincoln County, Oklahoma, USA, oil has

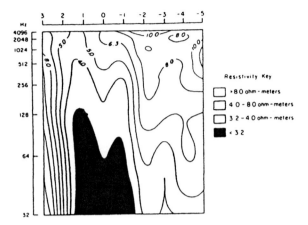

been recovered from the Prue Sand Unit since the 1930s. Brine injection has been used since the 1950s for enhanced oil recovery. At the same location, the Vamoosa Formation is the major source of drinking water and its base occurs at depths of between 45 m and 135 m. Test wells drilled in 1979 indicated significant brine concentrations.

CSAMT surveys in the area indicated the presence at depth of plumes of conductive material around abandoned injection wells. One such apparent resistivity section is shown in Figure 11.69 on which the conductive plume is self-evident. Some plumes were found to reach shallow depths. Following the CSAMT survey two additional test wells were constructed, and bromide/chloride ratios established that the source of the contamination was undoubtedly the Prue Sand brines, as previously suggested by the CSAMT surveys.

In a similar vein, the migration of conductive fluids for *in situ* mining tests as well as injection tests have been mapped using CSAMT methods. In one example, an *in situ* leaching project was losing 90% of the injected fluid and it was important, therefore, to identify where the remainder of the injected material was going. A CSAMT survey was instigated and the resistivity results processed to provide depth-level resistivity slices (Figure 11.70). Conductive areas are shown shaded.

It is evident from Figure 11.70 that the main body of conductive material was centred at the 1600 level (as the areal extent is greatest at this level). It appeared that the fluid had migrated more slowly than hydrologists had suspected. Two CSAMT surveys were undertaken over exactly the same ground but one month apart to map the migration of the conductive plume. It was found that in some areas the apparent resistivity was changing by as much as 1% per day.

As with many geophysical methods, the use of the appropriate method(s) can yield valuable information about the three-dimensional

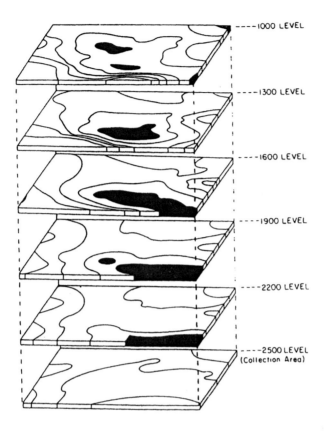

1000 LEVEL

1300 LEVEL

1600 LEVEL

1900 LEVEL

2200 LEVEL

2500 LEVEL
(Collection Area)

Figure 11.70 CSAMT depth-level resistivity slices; conductive zones are shown shaded. The largest areal extent of contamination can be seen at the 1900 level. From Zonge and Hughes (1991), by permission

spatial distribution of sub-surface features. In environmental applications particularly, the spatial changes as a function of time are increasingly important. Where appropriate, CSAMT along with other EM and electrical methods can provide valuable information cost-effectively and without any environmentally detrimental effects.

Chapter 12
Ground penetrating radar

12.1　INTRODUCTION

Since the mid-1980s, *ground penetrating radar* (GPR) has become enormously popular, particularly within the engineering and archaeological communities. However, radar has been used for geological applications since the 1960s, especially in connection with the development of *radio echosounding* of polar ice sheets. Glaciological applications of radar are now very well developed. Despite the recent upsurge in interest in the method, general experience in data processing and interpretation within the engineering community in particular has not kept pace with advances in technology or computer capabilities. There is enormous scope for the application of ground penetrating radar and it is extremely exciting to anticipate what might be achieved in the near future.

GPR applications can be divided into two virtually discrete classifications based on the main antenna frequencies. For geological applications, where depth penetration tends to be more important than very fine resolution, antennae with frequencies less than or equal to 500 MHz are used. For engineering or non-destructive testing (NDT) applications, antennae with frequencies of 500 MHz and greater are used, typically as high as 900 MHz or 1 GHz. A list of the range of applications of GPR is given in Table 12.1.

The first use of electromagnetic (EM) signals to locate remote buried objects is attributed to Hülsmeyer in a German patent in 1904, but the first published description of such investigations was by Leimbach and Löwy (1910), also in German patents. The systems used in these investigations employed continuous wave (CW) transmission. Hülsenbeck (1926) developed the first use of pulsed radar to investigate the nature of buried features.

Pulsed techniques were developed substantially over the following five decades. Its early civilian development was in radio echosounding of polar ice sheets (Cook 1960; Evans 1965; Swithinbank 1968). The first use of impulse radar for glaciological purposes was in the early 1970s (Watts *et al.* 1975). There is much pioneering research work being carried out in the glaciological field (for example, see papers by Wright *et al.* (1990), Hammond and Sprenke (1991) and Narod and Clarke (1994)). A useful review has been published by Bogorodsky *et al.* (1985). There has been wide acceptance of the radar method in certain areas of civil engineering, such as road pavement analysis and void detection behind tunnel linings. There has also been an expanding role for the method in geological applications, particularly in the rapid assessment of superficial deposits, location of swallow holes, etc. In archaeological studies, too, GPR has been used on many sites to identify potential excavation areas.

Many uses of ground penetrating radar have been described in the literature. These include the determination of permafrost thickness (Annan and Davis 1976); the detection of fractures in rock salt

Table 12.1   Range of applications of ground penetrating radar

Geological:
Detection of natural cavities and fissures
Subsidence mapping
Mapping sand body geometry
Mapping of superficial deposits
Soil stratigraphy mapping
Glacial geological investigations
Mineral exploration and resource evaluation
Peat thickness mapping and resource evaluation
Permafrost investigations
Location of ice wedges
Fracture mapping in rock salt
Location of faults, dykes, coal seams, etc.
Geological structure mapping
Lake and riverbed sediment mapping

Environmental:
Contaminant plume mapping
Mapping and monitoring pollutants within groundwater
Landfill investigations
Location of buried fuel tanks and oil drums
Location of gas leaks
Groundwater investigations

Glaciological:
Ice thickness mapping
Determination of internal glacier structures
Ice movement studies
Detection of concealed surface and basal glacier crevasses
Mapping water conduits within glaciers
Determination of thickness and type of sea and lake ice
Sub-glacial mass balance determination
Snow stratigraphy mapping

Engineering and construction:
Road pavement analysis
Void detection
Location of reinforcement (rebars) in concrete
Location of public utilities (pipes, cables, etc.)
Testing integrity of building materials
Concrete testing

Archaeology:
Location of buried structures
Detection and mapping of Roman Roads, etc.
Location of post-holes, etc.
Pre-excavation mapping
Detection of voids (crypts, etc.)
Location of graves

Forensic science:
Location of buried targets (e.g. bodies and bullion)

(Thierbach 1974; Unterberger 1978; Nickel *et al.* 1983; Olsson *et al.* 1983); and archaeological investigations (Bevan and Kenyon 1975; Imai *et al.* 1987; Bevan 1991). Examples of civil engineering and of other geological applications have been described by Darracott and Lake (1981), Leggo (1982), Ulriksen (1982), Leggo and Leech (1983), Davis and Annan (1989), Moorman *et al.* (1991), Doolittle (1993), and Huggenberger *et al.* (1994), among others. Cross-hole radar systems for use in crystalline rock have been described by Nilsson (1983), Wright and Watts (1982) and Olsson *et al.* (1990). GPR has also been used in police investigations to help locate buried bodies, such as in a double murder inquiry on Jersey, Channel Islands, in the 1980s, and in a gruesome search for human remains at two houses in Gloucester in 1994. In the latter, radar was instrumental in detecting where the corpses of 10 murdered women had been hidden within thick concrete inside the buildings, and in locating the remains of two other victims buried in a nearby field.

For regional and large-scale investigations, radar measurements have been made increasingly from aircraft and satellites. Such *remote sensing techniques* are beyond the scope of this chapter. Nevertheless, for sub-surface mapping in arid regions for hydrogeological purposes, for example, satellite radar imagery has been used to locate important features that would otherwise be extremely difficult to locate using ground-based surveys. An example of this is the identification of an ancient river drainage system now buried beneath desert sands in Africa and which was later proven to be an important source of potable water.

Ground radar was developed further by the US Army during the Vietnam War. Systems were constructed in order to locate labyrinths of tunnels excavated and used by the Viet Cong. At the end of the Vietnam War, the potential of GPR for civilian purposes was identified by Geophysical Survey Systems Inc. (GSSI), who are still the largest ground radar manufacturer internationally. In recent years other manufacturers have developed GPR systems, such as the PulseEKKO (Sensors & Software Ltd, Canada) and a range of systems produced by ERA Technologies Ltd, UK. Other companies are developing antennae to add on to existing radar systems, such as Radarteam AB (Sweden). Following the Falklands conflict in 1982, a radar system was developed with the aim of locating plastic mines that had been sown indiscriminately from the air.

12.2   PRINCIPLES OF OPERATION

A radar system comprises a signal generator, transmitting and receiving antennae, and a receiver that may or may not have recording facilities or hardcopy graphical output. Some advanced systems have an onboard computer that facilitates data processing both while acquiring data in the field, and post-recording.

The basic constituents of a radar system are shown in Figure 12.1. The radar system causes the transmitter antenna (Tx) to generate a wavetrain of radiowaves which propagates away in a broad beam. As radiowaves travel at high speeds (in air 300 000 km/s or 0.3 m/ns),

Figure 12.1 Simplified diagram of (A) the constituents of a radar system with (B) the interpreted section of (C) the radargram display. Adapted from Butler *et al.* (1991) and Daniels *et al.* (1988)

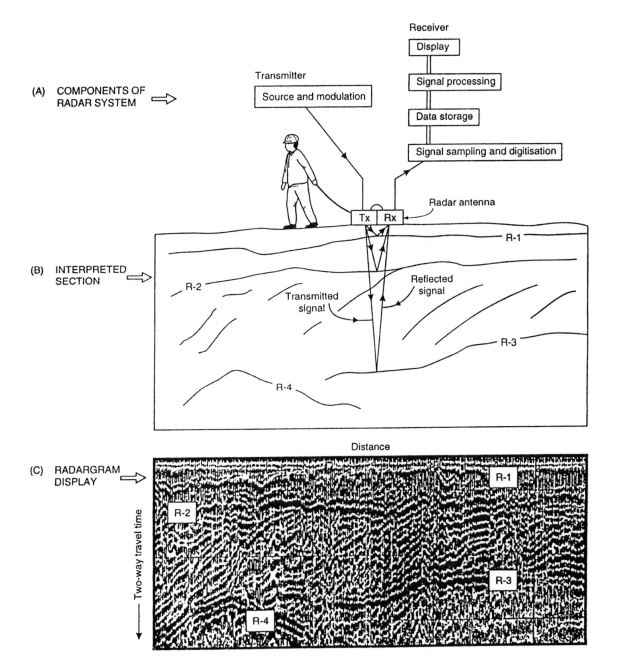

the travel time of a radiowave from instant of transmission through to its subsequent return to the receiving antenna (Rx) is of the order of a few tens to several thousand nanoseconds (ns; 10^{-9} seconds). This requires very accurate instrumentation to measure the transmit instant precisely enough for the final accuracy of the system to be reasonable with respect to the travel times in question. The antennae are used in either a monostatic or bistatic mode. *Monostatic mode* is when one antenna device is used as both transmitter and receiver, whereas *bistatic mode* is when two separate antennae are used with one serving as a transmitter and the other as a receiver. There are specific cases (such as in *wide-angle reflection and refraction* (WARR) measurements) when the bistatic mode is advantageous over the monostatic mode. The PulseEKKO system uses only bistatic antennae. For the majority of this chapter it can be assumed that any antennae are deployed in monostatic mode unless indicated otherwise.

The transmitter generates a pulse of radiowaves at a frequency determined by the characteristics of the antenna being used at a repetition rate of typically 50 000 times per second. The receiver is set to scan at a fixed rate, normally up to 32 scans per second, depending upon the system being used. Each scan lasts as long as the total two-way travel time range, which can be set from a few tens to several thousand nanoseconds. Each scan is displayed on either a video screen or a graphic recorder or both. As the antenna is moved over the ground, the received signals are displayed as a function of their two-way travel time, i.e. the time taken from instant of transmission to time of detection by the receiver, in the form of a *radargram*. This display is analogous to a seismic section (seismogram).

The pulse length of the transmitted radiowave should be short enough (typically <20 ns, depending upon antenna frequency and type) to provide resolvable reflections. It is important, therefore, that the shape and characteristics of the transmitted radiowave are both determinable and highly repeatable. The significance of this point will be discussed below (see Section 12.5). The manner in which the recorded signals are displayed on a graphic recorder, for example, are determined by the operator; a simplified output is illustrated in Figure 12.2. Signals with amplitudes greater than the set threshold are printed dark on the radar section as illustrated. In some cases, it may be most suitable to print both positive and negative, or when just positive or just negative. Displays can also be output in terms of *variable area wiggle* or wiggle trace only (just as in seismic data displays). Commonly, the more sophisticated digital recording systems display the amplitudes of the signals according to a grey scale or colour menu; for example, the strongest reflections can be picked out by the brightest colours.

Figure 12.2 (*opposite*) Schematic example of the translation of the received waveform (one scan) on to a graphic recorder output

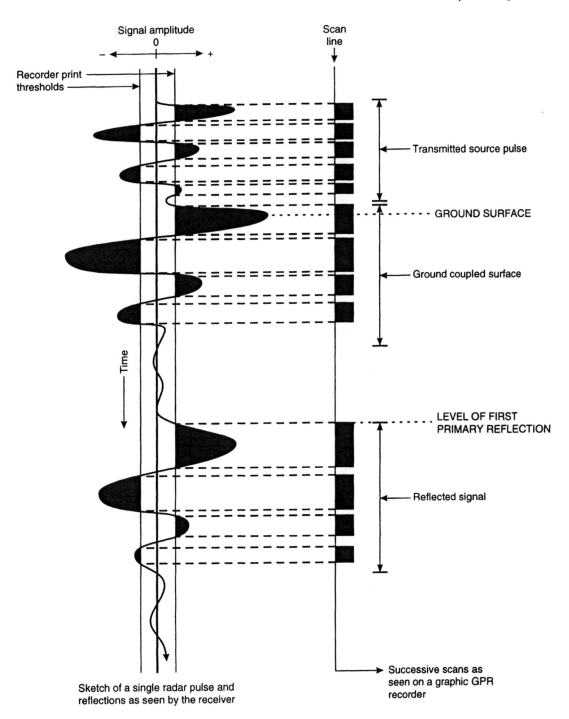

Signal amplitude
0
– ← → +

Recorder print thresholds

Scan line

Transmitted source pulse

GROUND SURFACE

Ground coupled surface

LEVEL OF FIRST PRIMARY REFLECTION

Reflected signal

Time

Sketch of a single radar pulse and reflections as seen by the receiver

Successive scans as seen on a graphic GPR recorder

Note that the source pulse consists of more than one wavelength and that it may have a complex waveshape. Ground coupling affects the shape and duration of the downgoing wavelet and thus the waveshape of any reflection is equally complex, but with a pulse-broadened duration due to attenuation of the higher frequency components of the signal. The reflection event consists of several wavelets, not just one, and it is imperative that this be borne in mind during the interpretation of radar data.

The measurement system should have sufficient dynamic range and sensitivity to be able to detect the low signal strengths associated with the returning radar pulses. It should also be able to produce printouts with adequate clarity for interpretation.

While the manufacturer's specifications may indicate the measurement accuracy of the timing within the instrument (e.g. to ± 1 ns), this should not be interpreted as being equivalent to the resolution capability of interpretation. Vertical and horizontal resolution are discussed in Section 12.5.

12.3   PROPAGATION OF RADIOWAVES

12.3.1   Theory

The electromagnetic properties of materials are related to their composition and water content, both of which exert the main control over the speed of radiowave propagation and the attenuation of electromagnetic waves in materials.

The speed of radiowaves in any medium is dependent upon the speed of light in free space ($c = 0.3$ m/ns), the relative dielectric constant (ε_r) and the relative magnetic permeability ($\mu_r = 1$ for non-magnetic materials) (see Box 12.1). The success of the ground radar method relies on the variability of the ground to allow the transmission of radiowaves. Some materials, such as polar ice, are virtually transparent to radiowaves. Other materials, such as water-saturated clay and seawater, either absorb or reflect the radiowaves to such an extent that they are virtually opaque to radiowaves. It is the contrast in relative dielectric constant between adjacent layers that gives rise to reflection of incident electromagnetic radiation. The greater the contrast, the greater will be the amount of radiowave energy reflected. The proportion of energy reflected, given by the *reflection coefficient* (R), is determined by the contrast in radiowave velocities, and, more fundamentally, by the contrast in the relative dielectric constants of adjacent media (see Box 12.2). In all cases the magnitude of R lies in the range ± 1. The proportion of energy transmitted is equal to $1 - R$. The equations given in Box 12.2 apply for normal incidence on a planar surface assuming no other signal losses and refer to the amplitude of a signal. The power reflection coefficient is equal to R^2.

Box 12.1   Speed of radiowaves

The speed of radiowaves in a material (V_m) is given by:

$$V_m = c/\{(\varepsilon_r\mu_r/2)[(1+P^2)+1]\}^{1/2}$$

where c is the speed of light in free space, ε_r is the relative dielectric constant, and μ_r is the relative magnetic permeability ($= 1$ for non-magnetic materials). P is the *loss factor*, such that $P = \sigma/\omega\varepsilon$, and σ is the conductivity, $\omega = 2\pi f$ where f is the frequency, ε is the permittivity $= \varepsilon_r\varepsilon_0$, and ε_0 is the permittivity of free space (8.854×10^{-12} F/m).

In low-loss materials, $P \approx 0$, and the speed of radiowaves, $V_m = c/\sqrt{e_r} = 0.3/\sqrt{\varepsilon_r}$.

It should always be remembered when dealing with ground radar that the radiation is electromagnetic and its propagation is described by Maxwell's equations with the electric (E) component orthogonal to the magnetic (H) component (Figure 12.3). The specific shape and size of the directivity pattern lobes are functions of the dielectric constant(s) of the host media. There is a danger in making the comparison of radargrams to seismograms that the vector nature of radar may be overlooked, so that incorrect assumptions are made about the way the radiowaves behave in geologic media. While seismic data processing can be used effectively in most cases, the electromagnetic polarisable characteristics of the radiowaves are more analogous to seismic S-waves than to P-waves.

Box 12.2   Amplitude reflection coefficient

The amplitude reflection coefficient is:

$$R = \frac{(V_1 - V_2)}{(V_1 + V_2)}$$

where V_1 and V_2 are the radiowave velocities in layers 1 and 2 respectively, and $V_1 < V_2$. Also:

$$R = \frac{\sqrt{\varepsilon_2} - \sqrt{\varepsilon_1}}{\sqrt{\varepsilon_2} + \sqrt{\varepsilon_1}}$$

where ε_1 and ε_2 are the respective relative dielectric constants (ε_r) of layers 1 and 2, applicable for incidence at right-angles to a plane reflector. Typically, ε_r increases with depth.

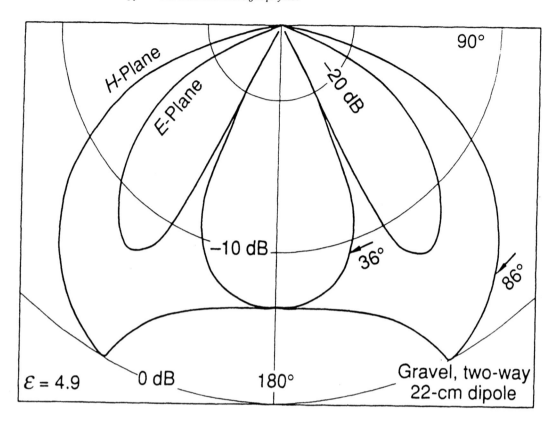

Figure 12.3 Theoretical *E*- and *H*-plane radar directivity patterns for a 22 cm resistively-loaded dipole situated over a medium with $\varepsilon = 4.9$. From Acrone *et al.* (1993), by permission

12.3.2 Energy loss and attenuation

Factors that result in a decrease in signal strength as radiowaves propagate through sub-surface media are illustrated schematically in Figure 12.4. Energy loss occurs as a consequence of reflection/ transmission losses about each interface and occur each time the radiowaves pass through a boundary. Furthermore, if there are objects with dimensions of the same order as the wavelength of the radar signal, these objects will cause scattering of energy in a random manner. This is known as *Mie scattering* and causes 'clutter' noise on the radar section. It is analogous to the noise seen on marine radar screens caused by the backscatter from seawaves in rough weather.

In addition to reflection/transmission losses at interfaces, energy is lost by *absorption* (turning the electromagnetic energy into heat). This is best pictured by analogy with a microwave oven which uses high-power radiowaves to cook food. A further loss of energy is caused by the geometrical spreading of the energy. The radar signal is transmitted in a beam with a cone angle of 90°. As the radio signals travel away from the transmitter, they spread out causing a reduction in energy per unit area at a rate of $1/r^2$, where r is the distance travelled.

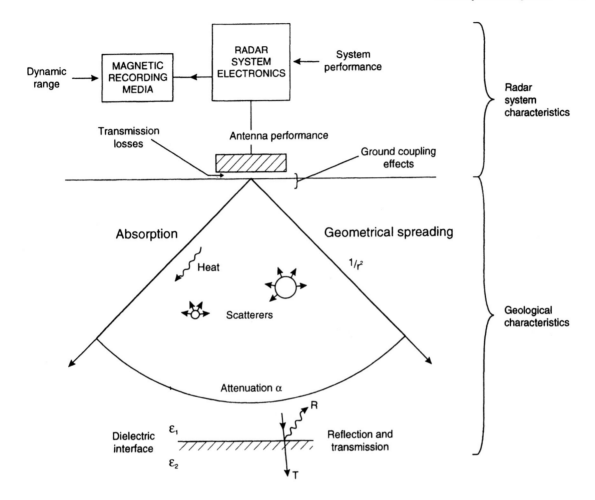

The diagram contains the following labels:

RADAR SYSTEM ELECTRONICS

MAGNETIC RECORDING MEDIA

Dynamic range

System performance

Radar system characteristics

Transmission losses

Antenna performance

Ground coupling effects

Absorption

Heat

Geometrical spreading

$1/r^2$

Scatterers

Geological characteristics

Attenuation α

R

Dielectric interface

ε_1

ε_2

Reflection and transmission

T

A fundamental cause of the loss of energy is *attenuation* which is a complex function of the dielectric and electrical properties of the media through which the radar signal is travelling. The attenuation factor (α) is dependent upon the electric (σ), magnetic (μ) and dielectric (ε) properties of the media through which the signal is propagating as well as the frequency of the signal itself ($2\pi f$). The bulk behaviour of a material is determined by the corresponding physical properties of the various constituents present and their respective proportional abundances.

As with other electromagnetic waves, the depth by which the signal has decreased in amplitude to $1/e$ (that is, to 37%) of the initial value is known as the skin depth (δ) and is inversely proportional to the attenuation factor (i.e. $\delta = 1/\alpha$). Mathematical definitions of the attenuation factor and skin depth are given in Box 12.3. Using the final term for the skin depth, and substituting typical values for seawater, it can be seen that the skin depth in seawater is only 1 cm, and for wet

Figure 12.4 Processes that lead to reduction in signal strength

clay it is only 0.3 m. Where fresh dry rock is encountered, the conductivity term decreases substantially and hence the skin depth increases, and much greater depth penetration is likely. The variation in skin depth is shown in Figure 12.5 as a function of ground resistivity at the extremes of expected *in situ* relative dielectric constants (McCann *et al.* 1988).

Box 12.3

If the peak electric field strength on transmission is E_0, and at a distance x away it has reduced to E_x, the ratio of the two amplitudes is given by:

$$E_0/E_x = \exp(-\alpha x)$$

where α is the attenuation coefficient;

$$\alpha = \omega \left\{ \left(\frac{\mu\varepsilon}{2} \right) \left[\left(1 + \frac{\sigma^2}{\omega^2\varepsilon^2} \right)^{1/2} - 1 \right] \right\}^{1/2}$$

where $\omega = 2\pi f$ where f is the frequency (Hz), μ is the magnetic permeability ($4\pi \times 10^{-7}$ H/m), σ is the bulk conductivity at the given frequency (S/m), and ε is the dielectric permittivity where $\varepsilon = \varepsilon_r \times 8.85 \times 10^{-12}$ F/m and ε_r is the bulk relative dielectric constant. The formula is valid for non-magnetic materials only.

The term ($\sigma/\omega\varepsilon$) above is equivalent to the *loss factor* (P), such that:

$$P = \sigma/\omega\varepsilon = \tan D.$$

Also, skin depth (δ) $= 1/\alpha$. When $\tan D \ll 1$, $\delta = (2/\sigma)(\varepsilon/\mu)^{1/2}$. Numerically:

$$\delta = (5.31\sqrt{\varepsilon_r})/\sigma, \text{ where } \sigma \text{ is in mS/m.}$$

It is important to remember that the simplified version of skin depth is valid only when the loss factor is considerably less than one. In order to determine when such conditions are valid, the graph shown in Figure 12.6 should be used. The figure shows the theoretical conductivity values (in mS/m) when the loss factor is equal to one. Thus the observed conductivity for the condition of being much less than unity to apply should be of the order of 0.05 of the theoretical conductivity. For example, if the observed true conductivity is 15 mS/m, then the loss factor needs to be considered in its full form in all cases other than when a 900 MHz antenna is being used, as long as the relative dielectric constant is greater than or equal to 6. If the full form of the attenuation factor is not used under these circumstances, the derived value of skin depth will be overestimated.

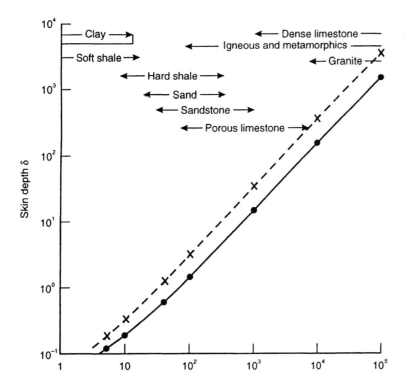

Figure 12.5 Variation of skin depth (δ) as a function of resistivity for $\varepsilon_r = 8$ and 40. After McCann *et al.* (1988), by permission

It should be noted that the skin depth does not equate to the depth of penetration of the ground radar. To determine radar range, instrumental factors also need to be taken into consideration in addition to those related to the sub-surface target and to the media through which the radiowaves travel. The total path loss for a given distance is made up of five terms: antenna losses; transmission losses between the air and the ground; losses caused by the geometrical spreading of the radar beam; attenuation within the ground as a function of the material properties; and losses due to scattering of the radar signal from the target itself. The radar range equation and definition of a radar system performance (Q) are given in Box 12.4 and the components affecting radiated and return power are illustrated schematically in Figure 12.7.

The system performance of modern radar equipment is between 120 and 160 dB, enabling a three-fold improvement in depth penetration under the same ambient conditions over that of a radar system with $Q = 80$ dB. The variation of radar signal range is shown in Figure 12.8 as a function of both attenuation and radar system performance (Q).

Within Box 12.4 are listed three types of target: smooth and rough plane reflectors and a point target. Of particular importance is the gF

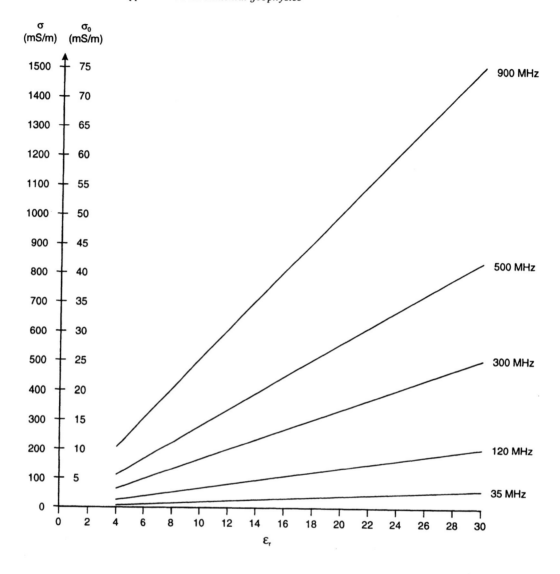

Figure 12.6 Conditions under which the loss factor (tan D) $\ll 1$

term in the first equation in Box 12.4. This product defines the power scattered by the target and also directed back to the receiver. The term g is the backscatter gain of the target and F is the target scattering cross-sectional area.

For a smooth plane reflector, the incident signal returned appears to be an image of the source, albeit reduced in power by the power reflection coefficient R ($=r^2$, where r is the amplitude reflection coefficient) of the interface, radiating upwards from a distance twice as far away as the boundary. The theory behind this is the same as in simple optics for a plane reflector.

Box 12.4   Radar range equation, and definition of Q
(Annan and Davis 1977)

Radar range equation (Ridenour 1947)

Q is the system performance (in decibels):

$$Q = 10\log\left\{\frac{E_{Tx}E_{Rx}G_{Tx}G_{Rx}V^2(gF)\exp(-4\alpha z)}{64\pi^3 f^2 z^4}\right\}.$$

The various terms are defined in Figure 12.7. Also:

$$Q = 10\log(P_{min}/P_s)$$

where P_{min} is the minimum detectable signal power, and P_s is the source power.

In low-loss materials the range of z is approximately $10D_2$. In high-loss materials the range is approximately D_2/D_1, where:

$$D_1 = 2A/(40-10B_2)$$

$$D_2 = \frac{\{-Q+10\log(S)+10\log V^2+10[B_1+(B_3-2)\log f]\}}{40-10B_2}$$

$$S = E_{Tx}E_{Rx}G_{Tx}G_{Rx}/64\pi^3$$

where B_1, B_2 and B_3 are as listed in the table below.

Type of target	gF	B_1	B_2	B_3
Smooth, plane reflector	$\pi z^2 R$	$\log(\pi R)$	2	0
Rough, plane reflector	$\pi(V^2/16f^2 + Vz/2f)R$	$\log(\pi V R/2)$	1	-1
Rayleigh point target	$(64\pi^5 a^6 f^4/V^4)R$	$\log(64\pi^5 a^6 f^4/V^4)$	0	4

For a rough, specular reflector, there is difficulty in defining the cross-sectional area of the target. Cook (1975) suggested that it equates to the area of the first Fresnel zone (see Figure 12.9). Consequently, where the wavelength of the roughness of the surface is greater than the diameter of the first Fresnel zone, the cross-sectional area, and hence the gF product, can be estimated. Where the wavelength of the surface roughness is less than the diameter of the first Fresnel zone, and especially when the amplitude of the roughness is greater than one-quarter wavelength, the actual cross-sectional target area is difficult to calculate. The power reflection coefficient would be reduced as a consequence of the greater scatter arising from such a surface roughness. The significance of the first Fresnel zone in terms of interpretation and resolution is discussed further in Section 12.5.

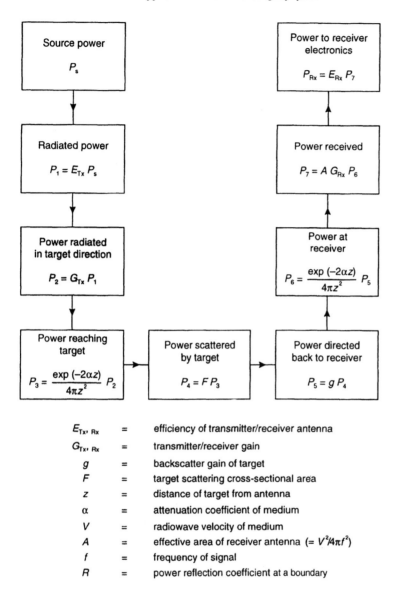

Figure 12.7 Block diagram illustrating radiated and return power for a radar system. After Annan and Davis (1977), by permission

For a point target, the characteristics of the returned energy is described by the Rayleigh Law of scattering in which the gF product is very strongly dependent upon frequency (to the fourth power). It is assumed for the expression in Box 12.4 for a point source, that the radius of the target (a) is much smaller than the wavelength of the incident radiation. In materials that consist of cobbles and gravel, for example, or where the geological units are severely distorted over distances shorter than the wavelength of the incident energy, then the

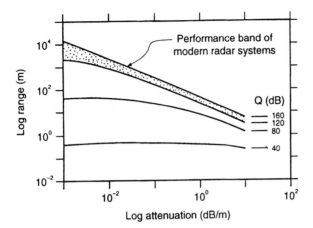

Figure 12.8 Radar range for radars with different system performances (Q) over a range of attenuation. From Davis and Annan (1989), by permission

amount of energy scattered is likely to be large and the resulting radargram is likely to show very few, if any, coherent reflection events associated with such materials. This characteristic can in itself be used indirectly during interpretation as being diagnostic of such material.

It has been shown (see Box 12.3) that attenuation is directly proportional to frequency. The higher the frequency, the greater will be the amount of attenuation. It is also evident that the bulk relative dielectric constant (ε_r) and bulk conductivity at the given frequency (σ) also affect attenuation significantly. Each of these properties is affected by the composition of the material and the electrical behaviour and relative abundance of each constituent. The loss factor (tan D in Box 12.3) is directly proportional to conductivity and inversely proportional to the relative dielectric constant and frequency. For saturated granular media, the conductivity and the relative dielectric constant of the saturating fluid will dominate over the respective matrix values. The bulk relative dielectric constant (ε_r) is roughly equal to the product of porosity (ϕ) and relative dielectric constant for the fluid (ε_f). The effect of this is that the more conductive the saturating fluid, and the greater the proportion of fluid present with a correspondingly high relative dielectric constant (remember: ε_r for water $= 81$), the greater will be the attenuation. Similarly, the greater the clay content, the greater will be the loss factor and hence attenuation. The importance of clay is that it possesses bound water within its lattice structure. Clay minerals also exhibit particular electrical properties as a result of their physicochemical structure, the details of which are beyond the scope of this section.

For both geological and engineering materials the electrical and dielectric properties, especially as functions of frequency, are still poorly understood. Furthermore, the petrophysical characteristics of such materials are largely unknown. The electrical and dielectric properties of materials are discussed in Section 12.4.

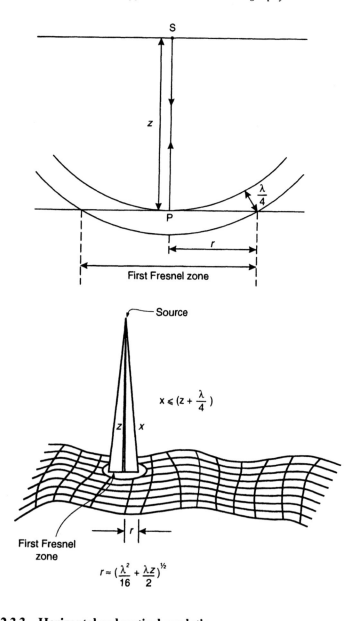

Figure 12.9   Reflection from a rough, specular interface; the target cross-sectional area is equivalent to the area of the first Fresnel Zone. After Annan and Davis (1977), by permission

12.3.3   Horizontal and vertical resolution

Vertical resolution is a measure of the ability to differentiate between two signals adjacent to each other in time. Simplistically, vertical resolution is a function of frequency. Each radar antenna is designed to operate over a range of frequencies (bandwidth) where the peak power occurs at the centre frequency of the antenna. It is the centre frequency that labels individual antenna; hence a 500 MHz antenna has a centre frequency of 500 MHz, for instance. The centre frequency is also inversely proportional to the pulse period (in nanoseconds). The

Table 12.2 Theoretical vertical resolution for two geological materials at three frequencies

	Antenna frequency (MHz)		
	120	500	900
Soil			
Wavelength (cm)	62.5	15	8
Resolution (cm)	15.6	3.75	2
Bedrock			
Wavelength (cm)	92	22	12
Resolution (cm)	23	5.5	3

500 MHz antenna, therefore, has a pulse period of 1/500 MHz = 2 ns, and for 35 MHz the pulse period is $1/35 \times 10^6$ or 28.6 ns. The equivalent length (in metres) of the pulse is the product of the pulse period and the radiowave velocity for the appropriate material. In a wet soil ($V = 0.06$ m/ns) and with a 100 MHz antenna (pulse period = 10 ns), the pulse (wave) length is 0.06×10 m or 0.6 m. Resolution can be taken as one-quarter of the wavelength (h) of incident radiation; $h = V/f$, where V is the radiowave velocity, and f is its frequency. In the last case, if the wavelength is 60 cm, the *theoretical* vertical resolution is 15 cm.

Examples of theoretical minimum resolutions for two different materials at three separate frequencies are listed in Table 12.2. The first example in the table is for a typical soil with $V = 0.075$ m/ns, and the second, a massive rock such as limestone with $V = 0.110$ m/ns. The vertical resolutions given are the very best that could be achieved theoretically. In reality, the resolution is less than these figures owing to the complex nature of the source waveform and the ground responses.

An antenna placed directly on the ground will produce a 'ground coupled' signal. That is, the transmitted waveform in air will not be reproduced when being transmitted into the ground. The material affects the shape, form and amplitude (power) of the downgoing source wavetrain and effectively filters it. The source pulse length decreases with increasing frequency, but describes the main pulse lobe only. With ground coupling, and depending upon the transmitter efficiency, the downgoing wavetrain is usually several times longer than the pulse length described in the manufacturer's literature for a given antenna. This complexity of source waveshape has serious consequences for interpretation.

If the downgoing radar wave has, for instance, three cycles with a total period of 25 ns, this means that a reflection from any interface will have equal if not greater complexity of shape and longer period. The lengthening is due to the loss of higher frequency components within the signal as higher frequencies are attenuated preferentially with respect to lower frequencies.

If two interfaces are separated by only a few tens of centimetres, for example, and the radiowave velocity of the material in between is such that the time interval between a reflection from the first (uppermost) interface and one from the second is shorter than the period of the source wavetrain, the onset of the second reflection will be masked by the tail of the first, and thus may not be resolved.

Another complexity is that the downgoing signal travels from the transmitter in a cone of radiation with a finite-sized footprint. The first Fresnel zone describes the minimum area in which features with smaller dimensions will not be imaged. The radius of the first Fresnel zone is indicated in Figure 12.9. The finite size of this footprint affects both the vertical resolution (when interfaces are steeply dipping or have high-amplitude surface roughness relative to the wavelength of the incident radiowaves), and the horizontal resolution. The larger the first Fresnel zone, the lower will be the horizontal resolution in discriminating between adjacent targets. Furthermore, spatial resolution is also affected by the conical beam width of the downgoing radiowaves (see Figure 12.10); the narrower the beam width, the greater will be the spatial resolution. Horizontal resolution is inversely proportional to $\sqrt{\alpha}$, where α is the attenuation coefficient (Daniels *et al.* 1988). Consequently, the horizontal resolution is actually better over a high-loss material than over a low-loss medium. Where radar systems permit horizontal stacking of adjacent scans to improve the signal-to-noise ratio, horizontal resolution is reduced as the amount of horizontal stacking is increased. There is a practical compromise to

Figure 12.10 Horizontal resolution due to beam width

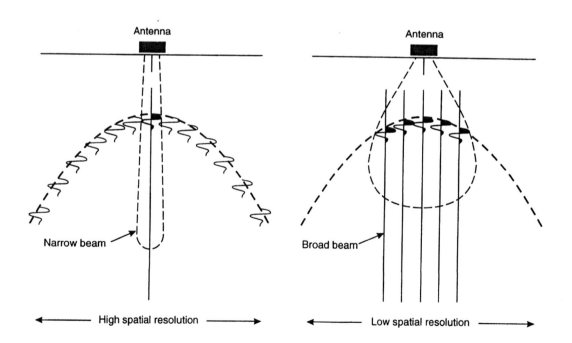

be reached between optimising return signal strengths by horizontal stacking and reducing horizontal resolution.

When *synthetic aperture radar* (SAR) is used, measurements are made by a single antenna at a number of different positions and the results combined to simulate a narrower beam than would have been achieved by using just an isolated antenna at one location. Details of the SAR or holographic radar are beyond the scope of this chapter.

12.4 DIELECTRIC PROPERTIES OF EARTH MATERIALS

The dielectric behaviour of a material is described in terms of its complex permittivity (ε^*) and complex conductivity (σ^*) which are interrelated (see Box 12.5). The high-frequency permittivity (ε_∞) is taken as the lowest real permittivity when the imaginary permittivity (ε'') is zero (see Figure 12.11). The real permittivity (ε') increases as frequency decreases. When the material is a non-conductor, the frequency–permittivity behaviour is described by a semicircle, the centre of which is located on the real permittivity axis half-way between the high-frequency and static permittivities (ε_∞ and ε_s respectively). The imaginary permittivity (ε'') indicates the absorption or energy loss within the dielectric material, and this in turn contributes to the absorption of radiowaves within the ground.

Box 12.5 Complex permittivity and conductivity

Complex permittivity ε^* of a non-conductive material is given by:

$$\varepsilon^* = \varepsilon' + i\varepsilon''.$$

When ε'' is plotted as a function of ε', the resultant graph is a semicircle. The plot is known as a Cole–Cole plot after its originators, Cole and Cole (1947).

If the material has a conductivity σ, then:

$$\varepsilon^* = \varepsilon' + i(\varepsilon'' + \sigma_s/\omega\varepsilon_0)$$

where σ_s is the static or DC conductivity, and ε_0 is the permittivity of free space. At low frequencies, the DC term dominates and produces a characteristic low-frequency tail (see Figure 12.11). The ε'' term is the frequency-dependent loss related to the relaxation response phenomena associated with water molecules (King and Smith 1981).

The complex conductivity σ^* is given by:

$$\sigma^* = \sigma' + i\sigma'' = j\omega\varepsilon_0\varepsilon^*.$$

(A)

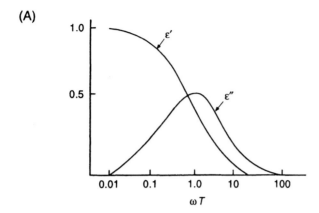

(B)

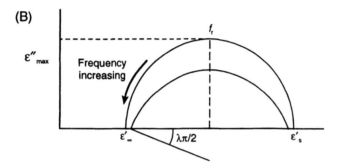

(C)

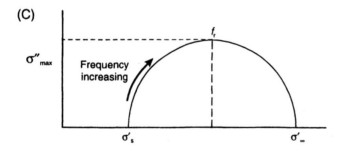

Figure 12.11   Cole–Cole plot of complex permitivity

If the material is conductive, then an appropriate additional term has to be included within the definition of the complex permittivity (Box 12.5). Conductivity also contributes to the loss within the material.

The relative dielectric constant (ε_r) varies from 1 in air through to 81 in water. For most geological materials, ε_r lies in the range 3–30. Consequently, the range of radiowave velocities is large (see Box 12.1), from around 0.06 to 0.175 m/ns (Figure 12.12). The speed of radiowaves in air is 299.8 mm/ns. In trying to estimate depths to any given target it is *essential* to have a detailed knowledge of the

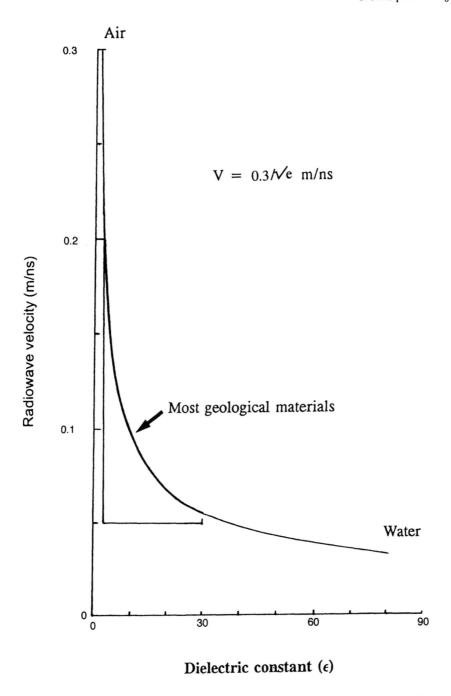

$$V = 0.3/\sqrt{e} \ m/ns$$

Air

Most geological materials

Water

Radiowave velocity (m/ns)

Dielectric constant (ϵ)

radiowave velocities through the sub-surface materials present. This aspect of radar interpretation will be dealt with in Section 12.6.

A list of the relative dielectric constants and associated radiowave velocities for a variety of geological and man-made materials is given

Figure 12.12 Radiowave velocities as a function of relative dielectric constant

Table 12.3 Table of relative dielectric constants and radiowave velocities for a range of geological and man-made materials

Material	ε_r	V(mm/ns)
Air	1	300
Water (fresh)	81	33
Water (sea)	81	33
Polar snow	1.4–3	194–252
Polar ice	3–3.15	168
Temperate ice	3.2	167
Pure ice	3.2	167
Freshwater lake ice	4	150
Sea ice	2.5–8	78–157
Permafrost	1–8	106–300
Coastal sand (dry)	10	95
Sand (dry)	3–6	120–170
Sand (wet)	25–30	55–60
Silt (wet)	10	95
Clay (wet)	8–15	86–110
Clay soil (dry)	3	173
Marsh	12	86
Agricultural land	15	77
Pastoral land	13	83
Average 'soil'	16	75
Granite	5–8	106–120
Limestone	7–9	100–113
Dolomite	6.8–8	106–115
Basalt (wet)	8	106
Shale (wet)	7	113
Sandstone (wet)	6	112
Coal	4–5	134–150
Quartz	4.3	145
Concrete	6–30	55–112
Asphalt	3–5	134–173
PVC, Epoxy, Polyesters	3	173

Data from Johnson *et al.* (1979), McCann *et al.* (1988), Morey (1974), Reynolds (1990b, 1991b)

in Table 12.3. It should be emphasised that the values of both relative dielectric constants and radiowave velocities should be taken only as guide figures. The lack of ranges for some materials is due to the paucity of measurements made and is not meant to imply that there is no variation within these materials. The ranges given are also not meant to be extremes. As more results are published the ranges listed may need to be extended as the true variability of both parameters becomes more widely realised.

Cook (1975) has produced a schematic illustration (Figure 12.13) to show the likely probing distances achievable for different geological

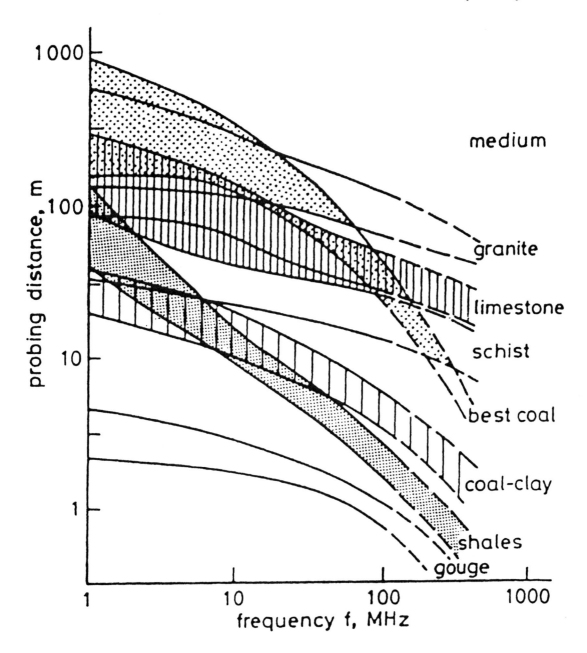

Figure 12.13 Probing distances as a function of frequency for different geological materials. From Cook (1975), by permission

materials over the frequency range from 1 to 500 MHz. Clay-rich materials have much shorter probing distances than more massive rocks such as granite and limestone.

Most materials, whether geological or man-made, are a complex mixture of components each of which is likely to have different electrical and dielectric properties. Grain size and even grain shape can affect bulk electrical and dielectric behaviour. Most rocks contain a degree of moisture, either as 'free' liquid contained within pore spaces, or 'bound' within the mineral lattice as with many types of clay minerals. Since the relative dielectric constant of water is high (81) relative to that of dry rock, even a small amount of water may increase the bulk permittivity of the rock. An example of the effect of soil moisture content of a variety of rocks on the relative dielectric constant is shown in Figure 12.13. Furthermore, the amount of water present within a rock will also affect the speed of propagation of radiowaves. The radar velocity through freshwater is 3.3×10^7 m/s (0.033 m/ns) whereas it is 1.2×10^8 m/s (0.12 m/ns) through a low-porosity sandstone (McCann *et al.* 1988).

The relative dielectric constant of a layered material has been demonstrated to be related to porosity (ϕ) by considering the proportion of constituents present and their respective relative dielectric constants. The relationships between constituent and bulk relative dielectric constants and porosity are given in Box 12.6; the variations of radiowave velocity with porosity for water-saturated and air-saturated porous media are illustrated in Figure 12.14. If the relative dielectric constant for each constituent of the material is known and that of the bulk material is measured or derived from the radiowave velocity, then the total porosity can be calculated.

Box 12.6 Relative dielectric constants and porosity
(Parkomenko 1967)

The relationship between bulk relative dielectric constant (ε_r) and porosity (ϕ) is:

$$\varepsilon_r = (1 - \phi)\varepsilon_m + \phi\varepsilon_w \qquad (1)$$

where ϕ is the porosity, ε_m and ε_w are the relative dielectric constants for the rock matrix and pore fluid water, respectively. This is valid when the external field is applied parallel to the bedding.

When the external field is applied perpendicular to the bedding, then:

$$\varepsilon_r = \varepsilon_m\varepsilon_w / [(1 - \phi)\varepsilon_m + \phi\varepsilon_w].$$

Using the simplified relationship that $V = c/\sqrt{\varepsilon_r}$, for low-loss materials, where c is the radiowave velocity in air, and substituting in equation (1) for ε_r, then:

$$V = c/[(1 - \phi)\varepsilon_m + \phi\varepsilon_w]^{1/2}.$$

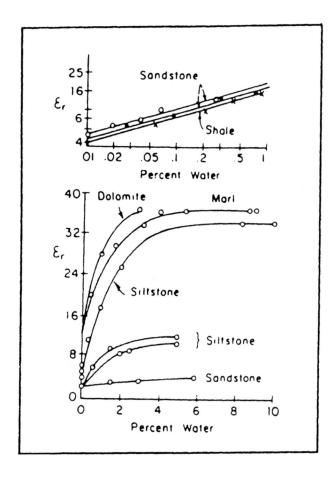

Figure 12.14 Effect of soil moisture content of rock on ε_r. From King and Smith (1981), by permission

From Figure 12.15 it can be seen that the radiowave velocity decreases with increasing soil moisture content. Consequently, wetter materials have a better vertical resolution than dry materials, although the attenuation in wetter materials is greater than for dry so depth penetration is likely to be smaller.

The determination of porosity assumes only a two-component system, i.e. made up of a matrix and pore spaces that are saturated with either air or another fluid of known relative dielectric constant. It also assumes that all the pore spaces are saturated with one fluid. This situation may not be achieved in many cases in nature.

Total porosity is the proportion of volume not filled by the solid constituents within a material and includes isolated pore/fracture space. The *effective porosity* is the porosity available to free fluids and excludes the isolated unconnected pores/fractures and space occupied by bound water in clays (Sheriff 1991). The isolated pore/fracture space (*residual porosity*) is thus the difference between the total and effective porosities. The permeability of a material is a measure of the

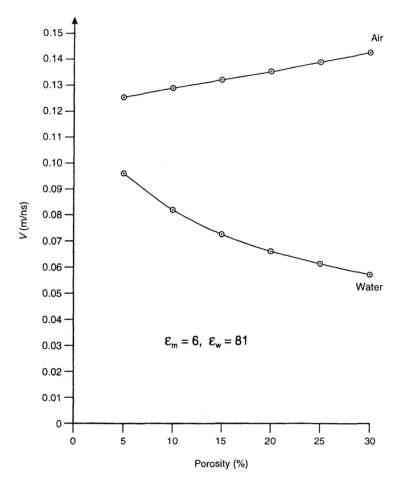

Figure 12.15 Radar velocities as a function of porosity for air- and water-saturated granular materials

ease with which a fluid can flow through the pore spaces within a given medium and thus is a function of the connectivity of the pore space, i.e. the effective porosity. Permeability is also a function of the viscosity of the fluid, the rate of fluid flow and the hydraulic pressure gradient causing the flow.

When electrical resistivity is used to derive porosity of clean granular rocks, such as by using Archie's Formula, it is the effective porosity which is being determined. Electrical continuity is provided by the electrolytes within the connected interstitial spaces. Dielectric measurements, however, are not dependent upon the connectivity and thus can be used to determine the total porosity. Conversely, values of porosity derived from the use of radiowave velocities are always likely to be overestimates of the effective porosity.

Microporosity – i.e. porosity at a scale of the order of microns but still large relative to the sensitivity of the electrical measuring system –

becomes especially significant in dielectric analysis. Electrical conductivity is affected by pore geometry and pore surface area. Clay not only affects the physical communication between pores and pore throats (affecting permeability as well as porosity), but the type of clay provides differing surface areas for double-ionic polarisation (Reynolds and Taylor 1992). For example, authigenic kaolinite occurs as disc-like 'booklets' whereas illite forms ribbons (Klimentos and McCann 1990; Klimentos 1991). The form of the clay, therefore, can affect the surface area within the pore space and it is probable that, at the scale of the order of microns, the microporosity has a measurable effect on the electrical properties. In contrast, ultrasonic acoustic methods appear to be less sensitive to this. This is not to say, however, that the microporosity does not influence the physical properties measured using acoustic methods, but that as yet the acoustic methods are not able to resolve the effects of microporosity. Indeed, electrical properties are being modelled by reference to the fractal nature of porosity (Ruffet *et al.* 1991). As it is the *effective* porosity that is directly related to permeability, the latter will only be determinable once the former can be derived accurately.

The significance of porosity, permeability and the dielectric properties of fluids is of particular importance in dielectric logging of hydrocarbon wells and in contamination mapping, for example.

Dielectric properties of concretes have also been demonstrated to exhibit a surprisingly large degree of variation (Reynolds 1991b; Reynolds and Taylor 1992). The relative dielectric constant can vary by more than 50% over a distance of less than 0.1 m within the same mix of concrete; the resulting change in radiowave velocity is of the order of 35%, with velocity decreasing with depth into the concrete. The effect is thought to be related to the amount of micro-cracking present within the concrete, with a greater amount of cracking present nearer the surface (hence more air present, thus the higher radiowave velocity). While this is but one isolated example, it does serve to demonstrate that even within a relatively controlled material like concrete, there is still a high degree of variability in the electrical properties.

12.5 MODES OF DATA ACQUISITION

There are three modes of deployment of radar systems; reflection profiling (using either monostatic or bistatic antennae); wide-angle reflection and refraction (WARR) or common-midpoint (CMP) sounding; and transillumination or radar tomography.

12.5.1 Radar reflection profiling

Figure 12.1 provides an example of radar being used to obtain a reflection profile. One or more radar antennae are moved over the

(A)

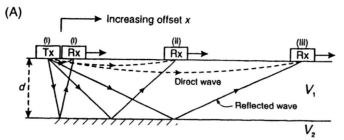

(B)

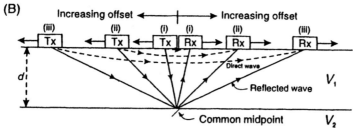

(C)

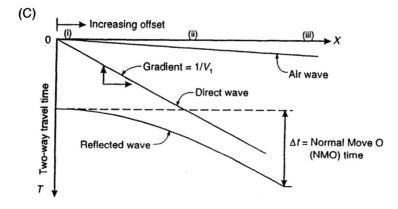

(D)

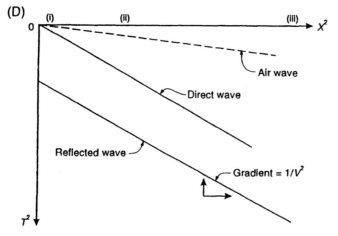

Figure 12.16 (A) WARR sounding and (B) CMP sounding with (C) a time-distance $(T–X)$ graph with normal moveout, and (D) the corresponding $T^2–X^2$ graph

ground surface simultaneously, with the measured travel times to radar reflectors being displayed on the vertical axis while the distance the antenna has travelled is shown on the horizontal axis. This mode of surveying is analogous to continuous seismic reflection profiling (see Chapter 6).

If the radiowave velocities have been measured independently (see next section) or reflections correlated with changes in ground characteristics observed from borehole data, then depths to the reflectors can be determined. See Section 12.7 for more details of interpretation techniques.

12.5.2 Wide-angle reflection and refraction (WARR) sounding

The WARR antenna configuration is shown in Figure 12.16A. The transmitter is kept at a fixed location and the receiver is towed away at increasing offsets. The location of a WARR sounding should be over an area where the principal reflectors are planar and either horizontal or dipping only at very shallow angles. It is also assumed that the material properties are uniform and that the reflector characteristics are the same over the sub-surface area over which the WARR sounding is undertaken. This assumption may not be true in all cases.

To avoid having to make this last assumption, an alternative and preferable deployment for the same analysis is the common midpoint (CMP) sounding. In this case, both the transmitter and receiver are moved away from each other so that the midpoint between them stays at a fixed location (see Figure 12.16B). In the CMP case, the point of reflection on each sub-surface reflector is used at each offset, and thus areal consistency at depth is not a requirement. The equivalent positions between the WARR and CMP soundings are given as (i), (ii) and (iii) in Figure 12.16.

12.5.3 Transillumination or radar tomography

The transillumination mode of deployment is where the transmitter and receiver are on opposite sides of the medium under investigation (Figure 12.17). The method is used underground within mines, for example, where the transmitter is located in one gallery and the receiver is either in a gallery to one side of the transmitter, or in a gallery above or below. Alternatively, the radar antennae can be located down boreholes and the radar signals are then propagated from one, through the medium in between, to the other.

The transillumination mode is also common in non-destructive testing (NDT) investigations of man-made structures, particularly using very high frequency and hence small antennae (e.g. 900 MHz centre frequency). Examples include testing concrete columns and masonry pillars.

712   *An introduction to applied and environmental geophysics*

(A)

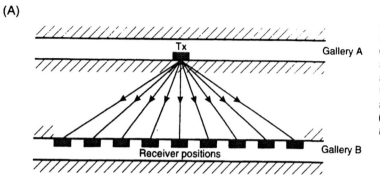

Gallery A

Receiver positions   Gallery B

Figure 12.17 Transillumination and cross-hole radar modes of data acquisition: (A) between galleries in a mine, (B) between boreholes or hole-surface, and (C) through a concrete pillar. In all cases the direct distance between transmitter (Tx) and receiver (Rx) antennae is known. Modes shown in (A) and (B) are also known as *radar tomography*

(B)

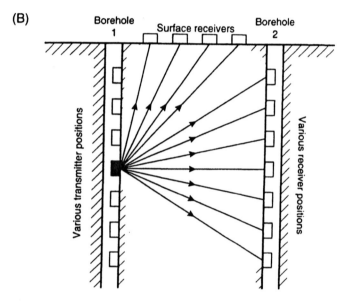

Cross-hole and hole-surface mode
(direct raypath from only one transmitter shown)

(C)

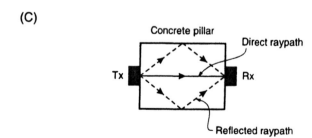

As the relative positions of the antennae are known at all times, and hence the distances between them, it is a simple matter to calculate the mean radiowave velocity of the appropriate raypath. If the signal amplitude is also measured, in addition to travel times, then attenuation can be determined. More details of this method have been given by Annan and Davis (1977). Sophisticated data-processing routines exist in order to produce tomograms that are analogous to seismic equivalents (see Chapter 6).

12.6 DATA PROCESSING

12.6.1 During data acquisition

All ground radar systems provide a means of filtering the data during acquisition. It is usually possible to set both highpass and lowpass filters to sharpen the signal waveform at the time of the survey. As with seismic filtering during acquisition, there is a significant element of qualitative feel to choosing appropriate filter settings. More sophisticated digital systems allow vertical and horizontal filtering as well as more powerful gain-setting options with which to optimise the data quality. As a rule of thumb, it is advisable to keep the filter settings as broadband as possible so that potentially valuable data are not excluded during the acquisition phase. It is far cheaper to filter broadband data after the field work has been completed than to realise that the data quality has been compromised by the use of filter settings which are too harsh, thereby necessitating a repeat of the fieldwork!

Digital systems have the function of stacking a limited number of adjacent traces in order to improve the signal-to-noise ratio. This works well in areas where the structure is largely parallel or subparallel to the ground surface. When steeper structures are present, horizontal stacking over too many adjacent scans can produce lateral smearing and a defocusing of the radar image.

12.6.2 Wide-angle reflection and refraction (WARR) sounding

If two separate antennae are used, one as a transmitter and the other as a receiver, in bistatic mode, it is possible to determine the vertical variation in radiowave velocity (and hence relative dielectric constant). If only one antenna is available in monostatic mode, it is not possible to undertake WARR sounding and hence velocity determination can only be by either direct correlation with adjacent borehole logs, targets at known depths of burial or by guesswork; the latter is the most commonly employed and may give depth estimates accurate to $\pm 20\%$.

In each of the WARR or CMP configurations three types of waves may be identified:

- the airwave, travelling from the transmitter to the receiver through the air at the speed of radiowaves in air (0.3 m/ns);
- the direct wave, travelling directly from the transmitter through the near-surface ground to the receiver at the speed of radiowaves in the near-surface medium (V_1);
- the reflected wave, travelling from the transmitter to the interface from which it is reflected to the receiver, also at the speed of radiowaves in the first layer (V_1).

The travel times for both the airwave and the direct wave plot as straight-line segments on the T–X graph, but those for the reflected wave plot on a curved (hyperbolic) line. The difference in travel time between zero offset and at finite offset is the *normal moveout* (NMO) time (Figure 12.16C). However, when these travel-time data are plotted on a T^2–X^2 graph, all the segments appear as straight lines (Figure 12.16D). The inverse gradients of each line are equal to the respective radiowave velocity squared. Further details of the velocity sounding techniques have been given by Arcone (1984).

The radiowave velocity determined for layer 1 is a time-averaged value over the interval from zero time to t_0, and is hence a root-mean-squared velocity (V_{RMS}). Where travel-time data are obtained for deeper reflections, the velocities determined from the above analysis for each layer are also RMS values. To determine a geologically more meaningful velocity for a particular layer, it is then necessary to use the Dix Formula to derive the 'interval velocity'. This analysis is exactly the same as for seismic reflection data and more complete details are given in Chapter 6.

12.6.3   Post-recording data processing

Only digital data can be processed post-recording. The degree of processing is often determined by (a) the budget available; (b) the time available; (c) data quality; (d) the available processing capability (software and hardware); (e) the requirement that the final interpretation justifies further analysis; and (f) the structural detail on the raw record meriting detailed quantitative data processing.

The first step is to filter the data in order to focus the image. For many applications this is sufficient in order to locate sub-surface features. For more detailed analysis, a wide range of processes are available, up to and including the same as for multifold seismic reflection data, including attribute analysis, details of which are given in Chapter 6.

The radar system produced by Geophysical Survey Systems Inc. has a suite of data-processing packages available called RADAN™

(Radar Data Analyser). The format of data recorded on to magnetic media, typically 2.5 Gbyte magnetic tape or magneto-optical disk, can be converted from SEG 2 to SEG Y format in order to be compatible with seismic industry-standard workstations. A program to convert from SEG 2 to SEG Y has been published by Bennett (1990). Datafile formats for radar data have been specified by the SEG Engineering and Groundwater Geophysics Committee (Pullan 1990). Similarly, Sensors & Software Ltd have a range of computer software designed for use with their PulseEKKO system, which produces data in a format compatible with seismic data-processing software.

Of particular significance is the ability to restore correct subsurface geometrical relationships through the process of migration. Diffraction hyperbolae can be migrated back to the apex from which the diffraction originated. Dipping planar surfaces can be corrected to their correct position relative to ground locations. Otherwise, significant errors can be made by believing that the location of a particular sub-surface feature on a raw radargram is exact, whereas it is only a virtual image and may be displaced from its actual position by significant horizontal and vertical distances. The principles of migration are discussed more fully in Chapter 6.

Other methods of quantitative analysis are available through *image analysis*. Rather than operate on the waveforms of the data, the radargram is scanned for analysis of trends. For example, trends such as reflections dipping in a particular direction can be picked out. Statistically significant trends can be identified from the entire radargram and displayed automatically as line interpretations for subsequent manual analysis.

12.7   INTERPRETATION TECHNIQUES

12.7.1   Graphic interpretation

From both analogue and digital radar data, hardcopy radargrams can be analysed in terms of identifying reflections and diffractions and measuring the two-way travel times to such identifiable events. By assuming, or having measured, a value for the appropriate relative dielectric constant – and hence obtained a realistic radiowave velocity – the two-way travel times to specific events can be translated into depths. Where radar data have been acquired over a regular grid, and reflections identified over significant areas, it is then possible to produce posted two-way travel time maps, or *isopachyte maps*, indicating the depth to, or thickness of, a particular layer, given a realistic measure of the radiowave velocity.

This approach is particularly prevalent in road pavement analysis, where the number of discrete layers is usually well constrained with up to four parallel to sub-parallel layers (bound layer of bitumen or concrete, granular layer, upper and lower sub-grades). The travel times to interpreted interfaces can be digitised off paper radargram records and, using an appropriate radiowave velocity for each of the discrete layers, the depths to each interface can be determined. There would need to be careful consideration of the accuracy of picking the onset of the various reflections on the radargrams as well as the likely reliability of the radiowave velocities used to derive depths. Local variability in radiowave velocities can occur within concretes (see Section 12.4; different mixes, even though all within specification), or due to changes in moisture content (see Section 12.4). Individual horizons, such as the sub-base, may have been prepared in layers, and these may be detected. In some cases the boundary between layers may be fuzzy – so where does the radar reflection come from? Is it always from the same relative position between layers? These are questions that need to be answered in road pavement work prior to the production of final drawings. In all cases, a statement as to the errors and limits of measurement should be made. There is no such thing as an absolute measure of depth using remote methods, particularly radar.

The interpreted data can be displayed in a wide range of ways using modern computer-aided design (CAD) systems and 3-D graphics software packages. While the final output may be extremely colourful and fancy, it should be remembered that the basic data analysis may be just that – basic! While the final drawings may indicate millimetre accuracy, is this justified? At present, the uncertainties and local variability in the dielectric properties of materials and the subjective nature of defining the onset position of reflections tend to make claims of such accuracy unjustified. As the electrical and dielectric properties of materials used in road pavements become better known, accuracies and reliabilities will improve.

In addition to interface mapping, it is possible to use the variations in character displayed on the radargram as an indication of sub-surface conditions. For example, areas of high attenuation may reveal zones with elevated conductivities which may be associated with pollution, or clay pockets. Sub-surface cavities may be evident by the resonance within the void space, indicated on the radargram as a series of large-amplitude pulses which are laterally very restricted. Zones of cobbles or severely distorted strata may be evident by the loss of coherency of primary reflection events. Delamination of road pavements may also be indicated by diagnostic character changes on radargrams (delamination is also evident when imaged using infrared thermography under appropriate conditions).

12.7.2 Quantitative analysis

Basic depth determination depends upon an adequate knowledge of the radiowave velocity and its vertical and lateral variation within a given survey area. Where WARR/CMP data have been acquired, then a detailed picture of the velocity field can be obtained. Consequently, geologically diagnostic values of the radiowave velocity, or more particularly of the relative dielectric constant, can be used to aid interpretation.

Where detailed quantitative attribute analysis has been undertaken of the recorded radiowave data – such as amplitude analysis, reflection coefficient determination, as well as variations in ε_r – then a much more comprehensive understanding of a site can be gained. Indeed, given adequate data quality, careful processing may yield more valuable information about the petrophysics of a given geological or engineering regime than would otherwise have been possible. High-level processing and analysis are becoming much more important where detailed discrimination is required, such as in hydrocarbon exploration and reservoir engineering, and in contaminated land investigations. The use of ground radar in both cases is likely to develop considerably over the next few years.

12.7.3   Interpretational pitfalls

The two commonest pitfalls in the interpretation of radar data are (a) not being able to identify the ground surface, and (b) misidentifying each black band on a black and white radargram as being caused by a discrete horizon. The easiest way to identify ground level, especially with antennae with centre frequencies $\geqslant 500\,\text{MHz}$, is to raise and lower the antenna above the ground surface. A distinctive cusp appears on the radargram and clearly indicates where the ground level is represented.

The over-identification of the number of layers highlights a real difficulty, especially with analogue radar data: how are the primary reflections to be identified from multiples, secondary events, and the tail of other primary reflections? Furthermore, when waveforms intersect each other, they cause interference which may give the appearance of a 'termination' of one dark band with respect to another. Geologically, this may be misinterpreted as one horizon abutting against another. If the geological conditions are such that the radargram is ambiguous in this regard, then there is justification for detailed quantitative analysis if the data have been recorded digitally. If the data have been produced as an analogue record, there is little that can be done to resolve the dilemma. The quantitative analysis can pick out likely multiple events (these are purely time repeats of earlier primary events), and by deconvolution, the shape of the downgoing wavetrain can be determined and hence primary reflection events identified. Subsequent migration can help to reduce diffraction hyperbolae and, by restoring some of the sub-surface geometry of primary reflections, can help to resolve geologically significant detail that otherwise would have been obscured.

The difficulties in distinguishing between geologically significant reflections and extraneous reverberations, multiples, noise, diffractions, off-section ghosts, etc., make the determination of soil and rock stratigraphy difficult in some cases. In others, the stratigraphy can be determined quite readily; one such example (from Best and Spies, 1990) is shown in Figure 12.18. An analogue record obtained in 1976

Figure 12.18 Radargram obtained in 1976 compared with one in 1991. From Best and Spies (1990), by permission

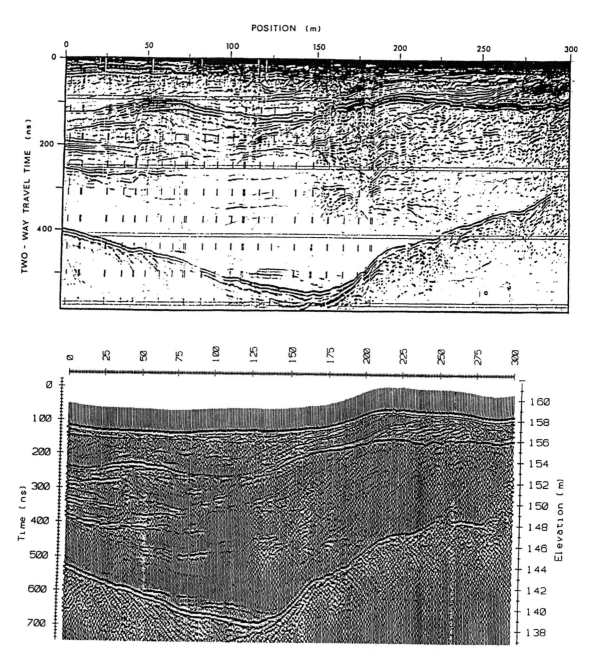

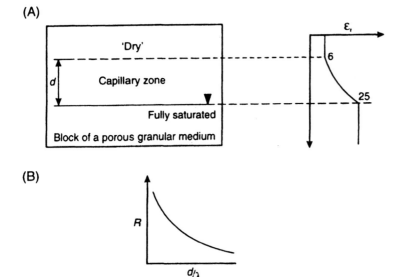

(A)

(B)

Figure 12.19 Effect of thickness of a capillary zone on the observed reflection strength arising from the water table. (A) A capillary zone of thickness d over the water table has a relative dielectric constant (ε_r) which increases to a maximum when fully saturated. (B). The amplitude reflection coefficient (R) decreases as the ratio of the thickness of the capillary zone to wavelength of incident radiowaves increases

is compared with a digital radargram produced over the same area more recently. While the gross structure is evident in the 1976 analogue record, the detail is much crisper in the digital record.

There are situations when the water table is detectable on a radargram and others where it is not. The reason for this is that the ratio of the thickness of the capillary zone to the wavelength of the incident radiowaves needs to be small (i.e. $d <$ wavelength) in order to provide sufficient contrast in relative dielectric constant between the unsaturated and saturated material to reflect a significant proportion of the energy (Figure 12.19). If the capillary zone is thick with respect to the wavelength, then the rate of change of relative dielectric constant with depth through this zone is small. The effect of this is that, for each incremental increase in the relative dielectric constant, a proportion of the incident energy is reflected so that the total reflected energy is smeared from the capillary zone, and hence the resulting reflection amplitude is too low to be detected with any clarity. In contrast, if the transition from dry to saturated is virtually instantaneous (e.g. the change in relative dielectric constant is from 6 to 25), then the amplitude reflection coefficient $|R|$ is 0.34 (using the expression in Box 12.2), which is a very strong return.

12.8   APPLICATIONS AND CASE HISTORIES

12.8.1   Sedimentary sequences

Ground penetrating radar has been demonstrated to be a valuable tool in mapping sediment sequences with a high degree of spatial

resolution on both land and through freshwater. An example of the
improvement in data quality in mapping soil stratigraphy was given
in Figure 12.18. Other examples are given in Section 12.8.7. A common failing of the analysis of radargrams acquired in stratigraphic
investigations is over-interpretation of the data. Too often apparently
coherent events are taken as indicating individual sedimentary interfaces without due regard for the physics of thin-bed interference,
vertical and horizontal resolution limits, the finite size of the first
Fresnel zone, migration effects, the complex form of the incident
wavetrain, etc. It is in these applications that seismic data processing
is likely to play an important role; for example, see the paper by
Huggenberger *et al.* (1994).

Ground radar can be deployed over frozen lakes and has been used
to investigate sub-lake sediments through freshwater up to 27 m deep;
an example of a through-ice survey is shown in Figure 12.20. The lake
ice provides a stable platform over which the radar was towed. The
freshwater within the lake is virtually transparent to radiowaves and
the lake sediments are clearly evident through 4.8 m of freshwater.
The resolution of the system (100 MHz antenna) is such that individual horizons within the sediment can be picked out. Note that the
reflection returns associated with the lake bed comprise at least four
bands owing to a ringy source. Furthermore, the period of the initial
wavetrain (around 70 ns) might be misinterpreted by some as indicating the presence of up to 6 m of ice (radiowave velocity through ice is

Figure 12.20 Radargram over an icecovered lake obtained using a 100 MHz antenna. Maximum water depth is 4.8 m; width of profile is 25 m; 500 ns two-way travel time range. From Mellett (1993), by permission

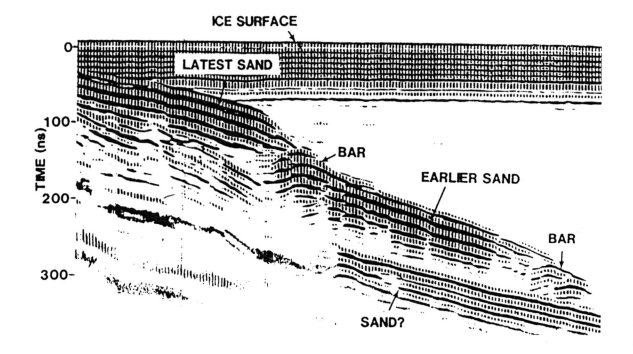

0.167 m/ns). The actual thickness was only 0.15 m. The two-way travel time through the ice layer would only have been around 2 ns. This shows effectively that, using a low-frequency antenna (100 MHz), near-surface features cannot be resolved at all as they are totally masked within the initial ground-coupled wavetrain. The radiowave velocity through the freshwater can be calculated knowing the depth of water (about 4.8 m) and the total travel time (around 300 ns). These values would give a radiowave velocity of 0.032 m/ns (from $2 \times 4.8/300$), ignoring the ice layer and assuming low-loss media. As a double check, the relative dielectric constant of water is 81, which thus gives a radiowave velocity of $0.3/\sqrt{81}$, or 0.033 m/ns (refer to Box 12.1 for the equation).

12.8.2 Hydrogeology and groundwater contamination

As environmental protection measures become more stringent, and the pressure on preserving the quality of groundwater sources increases, so the need to identify groundwater pollution grows. Davis and Annan (1989) have demonstrated how ground radar was used to locate and map out a plume of contaminated water leaking out from a landfill site; a schematic plan of the site area is shown in Figure 12.21A. Along the line of the radar transect shown in this figure, the soil consists of fine sand and overlies bedrock which occurs at a depth of about 20 m. A radar survey was undertaken using a PulseEKKO III radar system and the resulting radargram is shown in Figure 12.21B.

Where penetration of the radiowaves into the superficial sediments occurs, reflections are seen and are thought to be due to horizons with different grain size and density, and hence different soil moisture contents. It is also very clear on this section that there are areas where either only very weak reflections occur, such as at 150 m along the profile at around 400 ns two-way travel time, or the signals are completely attenuated. The presence of contamination, which has an associated high electrical conductivity, attenuates the radar signals severely.

Several boreholes had been constructed along the survey line and the conductivity of the groundwater was measured. The solid line joining a series of black dots on Figure 12.21B indicates the position below which the porewater conductivity is greater than 10 mS/m. As it was known that the superficial deposits were reasonably consistent in their properties over the area of the site, it is evident, therefore, that the pollution plume approaches the surface between 40 and 60 m (as proven by the borehole data) and that it also extends between 110 and 150 m along the profile line at a depth of about 6 m below ground level. The second part of the plume had not been expected, and thus the results from the radar survey were extremely useful in providing this additional information. A ground conductivity survey to

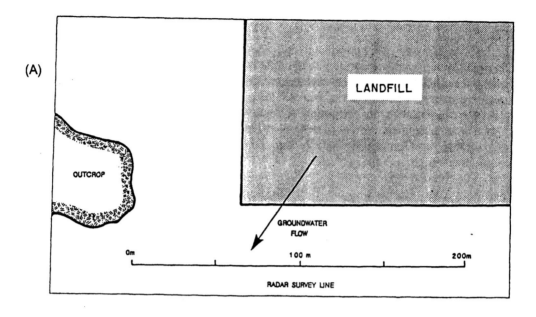

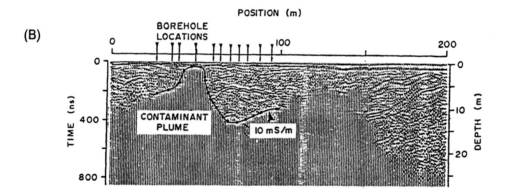

complement the radar work would have provided a non-invasive method of determining the spatial variation in sub-surface conductivity without having to drill extra boreholes, which in themselves may make the spread of the pollution worse.

Where it can be demonstrated that radar surveys would be useful under the ambient geological and ground conditions, and that the pollution can be detected by radar, then changes in the pollution plume can be mapped. By undertaking repeat surveys along the same ground transects, it is possible to detect changes as a function of time. Furthermore, where remediation measures are undertaken, the success of such treatment can be monitored using radar.

In certain cases, the actual pollution itself may not be detectable using radar, but the containers from which the pollution originated

Figure 12.21 Radar section showing the effect of a conductive pollution plume caused by leachate migrating from a landfill. From Davis and Annan (1989), by permission

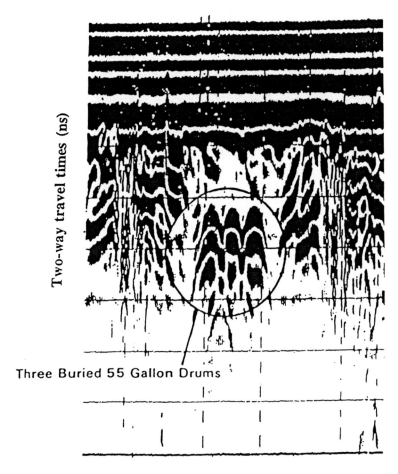

Figure 12.22 Radar record over three buried drums laid side by side

Two-way travel times (ns)

Three Buried 55 Gallon Drums

may be. Consider the case of buried 55-gallon steel drums which may have leaked their contents. Figure 12.22 shows a radar record over three buried drums. Note the characteristic diffractions arising from the drums which are located side-by-side. Also note that the incident radar waveform consists of more than one band, and hence the diffractions from the drums give the appearance of lower diffractions; these are the tails of the primary diffractions.

Ground penetrating radar is being used extensively in the Netherlands for hydrogeological assessment in groundwater management. Falling water tables and deteriorating water quality seriously affect agriculture and nature conservation and potable water supplies. Van Overmeeren (1994) has provided examples of types of hydrogeological applications current in the Netherlands where there are four main radar targets: (a) tectonic and sedimentary structures; (b) water tables within sandy deposits in push moraines, river terraces and sand dunes; (c) perched water tables as distinct from regional water tables; and (d) spatial extent and continuity of buried clay and peat layers within the superficial deposits.

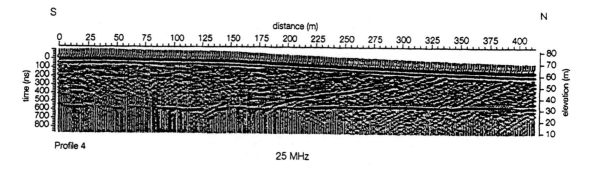

S   N

distance (m)

Profile 4

25 MHz

A radargram acquired using a PulseEKKO IV with 25 MHz antennae and station interval of 1 m is shown in Figure 12.23. The profile was obtained over push moraine comprising mainly sand. The section has been corrected for topography and has identical horizontal and vertical scales. The water table is evident as a horizontal reflection with a large amplitude. In the northern part of the section, the reflection is largely continuous whereas it is less so in the southern, most elevated, part. This suggests that the depth of penetration of the radar is at its limits here and that the greater depth below ground level to the water table in the southern part results in the poorer data quality. Using a radiowave velocity of 0.145 m/ns (derived from CMP measurements), the water table is still evident at 42 m below ground level. The radargram also reveals reflections from interfaces above the water table. The oblique reflections are associated with interfaces between sandy layers of different grain size or between sandy and clayey sediments. In both cases, the interface marks a change in moisture content and hence a contrast in relative dielectric constant.

A conspicuous reflector with a large amplitude and dipping southwards is evident in the middle of the section. There is a small vertical offset (around 2 m) in the water-table reflection where it is intersected by the inclined event. The step in the water table is thought to be caused by an inclined clay layer which gives rise to the strong oblique reflection.

Van Overmeeren (1994) has described another example of the hydrogeological usefulness of GPR, where a radargram (Figure 12.24A) was acquired using a PulseEKKO IV with 50 MHz antennae and a station interval of 0.5 m. The survey transect was over a valley with marine interglacial deposits passing into an uplifted ridge of ice-pushed sediments. A clay layer forms part of the marine formation. The objective of the radar survey was to map the lateral continuity of the clay layer and to locate its western limit (see Figure 12.24B which is a schematic interpretation of part of Figure 12.24A). The clay layer, which occurs at a depth of 15 m below ground level, sustains an artesian aquifer which is recharged by infiltration in the

Figure 12.23 Radargram acquired with 25 MHz antennae over a sandy ice-push moraine. The water-table reflection is evident at an elevation of 30 m. Note the slight vertical displacement in this reflection where an inclined reflection due to a clay layer intersects the water table at 185 m. From van Overmeeren (1994), by permission

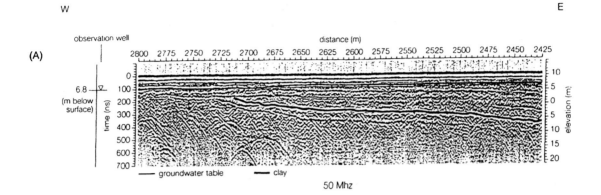

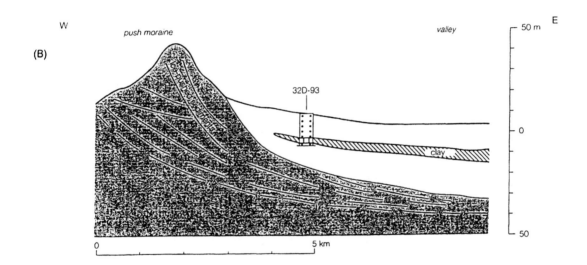

higher ice-pushed ridge. The precise boundaries of the clay layer were required for groundwater flow modelling.

In the radargram (Figure 12.24A), the vertical exaggeration is 2.5 times the horizontal scale. The regional groundwater level is at a depth of about 7 m, as determined from a nearby observation well. A radar CMP test near the well produced a value of the radiowave velocity of 0.115 m/ns for the sediments above the water table. In the area of the survey, the average radiowave velocity was found to be 0.075 m/ns for sediments above the clay band, which is in close agreement with a known value of 0.06 m/ns for sand saturated with freshwater. In contrast, dry sand has a radiowave velocity of 0.15 m/ns. The field-derived radiowave velocity is strongly indicative of freshwater-saturated sands above the clay layer, which can be

Figure 12.24 (A) Radargram acquired with 50 MHz antennae over a sandy ice-push ridge on to which interglacial marine sediments, including a clay layer, onlap. The clay layer is indicated by a solid black line. The water table occurs at a depth of about 7 m. (B) General interpretation of the part of the radargram shown in (A). The clay layer has been verified by a borehole (32D-93). From van Overmeeren (1994), by permission

identified on the radargram as a coherent reflection with large amplitude. The reflection from the clay layer persists clearly until station 2715 where it is cross-cut by an onlapping horizon.

The radargram shows that sensible reflection events arising from the sub-surface geology can be identified to an interpreted depth of 12 m, with the exception of several diffraction-like events in the western part of the radargram which occur at interpreted depths of about 16 m. Note in Figure 12.24A that the reflections are largely coherent across the section, but become increasingly incoherent or disturbed west of station 2675 where the ice-pushed ridge is encountered. The more chaotic nature of reflection events here, due to diffraction hyperbolae, is typical of ice-pushed moraine in this area.

Although the water table occurs at shallow depth below ground level (around 3–5 m), reflection events at significant depth are still evident on the radargram. Elsewhere in the Netherlands, the detection of buried clay layers below the water table has proved impossible, yet at this location it is obvious that the radar survey has provided extremely valuable information about the clay layer. This is attributed to the high electrical resistivity of the surface layers resulting in very little attenuation of the radar energy.

The transparency of freshwater saturated sediments to radiowaves has also been demonstrated in North America using a PulseEKKO IV with 100 MHz antennae (Figure 12.25). The reflection arising from the water table is clearly seen as a coherent reflection with a large amplitude between stations 375 and 625 at an elevation of around 500 ft (about 152 m). Diffraction hyperbolae arising from two cables/pipes above the water table are also obvious (arrowed). Note that there are many reflections evident from the sub-surface geology present to depths of around 33 ft (10 m). The data are published

Figure 12.25 Radargram acquired using 100 MHz antennae with a PulseEKKO IV system showing the water table (flat-lying reflection arrowed) and two diffraction hyperbolae from near-surface cables/pipes. Data courtesy of Michigan Department of Natural Resources, Environmental Response Division and Sensors & Software Inc.

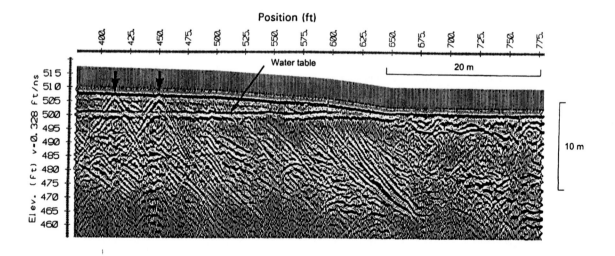

courtesy of the Michigan Department of Natural Resources, Environmental Response Division, and of Sensors & Software Inc.

In addition to investigations into natural groundwater resources, there is a growing need to map groundwater contamination (Greenhouse *et al.* 1993). Such pollution can arise from leachate migrating from a landfill, from saline water ingressing inland due to over-exploitation of freshwater sources, natural hydrocarbon contamination (from oil shales, etc.), through to chemical spillages (deliberate or otherwise) over timescales that can range from a few hours to many years. For example, could a chemical spill from an overturned railway tanker be monitored using geophysical methods? Or can petroleum products that have leaked from pipes at a refinery and which have ponded below ground at the water table be mapped?

Some chemicals can migrate from the source of contamination extremely quickly and would be difficult to detect by any means. However, an increasing amount of research is being undertaken to examine the protracted contamination of groundwater. The reason is that there are enormous quantities of carcinogenic organic groundwater contaminants, such as DNAPLs (colloquially pronounced as 'dee-napples') – DNAPLs are dense, non-aqueous phase liquids. The main chemical concerned is perchloroethylene (PCE) which is one of the main constituents of the dry-cleaning and metal-cleaning industries. PCE is but one of a type of liquids known as 'chlorinated organic solvents'. Other well-known chemicals are trichloroethylene (TCE) and dichloromethane (DCM) which are paint-strippers and metal-degreasers. In 1986, it is reported that in the USA alone, 120 million litres of PCE and 200 million litres of TCE were manufactured. Once used, a small but significant proportion of these volumes is disposed of underground, around dry-cleaning establishments, car-service garages, in landfills and waste lagoons, and as residues on old industrial sites.

As their name suggests, DNAPLs are dense (with a density of $1.623 \, \mathrm{Mg/m^3}$) and sink rapidly through the local groundwater leaving a residual trail of 5–20% of the pore volume, and eventually ponding for a time on a low-permeability layer such as a clay aquitard. DNAPLs also have low viscosity and low surface tension which allow the chemicals to migrate rapidly through porous media. They can even pass through very fine fractures in clay given adequate time. It is estimated that in the USA, the cost of cleaning up the existing DNAPL contamination to acceptable standards is in excess of one thousand billion dollars.

An excellent overview of a controlled experiment to examine the detectability of DNAPLs has been presented by Greenhouse *et al.* (1993). A schematic of the experimental site is shown in Figure 12.26. A $9 \times 9 \, \mathrm{m}$ cell was constructed by driving corrugated steel sheet piles, sealable at their joints, through the 3.3 m thick surficial aquifer into the underlying clay aquitard. Two concentric walls contained a 0.5 m

(A)

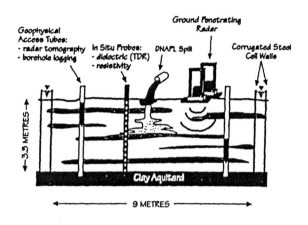

Figure 12.26 (A) Schematic cross-section of the 9×9m test cell at Borden. (B) Plan view of the Borden test cell showing access tubes AT1-9, resistivity probe locations RES-1 and RES-2, TDR locations TDR-1 and TDR-2, the 1 m radar grid and the point of PCE injection. From Greenhouse *et al.* (1993), by permission

(B)

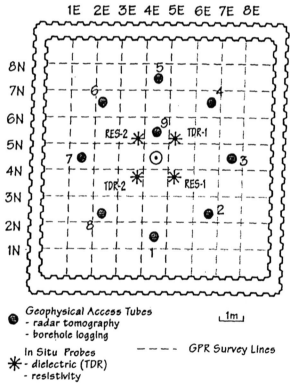

wide moat which effectively isolated the inner section hydraulically from its surroundings and allowed the interior water table to be maintained at a depth of 0.15 m below the ground surface. A tarpaulin covered the ground to restrict evaporation. The test cell was instrumented via nine vertical access tubes for radar tomography and borehole logging, and had four further vertical *in situ* probes to measure resistivity and relative dielectric constants (using time domain reflectometry; TDR). Surface ground radar traverses were undertaken repeatedly over two sets of orthogonal profiles with a 1 m line interval using a PulseEKKO IV radar system with 200, 300, 500 and 900 MHz antennae. Over a period of 70 hours, 770 litres of PCE were injected under a constant head of 2 m of water at the centre of the test cell at a point 0.6 m below the ground surface. Geophysical imaging of the test cell began several days before the contaminant injection and continued over 912 hours after the injection had been completed. Measurements were made throughout the period of the experiment so as to obtain time-dependent datasets. Geophysical measurements continued for several weeks after the main experiment had been completed in order to monitor the removal of the PCE from the test cell, which was undertaken by flushing the cell with surfactants.

Of the substantial body of data obtained from this experiment, a time series of radargrams obtained along transect 5N (Figure 12.26) using 200 MHz antennae is shown in Figure 12.27. Three radargrams are shown ranging from one obtained before any injection of PCE had occurred, and two acquired at 16 and 920 hours after injection. What is clear from comparing the two radargrams after injection is that PCE appears to pond at a depth of 1 m (iv on Figure 12.27B) and then drains downwards leaving a residual ((v) on Figure 12.27C) at 1 m depth with ponding above the clay aquitard evident at (vi).

These experiments demonstrated that PCEs can be imaged successfully using a variety of geophysical methods, and that GPR can be used to monitor both the migration and the subsequent remediation of the contaminated ground. Further aspects of the radar surveys have been reported by Brewster *et al.* (1992a, b), for example. A GPR survey at a site where DNAPL contamination was known to have occurred has been described by Carpenter *et al.* (1994), although no DNAPLs were imaged directly using radar.

12.8.3 Glaciological applications

Radar mapping of the polar ice sheets has been one of the most widely used geophysical methods in both Greenland and Antarctica. Radio echosounding has been developed substantially since its early use in the 1960s. Determination of ice thicknesses is now accurate to around 1% and has provided excellent agreement with values derived from both seismic and gravity surveys, as well as with borehole control.

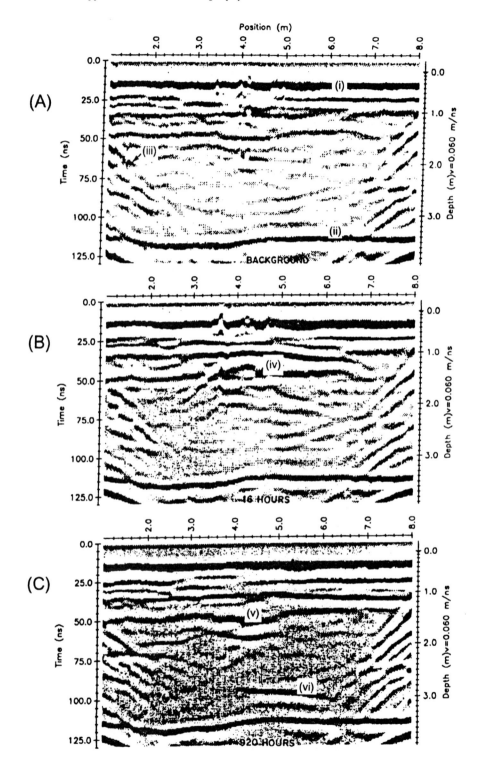

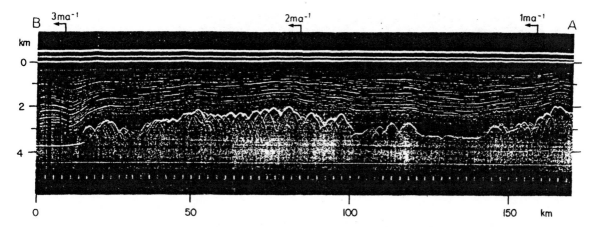

Figure 12.28 Radio echosounding record from along an ice flow line near Vostok, Antarctica. From Robin and Swithinbank (1987), by permission

An example of a radargram obtained using a 60 MHz antenna system over an ice flowline near the Russian research station Vostok, in Central Antarctica, is shown in Figure 12.28 (Robin and Swithinbank 1987). The total depth of penetration is around 3800 m. Three principal features are evident on this profile. The first is that there are a series of sub-parallel reflections within the ice itself. These horizons are thought to be due to elevated concentrations of sulphate (H_2SO_4) arising from large eruptions from volcanoes located outside Antarctica and transported in atmospheric aerosols to become incorporated within the polar snow, or from horizons which have undergone melting during the summer months (when the ice was at the surface) and subsequent refreezing. Ice has a thermal memory and the dielectric properties of the ice can be used to identify layers that have undergone melting and refreezing (Reynolds 1985). Some of the most prominent internal reflections can be traced over thousands of square kilometres and appear to be interfaces formed at the same time, and thus can be used as isochrons. These can then be used in the analysis of the dynamics of the flow of the ice sheet.

The second feature is that the bedrock surface can be seen very clearly and exhibits a range in elevation of about 1500 m. The bedrock topography is thus a hidden range of mountains. The third feature is located at the left-hand end of this profile at the base of the ice sheet. Note the flat, bright reflection at 3800 m depth. This is produced from what has been interpreted to be the surface of a large sub-glacial freshwater lake. The contrast in relative dielectric constants between the basal ice and the freshwater has given rise to this characteristic style of reflection. This type of feature can also be seen in another radio echosounding record shown in Figure 12.29 (Drewry 1986). Also note in both Figures 12.28 and 29 that there is a zone about 500 m wide immediately above the bedrock in which internal reflections are absent. This is possibly due to the basal deformation of the ice being such that the scale of deformation distorts the layering to such an extent that coherency is lost.

Figure 12.27 (*opposite*) Borden test cell imaged along profile line 5 N (see Figure 12.26) using 200 MHz antennae. Radargram (A) was acquired prior to PCE injection, and radargrams in (B) and (C) were obtained 16 and 920 hours after injection. (i) = cell surface; (ii) = top of clay aquitard; (iii) = reflections off the vertical cell walls; (iv) = pooled PCE at 1 m depth; (v) = a zone of residual PCE; and (vi) = PCE pooled on the clay aquitard. After Greenhouse *et al.* (1993), by permission

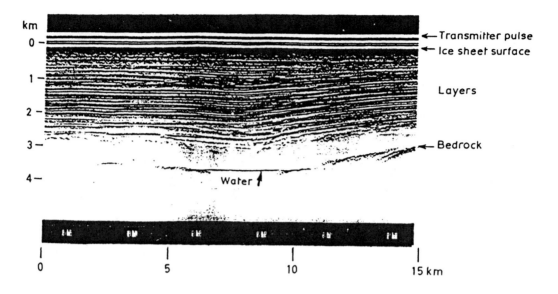

A radar system developed for radio echosounding (Wright *et al.* 1990) utilises antennae with much lower centre frequencies (1, 2, 4, 8 and 12.5 MHz) than those used previously ($\geqslant 60$ MHz). Records using this new system are shown in Figure 12.30. A 4 MHz centre-frequency antenna was used in this example which was obtained on Ice Stream B, near the Transantarctic Mountains. In Figure 12.30A, diffractions arising from near-surface crevasses and internal reflections are evident. The latter are also evident in Figure 12.30B, which shows the radar record for the lowermost 400 m of ice. In addition, a 'bright spot' is observed within the basal ice and is thought to be due to entrained basal rock debris or trapped water. Strong diffractions also occur at the base of the ice, just below the 400 m mark. These are thought to emanate from either bottom crevasses within the base of the ice or from within a saturated till layer sandwiched between the ice and the bedrock.

At the other extreme of range, Forster *et al.* (1991) have developed a very-high-frequency radar which transmits 200 mW at 11 GHz with a 2 GHz bandwidth to measure detailed snow stratigraphy (Figure 12.31A). This radar has an effective resolution in snow of around 5 cm. The normalised power return is plotted as a function of depth (Figure 12.31B), assuming a uniform radiowave velocity through the snow—which leads to an inaccuracy of about 2% in depth determination over the shallow range of depth penetration (< 3 m). The importance of a device like this is that it can be used to determine the effect of shallow sub-surface strata on measurements made by satellite remote sensors, such as synthetic aperture radar imagers, radar altimeters and passive microwave radiometers (Forster *et al.* 1991).

Figure 12.29   Radio echosounding record from a site in East Antarctica. From Drewry (1986), by permission

(A)

(B)

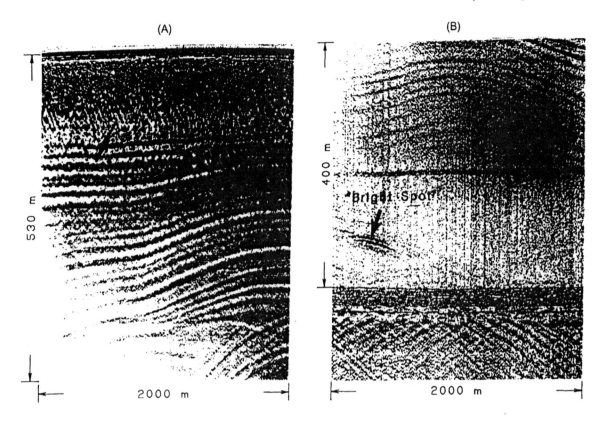

12.8.4   Engineering applications on man-made structures

Radar has enormous potential for use on engineered constructions. In these cases, the geometry and materials of each structure should be known. However, what can be at odds is whether the structure has been constructed according to the required specifications. Indeed, the construction methods may have left a legacy of subtle differences in physical properties within the structure. Radar can be used to ensure that, for example, reinforcement mesh has been placed at the correct level within concrete slabs. In some cases, the act of pouring the concrete can displace the mesh, so that instead of being located within the middle of a slab it is pushed to the bottom, and therefore cannot perform the function for which it was intended. Similarly, the location of reinforcement bars ('rebars') can be checked using higher-frequency radar (e.g. 500 MHz, 900 MHz or higher centre frequency) as shown in Figure 12.32.

An embankment made of fill material, which consisted largely of crushed dolerite, was thought to be uniform in its properties throughout. A radar survey using a 120 MHz antenna revealed sub-horizontal layering within the fill which were interpreted to be associated with compacted horizons. These layers had been produced by the

Figure 12.30 Low-frequency (4 MHz) records of sub-ice sheet environs: (A) diffractions from surface crevasses, and internal reflections; and (B) internal reflections and diffractions from basal crevasses and from a deformed sub-glacial sediment zone. From Wright *et al.* (1990), by permission

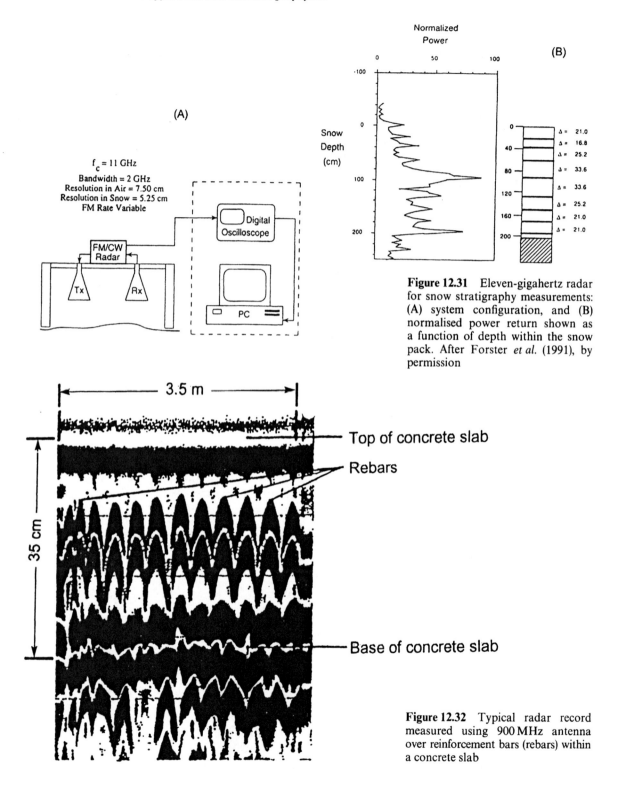

(A)

$f_c = 11$ GHz

Bandwidth = 2 GHz
Resolution in Air = 7.50 cm
Resolution in Snow = 5.25 cm
FM Rate Variable

FM/CW Radar

Tx   Rx

Digital Oscilloscope

PC

Normalized Power

(B)

Snow Depth (cm)

Figure 12.31 Eleven-gigahertz radar for snow stratigraphy measurements: (A) system configuration, and (B) normalised power return shown as a function of depth within the snow pack. After Forster *et al.* (1991), by permission

3.5 m

Top of concrete slab

Rebars

35 cm

Base of concrete slab

Figure 12.32 Typical radar record measured using 900 MHz antenna over reinforcement bars (rebars) within a concrete slab

contractor's vehicles driving over the fill along roadways. Distinct reflections were evident from these horizons and occurred as a result of the reduction in soil moisture content (and hence altered dielectric constant) of the compacted horizons (Reynolds and Taylor 1992).

Water-retaining dykes along Dutch rivers have been provided at various locations with granular filters. The purpose of these filters is to reduce the groundwater potential inside and below the dyke in order to increase the stability and prevent bursting of a clay layer at the lee side of the dyke (Figure 12.33A). The gravel-filled filter must be in hydraulic contact with the underlying sandy layers. The filters have a working life of several decades but suffer with becoming clogged by finer particles. Consequently, it is necessary to monitor the filters periodically, preferably using non-destructive methods. Ground penetrating radar has been used successfully in such investigations, as reported by de Feijter and van Deen (1990).

Radar surveys were carried out when the groundwater level was at its maximum and also at its minimum, over a time interval of several years. A sample radargram is shown in Figure 12.33B, which was obtained using a 300 MHz antenna with a GSSI radar system. The horizontal reflection arises from the water table (with its multiple). Reflection III is caused evidently by the filter–dyke interface. Note, however, that none of the horizontal layers within the dyke itself is imaged. Indeed, the zone to the right of the filter–dyke interface shows significant attenuation of the radar energy, most probably because of the presence of water-saturated sandy clay and clay.

The ground radar method lends itself very well to the investigation of road pavements and bridge decks as they are made up of discrete layers. Radar can be used to measure layer thicknesses and to detect areas of delamination, where one horizon separates from another causing weakness in the road. This can lead to rapid deterioration of the road surface. As the depth of investigation is small (usually < 1 m) and as fine vertical resolution is required, high-frequency antennae are used, typically 900 MHz centre frequency. One or more antennae are fixed to a frame on a vehicle so as to maintain a fixed and known geometry. The vehicle drives along the road being surveyed at speeds up to several tens of kilometres per hour. The graphical output is viewed for evidence of anomalous zones which might indicate potential problems within the road base. Radar can also be used to check that the specified thickness of sub-base material has been put down during construction or if there are areas where the sub-base is defective.

The radar method has considerable applications within stone and brick masonry. The method has also been used to investigate the internal composition of statues and of masonry façades of historic buildings.

In most of these applications, the radar method complements other non-destructive testing methods, such as acoustics (e.g. in bridge deck

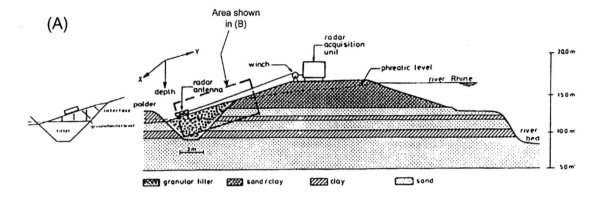

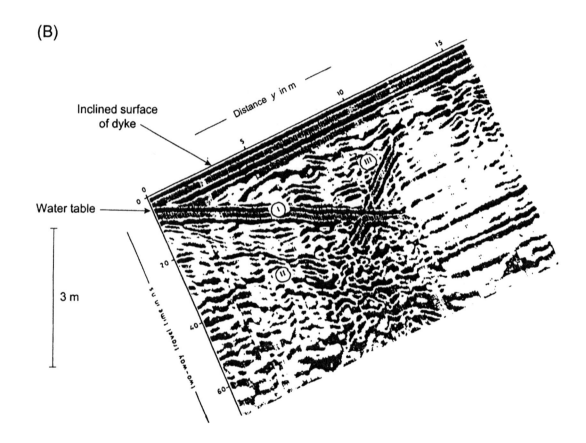

surveys), nuclear density measurements and thermal imaging (e.g. road studies). As with all geophysical methods, some ground-truth data are invaluable for correlation purposes. In road pavement studies, cores are used to provide point-to-point correlation with radargrams. The major advantage of the radar method is that it is entirely non-destructive.

Figure 12.33 (A) Cross-section of a dyke adjacent to the River Rhine, illustrating the engineered sub-structure, and the disposition of a radar transect shown in (B): 300 MHz radargram taken at highwater level. After de Feijter and van Deen (1990), by permission

12.8.5   Voids within man-made structures

One of the commonest applications of ground penetrating radar in the investigation of man-made structures is the detection and location of voids. Radar has been used extensively to inspect the condition of brick-lined tunnels and sewers. In the latter case, radar is often used in conjunction with closed-circuit TV (CCTV).

Two particular aspects of radar make the method particularly suited to such investigations. One is that the speed of radiowaves in air is around three times faster than in solid materials and thus produces a pronounced velocity 'pull-up' effect in association with a significant void. The second is the occurrence of resonance which happens when the wavelength of the incident radiowave energy is the same as, or shorter than, the dimensions of the void.

An example of a radargram acquired in a sewer survey is given in Figure 12.34A. This shows an extract of a radargram obtained where an air void was found above the crown of a sewer. Two features are evident on the radar section: one is the obvious difference in position of the crown of the sewer each side of the void (Figure 12.34B); the second is the obvious lozenge-shaped anomaly associated with the air void (picked out with white dashed lines in Figure 12.34A). The data were acquired using a vertical-looking 500 MHz antenna with a GSSI SIR-3 radar system.

Not all air voids occur as failures in construction. There are examples where radar has been used to locate hidden crypts in churches and in other historic buildings, or old Second World War air-raid shelters whose locations have been lost.

An excellent example of a radargram showing the location of a crypt has been provided by Stratascan from a survey at Worcester Cathedral (Figure 12.35). This particular survey profile was carried out at right-angles to the long axis of the Charnel House which has an arched roof. The stone–air interface of the roof of the Charnel House is obvious by the bright strong amplitude reflections, the apex of which is at 7.5 m along the profile. Note that at about 6 ns above the top of the Charnel House roof reflection, there is a much weaker reflection which has the same shape as the reflection from the crypt. This is interpreted to be from the top surface of a line of stones used to construct the ceiling arch of the Charnel House.

The bright domed reflection at the bottom of the radargram is not due to a heap of rubble on the floor of the Charnel House, but is the reflection arising from the floor itself. That it is domed is purely due to the velocity pull-up effect from the air-filled crypt. Given an air-filled void 2 m deep, the reflection from the floor would occur at a two-way travel time some 26 ns ahead of a reflection from a comparable depth within the adjacent soil. Note also that the reflection from the stone–air interface at the ceiling of the Charnel House has the form of a leading white then positive, etc., whereas the reflection from the

(A)

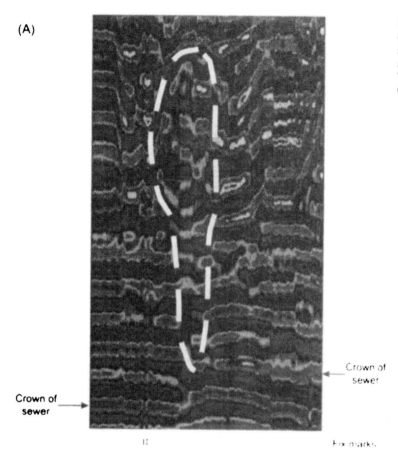

Crown of sewer

Crown of sewer

II   Fox marks

Figure 12.34 (A) Radargram through clay/hoggin material above a damaged sewer; a void is evident on the section and is delineated by dashed lines. The actual situation is shown in (B). Courtesy of Stratascan Ltd

(B)

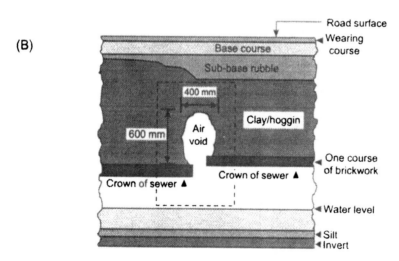

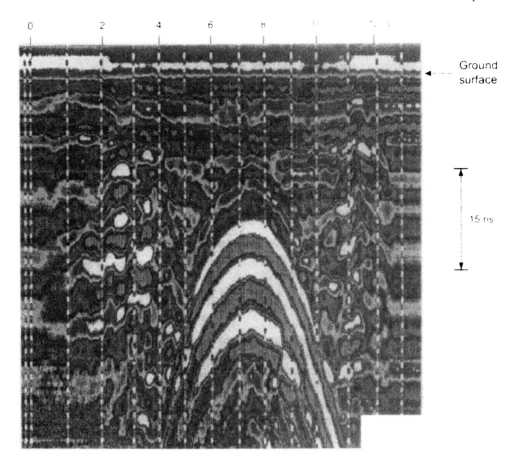

0   2   4   6   8   ...   ...

← Ground surface

15 ns

Figure 12.35 Worcester cathedral: radargram over a buried vault. The top of the vault is clearly identified as being at the apex of the strong white anomaly which dominates the section. Courtesy of Stratascan Ltd

floor (air–stone interface) has the form positive then negative white. The polarity change is attributable to the passage of radiowaves from (a) a slow to fast medium (stone to air) and then (b) from a fast to slow medium (air to stone). Consequently, the phase of the reflection changes as does the sign of the reflection coefficient (see Section 12.3.1).

In some ancient buildings, additional complications can arise from the dimensions of the stone used in the construction of the building in relation to the wavelength of the incident radiowaves. Using a 500 MHz antenna over stone blocks ($V \approx 0.1$ m/ns), the wavelength is 0.2 m which is comparable to the dimensions of the blocks. Consequently, adverse phase interference can arise from internal reflections from the faces of individual blocks, making the overall radargram much more chaotic than it would otherwise have been. Similarly, building materials used to fill-in behind stone walls can be of a wide variety of materials, ranging from stone rubble to soil which may compact to leave air-filled voids.

12.8.6 Archaeological investigations

Radar has many applications in archaeological no-dig investigations, especially as the depth of penetration required is usually small (commonly less than 3 m). Radar can be used as a first-look technique or as a fill-in method between areas of excavation. One example of the successful use of radar in imaging an archaeological feature is given in Figure 12.36. The radargram was obtained using a 300 MHz antenna (range setting = 40 ns) with a GSSI SIR-3 system over flat ground made up of alluvial silts, gravels and boulder clay at a site at Caersws, Powys, Wales. The radargram shows very clearly the reflections arising from the surface of a Roman road with ditches on each side. The swathe of the carriageway appears as a slightly depressed central zone.

The location of graves can be important for at least two reasons. As consecrated ground, there has to be considerable care exercised if the site is wanted for development. Also, there may be important archaeological information associated with the buried remains. It is doubly important, therefore, that the locations of graves can be determined non-destructively. One such survey of graves using GPR among other techniques has been described by Bevan (1991). Radargrams obtained in two orthogonal directions across a postulated grave are shown in Figure 12.37. To aid the interpretation of the radar results, an electromagnetic ground conductivity meter (an EM31) and a magnetometer were used for corroboration. As can be seen from Figure 12.37C and D, these other tools produced geophysical anomalies which are closed (form concentric contours) around the locations of the graves. It should be emphasised that it is the disturbed ground associated with the graves that tends to show up on radargrams, not the bodies themselves.

Figure 12.36 Example of a radargram over a buried Roman road. Courtesy of Stratascan Ltd

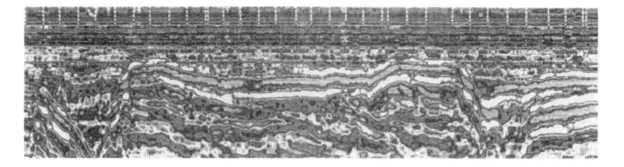

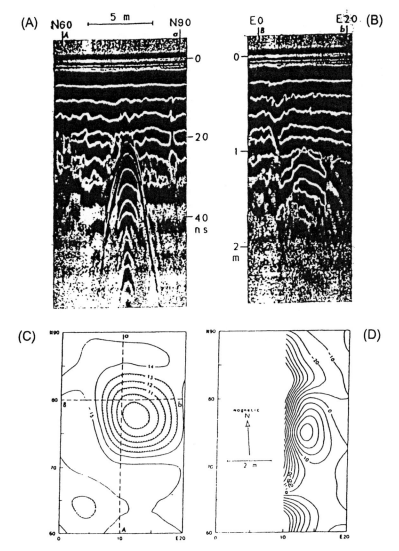

Figure 12.37 (A) North–south radar profile over a possible grave. (B) An east–west profile shows that the object is wider in this direction. (C) Apparent conductivity map with a contour interval of 1 mS/m; dashed lines indicate the orientations of the two radar profiles. (D) There is also buried iron or fired brick; the contour interval of this magnetic anomaly map is 10 nT. From Bevan (1991), by permission

Where human remains are thought to be located in specially protected sites – for example, those which may have been scheduled as 'Ancient Monuments' – radar can be particularly useful. The National Trust of Scotland has used radar in an attempt to locate the remains of the Duke of Northumberland's redcoats who died at the Battle of Culloden in 1746 after the historic routing of Bonnie Prince Charlie and his army. The investigation of the site is important as it is hoped that new information will be forthcoming about one of Scotland's greatest defeats at the hands of the English. It is thought that the remains of some of the English soldiers are buried inside a turf dyke.

12.8.7   Forensic uses of GPR

Radar has been used very successfully in the search not only for ancient graves but also for more recently buried bodies. It has been especially helpful where bodies have been concealed so as to avoid detection. Since the late 1980s, radar has been used increasingly by police in the search for disturbed ground which might indicate where a corpse has been hidden. For example, the States of Jersey, Channel Islands, employed radar to look for the bodies of a married couple which it was thought had been concealed in the back garden of their bungalow. Nothing was found. Radar also played an instrumental role in locating the remains of 12 corpses concealed at three locations in and around Gloucester, UK. A multiple murderer and his accomplice buried the bodies within the concrete foundations and inside newly plastered walls of their own house. The radar was able to identify anomalous targets within the concrete, thereby providing particular locations for excavation. Several more bodies buried in a nearby field were found with the help of radar.

In the USA, at Denton, Maryland, a high-school student Jamie Griffin was murdered, but neither his parents nor local police could locate his body for over seven years. Consequently, the person who had admitted perpetrating the crime could not be tried for murder. The body was thought to have been concealed somewhere within the 20 acres of Gunpowder Falls State Park, north of Baltimore. Previous searches using dogs, scuba divers and earth-moving equipment had been unsuccessful. It was only when a radar system was used that an anomaly was found which turned out to be due to Jamie Griffin's body. Not only was it then possible to convict the suspect of first-degree murder but the victim's family was able to hold a proper funeral for their son.

In addition to aiding the location of concealed corpses, radar has been used in the search for buried bullion following a major robbery in London in the late 1980s. With increasing publicity to such cases, so the forensic applications of radar increase. It is likely that, with the increasing reliability of radar, the applications for forensic investigations will become much more numerous.

12.8.8   Wide-aperture radar mapping and migration processing

Radar surveys are normally executed using one or two antennae in a profiling mode as a single-channel acquisition system. Fisher *et al.* (1992a) carried out a 40-channel GPR survey in a complicated fluvial/aeolian environment in the Ottawa River valley approximately 300 km north-northeast of Toronto, Canada. The receiver antenna was placed at each of 441 survey points at 1 m intervals along the profile line. For each of these receiver locations, data were recorded with the transmitting antenna at each of 40 separate

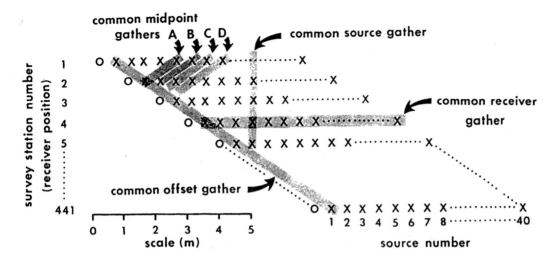

Figure 12.38 Survey geometry and data-gather definitions. A common-receiver gather contains all the traces recorded at one receiver position. A common-source gather contains all the traces generated at one source position. A common-offset gather contains all the traces with a fixed source-receiver separation. A common-mid-point gather contains all the traces with the same centre point between source and receiver. For the geometry used, there are four common-mid-point positions (cf. A, B, C, D) per receiver position. There are 40 traces in each common-receiver gather and 40 in each common-source gather. Except at the end of the survey, there are 10 traces in each common-mid-point gather. From Fisher *et al.* (1992a), by permission

positions along the line, at 0.5 m intervals between 0.5 m and 20 m from the receiver (Figure 12.38). The antennae were orientated parallel to each other with the *E*-plane of the dipoles perpendicular to the survey line. Instead of the usual seismic analogue with a single source recording into a collection of receivers ('common-source gathers'), the common-receiver gathers were acquired. Traces could be sorted into common-source, common-receiver, common-offset or common-midpoint gathers irrespective of the method by which the data were acquired in the field. By obtaining the radar data in this way, traditional seismic processing could be undertaken on the common-midpoint (CMP) gathers.

A PulseEKKO radar system (Sensors & Software Inc.) with 100 MHz antennae was used, producing usable energy within the 50–150 MHz bandwidth. For each recording, 64 source excitations of the 400 V pulser were stacked to improve the signal-to-noise ratio. A total of 1280 samples were recorded with a time sample interval of 800 ps for a total time record of 1024 ns. Given a typical radiowave velocity in soil of 0.065 m/ns, this time range provided a possible depth of penetration of up to 33 m.

The radar data were input into a standard seismic processing sequence (filtering, statics corrections, common-midpoint gathering, velocity analysis, normal- and dip-moveout corrections, stacking and depth migration). More details of the data processing are given in Fisher *et al.* (1992a), and of the migration procedures in Fisher *et al.* (1992b). While the common-receiver and common-midpoint gathers for the radar data look very much like their seismic analogues, the derived RMS velocity estimates from three representative CMP gathers are shown in Figure 12.39. Whereas seismic velocities tend to increase with depth, these velocity panels show that the reverse is true for radar waves.

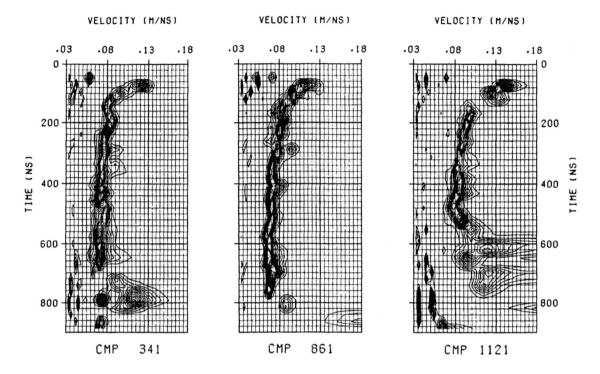

The results of the data processing are shown in Figures 12.40 and 12.41. Figure 12.40A shows a near-offset section; at each location the trace plotted is that recorded with the smallest (0.5 m) source–receiver separation. Figure 12.40B shows every fourth trace in the stacked time section; at each location, the trace plotted is the stack over the traces at the midpoint.

Note that the two panels in Figure 12.40 are shown as time sections. Compare those with the upper panel in Figure 12.41 which is a migrated depth image; the corresponding interpretation is provided in the lower panel. It is evident that significant detail has been obtained to a depth of over 25 m, as correlated with adjacent boreholes. The final migrated data are quite spectacular and show sub-metre vertical resolution even with a 100 MHz antenna to depths in excess of 20 m. Such success has to be tempered by the fact that, in the UK for example, sites where similar records could be obtained are rare, and comparable depths of penetration and similar resolution would be very difficult to achieve owing to the conductive nature of the soils. However, this example does demonstrate that, given the appropriate field conditions, equipment and processing facilities, exceptionally good results are possible with the method. Comparable results would not be possible using even high-resolution seismic reflection surveying.

Figure 12.39 RMS velocity estimates from three representative CMP gathers. Velocity generally decreases with increasing depth (i.e. with increasing travel time). Velocity estimates are not reliable at times greater than that at which the last coherent reflection occures. From Fisher *et al.* (1992a), by permission

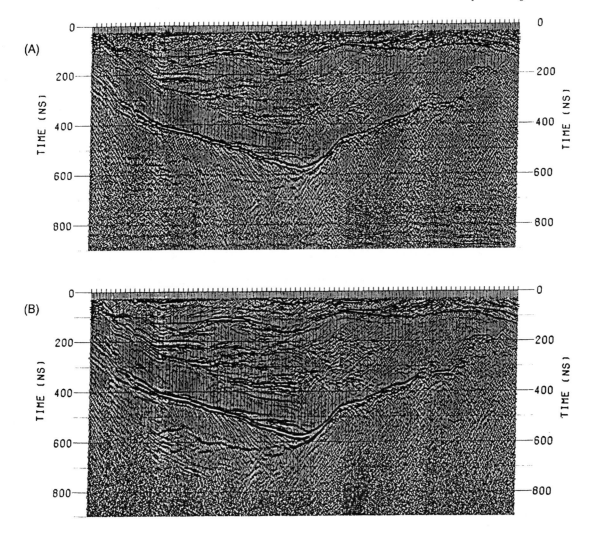

Figure 12.40 Radar sections from a 40-channel radar survey in the Ottawa River valley, Canada. (A) Near-offset section; at each location, the trace plotted is that recorded with the smallest (0.5 m) source–receiver separation. (B) Section displayed with every fourth trace of the stacked time section; at each location the trace plotted is the stack over the traces at that midpoint. Both (A) and (B) are plotted with AGC with a 200 ns window. From Fisher *et al.* (1992a), by permission

12.8.9   Borehole radar

In parallel with developments in surface ground penetrating radar, borehole systems have been in use since the early 1980s. One system called RAMAC was developed as part of the International Stripa Project in Sweden and is manufactured by Malå Geoscience AB and marketed by ABEM. The Stripa Project was undertaken in order to develop techniques suitable for use in underground nuclear fuel waste repositories. The short-pulse borehole radar was used primarily to obtain information about the structure and integrity of crystalline rock masses at a distance from tunnels and boreholes without affecting the rock in any way. In 1987, a RAMAC survey in Switzerland revealed fractures in granite 160 m away from the surveyed borehole.

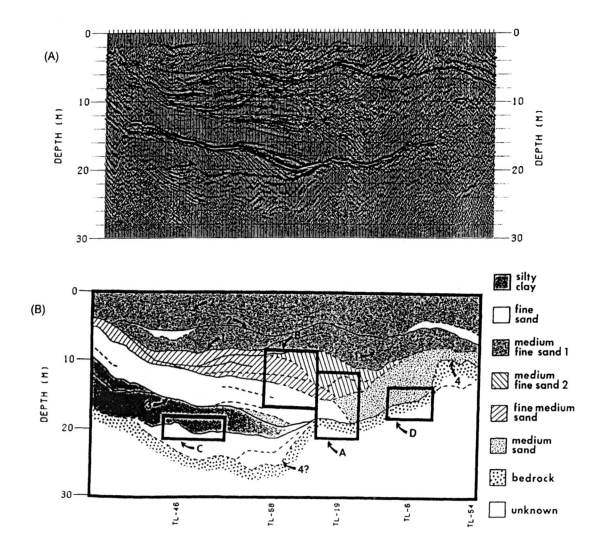

Since then, RAMAC has been used to record reflections arising from structures at distances greater than 300 m within rocksalt. The system is now well proven and has been used internationally in mining, hydrogeological and rock mechanical investigations. It is also being used to investigate sites for tunnels, dams and other construction type projects.

The basic system is illustrated schematically in Figure 12.42 with a simple radargram. A transmitter is used to generate the radiowaves with a pulse frequency of 43 kHz. A separate receiver is located a short distance further down the hole: typically 2–6 m when surveying in sedimentary rocks, and 5–15 m when in crystalline rocks. The receiver operates with a bandwidth of 10–200 MHz. Different frequency

Figure 12.41 (A) Migrated depth image and (B) its interpretation. The image is plotted with AGC and only every fourth trace is shown. Approximate locations of available drill cores are labelled at the bottom of (B). Reflector 1 is a garnet sand; 2 = silt layers; 3 = fine sand gravel; 4 = bedrock. From Fisher *et al.* (1992a), by permission

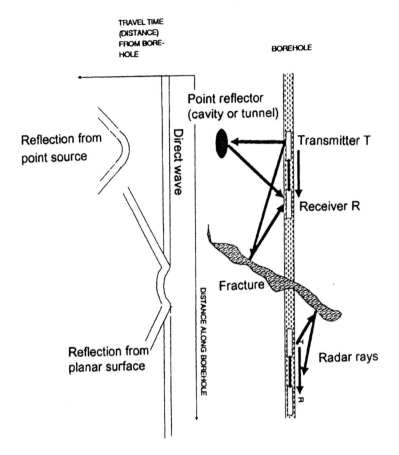

Figure 12.42 Basic arrangement for a borehole radar system with typical radar section images of plane and point reflectors. The cylindrical symmetry of dipole antennae creates V-shaped responses from planar reflectors. Hyperbolae arise from point sources. Courtesy of ABEM

antennae can be used, either 20 MHz or 60 MHz using the standard RAMAC system, or 50, 100, 200 or 400 MHz with a newer version released in 1994. Measurements are made at fixed intervals of 0.5 m or 1 m. It takes at most about 30 seconds at each location to make the required measurements.

Two modes of operation are possible. One uses an omnidirectional dipole antenna, and the other uses directional antennae so that reflections from discretely identified quadrants surrounding the borehole can be identified. The RAMAC system can also be used in cross-hole tomographic configurations with the transmitter antenna in one borehole and the receiver down another. Additionally a receiver can be placed down a borehole while the transmitter is located either in a mine tunnel or on the ground surface, as illustrated in Figure 12.43. Processing of radargrams can be accomplished using on-screen interpretational software.

A simple example of the types of radargrams generated using the RAMAC system is shown in Figure 12.44. The radargram was

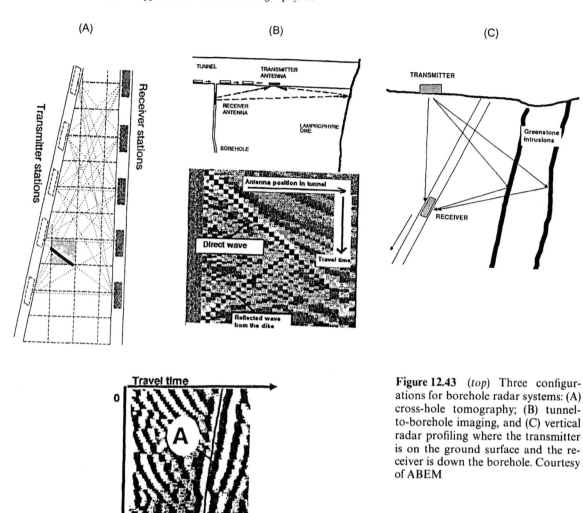

Figure 12.43 (*top*) Three configurations for borehole radar systems: (A) cross-hole tomography; (B) tunnel-to-borehole imaging, and (C) vertical radar profiling where the transmitter is on the ground surface and the receiver is down the borehole. Courtesy of ABEM

Figure 12.44 (*left*) Borehole radargram showing the location of an abandoned drill string (A) and a major fault zone (B) in which the drillbit was stuck. Courtesy of ABEM

obtained from one of three radar surveys made in three different boreholes to locate an abandoned drill string. Strong reflections from the drill string were observed on radargrams from all three boreholes. The radargrams also revealed the reason why the drill string became stuck – it had intersected an oblique major fault zone.

(A)

Travel time

Distance down borehole

Azimuth 250°

(B)

Travel time

Distance down borehole

Azimuth 160°

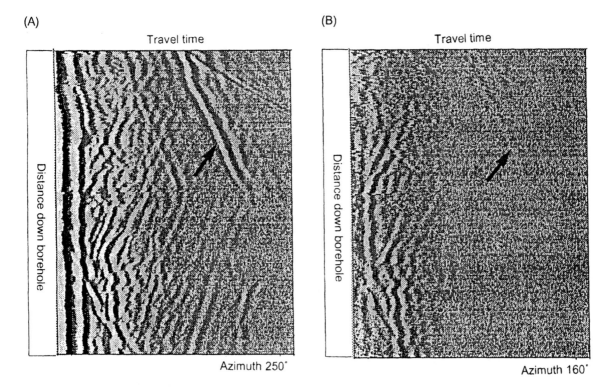

A second example presented in Figure 12.45 shows the effect of azimuth. In Figure 12.45A, the antennae are directed at 250° magnetic bearing and a plane reflection is evident (arrowed). The azimuth is rotated by 90° to 160° magnetic and the reflection is at a minimum (Figure 12.45B). Using such information, the dip and strike of planar features can be determined.

Figure 12.45 Borehole radargrams from the same location showing the effect of azimuth using directional antennae. (A) Azimuth is 250° magnetic with a major strong planar reflection (arrowed). (B) Azimuth is 160° magnetic and the planar reflection strength has decreased to its minimum value. Courtesy of ABEM

References

Abdoh, A., Cowan, D. and Pilkington, M. (1990) 3D gravity inversion of the Cheshire basin. *Geophysical Prospecting*, **38**(8): 999–1011.

Abbiss, C.P. (1981) Shear wave measurements of the elasticity of the ground. *Géotechnique*, **31**(1): 91–104.

Aldridge, D.F. and Oldenburg, D.W. (1992) Refractor imaging using an automated wavefront reconstruction method. *Geophysics*, **57**(3): 378–385.

Åm, K. (1972) The arbitrarily magnetized dyke; interpretation of characteristics. *Geoexploration*, **10**: 63–90.

Ackworth, R.I. and Griffths, D.H. (1985) Simple data processing of tripotential apparent resistivity measurements as an aid to the interpretation of subsurface structure. *Geophysical Prospecting*, **33**(6): 861–887.

Ahmed, H., Dillon, P.B., Johnstad, S.E. and Johnston, C.D. (1986) Northern Viking Graben multilevel three-component walkaway VSPs: a case history. *First Break*, **4**(10): 9–27.

Ahmad, M.U. (1961) A laboratory study of streaming potentials. *Geophysical Prospecting*, **12**(1): 49–64.

Airy, G.B. (1855) On the computation of the effect of the attraction of mountain masses, as disturbing the apparent astronomical latitude of stations in geodetic surveys. *Philosophical Transactions of the Royal Society*, Series B, **145**: 101–104.

Al-Chalabi, M. (1972) Interpretation of gravity anomalies by non-linear optimisation. *Geophysical Prospecting*, **20**(1): 1–16.

Allis, R.G. and Hunt, T.M. (1986) Analysis of exploitation-induced gravity changes at Wairakei Geothermal Field. *Geophysics*, **51**(8): 1647–1660.

Anderson, L.A. and Johnson, G.P. (1976) Application of the self potential method to geothermal exploration in Long Valley, California. *Journal of Geophysical Research*, **81**: 1527–1552.

Annan, A.P. and Davis, J.L. (1976) Impulse radar sounding in permafrost. *Radio Science*, **11**: 383–394.

Annan, A.P. and Davis, J.L. (1977) Radar range analysis for geological materials. *Reports of Activities: Part B*. Geological Survey of Canada, Paper 77-1B: 117–124.

Archie, G.E. (1942) The electrical resistivity log as an aid to determining some reservoir characteristics. *Trans. A.I.M.E.*, **146**: 389–409.

Arcone, S.A., Lawson, D.E. and Delaney, A.J. (1993) Radar reflection and refraction profiles of seasonal thaw over permafrost in Fairbanks, Alaska. *Advanced Group Penetrating Radar: Technologies and Applications, 26–28 Oct. 1993*. Ohio State University, Columbus, Ohio, USA, 241–256.

Arcone, S.A. (1984) Field observations of electromagnetic pulse propagation in dielectric slabs. *Geophysics*, **49**(10): 1763–1773.

Arkell, W.J. (1933) *The Jurassic System in Great Britain.* Oxford: Oxford University Press.

Ates, A. and Kearey, P. (1995) A new method for determining magnetization direction from gravity and magnetic anomalies: application to the deep structure of the Worcester Graben. *Journal of the Geological Society,* **152**(3): 561–566.

Baeten, G. and Ziolkowski, A. (1990) *The Vibroseis Source* (Advances in Geophysics 3). Amsterdam: Elsevier Science Publishers BV.

Baeten, G., Fokkema, J. and Ziolkowski, A. (1988) The marine vibrator source. *First Break,* **6**(9): 285–294.

Baoshan, Z. and Chopin, L. (1983) Shear wave velocity and geotechnical properties of tailings deposits. *Bulletin of the International Association of Engineering Geology* (26–27): 347–353.

Barbour, D.M. and Thurlow, J.G. (1982) Case histories of two massive sulphide discoveries in Central Newfoundland. In: *Prospecting in Areas of Glaciated Terrain – 1982. Can. Inst. Min. Metal.,* 300–321.

Barker, R.D. (1979) Signal contribution sections and their use in resistivity studies. *Geophysical Journal of the Royal Astronomical Society,* **59**(1): 123–129.

Barker, R.D. (1981) Offset system of electrical resistivity sounding and its use with a multicore cable. *Geophysical Prospecting,* **29**(1): 128–143.

Barker, R.D. (1990) Improving the quality of resistivity sounding data in landfill studies. In: Ward, S.H. (ed.), *Geotechnical and Environmental Geophysics. Vol. 2: Environmental and Groundwater.* Tulsa: Society of Exploration Geophysicists, 245–251.

Barker, R.D. (1992) A simple algorithm for electrical imaging of the subsurface. *First Break,* **10**(2): 53–62.

Barringer, A.R. (1962) A new approach to exploration: The INPUT airborne electrical pulse prospecting system. *Min. Congress Journal,* **48**: 49–52.

Barry, K.M. (1967) Delay time and its application to refraction profile interpretation. In: Musgrove, A.W. (ed.), *Seismic Refraction Prospecting.* Tulsa: Society of Exploration Geophysicists, 348–362.

Beaudoin, B.C., ten Brink, U.S. and Stern, T.A. (1992) Characteristics and processing of seismic data collected on thick, floating ice: results from the Ross Ice Shelf, Antarctica. *Geophysics,* **57**(10): 1359–1372.

Beck, A.E. (1981) *Physical Principles of Exploration Methods.* London: Macmillan.

Becker, A. and Telford, W.M. (1965) Spontaneous polarization studies. *Geophysical Prospecting,* **13**(2): 173–188.

Becker, A., Barringer, A.R. and Annan, P. (1990) Airborne electromagnetics 1978–1988. In: Fitterman, D.V. (ed.), Developments and applications of modern airborne electromagnetic surveys. US Geol. Surv. Bull. **1925**: 9–20.

Beer, K.E. and Fenning, P.J. (1976) *Geophysical Anomalies and Mineralisation at Sourton Tors, Okehampton, Devon.* Institute of Geological Sciences Report no. 76/1.

Bennett, B. (1990) A computer program to convert SEG-2 data to SEG-Y. *Geophysics,* **55**(9): 1272–1284.

Berkhout, A.J. (1984) Seismic migration. In: *Developments in Solid Earth Geophysics, 14B.* Amsterdam: Elsevier.

Bertelli, L., Mascarin, B. and Salvador, L. (1993) Planning and field techniques for 3D land acquisition in highly tilled and populated areas: today's results and future trends. *First Break,* **11**(1): 23–32.

Bertin, J. and Loeb, J. (1976) *Experimental and Theoretical Aspects of Induced Polarization,* Vols 1 and 2. Berlin: Gebrüder Borntraeger.

Best, M. and Spies, B. (1990) Recent developments in analysis and presentation of large data sets in mineral exploration. *Geophysics: The Leading Edge*, **9**(9): 37–43.

Bevan, B.W. (1991) The search for graves. *Geophysics*, **56**(9): 1310–1319.

Bevan, B.W. and Kenyon, J. (1975) Ground penetrating radar for historical archaeology. *MASCA Newsletter*, **11**(2): 2–7.

Bhattacharya, B.B. (1986) Reply to comment by N.S. Rajan, N.L. Mohan and M. Narasumha Chary. *Geophysical Prospecting*, **34**(8): 1294–1295.

Bhattacharya, B.B. and Roy, N. (1981) A note on the use of a nomogram for self-potential anomalies. *Geophysical Prospecting*, **29**(1): 102–107.

Bhattacharya, P.K. and Patra, H.P. (1968) *Direct Current Electrical Sounding*. Amsterdam: Elsevier.

Birch, F. (1960) The velocity of compressional waves in rocks to ten kilobars, Part. 1. *Journal of Geophysical Research*, **65**: 1083–1102.

Birch, F. (1961) The velocity of compressional waves in rocks to ten kilobars, Part 2. *Journal of Geophysical Research*, **66**: 2199–2224.

Black, R.A., Steeples D.W. and Miller, R.D. (1994) Migration of shallow seismic reflection data. *Geophysics*, **59**(3): 402–410.

Blizkovsky, M. (1979) Processing and applications of microgravity surveys. *Geophysical Prospecting*, **27**(4): 848–861.

Bloom, A.L. (1962) Physics of operation of the rubidium vapor magnetometer. *Applied Optics*, **1**: 61–68.

Bogorodsky, V.V., Bentley, C.R. and Gudmansen, P.E. (1985) *Radioglaciology*. Dordrecht: D. Reidel Publishing Co.

Bogoslovsky, V.A. and Ogilvy, A.A. (1970a) Natural potential anomalies as a quantitative index of seepage from water reservoirs. *Geophysical Prospecting*, **18**(2): 261–268.

Bogoslovsky, V.A. and Ogilvy, A.A. (1970b) Application of geophysical methods to studying the technical status of earth dams. *Geophysical Prospecting*, **18**(Suppl.): 758–773.

Bolt, B.A. (1982) *Inside the Earth*. San Fransisco: W.H. Freeman.

Boslough, J. (1989) Searching for the secrets of gravity. *National Geographic*, **175**(5): 562–583.

Bott, M.P.H. (1959) The use of electronic digital computers for the evaluation of gravimetric terrain corrections. *Geophysical Prospecting*, **5**(7): 45–54.

Bott, M.P.H. (1960) The use of rapid digital computing methods for direct gravity interpretation of sedimentary basins. *Geophysical Journal*, **3**: 63–67.

Bott, M.P.H. (1962) A simple criterion for interpreting negative gravity anomalies. *Geophysics*, **27**(3): 376–381.

Bowles, L.G. (1990) The seismic fleet and the marine environment: a compatible relationship. *Geophysics: The Leading Edge*, **9**(9): 54–56.

Breiner, S. (1981) Magnetometers for geophysical applications. In: Weinstock, H. and Overton, W.C. (eds), *SQUID Applications to Geophysics*. Tulsa: Society of Exploration Geophysicists, 3–12.

Brewster, M.L., Annan, A.P. and Redman, J.D. (1992a) GPR monitoring of DNAPL migration in a sandy aquifier. *Proceedings of the 4th International Conference of Ground Penetrating Radar*, Rovaniemi, Finland, June 1992.

Brewster, M.L., Redman, J.D. and Annan, A.P. (1992b) Monitoring a controlled injection of perchloroethylene in a sandy aquifier with ground penetrating radar and time-domain reflectometry. *Symposium on the Application of Geophysics to Engineering and Environmental Problems, 1992*. Englefield, USA: EEGS.

Briden, J.C., Clark, R.A. and Fairhead, J.D. (1982) Gravity and magnetic studies in the Channel Islands. *Journal of the Geological Society of London*, **139**(1): 35–48.

British Standards Institution (1981) *BS 5930: Code of Practice for Site Investigations*. London: BSI.

Broadbent, M. and Habberjam, G.M. (1971) A solution to the dipping interface problem using the square array resistivity technique. *Geophysical Prospecting*, **19**(3): 321–338.

Brock, J.S. (1973) Geophysical exploration leading to the discovery of the Faro deposit. *Canadian Institute of Mining and Metallurgy Bulletin*, **66**(738): 97–116.

Brown, G.C. and Mussett, A.E. (1981) *The Inaccessible Earth*. London: George Allen & Unwin.

Bukovics, C. and Nooteboom, J.J. (1990) Combining techniques in integrated 3D land, shallow water and deep channel seismic acquisition. *First Break*, **8**(10): 375–382.

Bullock, S.J. (1988) Future and present trends of navigation and positioning techniques in exploration geophysics. *Geophysical Journal*, **92**(3): 521.

Busby, J.P. (1987) An interactive Fortran 77 program using GKS graphics for 2.5D modeling of gravity and magnetic data. *Computers and Geosciences*, **13**(6): 639–644.

Buselli, G., Barber, C., Davis, G.B. and Salama, R.B. (1990) Detection of groundwater contamination near waste disposal sites with transient electromagnetic and electrical methods. In: Ward, S.H. (ed.), *Geotechnical and Environmental Geophysics. Vol. 2: Environmental and Groundwater*. Tulsa: Society of Exploration Geophysicists, 27–39.

Butler, D.K. (1984) Microgravimetric and gravity gradient techniques for detection of sub-surface cavities. *Geophysics*, **49**(7): 1084–1096.

Butler, D.K. and Llopis, J.L. (1990) Assessment of anomalous seepage conditions. In: Ward, S.H. (ed.), *Geotechnical and Environmental Geophysics. Vol. 2: Environmental and Groundwater*. Tulsa: Society of Exploration Geophysicists, 153–173.

Butler, D.K., Llopis, J.L., Dobecki, T.L., Wilt, M.J., Coewin, R.F. and Olhoeft, G. (1991) Comprehensive geophysics investigation of an existing dam foundation. *Geophysics: The Leading Edge*, **9**(9): 44–53.

Butler, K.E., Russell, R.D., Kepic, A.W. and Maxwell, M. (1994) Mapping of a stratigraphic boundary by its seismoelectric response. *Proceedings of the Symposium on the Application of Geophysics to Engineering and Environmental Problems, SAGEEP '94, Vol. 2*, Englefield, USA: EEGS, 689–699.

Cagniard, L. (1953) Basic theory of the magneto-telluric method of geophysical prospecting. *Geophysics*, **18**(3): 605–635.

Cahyna, F., Mazác, O. and Venhodová, D. (1990) Determination of the extent of cyanide contamination by surface geoelectrical methods. In: Ward, S.H. (ed.), *Geotechnical and Environmental Geophysics. Vol. 2: Environmental and Groundwater*. Tulsa: Society of Exploration Geophysicists, 97–99.

Cameron, G.W., Elliott, B.E. and Richardson, K.A. (1976) Effects of line spacing on contoured airborne gamma-ray spectrometry data. In: *Exploration for Uranium Ore Deposits*, Proc. Series, IAEA, Vienna, 81–92.

Campbell, W.H. (1986) An interpretation of induced electric currents in long pipelines caused by natural geomagnetic sources in the upper atmosphere. *Surveys in Geophysics*, **8**: 239–259.

Carmichael, R.S. and Henry, G. (1977) Gravity exploration for groundwater and bedrock topography in glaciated areas. *Geophysics*, **42**(4): 850–859.

Carpenter, E.W. (1955) Some notes concerning the Wenner configuration. *Geophysical Prospecting*, **3**(4): 388–402.

Carpenter, E.W. and Habberjam, G.M. (1956) A tripotential method of resistivity prospecting. *Geophysics*, **21**(2): 455–469.

Carpenter, P.J., Calkin, S.F. and Kaufman, R.S. (1991) Assessing a fractured landfill cover using electrical resistivity and seismic refraction techniques. *Geophysics*, **56**(11): 1896–1904.

Carpenter, P.J., Doll, W.E. and Phillips, B.E. (1994) Ground penetrating radar surveys over an alluvial DNAPL site, Paducah gaseous diffusion plant, Kentucky. *SAGEEP 1994*, 261–275.

Cartwright, K. and McComas, M.R. (1968) Geophysical surveys in the vicinity of sanitary landfills in north-eastern Illinois. *Ground Water*, **16**(5): 23–30.

Casten, U. and Gram, C. (1989) Recent developments in underground gravity surveys. *Geophysical Prospecting*, **37**(1): 73–90.

Caterpillar Tractor Company (1988) *Handbook of Ripping*, 8th edn.

Chave, A.D., Constable, S.C. and Edwards, R.N. (1991) Electrical exploration methods for the seafloor. In: Nabighian, M.N. (ed.), *Electromagnetic Methods in Applied Geophysics*, Vol. 2B. Tulsa: Society of Exploration Geophysicists, 931–966.

Christensen, E. (1992) Small vibrator development. *Abstracts of the EAEG Conference, 54th Meeting, Paris*. Netherlands: European Association of Exploration Geophysicists, 306–307.

Christensen, N.I. and Fountain, D.M. (1975) Constitution of the lower continental crust based on experimental studies of seismic velocities in granulites. *Bulletin of the Geological Society of America*, **86**: 227–236.

Cioni, R., Fanelli, G., Guidi, M., Kinyariro, J.K. and Marini, L. (1992) Lake Bogoria hot springs (Kenya): geochemical features and geothermal implications. *Journal of Volcanology and Geothermal Research*, **50**(3): 231–246.

Claerbout, J.F. (1976) *Fundamentals of Geophysical Data Processing*. New York: McGraw-Hill.

Claerbout, J.F. (1985) *Imaging the Earth's Interior*. Oxford: Blackwell Scientific.

Cole, K.S. and Cole, R.H. (1941) Dispersion and absorption in dielectrics. *J. Chem. Phys.*, **9**: 341–351.

Coleman, A.R. (1991) The use of the self-potential method in the delineation of a reclaimed landfill site. *Proceedings of the Conference on Planning and Engineering of Landfills*, 10–11 July 1991, University of Birmingham. The Midlands Geotechnical Society.

Collett, L.S. (1986) Development of the airborne electromagnetic technique. In: Palacky, G.J. (ed.), Airborne resistivity mapping. *Geol. Surv. Can. Paper* **86-22**: 9–18.

Collette, B.J. (1958) Structural sketch of the North Sea. *Geologie Minjnb*, **20**: 366–371.

Colley, G.C. (1963) The detection of caves by gravity measurements. *Geophysical Prospecting*, **11**(1): 1–9.

Cook, J.C. (1960) Proposed monocycle-pulse very-high-frequency radar for air-borne ice and snow measurement. *AIEE Comm. Electron.*, **51**: 588–594.

Cook, J.C. (1975) Radar transparencies of mine and tunnel rocks. *Geophysics* **40**(5): 865–885.

Cook, K.L. and van Nostrand, R.G. (1954) Interpretation of resistivity data over filled sinks. *Geophysics*, **19**(4): 761–790.

Cooper, N.J. and Swift, R. (1994) The application of TEM to Cypress-type massive sulfide exploration in Cyprus. *Geophysics*, **59**(2): 202–214.

Corwin, R.F. and Hoover, D.B. (1979) The self-potential method in geothermal exploration. *Geophysics*, **44**(2): 226–245.

Cosentino, M., Lombardo, G. and Privitera, E. (1989) A model for the internal dynamical processes on Mt Etna. *Geophysical Journal*, **97**(3): 367–379.

Crampin, S. (1985) Evaluation of anisotropy by shear-wave splitting. *Geophysics*, **50**(1): 142–152.

Crone, J.D. (1986) Field examples of borehole pulse EM surveys used to detect and outline conductive ore deposits. In: Killeen, P.G. (ed.), Borehole geophysics for mining and geotechnical applications. *Geological Surv. Can. Paper*, **85-27**: 59–70.

Crone, J.D. (1991) PEM case histories, Cigar and Winston Lakes, Canada. In: Nabighian, M.C. (ed.), *Electromagnetic Methods in Applied Geophysics. Vol. 2: Applications, Part A.* Tulsa: Society of Exploration Geophysicists, 490–495.

Cummings, D. (1988) Determination of depths to an irregular interface in shallow seismic refraction surveys using a pocket calculator. *Geophysics*, **44**(12): 1987–1998.

Dagley, P., Mussett, A.E., Wilson, R.L. and Hall, J.M. (1978) The British Teriary igneous province: palaeomagnetism of the Arran dykes. *Geophysical Journal of the Royal Astronomical Society*, **54**(1): 75–91.

Daily, W. and Owen, E. (1991) Cross-borehole resistivity tomography. *Geophysics*, **56**(8): 1228–1235.

Dam, van, J.C. and Meulenkamp, J.J. (1967) Some results of the geo-electrical resistivity method in ground water investigations in the Netherlands. *Geophysical Prospecting*, **15**(1): 92–115.

Dampney, C.N.G. (1977) Gravity interpretation for hydrocarbon exploration: a workshop manual, *Bulletin of the Australian Society of Exploration Geophysicists*, **8**(4): 161–178.

Daniels, D.J., Gunton, D.J. and Scott, H.F. (1988) Introduction to subsurface radar. *IEE Proceedings*, **135** (F,4): 278–320.

Darracott, B.W. and Lake, M.I. (1981) An initial appraisal of ground probing radar for site investigation in Britain. *Ground Engineering*, **14**: 14–18.

Darracott, B.W. and McCann, D.M. (1986) Planning engineering geophysical surveys. In: Hawkins, A.B. (ed.), *Site Investigation Practice: Assessing BS 5930*, Geological Society Engineering Geology Special Publication no. 2, 85–90.

David, A. (1995) *Geophysical Survey in Archaeological Field Evaluation*, Research and Professional Services Guideline no. 1. London: English Heritage.

Davis, J.L. and Annan, A.P. (1989) Ground-penetrating radar for high-resolution mapping of soil and rock stratigraphy. *Geophysical Prospecting*, **37**(5): 531–551.

de Feijter, J.W. and van Deen, J.K. (1990) Quality assessment of hydraulic engineering structures using ground radar. In: Ward, S. (ed.), *Geotechnical and Environmental Geophysics. Vol. III: Geotechnical.* Tulsa: Society of Exploration Geophysicists, 249–261.

de Witte, L. (1948) A new method of interpretation of self-potential field data. *Geophysics*, **13**(4): 600–608.

Dobrin, M.B. (1976) *Introduction to Geophysical Prospecting*, 3rd edn. New York: McGraw-Hill.

Dobrin, M.B. and Savit, C.H. (1988) *Introduction to Geophysical Prospecting*, 4th edn. New York: McGraw-Hill.

Docherty, P. (1992) Solving for the thickness and velocity of the weathering layer using 2-D refraction tomography. *Geophysics*, **57**(10): 1307–1318.

Dodds, A.R. and Ivic, D. (1990) Integrated geophysical methods used for groundwater studies in the Murray Basin, South Australia. In: Ward, S.H. (ed.), *Geotechnical and Environmental Geophysics. Vol. 2: Environmental and Groundwater.* Tulsa: Society of Exploration Geophysicists, 303–310.

Dohr, G. (1981) *Applied Geophysics*, New York: Halsted Press.

Doll, W.E. (1994) How can environmental geophysics be advanced? *Geophysics: The leading Edge*, **13**(10): 1035–1039.

Doolittle, J.A. (1993) Characteristics and monitoring the vadose zone with ground penetrating radar. *Advanced Ground Penetrating Radar: Technologies and Applications, 26–28 Oct. 1993*. Ohio State University, Columbus, Ohio, USA 105–119.

Douma, J., Den Rooijen, H. and Schokking, F. (1990) Anisotropy detected in shallow clays using shear-wave splitting in a VSP survey. *Geophysical Prospecting*, **38**(8): 983–998.

Drewry, D. (1986) *Glacial Geological Processes*. London: Edward Arnold.

Dunning, F.W. (1970) *Geophysical Exploration*. London: HMSO.

Dyck, A.V. (1991) Drill-hole electromagnetic methods. In: Nabighian, M.N. (ed.), *Electromagnetic Methods in Applied Geophysics*, Vol. 2B. Tulsa: Society of Exploration Geophysicists, 881–930.

Eaton, D.W.S., Stewart, R.R. and Harrison, M.P. (1991) The Fresnel zone for P–S waves. *Geophysics*, **56**(3): 360–364.

Edelmann, H.A.K. (1992) Circularly polarized shear waves used for VSP measurements. *Geophysics*, **57**(4): 643–646.

Edwards, R.N. and Nabighian, M.N. (1991) The magnetometric resistivity method. In: Nabighian, M.N. (ed.), *Electromagnetic Methods in Applied Geophysics*, Vol. 2A. Tulsa: Society of Exploration Geophysicists, 47–104.

Edwards, R.N., Law, L.K., Wolfgram, P.A., Nobes, D.C., Borne, M.N., Trigg, D.F. and DeLaurier, J.M. (1985) First results of the MOSES experiment: sea sediment conductivity and thickness determination, Bute Inlet, British Columbia, by magnetometric offshore electrical sounding. *Geophysics*, **50**(1): 153–161.

Edwards, R.N., Wolfgram, P.A. and Judge, A.S. (1988) The ICE-MOSES experiment: mapping permafrost zones electrically beneath the Beaufort sea. *Marine Geophysical Researches*, **9**: 265–290.

Eggers, A.A. (1987) Residual gravity changes and eruption magnitudes. *Journal of Volcanology and Geothermal Research*, **33**: 201–216.

Eiken, O., Degutsch, M., Riste, P. and Röd, K. (1989) Snowstreamer: an efficient tool in seismic acquisition. *First Break*, **7**(9): 374–378.

Elkins, T.A. (1951) The second derivative method of gravity interpretation. *Geophysics*, **16**(1): 29–50.

Eloranta, E. (1984) *Geoexploration*, **22**: 77–88.

European Association of Exploration Geophysicists (1991) *Standard Graphs for Resistivity Prospecting*. The Hague: EAEG.

Evans, R.B. (1982) Currently available geophysical methods for use in hazardous waste site investigations. In: Long, F.A. and Schweitzer, G.E. (eds), *Risk Assessment at Hazardous Waste Sites*. American Chemical Society Symposium Series no. 204, 93–115.

Evans, S. (1965) Dielectric properties of ice and snow: a review. *Journal of Glaciology*, **5**(42): 773–792.

Ewing, M., Worzel, J. and Pekeris, C.L. (1948) *Geological Society of America Memoir 27*.

Fajklewicz, Z. (1986) Origin of the anomalies of gravity and its vertical gradient over cavities in brittle rock. *Geophysical Prospecting*, **34**(8): 1233–1254.

Faust, L.Y. (1951) Seismic velocity as a function of depth and geologic time, *Geophysics*, **16**(2): 192–206.

Ferguson, C.C. (1992) The statistical basis for spatial sampling of contaminated land. *Ground Engineering*, **25**(1): 34–38.

Filloux, J.H. (1987) Instrumentation and experimental methods for oceanic studies. In: Jacobs, J.A. (ed.), *Geomagnetism*, Vol. 1. London: Academic Press, 143–246.

Fisher, E., McMechan, G.A. and Annan, A.P. (1992a) Acquisition and processing of wide-aperture ground penetrating radar data. *Geophysics,* **57**(3): 495–504.

Fisher, E., McMechan, G.A., Annan, A.P. and Cosway, S.W. (1992b) Examples of reverse-time migration of single-channel, ground-penetrating radar profiles. *Geophysics,* **57**(4): 577–586.

Fitterman, D.V. (1978) Electrokinetic and magnetic anomalies associated with dilatant regions in a layered earth. *Journal of Geophysical Research,* **83**: 5924–5934.

Fitterman, D.V. (1979a) Calculations of self-potential anomalies near vertical contacts. *Geophysics,* **44**(2): 195–205.

Fitterman, D.V. (1979b) *Journal of Geophysical Research,* **84**: 6031–6040.

Fitterman, D.V. (ed.) (1990) Developments and applications of modern airborne electromagnetic surveys. *US Geol. Surv. Bull.,* **1925**.

Fitterman, D.V. and Corwin, R.F. (1982) Inversion of self-potential data from the Cerro Prieto geothermal field, Mexico. *Geophysics,* **47**(6): 938–945.

Fitterman, D.V. and Stewart, M.T. (1986) Transient electromagnetic sounding for groundwater. *Geophysics,* **51**(4): 995–1005.

Fitzpatrick, F. (1991) *Studies of Sediments in a Tidal Environment.* Unpublished PhD thesis, University of Plymouth, UK.

Flinn, D. (1977) A geological interpretation of the aeromagnetic map of east-central Shetland. *Journal of the Geological Society of London,* **133**(2): 111–121.

Fofonoff, N.P. and Millard, R.C. (1983) Algorithms for computation of fundamental properties of seawater. *UNESCO Technical Papers in Marine Science,* 44.

Forster, R.R., Davis, C.H., Rand, T.W. and Moore, R.K. (1991) Snow-stratification investigation on an Antarctic ice stream with an X-band radar system. *Journal of Glaciology,* **37**(127): 323–325.

Fournier, C. (1989) Spontaneous potentials and resistivity surveys applied to hydrogeology in a volcanic area: case history of the Chaîne des Puys (Puy-de-Dôme, France). *Geophysical Prospecting,* **37**(6): 647–668.

Frangos, W. (1994) Electrical detection and monitoring of leaks in lined waste disposal ponds. *Proceedings of the Symposium on the Application of Geophysics to Engineering and Environmental Problems, SAGEEP '94, Vol. 2,* Englefield, USA: EEGS, 1073–1082.

Fraser, D.C. (1969) Contouring of VLF-EM data. *Geophysics,* **34**(6): 958–967.

Fraser, D.C., Keevil, N.B. and Ward, S.H. (1964) Conductivity spectra of rocks from the Craigmont ore environment. *Geophysics,* **29**(5): 832–847.

Frischknecht, F.C. (1987) Electromagnetic physical scale modelling. In: Nabighian, M.N. (ed.), *Electromagnetic Methods in Applied Geophysics,* Vol. 1. Tulsa: Society of Exploration Geophysicists, 365–441.

Frischknecht, F.C., Labson, V.F., Spies, B.R. and Anderson, W.L. (1991) Profiling methods using small sources. In: Nabighian, M.C. (ed.), *Electromagnetic Methods in Applied Geophysics. Vol. 2: Applications, Part A.* Tulsa: Society of Exploration Geophysicists, 105–270.

Frohlich, B. and Lancaster, W.J. (1986) Electromagnetic surveying in current Middle Eastern archaeology: application and evaluation. *Geophysics,* **51**(7): 1414–1425.

Garcia-Abdeslam, J. and Ness, G.E. (1994) Inversion of the power spectrum from magnetic anomalies. *Geophysics,* **59**(3): 391–401.

Gardener, R. (1992) Seismic refraction as a tool in the evaluation of rock quality for dredging and engineering purposes: case histories. In: Hudson, J.A. (ed.), *Eurock '92, Chester.* London: British Geotechnical Society, 153–158.

Gardner, L.W. (1939) An areal plan of mapping subsurface structure by refraction shooting. *Geophysics*, **4**(4): 247–259.

Gardner, L.W. (1967) Refraction seismograph profile interpretation. In: Musgrove, A.W. (ed.), *Seismic Refraction Prospecting*. Tulsa: Society of Exploration Geophysicists, 338–347.

Garland, G.D. (1965) *The Earth's Shape and Gravity*, Oxford: Pergamon.

Gay, S.P. (1963) Standard curves for the interpretation of magnetic anomalies over long tabular bodies. *Geophysics*, **28**(2): 161–200.

Gay, S.P. (1967) A 1800 millivolt self-potential anomaly near Hualgayoc, Peru. *Geophysical Prospecting*, **15**(2): 236–245.

Gay, S.P. (1986) The effects of cathodically protected pipelines on aeromagnetic surveys. *Geophysics*, **51**(8): 1671–1684.

Geological Society Engineering Group Working Party (1988) Engineering geophysics. *Quarterly Journal of Engineering Geology*, **21**(3): 207–271.

Ghosh, G.P. (1971) Inverse filter coefficients for the computation of apparent resistivity standard curves for a horizontally stratified earth. *Geophysical Prospecting*, **19**(4): 749–775.

Gluśko, V.T., Cerednicenko, V.P. and Usatenko, B.S. (1981) Reologija gornogo massiva, Kiev. Naukowa Dumka, 22–49.

Goldman, M., du Plooy, A. and Eckard, M. (1994) On reducing ambiguity in the interpretation of transient electromagnetic sounding data. *Geophysical Prospecting*, **42**(1): 3–25.

Goldstein, M.A. and Strangway, D.W. (1975) Audio-frequency magnetotellurics with a grounded electric dipole source. *Geophysics*, **40**(4): 669–683.

Goldstein, N.E., Benson, S.M. and Alumbough, D. (1990) Saline groundwater plume mapping with electromagnetics. In: Ward, S.H. (ed.), *Geotechnical and Environmental Geophysics. Vol. 2: Environmental and Groundwater*. Tulsa: Society of Exploration Geophysicists, 17–25.

Goree, W.S. and Fuller, M. (1976) Magnetometers using RF-driven SQUIDs and their applications in rock magnetism and palaeomagnetism. *Review of Geophysics and Space Physics*, **14**(4): 591–608.

Granser, H., Meurers, B. and Steinhauser, P. (1989) Apparent density mapping and 3D gravity inversion in the eastern Alps. *Geophysical Prospecting*, **37**(3): 279–292.

Grant, F.S. and West, G.F. (1965) *Interpretation Theory in Applied Geophysics*. New York: McGraw-Hill.

Green, R. (1960) Remanent magnetization and the interpretation of magnetic anomalies. *Geophysical Prospecting*, **8**(1): 98–110.

Green, W.B. (1983) *Digital Image Processing*. New York: Van Nostrand-Reinhold.

Greenhalgh, S.A. and Whiteley, R.J. (1977) Effective application of the seismic refraction method to highway engineering projects. *Australian Road Research*, **7**(1): 3–19.

Greenhouse, J., Brewster, M., Schneider, G., Redman, D., Annan, P., Olhoeft, G., Lucius, J., Sander, K. and Mazzella, A. (1993) Geophysics and solvents: the Borden experiment. *Geophysics: The Leading Edge*, **12**(4): 261–267.

Greenhouse, J.P. (1991) Environmental geophysics: it's about time. *Geophysics: The Leading Edge*, **10**(1): 32–34.

Greenhouse, J.P. and Slaine, D.D. (1983) The use of reconnaissance electromagnetic methods to map contaminant migration. *Ground Water Monitoring Review*, **3**(2): 47–59.

Griffiths, D.H. and King, R.F. (1981) *Applied Geophysics for Geologists and Engineers*. Oxford: Pergamon.

Griffths, D.H., Turnbull, J. and Olayinka, A.I. (1990) Two-dimensional resistivity mapping with a computer-controlled array. *First Break*, **8**(4): 121–129.

Grow, L.L. (1982) Induced polarization for geophysical exploration. *Geophysics: The Leading Edge of Exploration*, **1**(1): 55–56; 69–70.

Gubbins, D. (1990) *Seismology and Plate Tectonics*. Cambridge: Cambridge University Press.

Gupta, V.K. and Fitzpatrick, M.M. (1971) Evaluation of terrain effects in ground magnetic surveys. *Geophysics*, **36**(3): 582–589.

Habberjam. G.M. and Watkins, G.E. (1967a) The reduction of lateral effects in resistivity probing. *Geophysical Prospecting*, **15**(2): 221–235.

Haeberli, W. and Fisch, W. (1984) Electrical resistivity soundings of glacier beds: a test study on Grubengletscher, Wallis, Swiss Alps. *Journal of Glaciology*, **30**(106): 373–376.

Hagedoorn, J.G. (1959) The plus–minus method of interpreting seismic refraction sections. *Geophysical Prospecting*, **7**(2): 158–182.

Hahn, A., Kind, E.G. and Mishra, D.G. (1976) Depth estimation of magnetic sources by means of Fourier amplitude spectra. *Geophysical Prospecting*, **24**(2): 287–308.

Hale, D.I., Hill, N.R. and Stefani, J.P. (1991) Imaging salt with turning seismic waves. *Expanded Abstracts of 61st Annual International Meeting of SEG*, 1171–1174.

Hales, F.W. (1958) An accurate graphical method for interpreting seismic refraction lines. *Geophysical Prospecting*, **6**(3): 285–314.

Hall, D.H. and Hajnal, J. (1962) The gravimeter in studies of buried valleys. *Geophysics*, **27**(6, part 2): 939–951.

Hallof, P.G. (1967) *An Appraisal of the Variable Frequency IP Method after Twelve Years of Application*. Markham, Ontario: Phoenix Geophysics Ltd.

Hallof, P.G. (1974) The IP phase measurement and inductive coupling. *Geophysics*, **39**(5): 650–665.

Hallof, P.G. (1982) Reconnaissance and detailed geophysical results, Granite Mountain Area, Pershing County, Nevada. *Global Tectonics and Metallogeny*, **1**(4): 374–400.

Hallof, P.G. (1983) *An Introduction to the Use of the Spectral Induced Polarization Method*. Markham, Ontario: Phoenix Geophysics Ltd.

Hallof, P.G. and Klein, J.D. (1982) *Electrical Parameters of Volcanogenic Mineral Deposits in Ontario*. Markham, Ontario: Phoenix Geophysics Ltd.

Hallof, P.G. and Pelton, W.H. (1980) The removal of inductive coupling effects from spectral IP data. Presented at the SEG 50th Annual International Meeting, Houston, Texas, 16–20 November 1980 (Phoenix Geophysics Ltd, Ontario).

Hallof, P.G. and Pelton, W.H. (1981) *Recent and Future Advances in the Induced Polarization Method*. Markham, Ontario: Phoenix Geophysics Ltd.

Halpenny, J.F. and Darbha, D.M. (1995) Airborne gravity tests over Lake Ontario. *Geophysics*, **60**(1): 61–65.

Hammer, S. (1939) Terrain corrections for gravimeter stations. *Geophysics*, **4**(3): 184–194.

Hammer, S. (1945) Estimating ore masses in gravity prospecting. *Geophysics*, **10**(1): 50–62.

Hammer, S. (1963) Deep gravity interpretation by stripping. *Geophysics*, **28**(3): 369–378.

Hammer, S. (1982) Airborne gravity is here! *Oil & Gas Journal*, 11 January.

Hammer, S. (1984) Discussion on 'Airborne gravity is here!'. *Geophysics*, **49**(4): 471–472.

Hammond, W.R. and Sprenke, K.F. (1991) Radar detection of subglacial sulfides. *Geophysics*, **56**(6): 870–873.

Hansen, T., Kingston, J., Kjellesvik, S., Lane, G., l'Anson, K., Naylor, R. and Walker, C. (1989) 3-D Seismic surveys. *Oilfield Review*, **1**(3): 54–61.

Hardage, B.A. (1985) *Vertical Seismic Profiling. Part A: Principles*, 2nd (enlarged) edn. London: Geophysical Press.

Hatherley, P.J., Urosevic, M., Lamborne, A. and Evans, B.J. (1994) A simple approach to calculating refraction statics corrections. *Geophysics*, **59**(1): 156–160.

Hatton, L., Worthington, M.H. and Makin, J. (1986) *Seismic Data Processing: Theory and Practice*. Oxford: Blackwell Scientific.

Hawkins, A.B. (ed.) (1986) *Site Investigation Practice: Assessing BS 5930*, Geological Society Engineering Geology Special Publication no. 2.

Hayford, J.F. and Bowie, W. (1912) *US Coast and Geod. Survey*, Special Publication 10.

Haynes, R., Davis, A.M., Reynolds, J.M. and Taylor, D.I. (1993) The extraction of geotechnical information from high-resolution seismic reflection data. *Offshore Site Investigation and Foundation Behaviour*, **28**: 215–228, Society for Underwater Technology.

Heikes, R.R. and Ure, R.W. (1961) *Thermoelectricity: Science and Engineering*. New York: Interscience.

Heiskanen, W.A. (1938) Publ. Isostat. Inst., Int. Assoc. Geod., no. 1, Helsinki.

Henriet, J.P., Verschuren, M. and Versteeg, W. (1992) Very-high-resolution 3D seismic reflection imaging of small-scale structural deformation. *First Break*, **10**(3): 81–88.

Hermes, H.J. (1986) Calculation of pre-Zechstein Bouguer Anomaly in North-West Germany, *First Break*, **4**(11): 13–22.

Herrod, L.D.B. and Garrett, S.W. (1986) Geophysical fieldwork on the Ronne Ice Shelf, Antarctica. *First Break*, **4**(1): 9–14.

Hill, N.R. (1987) Downward continuation of refracted arrivals to determine shallow structure. *Geophysics*, **52**(9): 1188–1198.

Hinz, W.J. (ed.) (1985) *The Utility of Regional Gravity and Magnetic Anomaly Maps*. Tulsa: Society of Exploration Geophysicists.

Hird, G.A., Karwatowski, J., Jenkerson, M.R. and Eyres, A. (1993) 3D concentric circle survey: the art of going in circles. *Extended Abstracts 55th Meeting EAEG Stavanger Norway*, Paper A001.

Hoekstra, P. and Blohm, M.W. (1990) Case histories of time domain electromagnetic sounding in environmental geophysics. In: Ward, S.H. (ed.), *Geotechnical and Environmental Geophysics. Vol. 2: Environmental and Groundwater*. Tulsa: Society of Exploration Geophysicists, 1–15.

Hoekstra, P., Lahti, R., Hild, J., Bates, C.R. and Phillips, D. (1992) Case histories of shallow time domain electromagnetics in environmental site assessment. *Ground Water Monitoring Review*, Fall, 110–117.

Hohmann, G.W. (1973) Electromagnetic coupling between grounded wires at the surface of a two-layered earth. *Geophysics*, **38**(5): 854–863.

Hohmann, G.W. (1975) Three-dimensional induced-polarization and electromagnetic modeling. *Geophysics*, **40**(2): 309–324 (and discussion in *Geophysics*, **50**(11): 2279).

Hohmann, G.W. (1987) Numerical modeling for electromagnetic methods of geophysics. In: Nabighian, M.N. (ed.), *Electromagnetic Methods in Applied Geophysics*, Vol. 1. Tulsa: Society of Exploration Geophysicists, 313–363.

Hongisto, H. (1993) Self-potential interpretation using a surface polarization model. *Extended Abstracts of the 55th Meeting and Technical Exhibition, European Association of Exploration Geophysicists, Stavanger, Norway, 7–11 June 1993*, Paper D039.

Hood, P.J. (1981) Aeromagnetic gradiometry – a superior geological mapping tool for mineral exploration programs. In: Weinstock, H. and Overton, W.C. (eds), *SQUID Applications to geophysics*. Tulsa: Society of Exploration Geophysicists, 72–76.

Hood, P.J., Holroyd, M.T. and McGrath, P.H. (1979) Magnetic methods applied to base metal exploration. In: Hood, P.J. (ed.), *Geophysics and Geochemistry in the Search for Metallic Ores*, Geological Survey of Canada Economic Geology Report no. 31, 77–104.

Hovland, M. and Judd, A.G. (1988) *Seabed Pockmarks and Seepages*. London: Graham & Trotman.

Howell, B.F. (1990) *An Introduction to Seismological Research, History and Development*. Cambridge: Cambridge University Press.

Hülsenbeck & Co. (1926) German patent no. 489434.

Hülsmeyer, C. (1904) German patent no. 165546.

Huang, H. and Palacky, G.J. (1991) Damped least-squares inversion of time domain airborne EM data based on singular value decomposition. *Geophysical Prospecting*, **39**(6): 827–844.

Hubral, P., Schleicher, J., Tygel, M. and Hanitzch, Ch. (1993) Determination of Fresnel zones from travel time information. *Geophysics*, **58**(5): 703–712.

Huggenberger, P., Meier, E. and Pugin, A. (1994) Ground-probing radar as a tool for heterogeneity estimation in gravel deposits: advances in data processing and facies analysis. *Journal of Applied Geophysics*, **31**(1–4): 171–184.

Hunt, S. (1992) The use of cone penetration testing and sampling techniques in the investigation of contaminated land. In: Forde, M.C. (ed.), *Second International Conference on Polluted and Marginal Land*, 30 June–2 July, Brunel University, London. Engineering Technics Press.

Hunt, T.M. (1977) Recharge of water in Wairakei Geothermal field determined from repeat gravity measurements. *New Zealand Journal of Geology and Geophysics*, **20**(2): 303–317.

Hussein, A. (1983) Underground gravity surveys. In: Fitch, A.A. (ed.), *Developments in Geophysical Exploration Methods*, Vol. 5. Applied Science Publishers, 35–63.

IAGA Division I Working Group (1987) The International Geomagnetic Reference Field revision 1987. *Journal of Geomagnetism and Geoelectricity*, **39**: 773–779.

Imai, T., Sakayama, T. and Kanemori, T. (1987) Use of ground probing radar and resistivity surveys for archeological investigations. *Geophysics*, **52**(2): 137–150.

Ingham, M. (1992) Audiomagnetotelluric soundings in White Island volcano. *Journal of Volcanology & Geothermal Research*, **50**(3): 301–306.

Ivanov, A.G. (1939) Effect of electrization (sic) of earth layers by elastic waves passing through them. *Comptes Rendus (Doklady) de l'Académie des Sciences de l'URSS*, **24**(1): 42–45.

Jacobs, J.A. (1992) *Deep Interior of the Earth*. London: Chapman & Hall.

James, D.E. (1990) *The Encyclopedia of Solid Earth Geophysics*. New York: Van Nostrand Reinhold.

Jansen, J., Billington, E., Snider, F. and Jurcek, P. (1994) Marine SP surveys for dam seepage investigations: evaluation of array geometries through modeling and field trials. *Proceedings of the SAGEEP '94, 27–31 March, Boston, USA*. Englefield, USA: EEGS, 1053–1071.

Johnson, R.W., Glaccum, R. and Wojtasinski, R. (1979) Application of ground penetrating radar to soil survey. *Soil and Crop Science Society of Florida Proceedings*, **39**, 2–4 October: 68–72.

Jones, T.D. (1986) Pore fluids and frequency-dependent wave propagation in rocks. *Geophysics*, **51**(10): 1939–1953.

Kane, M.F. (1962) A comprehensive system of terrain corrections using a digital computer. *Geophysics*, **27**(4): 455–462.

Kawasaki, K., Osterkamp, T.E., Jurick, R.W. and Kienle, J. (1983) Gravity measurements in permafrost terrain containing massive ground ice. *Annals of Glaciology*, **4**: 133–140.

Kearey, P. and Brooks, M. (1991) *An Introduction to Geophysical Exploration*, 2nd edn. Oxford: Blackwell Scientific.

Kearey, P. and Vine, F.J. (1990) *Global Tectonics*. Oxford: Blackwell Scientific.

Keller, G.V. and Frischknecht, F.C. (1966) *Electrical methods in geophysical prospecting*. Pergamon Press.

Kellett, R., Bishop, J. and Van Reed, E. (1994) The effect of source polarization in CSAMT data over two massive sulfide deposits in Australia. *Geophysics*, **58**(12): 1764–1772.

Kennett, M., Laumann, T. and Lund, C. (1993) Helicopter-borne radio-echo sounding of Svartisen, Norway. *Annals of Glaciology*, **17**: 23–26.

Keppner, G. (1991) Ludger Mintrop, *Geophysics: The Leading Edge*, **10**(9): 21–28.

Ketelaar, A.C.R. (1976) A system for computer-calculation of the terrain correction in gravity surveying. *Geoexploration*, **14**: 57–65.

Ketola, M. (1979) On the application of geophysics in the indirect exploration for copper sulphide ores in Finland. In: Hood, P.J. (ed.), *Geophysics and Geochemistry in the Search for Metallic Ores*, Geological Survey of Canada, Economic Geology Report 31, 665–684.

Kick, J.F. (1985) Depth to bedrock using gravimetry. *Geophysics: The Leading Edge of Exploration*, **4**(4): 38–42.

Kilty, K.T. (1984) On the origin and interpretation of self potential anomalies. *Geophysical Prospecting*, **32**(1): 51–62.

King, E.C. and Jarvis, E.P. (1992) Short-offset seismic refraction results near Rothera Station, Antarctic Peninsula. *Antarctic Science*, **4**(4): 479–480.

King, R.W.P. and Smith, G.S. (1981) *Antennas in Matter*. New York: MIT Press.

King, W.C., Witten, A.J. and Reed, G.D. (1989) Detection and imaging of buried wastes using seismic wave propagation. *Journal of Environmental Engineering*, **115**(3): 527–540.

Kirchner, J.F. and Bentley, C.R. (1979) Seismic short-refraction studies on the Ross Ice Shelf, Antarctica. *Journal of Glaciology*, **24**(90): 313–319.

Klimentos, T. (1991) The effects of porosity-permeability-clay content on the velocity of compressional waves. *Geophysics*, **56**(12): 1930–1939.

Klimentos, T. and McCann, C. (1990) Relationships among compressional wave attenuation, porosity, clay content, and permeability in sandstones. *Geophysics*, **55**(8): 998–1014.

Knapp, R.W. (1991) Fresnel zones in the light of broadband data. *Geophysics*, **56**(3): 354–359.

Knight, M.J., Leanard, J.B. and Whiteley, R.J. (1978) Lucas Heights solid waste landfill and downstream leachate transport: a case study in environmental geology. *Bulletin of the International Association of Engineering Geology*, **18**: 45–64.

Koefoed, O. (1979) *Geosounding Principles*. Amsterdam: Elsevier.

Kohnen, H. (1974) The temperature dependence of seismic waves in ice. *Journal of Glaciology*, **13**(6): 144–147.

Korvin, G. (1982) *Geoexploration*, **19**: 267–276.

Kreitz, E. (1982) Seismic evaluation of the Mors salt dome. In: *Results of Geological Investigations for High-Level Waste Disposal in the Mors Salt Dome*, Proceedings of a Symposium, 18–19 Sept. 1981, Copenhagen, **1**: 74–98.

LaCoste, L.J.B. (1934) A new type of long period vertical seismograph. *Physics*, **5**: 178–180.

Ladwig, K.J. (1983) Electromagnetic induction methods for monitoring acid mine drainage. *Ground Water Monitoring Review*, Winter, 46–51.

LaFehr, T.R. (1982) Evaluation of surface and borehole gravity measurements at the Mors salt dome. In: *Results of Geological Investigations for High-Level Waste Disposal in the Mors Salt Dome*, Proceedings of a Symposium, 18–19 Sept. 1981, Copenhagen, **1**: 196–232.

Lakshmanan, J. (1991) The generalized gravity anomaly: endoscopic microgravity. *Geophysics*, **56**(5): 712–723.

Langhammer, J. and Landrø, M. (1993) Temperature effects on airgun signatures. *Geophysical Prospecting*, **41**(6): 737–750.

Lankston, R.W. (1990) High-resolution refraction data acquisition and interpretation. In: Ward, S.H. (ed.), *Geotechnical and Environmental Geophysics*. Vol. I: Review and Tutorial. Tulsa: Society of Exploration Geophysicists, 45–73.

Lankston, R.W. and Lankston, M.M. (1986) Obtaining multilayer reciprocal times through phantoming. *Geophysics*, **51**(1): 45–49.

Lee and Schwartz (1930) *Resistivity of Oil-Bearing Beds*. US Bureau of Mines Technical Paper 488.

Leech, C. and Johnson, R.M. (1992) Locating buried drums using a proton precession magnetometer and magnetic gradiometer. In: Forde, M.C. (ed.), *Proceedings of the Second International Conference on Construction on Polluted and Marginal Land, 30 June–2 July 1992*. Brunel University, 37–49.

Leggo, P.J. (1982) Geological applications of ground impulse radar. *Transactions of the Institute of Mining & Metallurgy; B: Applied Earth Sciences*, **91**: B1–5.

Leggo, P.J. and Leech, C. (1983) Subsurface investigations for shallow mine workings and cavities by the ground impulse radar technique. *Ground Engineering*, **16**: 20–23.

Leimbach, G. and Löwy, H. (1910) German patent no. 237944.

Lennox, D.H. and Carlson, V. (1967) Geophysical exploration for buried valleys in an area north of Two Hills, Alberta. *Geophysics*, **32**(2): 331–362.

Lines, L. and 18 co-authors (1993) Integrated reservoir characterization: beyond tomography. *Geophysics*, **60**(2): 354–364.

Logn, O. and Bölviken, B. (1974) Self potentials at the Joma pyrite deposit, Norway. *Geoexploration*, **12**: 11–28.

Lowrie, W. (1990) Magnetic analysis of rock fabric. In: James, D.E. (ed.), *The Encyclopedia of Solid Earth Geophysics*. New York: Van Nostrand Reinhold, 698–706.

Lugg, R. (1979) Marine seismic sources. In: Fitch, A.A. (ed.), *Developments in Geophysical Exploration Methods*, Vol. 1. London: Applied Science Publishers, 143–203.

MacGregor, F., Fell, R., Mostyn, G.R., Hocking, G. and McNally, G. (1994) The estimation of rock rippability. *Quarterly Journal of Engineering Geology*, **27**: 123–144.

Mackay, J.R. (1962) Pingos of the Pleistocene McKenzie delta area. *Geographical Bulletin*, **18**: 21–63.

MacLeod, I.N., Jones, J. and Dai, T.F. (1993) 3-D analytic signals in the interpretation of total magnetic field data at low magnitude latitudes. *Exploration Geophysics*, 679–688.

Madden, T.R. and Vozoff, K. (1971) *VLF Model Suite*, 2nd edn. 17 Winthrop Road, Lexington, MA 02173, USA.

Maillet, R. (1947) The fundamental equations of electrical prospecting. *Geophysics*, **12**(4): 529–556.

Maio, X-G., Moon, W.M. and Milkereit, B. (1995) A multioffset, three-component VSP study in the Sudbury Plain. *Geophysics*, **60**(2): 341–353.

Marsden, D. (1993a) Static corrections: a review. Part I. *Geophysics: The Leading Edge*, **12**(1): 43–49.

Marsden, D. (1993b) Static corrections: a review. Part II. *Geophysics: The Leading Edge*, **12**(2): 115–120.

Marsden, D. (1993c) Static corrections: a review. Part III. *Geophysics: The Leading Edge*, **12**(3): 210–216.

Marsh, G.R. (1991) Geophysics and environmentalism. *Geophysics: The Leading Edge*, **10**(8): 32–33.

Marshall, D.J. and Madden, T.R. (1959) Induced polarization: a study of its causes. *Geophysics*, **24**(4): 790–816.

Masero, W., Schnegg, P.-A. and Fontes, S.L. (1994) A magneto-telluric investigation of the Araguainha impact structure in Mato Grosso-Goiás, central Brazil, *Geophysical Journal International*, **116**(2): 366–376.

Masuda, H. (1981) Seismic refraction analysis for engineering study. *OYO Technical Note 10*, OYO Corporation.

Mazác, O., Benes, L., Landa, I, and Skuthan, B. (1990) Geoelectrical detection of sealing foil quality in light-ash dumps. In: Ward, S.H. (ed.), *Geotechnical and Environmental Geophysics. Vol. 2: Environmental and Groundwater.* Tulsa: Society of Exploration Geophysicists, 113–119.

McCann, C. and Till, R. (1974) Demonstration of the use of an on-line terminal for teaching geophysical data processing techniques. *International Journal of Mathematics in Education Science & Technology*, **5**: 729–737.

McCann, D.M., Jackson, P.D. and Fenning, P.J. (1988) Comparison of the seismic and ground-probing radar methods in geological surveying. *IEE Proceedings*, **135**(F, 4): 380–390.

McDowell, P.W. (1975) Detection of clay filled sink-holes in the chalk by geophysical methods. *Quarterly Journal of Engineering Geology*, **8**: 303–310.

McElhinny, M.W. (1973) *Palaeomagnetism and Plate Tectonics.* Cambridge: Cambridge University Press.

McGee, T.M. (1990) The use of marine seismic profiling for environmental assessment. *Geophysical Prospecting*, **38**(8): 861–880.

McGinnis, L.D. and Jensen, T.E. (1971) Permafrost-hydrogeologic regimen in two ice-free valleys, Antartica, from electrical depth sounding. *Quaternary Research*, **1**(3): 389–409.

McNeill, J.D., Bosner, M. and Snelgrove, F.B. (1988) A borehole induction logger for monitoring groundwater contamination. *Geonics Technical Note TN-25*, Geonics Ltd, Mississauga, Canada.

McNeill, J.D. (1980) *Electromagnetic Terrain Conductivity Measurement at Low Induction Numbers.* Technical Note TN-6, Geonics Ltd, Mississauga, Ontario.

McNeill, J.D. (1985) *EM34-3 Measurements at Two Intercoil Spacings to Reduce Sensitivity to Near-Surface Material.* Technical Note TN-19, Geonics Ltd, Mississauga, Ontario.

McNeill, J.D. (1990) Use of electromagnetic methods for groundwater studies. In: Ward, S.H. (ed.), *Geotechnical and Environmental Geophysics. Vol. 1: Review and Tutorial.* Tulsa: Society of Exploration Geophysicists, 191–218.

McNeill, J.D. and Labson, V.F. (1991) Geological mapping using VLF radiofields. In: Nabighian, M.C. (ed.), *Electromagnetic Methods in Applied Geophysics. Vol. 2: Applications, Part B.* Tulsa: Society of Exploration Geophysicists, 521–640.

McQuillin, R.M., Bacon, M. and Barclay, W. (1984) *An Introduction to Seismic Interpretation.* London: Graham & Trotman.

Meigh, A.C. (1987) *Cone Penetration Testing: Methods and Interpretation* (CIRIA ground engineering report: *in situ* testing), CIRIA. London: Butterworths.

Meinardus, H., Schleicher, K. and Sudhakar, V. (1993) A processing sequence for turning wave imaging, *Extended Abstracts of 62nd Annual International Meeting, Society of Exploration Geophysicists, New Orleans, Louisiana, USA, October 1992.*

Meiser, P. (1962) A method for quantitative interpretation of self potential measurements. *Geophysical Prospecting,* **10**(2): 203–218.

Mellet, J.S. (1993) Bathymetric studies of ponds and lakes using ground penetrating radar. *Advanced Ground Penetrating Radar: Technologies and Applications, 26–28 Oct. 1993.* Ohio State University, Columbus, Ohio, USA, 257–266.

Merrill, R.T. (1990) Rock magnetism. In: James, D.E. (ed.), *The Encyclopedia of Solid Earth Geophysics.* New York: Van Nostrand Reinhold, 946–950.

Meyer, de Stadelhofen, C. and Julliard, T. (1987) Truth is an instance of error temporarily appealing. *Geophysics: The Leading Edge of Exploration,* **6**(2): 38.

Miller, R.D., Anderson, N.L., Feldman, H.R. and Franseen, E.K. (1995) Vertical resolution of a seismic survey in stratigraphic sequences less than 100 m deep in southeastern Kanas. *Geophysics,* **60**(2): 423–430.

Miller, R.D., Pullan, S.E., Steeples, D.W. and Hunter, J.A. (1992) Field comparison of shallow seismic sources near Chino, California. *Geophysics,* **57**(5): 693–709.

Miller, R.D., Pullan, S.E., Steeples, D.W. and Hunter, J.A. (1994) Field comparison of shallow P-wave seismic sources near Houston, Texas. *Geophysics,* **59**(11): 1713–1728.

Miller, R.D., Pullan, S.E., Waldner, J.S. and Haeni, F.P. (1986) Field comparison of shallow seismic sources. *Geophysics,* **51**(11): 2067–2092.

Mills, T., Hoekstra, P., Blohm, M. and Evans, L. (1988) Time-domain electromagnetic soundings for mapping seawater intrusion in Monterey County, CA. *Ground Water,* **26**: 771–782.

Milsom, J. (1989) *Field Geophysics.* Milton Keynes: Open University Press.

Monier-Williams, M.E., Greenhouse, J.P., Mendes, J.M. and Ellert, N. (1990) Terrain conductivity mapping with topographic corrections at three waste disposal sites in Brazil. In: Ward, S.H. (ed.), *Geotechnical and Environmental Geophysics. Vol. 2: Environmental and Groundwater.* Tulsa: Society of Exploration Geophysicists, 41–55.

Mooney, H.M. and Wetzel, W.W. (1956) *The Potentials about a Point Electrode.* Minnesota: University of Minnesota Press.

Moorman, B.J., Judge, A.S. and Smith, D.G. (1991) Examining fluvial sediments using ground penetrating radar in British Columbia. *Current Research: Part A.* Geological Survey of Canada, Paper 91-1A, 31–36.

Morelli, C. (1971) *The International Gravity Standard Net 1971* (IGSN 71). Special Publication no. 4, International Association of Geodesy, Bureau Central de l'Association Internationale de Géodésie, 19 Rue Auber, 75009, Paris.

Morey, M. (1974) Continuous sub-surface profiling by impulse radar. In: *Subsurface Exploration for Underground Excavation and Heavy Construction,* ASCE Specialty Conference, Henneker, 213–232.

Morris, E. (1992) *Geophysical Lab Manual for Lotus 1-2-3 and Quattro Pro.* Regina, Canada: EcoTech Research Ltd.

Nabighian, M.N. (1984) Toward a three-dimensional automatic interpretation of potential field data via generalized Hilbert transforms: fundamental relations. *Geophysics*, **49**(6): 780–786.

Nabighian, M.N. (1987) *Electromagnetic Methods in Applied Geophysics*, Vol. 1. Tulsa, Society of Exploration Geophysicists.

Nabighian, M.N. and Macnae, J.C. (1991) Time domain electromagnetic prospecting methods. In: Nabighian, M.N. (ed.), *Electromagnetic Methods in Applied Geophysics*, Vol. 2A. Tulsa: Society of Exploration Geophysicists, 427–520.

Nafe, J.E. and Drake, C.L. (1963) Physical properties of marine sediments. In: Hill, M.N. (ed.), *The Sea*, Vol. 3. New York: Interscience Publishers, 794–815.

Nagata, T. (1961) *Rock Magnetism*, 2nd edn. Tokyo: Maruzen.

Narod, B.B. and Clarke, G.K.C. (1994) Miniature high-power impulse transmitter for radio-echo sounding. *Journal of Glaciology*, **40**(134): 190–194.

Nayak, P.N. (1981) Electromechanical potential in surveys for sulphides. *Geoexploration*, **18**: 311–320.

Nelson, R.G. and Haigh, J.H. (1990) Geophysical investigations of sinkholes in lateritic terrains. In: Ward, S.H. (ed.), *Geotechnical and Environmental Geophysics. Vol. 3: Geotechnical*. Tulsa: Society of Exploration Geophysicists, 133–153.

Nestvold, E.O. (1992) 3-D seismics: is the promise fulfilled? *Geophysics: The Leading Edge*, **11**(6): 12–19.

Nettleton, L.L. (1939) Determination of density for reduction of gravimeter observations. *Geophysics*, **4**(3): 176–183.

Nettleton, L.L. (1940) *Geophysical Prospecting for Oil*. New York: McGraw-Hill.

Nettleton, L.L. (1942) Gravity and magnetic calculations. *Geophysics*, **7**(3): 293–310.

Nettleton, L.L. (1954) Regionals, residuals and structures. *Geophysics*, **19**(1): 1–22.

Nettleton, L.L. (1971) *Elementary Gravity and Magnetics for Geologists and Seismologists*. Tulsa: Society of Exploration Geophysicists, Monograph Series no. 1.

Nettleton, L.L. (1976) *Gravity and Magnetics in Oil Exploration*. New York: McGraw-Hill.

New, B.M. (1985) An example of tomographic and Fourier microcomputer processing of seismic records. *Quarterly Journal of Engineering Geology* **18**(4): 335–344.

Newman, G.A. (1994) A study of downhole electromagnetic sources for mapping enhanced oil recovery processes. *Geophysics*, **59**(4): 534–545.

Nickel, H., Sender, F., Thierbach, R. and Weichart, H. (1983) Exploring the interior of salt domes from boreholes. *Geophysical Prospecting*, **31**(1): 131–148.

Nilsson, B. (1983) A new borehole radar system. In: *Borehole Geophysics for mining and Geotechnical Applications*. Geological Survey of Canada Paper 85-27, Ottawa, Canada.

Nissen, J. (1986) A versatile electromagnetic modelling program for 2D structures. *Geophysical Prospecting*, **34**(7): 1099–1110.

Noel, M. and Walker, R. (1990) Development of an electrical resistivity tomography system for imaging archaeological structures. In: Pernicka, E. and Wagner, G.A. (eds), *Archaeometry '90*. Birkhauser, Basel, pp. 767–776.

Noel, M. and Xu, B. (1991) Archaeological investigation by electric resistivity tomography: a preliminary study. *Geophysical Journal International*, **107**: 95–102.

Nostrand, van, R.G. and Cook, K.L. (1966) USGS Professional Paper no. 499.

Nunn, K.R. (1978) Geophysical surveys at the landfill sites in the West Midlands. In: *Proceedings of a Symposium on the Engineering Behaviour of Industrial and Urban Fill, University of Birmingham, April 1978*. Midland Geotechnical Society, C45–C53.

O'Neill, D.J. and Merrick, N.P. (1984) A digital linear filter for resistivity sounding with a generalized electrode array. *Geophysical Prospecting*, **32**(1): 105–123.

O'Reilly, W. (1984) *Rock and Mineral Magnetism*. Glasgow: Blackie.

Odins, J. (1975) *The Application of Seismic Refraction to Groundwater Studies of Unconsolidated Sediments*. Unpublished MSc thesis, University of New South Wales, Australia.

Ogilvy, A.A. and Kuzmina, E.N. (1972) Hydrogeologic and engineering geologic possibilities for employing the method of induced potentials. *Geophysics*, **37**(5): 839–861.

Ogilvy, A.A., Ayed, M.A. and Bogoslovsky, V.A. (1969) Geophysical studies of water leakages from reservoirs. *Geophysical Prospecting*, **17**(1): 36–62.

Ogilvy, R.D., and Lee, A.C. (1989) *KHFILTER: A Computer Program to Generate Current Density Pseudosections from VLF-EM In-Phase Data*. British Geological Survey Technical Report WK/89/6R.

Ogilvy, R.D., Cuadra, A., Jackson, P.D. and Monte, J.L. (1991) Detection of an air-filled drainage gallery by the VLF resistivity method. *Geophysical Prospecting*, **39**(6): 845–859.

Olhoeft, G.R. (1985) Low-frequency electrical properties. *Geophysics*, **50**(12): 2492–2503.

Olsson, O., Sandberg, E. and Nilsson, B. (1983) *The Use of Borehole Radar for the Detection of Fractures in Crystalline Rock*. Stripa Project report IR-83-06, SKB, Stockholm, Sweden.

Olsson, O., Falk, L., Forslund, O., Lundmark, L. and Sandberg, E. (1990) *Crosshole Investigations: Results from Borehole Radar Investigations*. Stripa Project technical report 87-11, Stockholm, Sweden.

Oppliger, G.L. (1984) Three-dimensional terrain corrections for mise-à-la-masse and magnetic resistivity surveys. *Geophysics*, **49**(10): 1718–1729.

Osterkamp, T.E. and Jurick, R.W. (1980) Detecting massive ground ice in permafrost by geophysical methods. *Northern Engineer*, **12**(4): 27–30.

Ostrander, W.J. (1984) Plane-wave reflection coefficients for gas sands at non-normal angles of incidence. *Geophysics*, **49**(10): 1637–1648.

Overmeeren, van, R.A. (1980) Tracing by gravity of a narrow buried graben structure, detected by seismic refraction, for ground-water investigations in North Chile. *Geophysical Prospecting*, **28**(3): 392–407.

Overmeeren, van, R.A. (1987) The plus–minus method for rapid field processing by portable computer of seismic refraction data in multilayer groundwater studies. *First Break*, **5**(3): 83–94.

Overmeeren, van, R.A. (1994) Georadar for hydrogeology. *First Break*, **12**(8): 401–408.

Overmeeren, van, R.A. and Ritsema, I.L. (1988) Continuous vertical electrical sounding. *First Break*, **6**(10): 313–324.

Owen, T.E. (1983) Detection and mapping of tunnels and caves. In: Fitch, A.A. (ed.), *Developments in Geophysical Exploration Methods*, Vol. 5. Applied Science Publishers: 161–258.

Palacky, G.J. (ed.) (1986) Airborne resistivity mapping. *Geol. Surv. Can. Paper* **86-22**.

Palacky, G.J. (1991) Application of the multi-frequency horizontal loop EM method in overburden investigations. *Geophysical Prospecting*, **39**(8): 1061–1082.

Palacky, G.J. and Stephens, L.E. (1992) Detection of sub-bottom ice-bonded permafrost on the Canadian Beaufort Shelf by ground electromagnetic measurements. *Geophysics*, **57**(11): 1419–1427.

Palacky, G.J. and West, G.F. (1991) Airborne electromagnetic methods. In: Nabighian, M.N. (ed.), *Electromagnetic Methods in Applied Geophysics*, Vol. 2B. Tulsa: Society of Exploration Geophysicists, 811–879.

Palmer, D. (1980) *The Generalised Reciprocal Method of Seismic Refraction Interpretation.* Tulsa: Society of Exploration Geophysicists.

Palmer, D. (1991) The resolution of narrow low-velocity zones with the generalised reciprocal method. *Geophysical Prospecting*, **39**(8): 1031–1060.

Pantze, R., Malmqvist, M. and Kristensson, G. (1986) Directional EM measurements in boreholes. In: Killeen, P.G. (ed.), Borehole geophysics for mining and geotechnical applications. *Geological Surv. Can. Paper* **85-27**: 79–88.

Parasnis, D.S. (1966) *Mining Geophysics.* Amsterdam: Elsevier.

Parasnis, D.S. (1967) Three-dimensional electric mise-à-la-masse survey of an irregular lead–zinc–copper deposit in Central Sweden. *Geophysical Prospecting*, **15**(3): 407–437.

Parasnis, D.S. (1973) *Mining Geophysics.* Amsterdam: Elsevier.

Parasnis, D.S. (1986) *Principles of Applied Geophysics*, 4th edn. London: Chapman & Hall.

Parasnis, D.S. (1991) Large-layout harmonic field systems: In: Nabighian, M.N. (ed.), *Electromagnetic Methods in Applied Geophysics*, Vol. 2A. Tulsa: Society of Exploration Geophysicists, 271–283.

Parker, R.L. and Zumberge (1989) An analysis of geophysical experiments to test Newton's Law of Gravity. *Nature*, **342**(6245): 29–32.

Parlowski, J., Lewis, R., Dobush, T. and Valleau, N. (1995) An integrated approach for measuring and processing geophysical data for the detection of unexploded ordnance. *Proceedings of SAGEEP-95*, Orlando, USA, 965–974.

Parra, J.O. (1984) Effects of pipelines on spectral induced-polarization surveys. *Geophysics*, **49**(11): 1979–1992.

Patella, D. and Schiavone, D. (1977) Comparative analysis of time-domain and frequency-domain in the induced-polarization prospecting method. *Geophysical Prospecting*, **25**(3): 496–511.

Paterson, N.R. and Ronka, V. (1971) Five years of surveying with the very low frequency electromagnetic method. *Geoexploration*, **9**: 7–26.

Peddie, N.W. (1982) International geomagnetic reference field: the third generation. *Journal of Geomagnetism and Geoelectricity*, **34**: 309–326.

Pelton, W.H. and Smith, P.K. (1976) Mapping porphyry copper deposits in the Philippines with IP. *Geophysics*, **41**(1): 106–122.

Pelton, W.H., Ward, S.H., Hallof, P.G., Sill, W.R. and Nelson, P.H. (1978) Mineral discrimination and removal of inductive coupling with multi-frequency induced polarization. *Geophysics*, **43**(3): 588–609.

Pelton, W.H., Sill, W.R. and Smith, B.D. (1983) Interpretation of complex resistivity and dielectric data, Part I. *Geophysical Transactions*, **29**(3): 297–330.

Pemberton, R.H. (1962) Airborne electromagnetics in review. *Geophysics*, **27**(5): 691–713.

Pendick, D. (1995) 'And here is the eruption forecast...'. *New Scientist*, **145**(1959): 26–29.

Peters, L.J. (1949) The direct approach to magnetic interpretation and its practical application. *Geophysics*, **14**(3): 290–320.

Petersen, N. (1990) Curie temperature. In: James, D.E. (ed.), *The Encyclopedia of Solid Earth Geophysics.* New York: Van Nostrand Reinhold 166–173.

Petrovski, A. (1928) The problem of a hidden polarized sphere. *Philosophical Magazine*, **5**: 334–358.

Phadke, S. and Kanasewich, E.R. (1990) The resolution possible in imaging with diffracted seismic waves. *Geophysical Prospecting*, **38**(8): 913–931.

Poldini, E. (1939) Geophysical exploration by spontaneous polarization methods. *Mining Magazine*, **60**: 22–27; 90–94.

Poster, C. and Cope, C. (1975) The location of backfilled quarries by high-resolution gravimetry. *The Arup Journal*, **10**: 7–9.

Powell, J.J.M. and Butcher, A.P. (1991) Assessment of ground stiffness from field and laboratory tests. *Proceedings of the 10th European Conference on Soil Mechanics and Foundation Engineering, Florence, May 1991*, 153–156.

Pratt, J.H. (1859) *Philosophical Transactions of the Royal Society*, **149**: 745.

Prentice, J.E. and McDowell, P. (1976) Geological and geophysical methods in the location and design of services. *Underground Services*, **4**: 15–18.

Primdahl, F. (1979) The fluxgate magnetometer. *Journal of Physics E: Soc. Instrum.*, **12**: 241–253.

Pullan, S.E. (Chairman) and subcommittee of the SEG Engineering and Groundwater Committee (1990) Recommended standard for seismic (/radar) files in the personal computer environment. *Geophysics*, **55**(9): 1260–1271.

Pullan, S.E. and Hunter, J.A. (1990) Delineation of buried bedrock valleys using the optimum offset shallow seismic reflection technique. In: Ward, S.H. (ed.), *Geotechnical and Environmental Geophysics. Vol. 3: Geotechnical*. Tulsa: Society of Exploration Geophysicists, 75–87.

Rampton, V.N. and Walcott, R.I. (1974) Gravity profiles across ice-cored topography. *Canadian Journal of Earth Sciences*, **11**(1): 110–122.

Rao, D.B., Praksh, M.J. and Babu, N.R. (1993) Gravity interpretation using Fourier transforms and simple geometrical models with exponential density contrast. *Geophysics*, **58**(8): 1074–1083.

Rao, C.V., Pramanik, A.G., Kumar, G.V.R.K. and Raju, M.L. (1994) Gravity interpretation of sedimentary basins with hyperbolic density contrast. *Geophysical Prospecting*, **42**(7): 825–839.

Ratcliff, D.W., Gray, S.H. and Whitmore, N.D. (1991) Seismic imaging of salt structure in the Gulf of Mexico. *Expanded Abstracts of 61st Annual International Meeting of SEG*, 1164–1167.

Rathor, B.S. (1977) Transient electromagnetic fields over a two-layer polarizable earth. *Geoexploration*, **15**: 137–149.

Redmayne, D.W. and Turbitt, T. (1990) Ground motion effects of the Lockerbie air crash impact. *Geophysical Journal International*, **101**(1): 293.

Redpath, B.B. (1973) *Seismic Refraction Exploration for Engineering Site Investigation*. National Technical Information Service.

Reeves, C.V. and MacLeod, I.N. (1986) A standard structure for geophysical data files on popular microcomputers. *First Break*, **4**(2): 9–17.

Regueiro, J. (1990a) Seam waves: what they are used for (part 1). *Geophysics: The Leading Edge*, **90**(4).

Regueiro, J. (1990b) Seam waves: what they are used for (part 2). *Geophysics: The Leading Edge*, **90**(8): 32–34.

Reid, A.B. (1980) Aeromagnetic survey design. *Geophysics*, **45**(5): 973–976.

Reid, A.B., Allsop, J.M., Granser, H., Millet, A.J. and Somerton, I.W. (1990) Magnetic interpretation in three dimensions using Euler deconvolution. *Geophysics*, **55**(1): 80–91.

Reilly, J.M. and Comeaux, L.B. (1993) 3D concentric circle survey: processing for steep dip imaging. *55th Meeting EAEG Stavanger Norway Extended Abstracts*, Paper A002.

Renata, D. (1977) Electromechanical phenomena associated with earthquakes. *Geophysical Surveys*, **3**: 157–174.

Renner, R.G.B., Sturgeon, L.J.S. and Garrett, S.W. (1985) *Reconnaissance Gravity and Aeromagnetic Surveys of the Antarctic Peninsula*. British Antarctic Survey Scientific Reports no. 110, Natural Environment Research Council.

Reynolds, J.M. (1982) Electrical resistivity of George VI Ice Shelf, Antarctica Peninsula. *Annals of Glaciology*, **3**: 279–283.

Reynolds, J.M. (1985) Dielectric behaviour of firn and ice from the Antarctic Peninsula. *Journal of Glaciology*, **31**(109): 253–262.

Reynolds, J.M. (1987a) The role of surface geophysics in the assessment of regional groundwater potential in northern Nigeria. In: Culshaw, M.G., Bell, F.G., Cripps, J.C. and O'Hara, M. (eds), *Planning and Engineering Geology*, Geological Society Engineering Group Special Publication no. 4, 185–190.

Reynolds, J.M. (1987b) Dielectric analysis of rocks: a forward look. *Geophysical Journal of the Royal Astronomical Society*, **89**(1): 457.

Reynolds, J.M. (1988) Surface geophysical anomalies associated with stratabound mineralisation at Sourton Tors, north-west Dartmoor. *Proceedings of the Usher Society*, 7.

Reynolds, J.M. (1990a) High-resolution seismic reflection surveying of shallow marine and estuarine environments. *Marine Geophysical Researches*, **12**: 41–48.

Reynolds, J.M. (1990b) Electrical properties of ice and the implications for ground penetrating radar. *Geophysical Journal International*, **101**(1): 268 [abstract].

Reynolds, J.M. (1991a) The need for recognized standards of applied geophysical software and the geophysical education of software users. *Computers & Geosciences*, **17**(8): 1099–1104.

Reynolds, J.M. (1991b) The determination of the degree of 'Alkali Silica Reaction' (Concrete Cancer) using dielectric analysis: preliminary results. *Geophysical Journal International*, **104**(3): 674 [abstract].

Reynolds, J.M. (1994) Resolution and differentiation of sub-surface materials using multi-method geophysical surveys. In: Forde, M.C. (ed.), *Polluted and Marginal Land – 94, University of Brunel*, London. Edinburgh: Engineering Technics Press.

Reynolds, J.M. (1995) Environmental geophysics: towards the new millenium. *Geoscientist*, **5**(1): 21–23.

Reynolds, J.M. and McCann, D.M. (1992) Geophysical methods for the assessment of landfill and waste disposal sites. In: Forde, M.C. (ed.), *Proceedings of 2nd International Conference on Construction on Polluted and Marginal Land*, 30 June–2 July 1992, Brunel University, London, 63–71.

Reynolds, J.M. and Paren, J.G. (1980) Recrystallization and the electrical behaviour of glacier ice. *Nature*, **283**(5742): 63–64.

Reynolds, J.M. and Paren, J.G. (1984) Electrical resistivity of ice from the Antarctic Peninsula. *Journal of Glaciology*, **30**(106): 289–295.

Reynolds, J.M. and Taylor, D.I. (1992) The use of sub-surface imaging techniques in the investigation of contaminated sites. In: Forde, M.C. (ed.), *Proceedings of 2nd International Conference on Construction on Polluted and Marginal Land*, 30 June–2 July 1992, Brunel University, London. Edinburgh: Engineering Technics Press, 121–131.

Reynolds, J.M. and Taylor, D.I. (1995) The use of geophysical surveys during the planning, construction and remediation of landfills. In: Bentley, S. (ed.), *Engineering Geology of Waste Storage and Disposal*. Geological Society Special Publication no. 11, 93–98.

Richter-Bernburg, G. (1982) Interior structures of salt domes. In: *Results of Geological Investigations for High-Level Waste Disposal in the Mors Salt Dome*, Proceedings of a Symposium, 18–19 Sept. 1981. Copenhagen, **1**: 50–73.

Rikitake, T. (1976) *Earthquake Prediction*. Amsterdam: Elsevier.

Rimbert, F., Erling, J-C. and Lakshmanan, J. (1987) Variable density Bouguer processing of gravity data from Herault, France. *First Break*, **5**(1): 9–13.

Roberts, G.A. and Goulty, N.R. (1990) Directional deconvolution of marine seismic reflection data: North Sea example. *Geophysical Prospecting*, **38**(8): 881–888.

Roberts, R.L., Hinze, W.J. and Leap, D.I. (1990) Data enhancement procedures on magnetic data from landfill investigations. In: Ward, S.H. (ed.), *Geotechnical and Environmental Geophysics, Vol. II: Environmental and Groundwater* Tulsa: Society of Exploration Geophysicists, 261–266.

Robin, G. de Q. and Swithinbank, C. (1987) Fifty years of progress in understanding ice sheets. *Journal of Glaciology* (special issue), 33–47.

Robinson, E.S. and Coruh, C. (1988) *Basic Exploration Geophysics*. New York: John Wiley.

Roest, W.R., Verhoef, J. and Pilkington, M. (1992) Magnetic interpretation using the 3-D analytic signal. *Geophysics*, **57**(1): 116–125.

Rollin, K.E. (1986) Geophysical surveys on the Lizard Complex, Cornwall. *Journal of the Geological Society of London*, **143**(3): 437–446.

Ruffet, C., Gueguen, Y. and Darot, M. (1991) Complex conductivity measurements and fractal nature of porosity. *Geophysics*, **56**(6): 758–768.

Russell, A. and Gee, R. (1990) The use of the dynamic probe on polluted and marginal sites. In: Forde, M.C. (ed.), *Proceedings of the International Conference on Construction on Polluted and Marginal Land*, 28–29 June 1990, Brunel University, London. Engineering Technics Press, 23–28.

Rymer, H. (1993) Predicting volcanic eruptions using microgravity, and the mitigation of volcanic hazard. In: Merriman, P.A. and Browitt, C.W.A. (eds), *Natural Disasters: Protecting Vulnerable Communities*. London: Thomas Telford, 252–269.

Rymer, H. and Brown, G.C. (1986) Gravity fields and the interpretation of volcanic structures: geological discrimination and temporal evolution. *Journal of Volcanology and Geothermal Research*, **27**: 229–254.

Rymer, H. and Brown, G.C. (1987) Causes of microgravity change at Poás volcano, Costa Rica: an active but non-erupting system. *Bulletin of Volcanology*, **49**: 389–398.

Rymer, H. and Brown, G.C. (1989) Gravity changes as a precursor to volcanic eruption at Poás volcano, Costa Rica. *Nature*, **342**(6252): 902–905.

Rymer, H., Murray, J.B., Brown, G.C., Ferrucci, F. and McGuire, W. (1993) Magma eruption and emplacement mechanisms at Mt Etna 1989–1992. *Nature*, **361**: 439–441.

Saint-Amant, M. and Strangway, D.W. (1970) Dielectric properties of dry, geologic materials. *Geophysics*, **35**(4): 624–645.

Sanderson, T.J.O., Berrino, G., Corrado, G. and Grimaldi, M. (1983) Ground deformation and gravity changes accompanying the March 1981 eruption of Mount Etna. *Journal of Volcanology and Geothermal Research*, **16**: 299–315.

Sato, M. and Mooney, H.M. (1960) The electrochemical mechanism of sulphide self-potentials. *Geophysics*, **25**(1): 226–249.

Saxov, S. (1956) Some gravity measurements in Thy, Mors and Vendsyssel. *Geodet. Inst. Skrifter, Ser. 3*, no. 25.

Schenck, F.L. (1967) Refraction solutions and wavefront targeting. In: Musgrove, A.W. (ed.), *Seismic Refraction Prospecting*. Tulsa: Society of Exploration Geophysicists, 416–425.

Schiavone, D. and Quarto, R. (1984) Self-potential prospecting in the study of water movements. *Geoexploration*, **22**: 47–58.

Schufle, J.A. (1958) Cation exchange and induced electrical polarization, *Geophysics*, **24**(1): 164–166.

Scollar, I., Weidner, B. and Segeth, K. (1986) Display of archaeological magnetic data. *Geophysics*, **51**(3): 623–633.

Scott, W.J., Sellman, P.V. and Hunter, J.A. (1990) Geophysics in the study of permafrost. In: Ward, S.H. (ed.), *Geotechnical and Environmental Geophysics. Vol. 1: Review and Tutorial*. Tulsa: Society of Exploration Geophysicists, 355–384.

Seigel, H.O., Hill, H.L. and Baird, J.G. (1968) Discovery case history of the Pyramid ore bodies, Pine Point, Northwest Territories, Canada. *Geophysics*, **33**(4): 645–656.

Semenov, A.S. (1980) *Elektrorazvedka metodom Estectvennogo Elektricheskogo Polya*. Leningrad: Nedra.

Shabtaie, S., Bentley, C.R., Blankenship, D.D., Lovell, J.S. and Gassett, R.M. (1980) Dome C geophysical survey. *Antarctic Journal of the United States*, **15**(5): 2–5.

Shabtaie, S., Thyssen, F. and Bentley, C.R. (1982) Deep geoelectric and radar soundings at Dome C, East Antarctica. *Annals of Glaciology*, **3**: 342.

Sharma, P.V. (1986) *Geophysical Methods in Geology*, 2nd edn. New York: Elsevier Science.

Sheriff, R.E. (1991) *Encyclopedic Dictionary of Exploration Geophysics*, 3rd .edn. Tulsa: Society of Exploration Geophysicists.

Sheriff, R.E. and Geldart, L.P. (1982) *Exploration Seismology*, Vol. 1. Cambridge: Cambridge University Press.

Sherrell, F. (1987) I'll sue if I can...aspects of professional liability. *British Geologist*, **13**(2): 66–71.

Shima, H. (1990) Two-dimensional automatic resistivity inversion technique using alpha centers. *Geophysics*, **55**: 682–694.

Shuey, R.T. and Johnson, M. (1973) On the phenomenology of electrical relaxation in rocks. *Geophysics*, **38**(1): 37–48.

Sieck, H.C. and Self, G.W. (1977) Analysis of high-resolution seismic data. In: Payton, C.E. (ed.), *Memoir 26: Seismic Stratigraphy: Applications to Hydrocarbon Exploration*. Tulsa: American Association of Petroleum Geologists, 353–385.

Sill, W.R. (1983) Self-potential modeling from primary flows. *Geophysics*, **48**(1): 76–86.

Singh, S.K. (1976) Fortran IV program to compute apparent resistivity of a perfectly conducting sphere buried in a half-space. *Computers & Geosciences*, **1**: 241–245.

Sinha, A.K. (1990) Interpretation of ground VLF-EM data in terms of inclined sheet-like conductor models. *Pure & Applied Geophysics*, **132**: 733–756.

Sinha, A.K. and Stephens, L.E. (1983) Permafrost mapping over a drained lake by electromagnetic induction methods. In: *Current Research, Part A, Geological Survey of Canada, Paper 83-1A*, 213–220.

Sjögren, N.A., Öfthus, A. and Sandberg, J. (1979) Seismic classification of rock mass qualities. *Geophysical Prospecting*, **27** (2): 409–442.

Sjögren, B. (1984) *Shallow Refraction Seismics*. London: Chapman & Hall.

Slaine, D.D. and Greenhouse, J.P. (1982) Case histories of geophysical contaminant mapping at several waste disposal sites. *Proceedings of the Second National Symposium on Aquifer Restoration and Ground Water Monitoring 26–28 May Columbus, Ohio, USA*, 299–315.

Slaine, D.D., Pehme, P.E., Hunter, J.A., Pullan, S.E. and Greenhouse, J.P. (1990) Mapping overburden stratigraphy at a proposed hazardous waste facility using shallow seismic reflection methods. In: Ward, S.H. (ed.), *Geotechnical and Environmental Geophysics. Vol. 2: Environmental and Groundwater*. Tulsa: Society of Exploration Geophysicists, 273–280.

Smith, R.A. (1959) Some depth formulae for local magnetic and gravity anomalies. *Geophysical Prospecting*, **7**(1): 55–63.

Smith, R.A. (1960) Some formulae for interpreting local gravity anomalies. *Geophysical Prospecting*, **8**(4): 607–613.

Soininen, H.T. and Vanhala, H. (1992) Spectral induced polarization method in mapping soils polluted by organic chemicals. *Extended Abstracts, 54th Meeting of EAEG, Paris, France*, 366–367.

Song, L. and Vozoff, K. (1985) The complex resistivity spectra of models consisting of two polarizable media of different intrinsic properties. *Geophysical Prospecting*, **33**(7): 1029–1062.

Sorgenfrei, T. (1971) On the granite problem and the similarity of salt and granite structures. *Geologiska Föreningens Förhandlingar*, **93**(2; 545): 371–435.

Sörensen, K. (1994) Pulled Array Continuous Electrical Profiling. Symposium on the Application of Geophysics to Engineering and Environmental Problems SAGEEP '94, **2**: 977–983.

Sowerbutts, W.T.C. (1987) Magnetic mapping of the Butterton Dyke: an example of detailed geophysical surveying. *Journal of the Geological Society of London*, **144**(1): 29–33.

Sowerbutts, W.T.C. (1988) The use of geophysical methods to locate joints in underground metal pipelines. *Quarterly Journal of Engineering Geology*, **21**: 273–281.

Sowerbutts, W.T.C. and Mason, R.W.I. (1984) A microcomputer-based system for small-scale geophysical surveys. *Geophysics*, **49**(2): 189–193.

Spector, A. and Grant, F.S. (1970) Statistical models for interpreting aeromagnetic data. *Geophysics*, **35**(2): 293–302.

Spector, A. and Parker, W. (1979) Computer compilation and interpretation of geophysical data. In: Hood, P.J. (ed.), *Geophysics and Geochemistry in the Search for Metallic Ores*, Geological Survey of Canada, Economic Geology Report 31, 527–544.

Spies, B.R. and Frischknecht, F.C. (1991) Electromagnetic sounding. In: Nabighian, M.N. (ed.), *Electromagnetic Methods in Applied Geophysics*, Vol. 2A. Tulsa: Society of Exploration Geophysicists, 285–425.

Stacey, F.D. and Banerjee, S.K. (1974) *The Physical Principles of Rock Magnetism*. Amsterdam: Elsevier.

Steeples, D. (1991) Uses and techniques of environmental geophysics. *Geophysics: The Leading Edge*, **10**(9): 30–31.

Steeples, D.W. and Miller, R.D. (1994) Pitfalls in shallow seismic reflection. *Proceedings SAGEEP '94*. Englewood, USA: EEGS.

Stephan, A., Schniggenfittig, H. and Strack, K.-M. (1991) Long-offset transient EM sounding north of the Rhine–Ruhr coal district, Germany. *Geophysical Prospecting*, **39**(4): 505–525.

Stephenson, S.N. (1984) Glacier flexure and the position of grounding lines: measurements by tiltmeter on Rutford Ice Stream, Antarctica. *Annals of Glaciology*, **5**: 165–169.

Stephenson, S.N. and Doake, C.S.M. (1982) Dynamic behaviour of Rutford Ice Stream. *Annals of Glaciology*, **3**: 295–299.

Stern, W. (1945) Transactions *AIME*, **164**: 189.

Stewart, R.R. (1991) *Exploration seismic tomography: fundamentals*. Society of Exploration Geophysicists, Tulsa, Oklahoma, USA. SEG Continuing Education Short Course notes.

Stolt, R.H. (1978) Migration by Fourier transform. *Geophysics*, **43**(1): 23–48.

Stoyer, C. (1989) *EMIX34 Plus™ User Manual*. Golden, Colorado: Interpex Ltd.

Sumner, J.S. (1976) *Principle of Induced Polarization for Geophysical Exploration*. Amsterdam: Elsevier.

Sundberg, K. (1931) *Gerlands Beitr. Geoph., Ergänzungs-Hefte*, **1**: 298–361.

Sundberg, K. and Hedström, F.H. (1934) Structural investigations by electromagnetic methods. *Proceedings of the World Petroleum Conference*, **V**, **B**, (4): 102–110. Institute of Petroleum Technologists.

Sunderland, J. (1972) Deep sedimentary basin in the Moray Firth. *Nature*, **236**: 24–25.

Swift, C.M. (1988) Fundamentals of the electromagnetic method. In: Nabighian, M.N. (ed.), *Electromagnetic Methods in Applied Geophysics*, Vol. 1. Tulsa: Society of Exploration Geophysicists, 5–10.

Swithinbank, C.S.M. (1968) *Radio Echosounding of Antarctic Glaciers from Light Aircraft*. Int. Ass. Sci. Hydrol., Publication 79: 405–414.

Syberg, F.J.R. (1972) A Fourier method for the regional-residual problem of potential fields. *Geophysical Prospecting*, **20**(1): 47–75.

Talwani, M. (1965) Computation with the help of a digital computer of magnetic anomalies caused by bodies of arbitrary shape. *Geophysics*, **30**(5): 797–817.

Talwani, M. and Ewing, M. (1960) Rapid computation of gravitational attraction of three-dimensional bodies of arbitrary shape. *Geophysics*, **25**(1): 203–225.

Talwani, M., Worzel, J.L. and Landisman, M. (1959) Rapid gravity computations for two-dimensional bodies with application to the Mendocino submarine fracture zone. *Journal of Geophysical Research*, **64**: 49–59.

Talwani, M., Le Pichon, X. and Ewing, M. (1965) Crustal structures of the midocean ridges. 2: Computed models from gravity and seismic refraction data. *Journal of Geophysical Research*, **70**: 341–352.

Tanner, J.G. and Gibb, R.A. (1979) Gravity method applied to base metal exploration. In: Hood, P.J. (ed.), *Geophysics and Geochemistry in the Search for Metallic Ores*, Geological Survey of Canada, Economic Geology Report 31, 105–122.

Tarling, D.H. (1983) *Palaeomagnetism*. London: Chapman & Hall.

Tauxe, L. (1990) Magnetostratigraphy. In: James, D.E. (ed.), *The Encyclopedia of Solid Earth Geophysics* New York: Van Nostrand Reinhold, 740–746.

Taylor, D.I. (1992) Nearshore shallow gas around the UK coast. *Continental Shelf Research*, **12**(10): 1135–1144.

Telford, W.M., Geldart, L.P., Sheriff, R.E. and Keys, D.A. (1990) *Applied Geophysics*, 2nd edn. Cambridge: Cambridge University Press.

Thiel, E.C., Crary, A.P., Haubrich, R.A. and Behrendt, J.C. (1960) Gravimetric determination of the ocean tide, Weddell and Ross seas, Antarctica. *Journal of Geophysical Research*, **65**(2): 629–636.

Thierbach, R. (1974) Electromagnetic reflections in salt deposits. *Journal of Geophysics*, **40**: 633–637.

Thimus, J.Fr. and van Ruymbeke, M. (1988) Improvements in field technique for use of the LaCoste and Romberg gravimeter in microgravity surveys. *First Break*, **6**(4): 109–112.

Thompson, A.H. and Gist, G.A. (1993) Geophysical applications of electrokinetic conversion. *The Leading Edge*, **12**(12): 1169–1173.

Thompson, D.T. (1982) EULDPH: a new technique for making computer-assisted depth estimates from magnetic data. *Geophysics*, **47**(1): 31–37.

Thompson, R.R. (1936) The seismic electric effect. *Geophysics*, **1**(3): 327–335.

Thornburgh, H.R. (1930) Wavefront diagrams in seismic interpretation. *Bulletin of the American Association of Petroleum Geologists*, **14**: 185–200.

Thurston, J.B. and Brown, R.J. (1994) Automated source-edge location with a new variable pass-band horizontal gradient operator. *Geophysics*, **59**(4): 546–554.

Timco, G.W. (1979) An analysis of *in situ* resistivity of sea ice in terms of its microstructure. *Journal of Glaciology*, **22**(88): 461–471.

Todd, B.J., Pullan, S.E. and Hunter, J.A. (1991) Electromagnetic studies across lakes and rivers in permafrost terrain, Mackenzie River Delta, Northwest Territories, Canada. In: *Expanded Abstracts of 61st Annual International Meeting, 10–14, Nov.* Tulsa: Society of Exploration Geophysicists, 565–568.

Tonks, D.M., Hunt, S. and Bayne, J.M. (1993) Use of the conductivity probe to evaluate groundwater contamination. *Ground Engineering*, **26**(9): 24–29.

Trorey, A.W. (1970) A simple theory for seismic diffractions. *Geophysics*, **35**(5): 762–784.

Tucker, P.M. (1982) Pitfalls revisited. *Geophysical Monograph no. 3.* Tulsa: Society of Exploration Geophysicists.

Tucker, P.M. and Yorston, H.J. (1973) Pitfalls in seismic interpretation. *Geophysical Monograph no. 2.* Tulsa: Society of Exploration Geophysicists.

Tura, M.A.C., Greaves, R.J. and Beydoun, W.B. (1994) Crosswell seismic reflection/diffraction tomography: a reservoir characterization application. *Geophysics*, **59**(3): 351–361.

Ulriksen, C.P.F (1982) *Application of Impulse Radar to Civil Engineering.* North Salem, NH: Geophysical Survey Systems Inc.

Unterberger, R.R. (1978) Radar propagation in rock salt. *Geophysical Prospecting*, **26**(2): 312–328.

Vacquier, V., Steenland, N.C., Henderson, R. and Zietz, I. (1951) Interpretation of aeromagnetic maps. *Geological Society of America Memoir 47*, 151.

Vacquier, V., Holmes, C.R., Kintzinger, P.P. and Lavergne, M. (1957) Prospecting for ground-water by induced electrical polarization. *Geophysics*, **22**(3): 660–687.

Vanhala, H. and Peltoniemi, M. (1992) Spectral IP studies of Finnish ore prospects. *Geophysics*, **57**(12): 1545–1555.

Varela, C.L., Rosa, A.L.R. and Ulryck, T.J. (1993) Modeling of attenuation and dispersion. *Geophysics*, **58**(8): 1167–1173.

Vermeer, G.J.O. (1991) Symmetric sampling. *Geophysics: The Leading Edge*, **10**(11): 17–27.

Vidale, J. (1990) Finite-difference calculation of travel times in three dimensions. *Geophysics*, **55** (6): 521–526.

Vozoff, K. (ed.), (1986) *Magneto-telluric Methods.* Tulsa: Society of Exploration Geophysicists.

Vozoff, K. (1991) The magneto-telluric method. In: Nabighian, M.C. (ed.), *Electromagnetic Methods in Applied Geophysics. Vol. 2: Applications, Part B.* Tulsa: Society of Exploration Geophysicists, 641–711.

Wagner, C.A., Lerch, F.J., Brown, J.E. and Richardson, J.A. (1977) Improvement in the geopotential derived from satellite and surface data (GEM 7 and 8). *Journal of Geophysical Research*, **82**: 901–914.

Ward, S.H. (1990) *Geotechnical and Environmental Geophysics.* Tulsa: Society of Exploration Geophysicists.

Wartnaby, J. (1957) *Seismology*, Geophysics Handbook no. 1. London: HMSO.

Waters, K.H. (1978) Reflection Seismology: A Tool for Energy Resource Exploration, 3rd edn. New York: John Wiley.

Watts, R.D., England, A.W., Vickers, R.S. and Meier, M.F. (1975) Radio-echo sounding on South Cascade Glacier, Washington, using a long-wavelength, mono-pulse source. *Journal of Glaciology*, **15**(73): 459–461.

Webb, S., Constable, S.C., Cox, C.S. and Deaton, T.K. (1985) A seafloor electric field instrument. *J. Geomagn. Geoelectr.* **37**: 1115–1129.

Wenner, F. (1912a) The four-terminal conductor and the Thompson Bridge. *US Bureau of Standards Bulletin*, **8**: 559–610.

Wenner, F. (1912b) A method of measuring earth resistivity. *US Bureau of Standards Bulletin*, **12**: 469–478.

Whitmore, N.D. and Garing, J.D. (1993) Interval velocity estimation using iterative prestack depth migration in the constant angle domain. *Geophysics: The Leading Edge*, **12**(7): 757–762.

Witten, A.J. and King, W.C. (1990a) Acoustic imaging of subsurface features. *Journal of Environmental Engineering*, **116**(1): 166–181.

Witten, A.J. and King, W.C. (1990b) Sounding out buried waste. *Civil Engineering*, **60**: 62–64.

Witten, A.J. Gillette, D.D., Sypniewski, J. and King, W.C. (1992) Geophysical diffraction tomography at a dinosaur site. *Geophysics*, **57**(1): 187–195.

Wolfgram, P.A., Edwards, R.N., Law, L.K. and Bone, M.N. (1986) Polymetallic sulfide exploration on the deep sea floor: the feasibility of the MINI-MOSES technique. *Geophysics*, **51**(9): 1808–1818.

Wong, J. (1979) An electrical model of the induced polarization phenomenon in disseminated sulphide ores. *Geophysics*, **44**(7): 1245–1265.

Woollard, G.P. (1950) The gravity meter as a geodetic instrument (surveys). *Geophysics*, **15**(1): 1–29.

Woollard, G.P. (1959) *Journal of Geophysical Research*, **64**: 1521.

Woollard, G.P. (1975) Regional changes in gravity and their relation to crustal parameters. *Bur. Grav. Int. Bull. Inf.*, **36**: 106–110.

Woollard, G.P. and Rose, J.C. (1963) *International Gravity Measurement*. Tulsa: Society of Exploration Geophysicists.

Wright, D.L., Hodge, S.M., Bradley, J.A., Grover, T.P. and Jacobel, R.W. (1990) A digital low-frequency, surface-profiling ice-radar system. *Journal of Glaciology*, **36**(122): 112–121.

Wright, D.L. and Watts, R.D. (1982) A single-hole shortpulse radar system. In: *Geophysical Investigations in Connection with Geological Disposal of Radioactive Waste*, OECD/NEA, Ottawa, Canada.

Wu, C. (1990) Adequate sampling in magnetic surveys, the resolution of closely-spaced sources, and their importance to data processing, image-enhancement techniques and interpretation (MSc thesis, ITC Delft), *ITC Journal 1990–2*, 176–179.

Wyllie, M.R., Gregory, A.R. and Gardner, G.H.F. (1958) An experimental investigation of factors affecting elastic wave velocities in porous media. *Geophysics*, **23**(3): 459–493.

Wynn, J.C. and Zonge, K.L. (1975) EM coupling: its intrinsic value, its removal and the cultural coupling problem. *Geophysics*, **40**(5): 831–850.

Wynn, J.C. and Zonge, K.L. (1977) Electromagnetic coupling. *Geophysical Prospecting*, **25**(1): 29–51.

Wyrobek, S.M. (1956) Application of delay and intercept times in the interpretation of multi-layer refraction time distance curves. *Geophysical Prospecting*, **4**(2): 112–130.

Xia, J. and Sprowl, D.R. (1992) Inversion of potential-field data by iterative forward modelling in the wavenumber domain. *Geophysics*, **57**(1): 126–130.

2222

222

22222

Xu, B. and Noel, M. (1993) On the completeness of data sets with multielectrode systems for electrical resistivity survey. *Geophysical Prospecting,* **41**(6): 791–801.

Yaghoobian, A., Boustead, G.A. and Dobush, T.M. (1992) Object delineation using Euler's inhomogeneity equation – location and depth determination of buried ferro-metallic bodies. *Proceedings of SAGEEP-92, San Diego, California.* Englewood, USA: Environment and Engineering Society.

Yilmaz, O. (1987) *Investigations in Geophysics. Vol. 2: Seismic Data Processing.* Tulsa: Society of Exploration Geophysicists.

Yüngül, S.H. (1950) Interpretation of spontaneous-polarization anomalies caused by spheroidal orebodies. *Geophysics,* **15**(2): 237–246.

Yüngül, S.H. (1954) Spontaneous-potential survey of a copper deposit at Sariyer, Turkey. *Geophysics,* **19**(3): 455–458.

Yuanxuan, Z. (1993) Radon transform application to the improved gridding of airborne geophysical survey data. *Geophysical Prospecting,* **41**(4): 459–494.

Zachos, K. (1963) Discovery of a copper deposit in Chalkidiki Peninsula, northern Greece (in Greek with English summary). *Inst. Geol. Subsurface Res. Publ.,* **8**(1): 1–26.

Zalasiewicz, J.A., Mather, S.J. and Cornwell, J.D. (1985) The application of ground conductivity measurements to geological mapping. *Quarterly Journal of Engineering Geology,* **18**(2): 139–148.

Zanzi, L. (1990) Inversion of refracted arrivals: a few problems. *Geophysical Prospecting,* **38**(4): 339–364.

Zanzi, L. and Carlini, A. (1991) Refraction statics in the wavenumber domain. *Geophysics,* **56**(10): 1661–1670.

Zinni, E.V. (1995) Sub-surface fault detection using seismic data for hazardous-waste-injection well permitting: an example from St John the Baptist Parish, Louisiana. *Geophysics,* **60**(2): 468–475.

Zohdy, A.A.R. (1974) Electrical methods. In: *Applications of Surface Geophysics to Groundwater Investigations,* Book 2, Chapter D1. US Department of the Interior, 5–66.

Zohdy, A.A.R. (1989) A new method for the interpretation of Schlumberger and Wenner sounding curves. *Geophysics,* **54**(2): 245–253.

Zohdy, A.A.R., Anderson, L.A. and Muffler, L.J.P. (1973) Resistivity, self-potential and induced-polarization surveys of a vapor-dominated geothermal system. *Geophysics,* **38**(6): 1130–1144 (and discussion by Roy, A., *Geophysics,* **40**(3): 538; and by Zohdy, A.A.R., *Geophysics,* **40**(3): 538).

Zonge, K.L. and Hughes, L.H. (1991) Controlled-source audio-frequency magnetotellurics. In: Nabighian, M.C. (ed.), *Electromagnetic Methods in Applied Geophysics. Vol. 2: Applications, Part B.* Tulsa: Society of Exploration Geophysicists, 713–809.

Zonge, K.L. and Wynn, J.C. (1975) Recent advances and applications in complex resistivity measurements. *Geophysics,* **40**(5): 851–864.

Index

Note: Page references in **bold** refer to Figures, page references in *italic* refer to Tables and Boxes.